The Hawker Hunter

The Hawker Hunter

Tim McLelland

Crécy Publishing

Published in 2008 by Crécy Publishing Limited, Manchester

A CIP record for this book is available from the British Library

ISBN 978 0 85979 123 6

Printed and bound in Great Britain
by Cromwell Press

Crécy Publishing Limited
1a Ringway Trading Estate, Shadowmoss Road, Manchester M22 5LH
www.crecy.co.uk

www.hunterteam.com

Frontispiece: Pulling up into the vertical, Hawker Hunter Aviation's XF995, pictured whilst on detachment to the Empire Test Pilots School at Boscombe Down. *HHA*

Rear cover (top): Refurbished Hunter prior to delivery to Jordan. *Paul Tomlin*

Rear cover (bottom): Hunter F Mk 6 XJ637 resplendent in the markings of No 79 Squadron, one of No 229 Operational Conversion Unit's component squadrons at Chivenor. *Paul Tomlin*

Contents

Abbreviations

ACMI	Air Combat Manoeuvring Instrumentation
ADTU	Air Direction Training Unit
AFDS	Air Fighting Development Squadron
AFTS	Advanced Flying Training School
APS	Armament Practice School
ASM	Air-to-Surface Missile
AWA	Armstrong Whitworth Aircraft
CA	Certificate of Airworthiness
CAACU	Civilian Anti Aircraft Co-operation Unit
CAT1-5	Category 1 is damage that is repairable within established first-line resources Category 2 is damage that is repairable within established second-line resources Category 3 is damage that is repairable on site, but is beyond unit technical resources Category 4 is damage that is not repairable on site because special facilities or equipment is needed Category 5 is damage beyond economic repair or the aircraft is missing
CDTC	Controlled Descent Through Cloud
CEPE	Central Experimental & Proving Establishment
CFE	Central Fighter Establishment
CFS	Central Flying School
CWP	Contractor's Working Party
DBR	Damaged Beyond Repair
DFCS	Day Fighter Conversion Squadron
DFLS	Day Fighter Leading School
DTEO	Defence Test and Evaluation Organisation
ETPS	Empire Test Pilots School
ff	first flown
FRADU	Fleet Requirements & Air Direction Unit
FRU	Fleet Requirements Unit
FTS	Flying Training School
FTU	Ferry Training Unit
FWS	Fighter Weapons School
GCA	Ground Controlled Approach
GGS	Ground Gunnery School
GIA	Ground Instructional Airframe
HAL	Hawker Aircraft Ltd
HAS	Hardened Aircraft Shelter
HHA	Hawker Hunter Aviation
HSA	Hawker Siddeley Aviation
IAS	Indicated Air Speed
JPT	Jet Pipe Temperature
OAT	Outside Air Temperature
OCU	Operational Conversion Unit
OFMC	Old Flying Machine Co
PEE	Proof & Experiental Establishment

QFI	Qualified Flying Instructor
QGH	Ground-interpreted radar let-down procedure
RNEC	Royal Naval Engineering College
RRE	Royal Radar Establishment
SAH	School of Aircraft Handling
SNEB	Societe Nouvelle des Etablissements Edgar Brandt
SOC	Struck Off Charge
SoTT	School of Technical Training
TAF	Tactical Air Force
TEE	Test & Evaluation Establishment
TWU	Tactical Weapons Unit
W/O	Written off
WFU	Withdrawn from use

Introduction

Although I have never served in the Royal Air Force, I have spent an almost indecent proportion of my life writing about, talking about, photographing and looking at military aircraft. Having reached the status of 'middle age' I can at least be thankful that I was able to witness the RAF's 'Cold War' years when Britain's skies were seemingly full of warplanes, pursuing their various roles in preparedness for what would undoubtedly have been the RAF's last conflict, had it ever taken place.

I grew up during the 'sixties and 'seventies, eagerly reading every magazine article and watching every television clip that gave a fleeting insight into the seemingly distant world of military flying. On a few, greatly relished occasions I was able to spend many hours lurking amongst the weeds outside the RAF's airfields, watching the warplanes go about their business, and on a few even more memorable days I was able to get even closer to the objects of my fascination, at various air shows around the country. It was here that I first marvelled at the magnificent Hunter, a sleek, purposeful machine that looked every inch a potent fighting machine. One of the highlights of many air shows was the 'set piece' – a display of raw military power when the RAF ranged its assets against a suitable target, usually a cloth-covered tent or a pile of oil drums. The explosions are unashamedly fake (and often erupted before the aircraft had even arrived over the target) but the effect was captivating. The stars of the show? Of course it was the Hunters, streaking in low, the unmistakable 'blue note' howl making one's hair literally stand on end as they roared over the airfield, often surrounded by clouds of thick, white water vapour. It was a joy to behold. In later years it was the Navy that kept the Hunter's spirit alive, with its team of beautiful white-bellied machines performing graceful loops and rolls; they lacked the thrill and spectacle of the ground attack demonstrations, but still kept the observer spellbound. Then there was Greenham Common and the Hunter's anniversary – a long, long line of Hunters led by the scarlet prototype, flanked by aircraft clothed in dowdy camouflage or eye-catching trainer and test flying colours. Nobody could claim to be anything but impressed.

Shortly before the Hunter finally retired from Fleet Air Arm operations, I was given an opportunity to experience the aircraft first-hand, courtesy of FRADU, which offered me a seat in one of its few remaining twin-seat Hunters. A five-hour drive seemed like a small price to pay for such a unique opportunity, and even when I found myself in a flight safety room, hanging blindfolded from a parachute harness and being spun around like a top, the prospect of a seat in a classic fighter made the ordeal still seem like a bargain offer. When I was finally introduced to my aircraft, the first impression was one of size. Like most civilians, it is often easy to underestimate the sheer bulk of an aircraft until you are attempting to climb into it. Conversely, the feeling of size is then immediately exchanged for one of confinement when one realises just how little space there is inside the cockpit, with two very solid-looking ejection seats crammed under a weighty Perspex canopy, just inches from an instrument panel packed with a bewildering array of dials and switches. Once the canopy clunks into its locked position, the world inside the Hunter seems astonishingly small; everything is literally no more than an arm's length away, and with the weight of a life jacket, 'bone dome' helmet, boots, and a variety of tight-fitting straps, the feeling is one of claustrophobia, particularly when the malodorous rubber oxygen mask is clipped into place and the helmet visor is pulled down.

The background buzz of electronics is interrupted by the crackling voice of my pilot as he prepares to start up the Hunter's Avon engine – a simple task that results in a hiss of cartridge exhaust and a familiar plume of sooty smoke that briefly surrounds the canopy. As it clears, the satisfying rumble of the turning Avon engine feeds through the airframe and the Hunter is alive, the switches and dials slowly starting to make sense as I carefully remove the ejection seat pins, rendering the seat 'live' and instantly ready to punch me out of the aircraft with such a force that I would probably find myself in hospital as a result. If the seat should fire before we achieve some forward speed and altitude, I would be lucky to even make it to hospital – unlike modern ejection seats, the Hunter's escape system offers no chance of successful ejection from a stationary start, so I make a conscious effort to keep my hands well away from the black-and-yellow-striped D-ring handle wedged between my legs.

With a gentle surge of engine power, the Hunter lurches forward, performing a brief curtsey as my pilot checks the wheel brakes for effectiveness. Out we go, across the apron where V-Bombers were once parked on dispersal. The mighty bombers are long gone, but the Hunter is still very much in business, sleek and purposeful, perfectly at home amongst the assorted Sea Harriers from a much more modern era that share the taxiway. We might be occupying a Hawker machine from a distinctly older generation but, given a choice between the two, the Sea Harrier pilots would be equally delighted to take my seat in the Hunter.

With clearance from the tower and a last-minute run-through of our escape drills – just in case our departure doesn't go quite as planned – we are on to the 'piano keys' at the threshold of Yeovilton's runway, the crisp spring sunlight interrupted by the twinkle of the runway lights stretching ahead of us, and, with everything 'good to go', the Avon spools up to full power with a solid and thunderous roar. Even through the thick canopy and the bulky, insulated 'bone dome', there is no doubt that the Avon is spooling up to speed, making its familiar noisy presence known to the Yeovilton locals who have lived with the Hunters as part of their everyday lives for decades. Brakes off, and the cockpit lurches upwards as the Hunter leaps forwards and begins to race off along the runway. I have been lucky to take a passenger seat in many different combat aircraft, but the memory of the Hunter's take-off performance is still a vivid one that stays with me.

Maybe it's the relative lightness of the airframe, the power of the engine (which doesn't even match that of its single-seat counterparts but still provides quite a punch), or maybe it's the side-by-side seating that gives you an uninterrupted forward view of the runway, but the sensation is one of sheer speed as the 1,000-foot markers stream by with alarming haste, the line of red runway threshold lights seemingly racing towards us. A long runway suddenly seems to be incredibly short when the Hunter is getting airborne. Effortlessly, as we accelerate through 100 knots, my pilot gently eases the control column backwards and the Hunter lifts gently into the air, the undercarriage clunking into the wheel wells as we rush out over the perimeter road, with a handful of hardy plane-spotters busy with their cameras below us. They have probably been waiting in the cold for a long time, but the sight and sound of Hawker's classic fighter streaking over their heads must surely have made the cold worthwhile. In the cockpit, the temperature is already warming up and the Hunter's mighty roar is behind us now, replaced by a gentle hiss of airflow, periodically interrupted by crackling chatter on the radio.

En route to our tasking, there is time to consider my surroundings. Once in the air, the all-round view from the Hunter's cockpit is good, with a superbly clear forward view and excellent visibility out to the side, where, with a good strain against the seat harness, the dark grey starboard wing can be seen, reaching out to the horizon with two slightly chipped and battered external fuel tanks visible hanging from the underside, their glossy grey paintwork contrasting against the greens and yellows of the Somerset countryside far below us. In a matter of minutes the rolling fields are replaced by greenish-grey seas and my pilot begins the serious business for which the Hunter's performance has proved to be so ideal. Somewhere, far away on the horizon behind a wall of brown haze, is a German frigate, waiting passively, her radar screens carefully searching for signs that they might be under attack. Today we are the attackers and, although we carry no live ordnance, the Hunter's sprightly performance is to be used as a mimic, replicating the speeds and altitudes of an incoming anti-ship missile. On the frigate's radar screen, the effect is convincing – the radar return has all the characteristics of an enemy attack and provides the ideal training situation for their radar interpreters.

At 10,000 feet above the coast, we are in the wrong place to imitate a sea-skimming missile, so the Hunter has to be repositioned to begin our attack profile. In a gentle

XL601 (the aircraft flown by the author) is seen shortly after the beginning of its service with the Royal Air Force, proudly wearing the markings of No 1 Squadron on its nose (and – unusually – on the underwing tank).

During the early 1970s XL601 was allocated to the Harrier Conversion Unit at Wittering, being used as a continuation, chase and conversion trainer for the fledgling Harrier force. *Paul Tomlin*

descent from altitude, the aircraft gives no impression of movement. The outside world appears still with not so much as a twitch from the rock-steady airframe and the only perceptible change is a gradual darkening of the forward view as the crisp and clear upper air slowly gives way to the dull, hazy blanket of mist and salt spray that is spread over the South Western approaches. The instrument panel gives a better impression of what is really going on; the altitude is rapidly winding down and, with an audible rumble of power from behind us, the speed starts to build up, winding rapidly through 400 knots, 430 knots, 480 knots, and outside the sea surface now seems to be slowly wrapping itself around the Hunter's canopy.

Lower and lower we go, the rough surface air starting to create turbulence that gently kicks the Hunter up and down, left and right, the cockpit instrument panel rattling against its mountings. We are down to almost wave-top height now, little higher than a double-deck bus as we rush ahead towards the target, shaking, rattling, thundering on across the grey sea, with salt spray rolling up across the Hunter's windscreen until, with a sudden hard push of teeth-rattling g-forces, we are climbing, up through forty-five degrees, still climbing, and as we enter a lazy climbing turn I catch a glimpse of the hapless German frigate, far below us in a tight spiral evasive turn. The frigate's crew have had their rude awakening and our task is over.

Back to the sunshine and across the rolling countryside, and the clear, blue skies of Yeovilton. The Hunter's work is done for today, but with four external tanks under our wings and a relatively short 'hop' out to the English Channel, the Hunter's fuel reserves are far from exhausted. My pilot makes a call to Yeovilton's tower to announce that we will remain in the airfield overhead while we burn off some fuel to lower our landing weight. There was no need – the aircraft was already light enough to safely recover – but it provides a semi-legitimate excuse to demonstrate yet more of the Hunter's brilliant manoeuvrability.

On with the power and up we go into a lazy loop, slowly over the top, inverted, straps hanging from the seat harness as we bleed off the airspeed, then down towards the fields. Back comes the rush of airflow noise and on comes the g-force as we power back into the sky again, a roar of power from the Avon as we effortlessly punch back up into the blue and gently roll out off the top, slightly shaken and more than a little stirred. Had the oxygen masks been removed, there would have been two broadly smiling faces under the Hunter's cockpit canopy. Another roll over onto our backs and we are heading back down again towards the airfield, the air traffickers giving us clearance to join the airfield circuit.

Down towards the runway and back at low level there is time for one more burst of power to give us a sprightly push into a streaking arrival over Yeovilton, pulling hard to port as the aircraft rattles down to circuit speed, the huge under-fuselage air brake fully deployed. Settled on the downwind leg, the flaps and undercarriage rumble down and clunk into position, and round we go onto final approach, sweeping in over the approach lights at what seems like a surprisingly high speed even for a 'fast jet', before making a firm thump onto the runway as the brake parachute cracks open and tugs us to a gentle taxi speed.

This, then, is a taste of the Hunter. A truly outstanding example of British design and manufacturing expertise, it is a true 'classic' that served the RAF and a variety of overseas customers flawlessly for decades. Fast, and manoeuvrable, built with simplicity in mind but with a rugged construction, the Hunter was a capable interceptor and outstanding ground-attack aircraft. In service around the world, the Hunter remained in business long after all of its contemporaries had long since retired into museums or scrapyards. Quite simply, there was nothing able to match it.

In this book I have attempted to create a comprehensive look at the magnificent Hunter, tracing the aircraft's background, design and development, and service history. Of course the Hunter's story is well-known, therefore I've resisted the temptation to dwell unnecessarily on too many details of its development and operational use, as this information is readily available through countless other books, magazine features and websites. Instead of simply producing just another dull history book, I've drawn upon the experiences of former Hunter people to provide a more 'hands-on' look at the aircraft, describing what it was like to fly it, maintain it and live with it during its service years and beyond. In this way I hope that I (and the many people who have contributed) have provided a much more interesting and readable account of the Hunter's fascinating history. Facts and figures are fine, but it is people – Hunter people – who make a story, and only they can give a proper description of what the Hunter is all about.

For the more technically minded, I have also traced the histories of each and every Hunter that was operational in the UK, and although it would have been fascinating to have done likewise for all of the many Hunters that were exported overseas, there simply is not sufficient space, even within a book of this size, to do the job properly. It would have had to run into a series of volumes to cover every aspect of the Hunter's career! I have also included a reproduction of the Hunter Pilot's Notes – in effect the user's manual – which will hopefully give the reader a fairly good account of the more technical aspects of Hunter operations and thereby create a mental escape into the pilot's seat. Naturally, there is a great deal more that could have been included, but ultimately a book of this nature has to be a choice between conflicting possibilities. I hope that I have managed to strike a good balance!

I must take this opportunity to thank the many individuals who have given so much time and effort to this project. I have been constantly amazed at the enthusiasm and affection that the Hunter inspires in so many people, and I must thank them all for their co-operation. Of course it is impossible to mention every individual who has helped with this book, but to everyone who provided advice, information, contributions, photographs, drawings and encouragement, I must extend a very big thank you. I must also make special thanks to some individuals whose helpfulness was vital to the success of this project: Ray Deacon; Allan Pollock; Ian Hartley; Craig Jackson; Paul Constable; Glenn Sands; David Lockspeiser; Steve Legassick (Northern Lights); David Griffin; Michael Elliot (HHA); Vatche Mitilian; Richard Caruana; Lennart Berns; Mike McEvoy; Erik Johansson; Paul Tomlin; Coen van den Heuvel; Terry Kingsley; Derek Bell; Alick Nicholson; Chris Bolton; David Flower; Andrew Lawson; Stan Lawson; Tony Ogilvy; Bryan Kelly; Mat Potulski (HHA); Mick Freer; Martin Johnson; Sergio de la Puente; Keith Hartley; Ivan Pennell; Michael Doust; and Brian Irvine.

Chapter 1

Genesis of the P.1067

As the Second World War entered its final phase, Britain's armed forces undoubtedly possessed a vast array of formidable weaponry, the Royal Air Force's inventory including a considerable number of piston-engined fighter aircraft types that were, at the time, amongst the very best examples of advanced aircraft design and manufacturing available. The legendary Spitfire, having earned its place in history during the Battle of Britain, still formed a significant part of the Royal Air Force's Order of Battle, even though the basic design had been developed to such an extent that it bore very little resemblance to the aircraft that had fought the Luftwaffe over South East England just a few years previously. Advances in airframe design and power plant development had already resulted in much more than the series of powerful Griffon-engine Spitfires; the very capable Typhoon ground-attack aircraft and other equally impressive machines such as the Tempest and Hornet were also eventually developed, shifting the emphasis from traditional fighter and bomber production, partly in response to an ever-growing requirement for more versatile tactical fighter-bombers that could support troops on the ground, as fighting continued across Europe and beyond. Although night-fighters were still regarded as being a vital part of the RAF's inventory, it is fair to say that the relatively uncomplicated, light and agile day-interceptor became much less of an important asset, moving further and further down the Air Ministry's list of priorities as the war progressed.

Undoubtedly, one of the most important technological developments to emerge during the Second World War (even though its significance often did not seem quite so obvious at the time) was the gas-turbine 'jet' engine, a revolutionary concept that Frank Whittle pursued for some considerable time, often with little more than lukewarm interest and support from the British Government, which often seem to be almost scornful of the whole idea. However, even the Air Ministry eventually grasped the potential of the jet engine and did finally place a contract with Whittle's Power Jets company in March 1938 for the development and construction of a flight-standard engine, but it was not until February 1940 that the Government took responsibility for the project and placed a contract with Gloster Aircraft to produce an aircraft to carry the engine, under Specification E.28/39. By this stage Germany had already test-flown a jet engine (in the Heinkel He.178) and development of an operational Luftwaffe jet fighter was under way.

Britain's Gloster E.28/39 Pioneer aircraft was something much less ambitious and was primarily intended to act simply as a test airframe for the experimental engine, but it was also specified that the aircraft should be designed with the potential to be developed into an interceptor-fighter with fixed gun armament, within the very clear limitations of the new low-powered engine. The fact that there was really no urgent Air Ministry interest in the production of new 'day-fighter' designs may well have contributed to the painfully slow development of a British jet engine, but the revolutionary (and temperamental) power plant appeared at a time when both the Air Ministry and Britain's aircraft manufacturers were understandably reluctant to allocate precious resources to what must have looked like a seemingly fanciful concept, when so much experience (and success) had been achieved with the development of piston engines.

Despite the almost grudging support of Whittle's efforts, the Air Ministry's interest in the project did gradually begin to grow (and a constant trickle of information on the continuing German activities in this field undoubtedly served to starkly illustrate the jet engine's potential), and in December 1940 a specification (F.9/40) was issued, based around a Gloster proposal for an operational fighter aircraft armed with six 20mm cannon, powered by a pair of 2,000lb-thrust jet engines. As development of the jet engine's potential slowly got under way, an initial order for twelve aircraft was placed in February 1941, and on 15 May of the same year the small and rather oddly proportioned E.28/39 Pioneer made the first jet-powered flight in the UK, appropriately from the RAF's historic airfield at Cranwell, although it had already made a number of unofficial 'hops' as part of high-speed taxi trials at the company's Hucclecote airfield during the previous month.

Although the aircraft could not be described as breathtaking (indeed, it would have escaped any casual observer's attention, had it not been for the fact that it was conspicuously lacking a propeller), the Pioneer's performance was enough to convince the Air Ministry that the jet engine worked, and that a prop-less fighter was a practical (and very desirable) proposition. The Rover W.2B engine fitted to the second E.28/39 prototype (which crashed after entering an unintentional flat spin during a test flight) was selected as the power plant for the new twin-engine Gloster fighter design, but when the first aircraft had been completed and commenced taxiing trials, the engines were found to be only capable of delivering a meagre 1,000lb of thrust, which was insufficient to even achieve flying speed. Consequently, a pair of de Havilland Halford H.1 (later named Goblin) turbojets was substituted while the Rover engines were modified. However, even these replacement engines could achieve only 1,500lb of thrust, but this was sufficient to get the first F.9/40 fighter prototype into the air on 5 March 1943.

Meanwhile, the W.2B engine was developed by Rolls-Royce and finally emerged as the 1,700lb Welland that powered the first production variant of Gloster's fighter design – the Meteor F Mk 1. The Meteor's emergence was swiftly followed by de Havilland's own design (using its Goblin engine), the single-engine Vampire, which first flew in September 1943. Progress had been slow, but finally Britain had two new fighters, which, although scarcely superior to the piston-engine fighters they were designed to replace, at least demonstrated that jet propulsion was the way forward and that there was a very clear prospect of creating new aircraft based around the new power plant that would offer a performance far beyond that possessed by any aircraft that had preceded them.

When the Second World War finally ended, the Royal Air Force entered a period of rapid and drastic contraction, as countless squadrons disbanded and huge quantities of aircraft were suddenly declared redundant. Surplus second-line aircraft, redundant training airframes and obsolescent bombers and fighters were rapidly scrapped, as the emphasis shifted towards a consolidation policy based on the more modern front-line fighters and bombers that would form the basis of Britain's post-war defences. By this stage the various advanced piston-engine fighter-bomber aircraft such as the Tempest and Hornet had also effectively reached the very end of their development potential and there was little appetite for developing them still further, when even the most ambitious and costly efforts would inevitably produce only modest improvements in speed. Of course, there was also the important question as to whether any drastic improvements in fighter aircraft performance were even required at this time, when a brutal war had just ended and another conflict seemed unlikely – at least, if one chose to ignore any signs to the contrary. With the benefit of hindsight, it is clear to see that the seeds of a new Cold War had already been sown, but the post-war British Government had no appetite for rearmament. Indeed, there was a prevailing attitude of austerity, and with a country crippled almost to the extent of bankruptcy, the prospect of embarking on any major new military expenditure programmes was a proverbial pill that the Government was far from keen to swallow. Consequently, there was a very obvious eagerness to proclaim that nothing but peaceful times were ahead, even though there were often some disturbingly ominous signals that the post-war era might not be quite as secure as the Government might have been hoping.

Sadly, the fundamental flaw in this posture was the complete ignorance of the time scales involved in the design and production of new fighter and bomber aircraft, especially under peacetime conditions. For example, some four years were to elapse before the Meteor transitioned from design to operational status, and without the impetus and resources of wartime effort and urgency this time scale could be expected to double in the relaxed and cost-conscious post-war environment. In effect, the Air Ministry's thinking on fighter developments had concluded that the most sensible way forward was to assume that no urgent requirement for new fighters existed, and that it would be acceptable to simply concentrate on the introduction of more advanced versions of the Meteor and Vampire fighter jets, both types armed with four wartime-standard Hispano 20mm cannon, until more significant design proposals emerged, or a more obvious military requirement became clear. In principle, this posture seemed practical, but proverbially sitting on one's hands while other countries pursue their own less-benign agendas is, of course, an act of supreme folly that ought to have recognised by the Air Ministry.

These early post-war years of relative stagnation saw many potentially successful design projects simply crippled by the lack of Governmental support for research work. Having emerged from a long and costly war, the country could barely survive and there was virtually no money available to fund research and development projects, so the responsibility for almost all of this work now fell squarely on the shoulders of the British aircraft industry. Although a great deal of encouraging work was still being undertaken, the industry continually suffered from a woeful lack of funding while, in stark contrast, significant progress was being made elsewhere, particularly in America, where much of Britain's technical know-how was obligingly and willingly exported, often in exchange for promised information and co-operation that never materialised. British manufacturers were continually given plenty of information on Frank Whittle's engine work, and a variety of design concepts were privately studied for different aircraft, using the new jet engine as a basis. Initially, in much the same way as the Air Ministry had shown distinct indifference to the jet engine, manufacturers also did not share much enthusiasm either for the new engine (hardly surprising when one considers the almost pitiful amount of thrust that it could deliver) and there was an understandable reluctance to divert vital resources from more conventional design work to pursue a concept that did not appear (at that time) to offer any significant advantages over the piston engine.

Meanwhile, the Air Ministry's Operational Requirements Branch, having concluded that major advances in the top speeds of future (piston or jet-engine) fighters were unlikely, had turned its attention to the only other realistic option to create more modern and effective aircraft, which was simply to improve the aircraft's armament. At this time the 20mm Hispano cannon was the only reliable weapon available although other potential gun armament schemes were studied, including use of the hefty 57mm Molins gun, which had been used by the Mosquito on wartime anti-shipping attacks. The logic behind this 'bigger guns' thinking was that the gradual increases in speed performance had effectively meant that, in a combat situation, fighters could only be realistically expected to make one fast firing pass during an interception, so this 'one chance' pass would have to be a pretty lethal one. The concept was sound, but of course a multi-purpose day and night fighter carrying a heavy cannon armament would probably perform little better than the fighter types it was intended to replace.

Despite this inescapable fact, two specifications were eventually issued (F.43/46 and F.44/46), which (with a hint of hope rather than anticipation) called for single- and twin-seat fighters, intended ultimately to replace the Meteor and Vampire in the day-fighter role, and also the Mosquito in the night-fighter role. Companies did submit some fairly optimistic tenders for this requirement, but without any obvious leaps in predicted performance, and the Air Ministry's evidently poor understanding of what was actually needed, these project studies made little progress and drifted around the Air Ministry and industry desks for some time without any clear direction.

There was also another very significant factor that may well have contributed to the Air Ministry's sometimes rather confused thinking during this time period, and it was one that has often escaped historical accounts of this period: while work on the early jet designs gradually progressed, another (and much less well-known) project was being secretly pursued under Specification E.24/43, which described a research aircraft capable of flying at an altitude of 36,000 feet, at an astonishing top speed of 1,000mph. Even to this day it is still unclear why such an ambitious aim was even considered in 1942, and in many respects the Air Ministry's interest displays all the signs of having simply been a rather naïve aspiration rather than a practical proposition. However, it is also quite possible that research data recovered from Germany (or at least the data that had not been ushered away by America) may have provided sufficient knowledge and confidence to warrant consideration of such an ambitious project, so perhaps the concept was not quite as bizarre as it might appear in retrospect.

Miles Aircraft received instruction to proceed with the design of an aircraft based around this requirement in conditions of extreme secrecy, requiring the company to even set up its own metal foundry and wind tunnel so that the project could continue without any undue external attention being drawn to it. The M.52 was a revolutionary design featuring a bullet-shaped fuselage and small bi-convex wings combined with an annular air intake with the cockpit situated within the central nose shock cone. The engine was to have been a Power Jets W.2/700 developing only 2,000lb of thrust, but using a rudimentary afterburner system. Development of the M.52 progressed surprisingly well, and by the end of 1945 the design work was ninety per cent complete, with jigs ready for the assembly of three prototypes. By this stage both Miles Aircraft and the Air Ministry believed that the project had the potential to create an RAF fighter capable of achieving a staggering Mach 1.5 and an altitude of 40,000 feet. Undoubtedly, this was pretty impressive for an aircraft that could have been in service in the early 1950s, but in February 1946, and without any warning, the Ministry of Aircraft Production decided that all work on the project should stop. No plausible reason was given for this devastating decision other than a general view that the programme was likely to be prohibitively expensive and that the design might not even achieve supersonic flight, even though Miles Aircraft confidently (and rightly) believed the aircraft would go supersonic within a matter of months. The programme was simply abandoned while the jigs were unceremoniously destroyed, and even sixty years later there is no clear reason why the programme was terminated so abruptly.

The wind tunnel aerodynamic test model of the M.52.

The post-war avalanche of captured German research data (including that which was taken to America but later re-released to Britain after it had been digested by American manufacturers) could possibly have convinced the Ministry that Britain's grasp of supersonic aerodynamics was woefully (and fatally) inadequate, and this might have planted the seeds of doubt that led to the project's termination, but given the continual exchange of information between Britain and America on supersonic research, there is no reason to suppose that sufficient confidence in the design was not already well established. In retrospect, it does appear as if the project may well have been terminated due to political (and financial) pressure from America, possibly in exchange for other important transatlantic co-operation, details of which still have not been revealed. But whatever the true reasons behind its cancellation, the potentially world-beating M.52 was virtually dead and buried.

Eventually further research on the stillborn project was conducted by Vickers utilising a series of unmanned test models, following a great deal of persistent lobbying by Dr Barnes Wallis (of 'Dambusters' fame), who believed that on both cost and safety grounds the project should have been confined to unmanned experiments all along. Plausible though his attitude may have seemed, some commentators believe that his fervent belief in the need to conduct unmanned trials was not entirely altruistic and that his lobbying for unmanned tests to be conducted by Vickers was merely an effort to secure more work for his company, and actually did little more than kill off the main M.52 programme that these tests were meant to support. It was therefore with some irony that the tests conducted by Vickers proved that the M.52 would probably have flown very successfully, and one of the models actually achieved Mach 1.5 during a test flight, proving the soundness of the concept, but with the manned programme now abandoned (and no money or appetite to re-start it), these encouraging results were largely academic and did little more than generate income for Vickers. It was of course America's Bell X-1 that ultimately achieved fame as the first aircraft to break the sound barrier, although it was certainly no accident that the X-1 bore more than a passing resemblance to the abandoned M.52. This unhappy episode may well have been more complex than it might first appear, and further details of the programme may still be buried in obscure files in dusty cabinets somewhere, but it is clear that the M.52 was the victim of dark international political manoeuvring and industry self-interest. However, it is also just as likely that it was simply a potentially world-beating project that Britain just could not afford.

The Bell X-1 earned worldwide fame as the aircraft that first broke the sound barrier. It was no coincidence that it bore more than a passing resemblance to the abandoned Miles M.52.

Nonetheless, shortly before the end of the Second World War design staff at Hawker's Kingston factory had already embarked upon private preliminary studies of the potential use of the new jet engine as a power source for new fighters and bombers. During work on the development of a new high-speed bomber design ordered under Specification B.11/41 (which was subsequently cancelled), the Hawker team had considered the possibility of substituting a pair of Power Jet engines to replace the Napier Sabre pistons that had originally been specified. The idea did not progress far, but the idea of using jet power was already beginning to infiltrate significantly the minds of Hawker's design team. Other schemes were also considered, including the concept of fitting a Power Jets engine to an existing Hawker piston-engine design such as the Sea Fury and Tempest. Fascinating though these relatively simple proposals may have been, little serious progress was made (not least because the re-engined designs were unlikely to perform any better than the piston-engine variant) until Hawker began to learn more about the gas-turbine engines being developed by Rolls-Royce, particularly the new B.41 engine (which became the Nene).

More design studies were initiated, including the P.1031, which comprised a Sea Fury airframe fitted with a B.40 jet engine in the nose section, replacing the Centaurus piston engine and exhausting under the forward fuselage; this produced an aircraft with an unusual outline, but a performance scarcely better than the standard piston-engine Sea Fury. Rather more advanced was the P.1035 initiated in 1944, which was also based on the Fury airframe, but with a B.41 jet engine mounted amidships, and with the exhaust in the tail section of the fuselage. The air intakes for the engine were positioned in the wing roots (the B.41's double-sided centrifugal compressor enabled air to be fed easily into it from both sides of the fuselage) and, as the design was refined, the exhaust was also similarly modified so that it fed through bifurcated (split) pipes emerging on either side of the fuselage behind the wing trailing edge, thus avoiding the predicted loss of valuable thrust caused by a long jet pipe to the tail. This development was given the designation P.1040, and in 1945 Hawker's Project Office continued to develop the design until it reached a stage where it demonstrated a clear potential for significant improvements in performance over earlier designs, and a decision was made to manufacture an unarmed prototype, funded entirely by the company.

Although Hawker was sufficiently confident in the design to allocate funds to a prototype, the Air Ministry remained distinctly sceptical, believing that the design would offer little improvement over existing Meteors, which were still being produced. However, a much more favourable response was received from the Admiralty, which concluded that Hawker's design would certainly perform better than the Supermarine Attacker that was entering FAA service (the Attacker was a relatively simple design that, in much the same way as Hawker had considered a jet-powered Fury, was in essence a jet-powered Spiteful, even retaining the latter design's tailwheel undercarriage layout). The Admiralty quickly issued Specification N.7/46 around the new design, and this eventually led to the production of the elegant (and remarkably successful) Sea Hawk. Stung by the Air Ministry's lack of enthusiasm for what proved to be an excellent aircraft, Hawker's Sydney Camm later commented in a company memo, 'Thank God for the Navy.' Production of the Sea Hawk was eventually transferred to Armstrong Whitworth so that Hawker could devote its production resources to other projects, but the P.1040 was undoubtedly an important developmental link in the process that had begun with the jet-powered Sea Fury concept, and which ended with Hawker's ultimate fighter creation, the magnificent Hunter.

Although the Sea Hawk did not make its first flight until 2 September 1947, Camm's design team was already looking at more advanced developments of the P.1040 design by this stage. A year previously, while looking at studies that might meet the requirements of the Air Ministry Specification E.38/46 (to undertake studies into the aerodynamic properties of swept wings), Camm had approved the concept of mating the P.1040 fuselage with a new swept wing under the designation P.1052 (which also initially anticipated the use of a rocket engine as a means of propulsion, although this project was ultimately allocated to part of the continuing P.1040 programme), and research on this proposal captured the interest of the Royal Aircraft Establishment at Farnborough. Although a great deal of research data on swept-wing designs was already available by this stage, the bulk of the captured German information was still jealously guarded by America while, by comparison, Britain's technical knowledge of the subject was undoubtedly sketchy, to say the least. Hawker and the RAE exchanged a great deal of information on the proposed wing layout before settling on a pressure-plotted 31ft 6in span and a sweep of some 35 degrees, but with an un-swept tail unit. The power plant was to be a 5,000lb Nene 2, and Hawker expected the aircraft to be capable of reaching a speed of 560mph at 36,000 feet. The first prototype of the new P.1052 flew on 13 April 1949 and quickly demonstrated that it was more than capable of achieving its predicted performance figures.

As development of the P.1052 progressed, Rolls-Royce was also making further progress with the Nene, and with the increased use of magnesium alloy (to reduce weight) and the installation of an afterburner (reheat) unit, a further twenty-five per cent increase in thrust was predicted. This development (the Tay engine), together with a suitably modified P.1052 airframe, captured the attention of the Australian Government, where a requirement was emerging for a new fighter aircraft to replace the existing force of Meteors. The second P.1052 prototype was selected for modification, and following trails with the first prototype, which had been conducted from Boscombe Down (including combat exercises flown against Meteors), the opportunity was also taken to introduce swept tail surfaces, as these early P.1052 flight trials had revealed a lack of adequate elevator control. In all other respects the aircraft had performed remarkably well, apart from a tendency to 'Dutch roll' (an out-of-sequence yawing and rolling movement sometimes described by pilots as something similar to pedalling a bicycle backwards!). In addition to the new swept tail surfaces, the Tay engine's reheat system was obviously not suitable for a bifurcated exhaust layout, so the rear fuselage was also modified to incorporate a straight-through jet pipe (which, with some irony, was taken from an Attacker) emerging under the tail.

VX279, complete with the revised swept tailplane and fin, in company with Supermarine's competing design, on show at Farnborough.

Unfortunately, the development of the Tay engine was rather more protracted than Rolls-Royce had anticipated and eventually Hawker was forced to continue work on the new aircraft (designated P.1081) based around the earlier (and less powerful) Nene power plant. Not surprisingly, Australia eventually began to lose interest in the project as the delays mounted, and after a North American Aviation mission visited the country to look at the possibility of selling Australia the F-86 Sabre, the choice was finally made to abandon the P.1081 in favour of the Sabre (albeit fitted with a Rolls-Royce engine). The prototype P.1081 was eventually completed and showed considerable potential, but without any prospect of a production order it was transferred to the RAE at Farnborough to conduct high-speed research flights. Sadly, the aircraft crashed on 3 April 1951 and, although the cause of the accident is still unclear (it may well have inadvertently gone

The second prototype P.1052, illustrating the straight tailplane inherited from the Sea Hawk. *Hawker*

supersonic and, without adequate elevators, control may have simply been lost), Hawker's Chief Test Pilot, Trevor Wade, was killed after having attempted to operate his ejection seat at an altitude that was judged to be too low for successful abandonment. The aircraft, possibly slowed by the drag of an open cockpit, survived the crash almost intact (which says a great deal about the rugged nature of Hawker's construction), but the P.1081 programme effectively ended with this tragic incident.

It is interesting to note that even the sometimes cynical Air Ministry was suitably impressed by the P.1081's performance, and towards the end of 1948 serious consideration was given to a programme to replace RAF Meteor fighters with the new design. However, development of the jet engine was now beginning to progress quite rapidly, and Hawker's main interest had already turned towards even more powerful engines, particularly the new AJ.65 axial-flow turbojet that was being developed by Rolls-Royce. With an anticipated thrust of some 6,500lb and potential for even more power (effectively the same as the combined thrust of the Meteor Mk 4's two engines), the new engine would enable Hawker to produce a fighter design that would meet the requirements of the still-outstanding F.43/46 requirement, using just a single engine (of much-reduced diameter, thus reducing drag) instead of the twin power plants envisaged in the original specification. Unfortunately, the development of the AJ.65 (which became the Avon) was a difficult and protracted process that extended to some seven years, but at the time of Hawker's initial interest there was no reason to imagine that the engine would not be available within a much shorter time scale, even though many people felt that the Air Ministry's (and Hawker's) interest in the engine was premature. In many respects the Armstrong-Siddeley Sapphire (developed from the wartime-era Metrovick F9) was a better, more reliable, less expensive and more 'mature' design, but Camm's traditional enthusiasm for Rolls-Royce and its Avon was undiminished, so Camm's design team embarked upon a new fighter design based around the new engine. (In fact, Hawker produced no fewer than twelve different design studies before settling on a 'definitive' choice.)

As the project progressed, conferences were held at the Ministry of Supply, and it became clear that the long-standing Specification F.43/46 had effectively been abandoned by default, so a new Specification (F.3/48) was issued, directly based on Hawker's plans for a fighter that would be capable of intercepting aircraft with the performance of the then new high-altitude jet bombers emerging from both the USA and the Soviet Union. Issued to Hawker in March 1948, F.3/48 called for a single-seat land-based day-interceptor and fighter, able to achieve Mach 0.94 (620mph at 36,000 feet or 724mph at sea level), armed with either four 20mm Hispano cannons or two of the new 30mm Aden guns. A radar-ranging gunsight would be incorporated together with an ejection seat system for the pilot, and fuel endurance would be some sixty minutes (plus climb-to-height and ten minutes of combat power). Operational Requirement OR.228/3 (drawn up around the design) specified that the power plant would be either the Rolls-Royce Avon or the Armstrong-Siddeley Sapphire (both types wisely being specified in case further development or production difficulties were encountered with either power plant).

Based on this requirement, a time-to-height (45,000 feet) of less than six minutes from engine start would be necessary, so that the aircraft would have sufficient time to intercept an incoming bomber at a relatively safe distance, after having been identified on ground radar (AEW radar for the RAF still being some way off). At a higher altitude of 50,000 feet the aircraft was expected to be able to achieve a climb rate of 1,000 feet per minute and be capable of reaching 547 knots at more at 45,000 feet. The specification also covered the aircraft's airfield performance, and a take-off distance of 1,200 yards was required to clear a 50-foot obstacle. This would ensure that the aircraft was compatible with the length of typical RAF runways, although in subsequent years a number of fighter stations were given runway extensions to enable operations to be conducted rather more comfortably and safely.

For armament, the P.1067 was initially studied for compatibility with a single 4.5-inch recoilless gun, but when development of this weapon was abandoned Hawker concentrated on the 30mm Aden Cannon (Aden being an acronym derived from the Armament Development Establishment development of the German Mauser MG312 cannon, manufactured at Enfield), studying the use of two such weapons with 10 seconds of firing time (200rpg) or four guns with 7.5 seconds firing time (150rpg). As an alternative option, the old Hispano 20mm cannon was still incorporated into the design as an alternative weapons fit, and likewise the provision for air-to-air rocket batteries or beam-riding air-to-air missiles was also established at an early stage, even though such concepts were still in only an embryonic state.

Hawker started work on its finalised design choice – the P.1067 – which ultimately (and perhaps unsurprisingly) bore a distinct external similarity to the earlier P.1081, but only after a series of fairly radical changes had been made to the aircraft's proposed layout. Initially, the design featured a T-tail (the tailplane mounted on top of the fin), twin underwing intakes and a fattened elliptical fuselage. This rather unrefined creation was gradually developed so that the intake was repositioned to the nose (with a simple pitot-style design rather like the French Mystere), and the fuselage was switched to a more conventional circular cross-section. The four cannon in the nose were repositioned so that two were now fixed in the wing roots, but concern was expressed by avionics experts that the revised intake had resulted in a centrally mounted shock cone that would be too small to accommodate the projected ranging radar that was to be fitted inside it. Likewise, Rolls-Royce expressed its own concern that the long intake ducting would reduce the engine's efficiency, and suggested that Hawker look at shifting the intakes to the wing roots. Sydney Camm resisted the pressure to reposition the intake again (a move that would effectively lose the two wing-root-mounted guns), but having also been concerned at the increasing lack of space in an already cramped cockpit (where the Air Ministry gradually specified the addition of more and more equipment), it eventually seemed clear to Camm that wing-root intakes would be the only practical option, especially when it became obvious that the revision of the intakes would free up more vital fuselage space for fuel capacity. Consequently, a pair of P.1081-style intakes was incorporated and fifty per cent additional space was created in the cockpit at a stroke, albeit at the expense of losing two of the aircraft's cannon armament.

The aircraft was now approaching what would be its finalised form, and it proceeded to a mock-up stage (the P.1067/1) within the experimental shop at the company's Kingston factory. Considerable attention was then given to the tail area of the design, after wind tunnel tests suggested the possibility of control problems associated with the high-set tailplane (something that was eventually to cause Gloster plenty of difficulties with its Javelin design). Ultimately, the wing sweep of the tailplane was increased and its position shifted to a location further down the fin, leaving Camm's team to redesign the upper portion of the fin with a graceful sweeping curve that would become a trademark characteristic of the completed fighter.

Having reduced the armament to just two cannon, Sydney Camm remained confident that it was more than sufficient for the new fighter; indeed, it was also the chosen option for the new Swift fighter, which was being designed by Supermarine at the same time. However, C. F. G. Gray (one of the Hawker installation design staff) suggested to Camm that no fewer than four guns, together with their magazines, could be incorporated into a removable pack if it was positioned directly behind the cockpit. Camm had always wanted the P.1067 to be an adaptable and simple design incorporating easily removable components that would be capable of interchangability wherever possible, so the concept of a removable gun pack (which could be rapidly replaced by a pre-loaded pack, thus improving aircraft turn-round times significantly) appealed to him immensely. The cannon would have only a seven-second firing time (at 135rpg) in order to give a more reliable ammunition feed, but the pilot would be able to select only one of the two pairs if necessary, to extend the firing time. Together, the four guns gave the new Hawker design a blistering destructive power that was (at the time) unsurpassed by any other operational fighter.

The mock-up of the original P.1067 nose intake and ducting, illustrating the restricted area available for the cockpit between the intake trunks. *Hawker*

The full-size wooden mock-up of the P.1067. By this stage in the design process the engine air intakes had been relocated at the wing roots, but the tailplane was still positioned on top of the fin. *Hawker*

By February 1949 the basic shape of the P.1067 had been established. Only the tailplane position was subsequently revised after wind tunnel tests revealed that a lower position on the fin would create better aerodynamic properties. *Hawker*

Most significantly, the P.1067 compared very favourably with the North American F-86A Sabre, which was being produced in large numbers across the Atlantic. Although the F-86 was a fast and agile fighter, its armament was rather less impressive, comprising just six 0.5-inch machine-guns, which were completely outclassed by Hawker's new concept. On the other hand, the F-86E was to be fitted with an 'all flying tail' – a powered tailplane with fully boosted linked elevators that would inevitably give the Sabre a much better manoeuvrability. With Canadian-built Sabres being ordered for delivery to the RAF as a 'stop-gap' pending the arrival of the new Hawker and Supermarine fighters, it is perhaps surprising that more attention was not given to this deficiency in the P.1067. However, despite this, in many respects Hawker's new fighter would still prove be superior to the Sabre, particularly in terms of armament (Sabre pilots reported that on some early gunnery practice sorties, it was sometimes difficult to determine whether the holes in banner targets had been created by the Sabre's guns or runway tears!). The P.1067 would also possess more stable and pleasant handling characteristics, many of which were due to the its relatively thick wing roots (necessitated by the position of the intakes), which gradually tapered with the wing sweep. It is debatable whether such aerodynamic refinements were inherited by design or good fortune, and although it seems only fair to credit Hawker with some technically sound foresight, other subsequent aspects of the P.1067's development programme illustrated that not all of the aerodynamic qualities of the new design were anticipated by the design staff.

Chapter 2

Teething troubles

By April 1949 the basic design of the P.1067 had been fixed, and assembly of the construction jigs for the prototype airframes got under way towards the end of the same year. Of the three planned prototypes, Hawker proceeded to complete one of the aircraft with an Armstrong-Siddeley Sapphire engine, largely as part of the 'insurance' policy against possible delays or deficiencies with the continuing Avon programme, although it was agreed that if both engines lived up to their potential, aircraft powered by both types would be ordered into production. The construction jigs were 'hand-built', based on the design staff's figures and practical experience with the full-scale mock-up. Although by modern standards this sort of practice might seem expensive, outdated and impractical, at the time of the new fighter's development Hawker (like every other British aviation manufacturer) was still very much set in the ways of wartime production procedures; indeed, Sea Fury aircraft were still under production at Kingston, and a great deal of the company's resources were being devoted to the gradual transfer to Sea Hawk assembly. Ultimately, when large-scale manufacture of the Hunter commenced, Hawker transferred Sea Hawk production to Armstrong Whitworth at Coventry, and only thirty-five Sea Hawks were actually completed at Kingston, despite it being every inch a Hawker product.

The Hunter F Mk 1.

As the P.1067 was a completely new aircraft, the company had to start from the very basics and build a completely new production set-up at a time when production of other types of aircraft had been allowed to run down, with the loss of a considerable amount of skilled labour, which sometimes seemed almost impossible to replace. An even more serious aspect of Hawker's predicament was the need to find firms that could make large jigs and tools in the numbers required – 3,250 tool designs and about 40,000 jigs, tools and fixtures. The Government's new 'super-priority' scheme (more of which later) only partially produced satisfactory results, chiefly because priority had to be allocated to so many items, and this meant that suppliers and sub-contractors were often unable to give any major preference even if they had been instructed so to do. Additionally there was soon to be a continuing demand for exports (see Chapter 10).

Eventually, tool-making difficulties were overcome by a combination of two methods. The first took care of initial detail and sub-assembly work and involved a compromise – the development and use of 'rough tools', prepared not by the virtually unobtainable toolmakers but by skilled fitters. Although capable of manufacturing several hundred parts, these tools were intended as a temporary measure. Such methods obviously could not be employed for major assembly jigs, and in this case it was a matter of searching far and wide until someone with the capacity to accept orders could be found. Eventually a large number of jigs were procured from such Italian firms as Macchi, Breda and Fiat.

Conforming to Second World War practice, Hawker had decided that the P.1067 should be divided into major units that could be completed individually and would be available as fully equipped spares if needed. Their size depended partly upon structural design and partly upon service requirements for transportation, but these units should not be confused with manufacturing sub-assemblies, into which the airframe was further sub-divided. It was broken down into six fully interchangeable major components: front fuselage, which carried the cockpit, armament pack, nose wheel and the cowling for the radar-ranging sight and camera gun; centre fuselage, with integral wing roots, engine supports and air intake ducts; detachable rear fuselage, with integral fin base and removable jet pipe/tail cone unit; tail unit assembly; and main planes, complete with the main undercarriage legs. For final assembly it was necessary only to connect together the structural members and to plug the pipes and leads together, and one reason for this method was the difficulty inherent in building a relatively complex aircraft in a number of factories some distance apart. During the early days of the P.1067 programme, the resources of the entire Hawker Siddeley Group were pooled for difficult items. Hawker Aircraft itself was far from being housed all under one roof. Production plans, therefore, called for the detail parts and main assemblies to be made in several places, with Kingston-upon-Thames as headquarters. There were two other factories in the south of England and assemblies were fed to the final assembly plant and airfield about thirty miles from Kingston. Certain specialised components were sent some 250 miles to Kingston from Hawker Aircraft (Blackpool) Ltd at Squire's Gate.

A production line image taken at Kingston, showing the forward bulkhead where the radar dish and nose cone were subsequently attached. *Hawker*

The Kingston production line, illustrating the wing construction jigs and the Hunter F.1 straight wing design. *Hawker*

Assembly work at Hawker's Blackpool factory, with a Hunter fuselage (mounted on temporary factory manoeuvring wheels) receiving surface finish attention. *Hawker*

Blackpool's assembly hall, showing a Hunter nearing completion. In the background is the aircraft's internal jet pipe ducting, mounted on a trolley. *Hawker*

Factory workers give a good indication of the size of the Hunter's internal air intake ducting as they complete rivet work on the inner face of the fuselage, at the forward transport joint. *Hawker*

A relatively rare image of the final stages of Hunter production as the aircraft are assembled for final surface finishing and air tests. This image illustrates the way in which the Hunter's construction was divided into a series of interchangeable sub-sections. *Hawker*

By April 1950 the first prototype's airframe had been completed and work had commenced on the second aircraft. Although the initial construction process went smoothly, there was one significant hold-up when, in May 1950, the Ministry of Supply suddenly announced that the Aden cannon (intended as the aircraft's primary armament) would be scrapped, claiming that it would be prohibitively heavy and could also affect the fighter's development programme. The Ministry suggested that the older Hispano cannon should be substituted, but after a great deal of heated communication between Sydney Camm, Hawker's Board of Directors and the Ministry of Supply, it became evident that the decision had merely been based on short-sighted cost concerns, and eventually the Ministry reluctantly agreed to proceed with the Aden. Even so, the first prototype was completed with nose ballast in place of the cannons.

Construction and flight testing of the P.1067 would have taken place at Hawker's Langley airfield (famous as the home of the wartime Hurricane), but although the airfield

A historic image of the Hunter prototype, WB188, at Kingston early in 1950, shortly after the engine had been fitted and the rear fuselage attached. *Hawker*

had been perfectly suitable for piston-engine designs, it was considered unsuitable to safely accommodate jet operations, especially when the ever-growing Heathrow airport was not too far away. (Likewise, much of the Sea Hawk's test programme was conducted from temporary facilities at Farnborough.) Consequently a disused wartime airfield at Dunsfold in Surrey was purchased by Hawker as a new home for flight testing (although Langley remained in business for a few more years). While Dunsfold was being prepared for test flying, the completed P.1067 prototype (WB188) was dismantled at Kingston and transported by road to Boscombe Down, where it was re-assembled and prepared for engine runs and test 'hops' on the airfield's huge runway.

Emerging from its hangar in June 1951, WB188 certainly looked every inch a modern jet fighter. Unusually, the airframe had been painted an attractive duck-egg green, applied by a Kingston-based paint finishing company that had identified the new aircraft's potential for publicity, and had offered to supply free paint for the prototype. The first engine run was conducted on 1 July and the aircraft was weighed (and the centre of gravity established) a couple of days later. Taxi runs were then commenced, and although these were conducted without any major difficulties, the prototype's differential braking system developed a fault on 8 July during a high-speed taxi run, causing the brakes to burn out and causing WB188 to veer off the runway. Dunlop rectified the problem in just ten days and the programme continued without any further delay.

The prototype, WB188, lifts off from Boscombe Down on its first flight. *Hawker*

A well-known but nonetheless magnificent publicity image of Hunter prototype WB188. *Hawker*

Following the tragic death of Chief Test Pilot Trevor 'Wimpy' Wade in P.1081, the initial flight testing of the P.1067 was assigned to Neville Duke, who had hitherto been Wade's number two, and was himself a hugely experienced pilot who had done a great deal of test flying in Hawker's earlier jet aircraft. Duke spent a great deal of time familiarising himself with the cockpit's layout, using the mock-up at Kingston, and he later recalled that

> '…a mass of instruments and controls was stipulated by service requirements, and we test pilots were told in effect that "these all have to go somewhere, so see what you can do with them". I also flew a Canberra at Boscombe Down to gain experience with the Rolls-Royce Avon engine, an experimental Firefly Mk 4 with powered controls similar to those fitted to the P.1067 and, from Hawker's Langley airfield, I also flew a Sea Fury fitted with a powered elevator system.'

Satisfied that the prototype was ready for flight, after a series of high-speed runs along the runway the P.1067 was topped up with 220 gallons of fuel and Duke took WB188 into the air for the first time on the afternoon of 20 July for a forty-seven-minute flight over Wiltshire, Hampshire and Dorset. Duke later commented that

> '…the Hunter seemed a pilot's aeroplane all the way from drawing board to roll-out; it looked right both on paper and in its finished form. The cockpit felt right and was just small enough to make one feel entirely a part of the aeroplane. It was slightly narrow at the shoulders, giving the snug effect common to British fighters, and the position of the seat gave a commanding view and a feeling that the aeroplane was an extension of oneself.'

Chief Test Pilot Neville Duke poses for the camera in a Hunter. This image clearly illustrates the canopy rails, the windscreen design, and the later boundary air duct on the upper surface of the port wing. *Hawker*

The first test flight included a climb to 32,000 feet where a series of fairly gentle handling manoeuvres was conducted, before Duke descended to 10,000 feet, where he completed a series of stall tests at various flap settings, with the undercarriage both raised and lowered. Overall, speed was kept subsonic, not exceeding Mach 0.88. After completing the flight, and removing his American-style 'bone dome' helmet (possibly the first time such gear had been worn by a British pilot), Duke reported that the aircraft had handled extremely well although there was some post-flight concern that it had consumed rather more fuel than Rolls-Royce had predicted; this matter was not resolved until further test flights had been completed, at which stage it was discovered that the rear fuselage fuel tank had a small leak, and had been losing fuel during each flight.

For the first flight, the elevator hydraulic boost had been disconnected, although the aileron's system was active (caution was exercised because so little practical knowledge of such systems was available at the time), but the result of this configuration was that the elevator forces were unreasonably heavy, while in comparison the ailerons were extremely light. Another minor problem was that the tailplane trimmer (electrically actuated) was set within an unacceptable range and Duke had to exercise a considerable amount of brute force to keep the aircraft set up for the first landing. Duke later commented that

> '...the UK was particularly backward in powered flying controls. This is why the Hunter first flew – as the P.1067 – with the elevators in manual mode. Any Hunter pilot will tell you that manual really does mean manual and it is the emergency means of control should the normal system be lost. This manual control, plus the all-moving trimming tailplane, which had inadequate trim range for landing, resulted in a very considerable two-handed heave-to round-out for the first landing.'

However, in overall terms the aircraft had displayed no significant vices and further test flights were made until 10 August, when it was transferred to Farnborough, where a further four weeks of test flying was conducted (Farnborough being selected as a suitable temporary base while Dunsfold was still being brought up to standard). Duke said of this period that

> '...these were the days of the "super-priority" production programme, which was a mixed blessing under which development was expedited, while the incorporation of improvements and modifications found necessary during flight testing was affected by the need to freeze production drawings. The Hunter was one of the few aircraft ordered in large quantities from the drawing board. The flight-test programme proceeded with few snags or delays, and the aircraft was being flown at speeds of more than 700mph within a month of the first flight. At first we were not

entirely sure of achieving supersonic flight. Mach number indicator [MNI] readings were unreliable and grossly in error at extreme high speed. But the supersonic boom soon provided confirmation and heralded the era of the phenomenon at Farnborough-type public displays (until people got fed up with it).

Needless to say, among the recipients of our early bangs were our friends and rivals Supermarine and de Havilland at South Marston and Hatfield. We wasted no time in letting them know of our progress by means of a couple of sharp detonations. They were struggling with the Swift and the DH.110 and weren't yet supersonic. After five sorties from Boscombe Down the P.1067 was flown on August 10 to Farnborough, where company testing was carried out until we moved our test facility from Langley to Dunsfold in Surrey on September 7, 1951. After only 11 hours' total flight test time the aircraft was being demonstrated at the Farnborough Air Show at speeds of up to 700mph, so development had progressed at a fairly satisfactory pace.'

In fact, it was not until April 1952 that the P.1067 officially broke the sound barrier. After completing the initial series of flight tests from Farnborough, WB188 was grounded while a production-standard version of the Rolls-Royce Avon RA.7 was installed. When the aircraft re-emerged, it had officially been named 'Hunter' and flight testing began from Hawker's new base at Dunsfold. The aircraft's flight envelope was gradually expanded and climbs up to 40,000 feet were quickly explored. However, much to the Hawker team's surprise, the Hunter did not seem to handle particularly well as it approached supersonic speed. Severe rudder vibration and airframe buffeting became pronounced as the speed approached 0.97M and, although Duke took WB188 to Mach 1.3 in a test dive during April, the instrument readings could not be relied upon, thanks to the effect of shock waves on the pitot head. For some time it was feared that the Hunter's safe limiting speed may be lower than hoped, but a relatively simple 'fix' was finally found, as Duke explained:

'Many fixes were tried but the remaining buffet precluded pushing beyond 0.97M. In an endeavour to cure these troubles various modifications were tried: spoilers on the fin, a combination of fin and tail turbulators, a short rudder, an additional dorsal fin, and modified rear-end shapes around the jet pipe area. Our efforts finally bore fruit on June 6, when we were rewarded with a first vibration-free flight after fitting a bullet fairing to the fin and tailplane trailing-edge junction.

Having researched photographs of certain Russian and American aircraft, I noted some fitted with a rear-facing bullet fairing. Shortly after submitting these pictures to Kingston, a similar bullet was incorporated (I believe on the basis of trying anything) and we went supersonic immediately – not, as sometimes stated, after encouragement by Ministry of Supply test pilots.'

Indeed, with the new tail fairing fitted, the Hunter quickly achieved 1.06M in a thirty-degree dive from 45,000 feet, although no indisputable proof of the Hunter's supersonic ability was available until 24 June when a test dive from 30,000 down to 15,000 feet produced a very audible (and for the Hawker team, a very satisfying) sonic boom.

As mentioned previously, by this stage the Hunter had become part of the Government's 'super-priority' production programme. After years of lethargy, the new Churchill Government that came into power during October 1951 had immediately turned its attention to the sorry and directionless state of Britain's defences. The Korean War was already in progress and the British contribution to the United Nations force in the theatre was embarrassingly small, comprising mostly naval air power that was only capable of conducting fighter-bomber duties under the protection of American fighter cover. The RAF simply did not have a fighter aircraft that was capable of taking on communist-flown MiG-15s, and, following the experience of the RAAF Meteors that operated in Korea, it was evident that Gloster's first-generation jet design was already nearing obsolescence.

This promotional image clearly illustrates the revised fin shape that appeared during Hawker's attempts to improve aerodynamic stability during 1952. *Hawker*

Churchill's new Government put into place a programme that bestowed priority status on various projects in order to expedite progress. Although the concept had some merit, in practice it did not achieve much, chiefly because so many years of relative inactivity had drained Britain's aviation industry of both technical and practical capacity, and it is questionable whether it actually did much to hasten the completion of the Hunter's development programme. It is probably not worth contemplating how long the aircraft might have taken to reach RAF service had it not been part of the Government's programme.

However, progress was certainly made, and on 5 May 1952 the second prototype Hunter (WB195) made its first flight from Dunsfold in the hands of Neville Duke, complete with both a production-standard Avon engine and a working Aden gun pack (together with a gun-ranging radar in the nose cone). Much of the aircraft's early work was centred on development of a suitable air brake. The Hunter's sleek lines and relatively powerful engine had clearly given the aircraft an outstanding performance, but keeping its deceleration within combat limits proved to be a complicated task that (some would say) was never properly resolved.

WB195 on display at Farnborough, sharing the limelight with Supermarine's competing design that eventually entered RAF service as the Swift. Unlike the Hunter, the Swift's service career was relatively short and continually plagued with problems.

Originally, the P.1067 design employed the use of small brake flaps fitted into the upper surfaces of the wing – similar in nature to those fitted on earlier Hawker jet designs. However, these were intended to work in company with strengthened landing flaps, but on the Hunter this system quickly proved to be inadequate. Various modified versions of the air brake layout were tested but none proved satisfactory, and generally caused the aircraft to pitch down quite seriously when they were deployed. The Air Fighting Development Squadron at West Raynham (tasked with the evaluation of new fighters prior to their entry into service) sent pilots to Dunsfold to fly sorties in a Hunter, and (as one pilot succinctly put it) when the air brakes were operated 'the pilot headed for the canopy'. Neville Duke recalled that

> '...the P.1067 was designed with an air brake system. It consisted of landing flaps coupled with small dive-recovery flaps on the upper inboard wing surfaces, and had been tested satisfactorily on the P.1052 and P.1081. But the P.1067 dive-recovery flaps did nothing and the anticipated nose-down effect from the very powerful landing flap/air brake was quite unacceptable. During tests we nearly got the answer with third-span landing flaps, which were extremely successful and powerful up to a maximum speed of 610kt (or Mach number in excess of 1.0). Only in a small speed range around 550kt/0.94M were there unacceptable trim changes.'

The air brake saga proved to be one of the most persistent difficulties in the Hunter development programme and ultimately resulted in an airframe modification that inevitably had all the appearance of an afterthought, and marred the elegant lines of the fighter throughout its operational career.

The third of the Hunter prototypes – painted in a more traditional overall high-speed silver finish – was WB202, first flown on 30 November 1952, again in the hands of Neville Duke. Unlike the two earlier prototypes, which had the 6,500lb Avon 103 and 7,550lb

As part of the early air brake trials, perforated split flaps were flown on test flight, but proved unsatisfactory. *Hawker*

The Richmond Road experimental shop with WB195 in the background and the mock-up of WB202 for Sapphire engine installation. In the centre is the fuselage of WB202 under construction. *Hawker*

Avon 107 respectively, the third prototype was powered by an Armstrong-Siddeley Sa.6 Sapphire engine rated at 8,000lb, and was in effect the prototype Hunter Mk 2. Unusually, production Hunters had already been effectively ordered 'off the drawing board', with the first ITP (Instruction to Proceed) having been issued on 20 October 1950, leading to a contract for the production of 198 aircraft at a unit cost (including power plant) of some £172,000.

By the time of WB202's first flight, it had become clear that a significant number of early production Hunters from this initial batch would need to be assigned to various trials programmes before the Hunter could be cleared for entry into RAF service. In retrospect, there is no doubt that the construction of just three prototypes was wholly inadequate for an aircraft that was to form such an important part of the RAF's post-war structure. If more prototypes (or a batch of pre-production aircraft) had been ordered, the Hunter may well have entered RAF service perhaps more than a year before it finally reached operational status. Hawker did do a great deal to hasten the production effort, and while trials with the prototypes continued, the company's newly acquired factory at Blackpool was prepared to begin production of 26 aircraft, while the Armstrong Whitworth factory at Coventry (Baginton) was prepared to commence construction of forty-five Sapphire-powered Mk 2 aircraft.

In many respects the production facilities at the AW factory were actually superior to those at Kingston, where sub-sections had to be fabricated and transported to Langley for assembly before being transported to Dunsfold for flight testing (and sometimes transported back to Kingston again for modifications). At Coventry the assembly and flight testing could all be conducted at the one site, and even the Hunter Mk 2's power plant was also part of the same Armstrong Whitworth set-up.

No 257 Squadron operated the Hunter F Mk 2 for only three years, and pictures of the unit's markings are fairly rare. This well-known image was taken at Colerne at a stage when WN907 had been retired.

Langley's assembly line of Hunter cockpit sections, with semi-complete airframes just visible in the distance. *Hawker*

Langley's final assembly shed, with completed airframes visible in the distance. *Hawker*

As the Hunter test programme continued, generally with few truly serious difficulties, the Air Ministry's Operational Requirements Branch was already looking at future developments of the aircraft, particularly the possibility of squeezing more speed, range and endurance from the airframe. Hawker addressed this future requirement by studying a truly supersonic version of the Hunter – the P.1083 – which featured a fifty-degree swept wing and a Rolls-Royce RA.14 Avon with an afterburner unit. Manufacture of the new wings for this project was commenced at Kingston in October 1952 while the early Hunter prototypes were busy conducting their pre-service tests from Dunsfold. In anticipation of the new design, WB188 was withdrawn from the test flight programme, and while WB195 transferred to Boscombe Down to begin initial operational trials, WB188 was fitted with an afterburning Avon and internal wing fuel tanks. The difficulties surrounding the design of a suitable air brake prompted Hawker to completely abandon its attempts to refine the various wing brake configurations, and instead the company opted for a completely new air brake fixed within the rear fuselage. Consequently, WB188 reappeared with twin clam-shell brakes, one on each side of the rear fuselage, powered by tough hydraulic rams. Removing WB188 from the main Hunter flight test programme was possibly not the most sensible decision, but perhaps Hawker was already anticipating the need to avoid future delays with the proposed P.1083, and made the judgement that it was better to divert resources from the early Hunter in order to pave the way for the later, more capable variant.

WB188's unique air brake installation. *Hawker*

However, regardless of Hawker's motives, it was the Air Ministry that ultimately threw Hawker's plans into chaos when senior staff visited Kingston and announced that, with the end of the Korean War now a reality, future policy appeared to be drifting back towards an almost traditional obsession with short-term economy. It had been decided that work on the increased swept wing would no longer be necessary, and while the Directorate of Military Aircraft Research was keen to explore the potential of larger 'dry' Avon engine developments, and the Directorate of Operational Requirements wanted to continue exploring the reheated Avon, there was no longer any real requirement for the P.1083 as a stand-alone programme. Consequently, construction of the prototype (WN470) was halted on 13 July 1953 when the project was officially terminated. It is certainly true that, with a new Avon engine emerging that could deliver the same 'dry' thrust as the reheated RA.19, the concept of using a heavy and thirsty reheat system did not seem quite so attractive, but with the luxury of no obvious immediate military threat to occupy their thoughts, it is little wonder that the Air Ministry's appetite for ambitious developments had disappeared again. Of course, this decision also left Hawker with WB188, which had been extensively modified and was now effectively redundant.

Despite this major setback the Hawker design team opted to use the aircraft to study reheat developments still further, and with a surprisingly modern instinct for good publicity they also seized the opportunity to make an attempt on the World Air Speed record, which was currently held by an American F-86D (at 715.15mph). Rolls-Royce prepared a 'racing' version of the Avon – the RA.7R – which delivered a thrust of 6,750lb, rising to 9,500lb with reheat (complete with a two-stage nozzle), and with this engine installed WB188 became the sole Hunter Mk 3. The nose section of the aircraft was

The all-red WB188 pictured after conversion to Mk 3 standard, prior to embarking on high-speed record test flights. *Hawker*

modified to incorporate a sharper cone, while the canopy windscreen was redesigned to create a more aerodynamically smooth contour (and a better view for the pilot). Strangely, however, Hawker opted to remove the clam-shell air brakes (restoring them after the record attempt), and the doors were skinned over (this was done because of the recognised potential for unnecessary drag created by the brake actuating fairings that emerged on the fuselage sides). Painted in an eye-catching red colour scheme, WB188 was ferried to Tangmere in Sussex and Neville Duke flew his first practice run on 31 August.

Duke was familiar with the designated high-speed course that had been established off the South Coast between Bognor, Littlehampton and Worthing. He had already served with the RAF's High Speed Flight in 1946 – a unit that had already established an earlier record of 616mph with a Meteor, using the same course. Duke attained 722.5mph on his first practice run despite less than ideal weather conditions. On the second day the run had to be abandoned thanks to an undercarriage malfunction, as Duke described:

> 'The starboard undercarriage leg came out at 550-600kt at low level on a run in to the measured distance by Bognor Pier. It says all that is necessary about the structural integrity and controllability of the Hunter in extremis that this incident and the ensuing landing kept WB188 out of the air for only six days. The record was taken on the day it next flew.'

It was on 7 September 1953 that Duke successfully achieved a speed of 727.6mph during three runs over the course, and even to this day this figure still stands as a British record for speed at low altitude. (A later – and successful – attempt by Supermarine with its new Swift was something of a 'fix' as it was flown over Tripoli harbour where ambient conditions made the speed more easily attainable, and when the Fairey FD2 flew even faster it was at 38,000 feet). With the record attempts completed, WB188 was again redundant and, although it could have made further valuable contributions to the Hunter development programme, restoring it to Mk 1 or Mk 2 standard would have been prohibitively expensive, so this historically significant aircraft was withdrawn from use.

The Hunter Mk 3 was a one-off and, with a valuable prototype lost from the trials fleet, the remaining aircraft were worked even harder. The third prototype, as described previously, was WB202, the first Hunter to be fitted with an 8,000lb-thrust Sa.6 Sapphire engine and certainly the first that had the outward appearance of a traditional fighter aircraft, being finished in standard Fighter Command high-speed silver. Assigned to gun-firing trials, the aircraft began ground firing tests in February 1953 and airborne tests commenced from Boscombe down during the following month. From the outset, the firing trials looked encouraging, although there had been some initial concern based on experience with a series of test flights that had been made with an Aden gun installed in a Beaufighter, which suggested that there was potential for damage to the Hunter's undersides from spent cartridge cases and links. However, no significant problems were encountered with the Aden and WB202 was eventually re-tasked with Hawker's seemingly insurmountable problem of how to fit the aircraft with a suitable air brake.

Experience with the now abandoned F.3 indicated that fuselage clam-shell brakes were not an ideal solution as they also created a significant change in pitch when they were operated. Having explored all the options (using WT566, the twelfth production Hunter F.1, and WB202), it became clear that the only practical location for an air brake that would be satisfactory for the Air Ministry's performance requirements would be under the rear fuselage, and this meant that a neat (and aerodynamically clean) built-in air brake would not be possible – it would have to be literally added on to the outer skin. In accordance with this decision, WB202 was equipped with a vernier-mounted ventral air brake, which was flown on eight test flights, each with a different brake deflection angle set on the adjustment rails. Even at this stage the problem was still not solved, and it was not until further flights were conducted with WT566 that a production-standard air brake was tested and eventually fixed at an optimum deflection angle of sixty degrees, with further refinements being made to the aerodynamic fairing fixed ahead of the air brake mechanism on WT573, enabling the brake deflection to be set at a maximum of sixty-seven degrees. It was not until June 1954 that the air brake saga was finally resolved, but it had ultimately proved impossible to perfect an ideal system for the Hunter in this retrospective post-production manner.

The finalised air brake looked like (and was) an afterthought, and was far from ideal. Most inconveniently, its ventral position meant that it could not be used on a landing approach and it was automatically locked up when the landing gear was extended. Hawker's frustration at the rather 'fudged' nature of the air brake outcome was compounded by the fact that early production Hunters were now already emerging from the production lines, only to be rendered unfit for service acceptance until an air brake was fitted. Consequently, early Hunters left the three production lines only to be stored at Dunsfold and Kemble until air brakes could be fitted – hardly an ideal situation for a 'super-priority' programme.

On 16 May 1953 the first production Hunter F.1 (WT555) made its first flight from Dunsfold, illustrating how the flight test programme lagged behind the completion of what were meant to be production-standard airframes. The first few production aircraft (including the first sixteen) were used for the necessary trials work, which would have been performed by additional prototypes or pre-production aircraft had they been ordered. Gun firing, spinning and other expansions of the aircraft's envelope were conducted while other aircraft were flown on more specific test and evaluation tasks, including preparation of the Pilot's Notes, which would be vital before service entry could commence. WT558 flew trials with a wing fence design (subsequently dropped), WT559 flew canopy-jettisoning trials, WT562 was fitted with one-third-span landing flaps, and WT570 was equipped with full-power ailerons.

An early photograph of a Hunter F.1 allocated to short-term storage at Kemble. The clean lines of the aircraft have yet to be interrupted by gun link collectors, and the standard camouflage finish applied at Kingston has yet to receive unit markings.

The first production Hunter F Mk 1, WT555. *Hawker*

A limited Certificate of Release was finally issued on 1 July, and this enabled the first aircraft to be transferred from No 5MU at Kemble (where aircraft were held after being ferried from Dunsfold) to the Central Fighter Establishment at West Raynham, and during the same month a dozen aircraft (starting with WT577 and WT578) were delivered to the RAF, immediately embarking on evaluation flights with the unit's experienced pilots. Indeed, Hunters participated in Exercise 'Dividend' during their first month with the CFE and performed well despite their limitations. Successful intercepts on USAF B-45 and B-47 bombers were made, and for Fighter Command it was refreshing to hear of RAF Canberra pilots expressing some surprise at suddenly finding themselves vulnerable to interception from a brace of Hunters that were actually above their hitherto unattainable altitude.

A more minor difficulty experienced during the exercise was the unserviceability of aircraft left on West Raynham's exposed ORP (Operational Readiness Platform) overnight, caused by condensation affecting the aircraft's radio and radar equipment. But the Hunter's more serious deficiencies were also becoming clear, and the very fact that they were only able to operate to within 80 miles of their base during their first exercise illustrated how short on range the aircraft was. The originally specified range and endurance figures had largely been ignored as the Hunter programme progressed, and an internal fuel capacity of just 80 imperial gallons was patently inadequate by any standards.

Although Hawker had identified the glaring range deficiency, no provision for additional internal fuel had been attempted, and external fuel tanks were only just being investigated at this stage; only WB202 had been fitted with dummy drop tanks to begin an exploration of their aerodynamic qualities. The air brake problem had been all too evident (not rectified until the nineteenth production Hunter finally entered service with a ventral air brake), and the original requirement that combat damage from a single bullet would not lead to the loss of more than twenty per cent of remaining fuel had simply been overlooked, as had another requirement for gas purging of the fuel tanks and even the sun blinds for the pilot, although the early experiences of CFE pilots demonstrated that it was actually high-altitude misting and icing that was the main (and most dangerous) visibility deficiency within the cockpit. The 'all-flying tail' had yet to be incorporated, despite being a standard fit on all emerging American designs, but perhaps most importantly the Avon-powered Hunter proved to be far from compatible with the Aden guns with which most production aircraft would be equipped.

Wingtip-mounted fuel tanks were an idea that had some merit, but flight trials (conducted by Hunter F.6 XG131 seen here at Farnborough) demonstrated that the tanks created severe buffeting problems, and the idea was ultimately abandoned.

Hunter F.4 WT798 was used by Hawker to fly trials with dummy 230-gallon tanks. The project was company-funded but proved to be a successful project. *Hawker*

Although trials with the Aden gun had progressed relatively smoothly with the first Sapphire-powered prototype, nobody had predicted that the situation would be very different when pilots of the Avon-powered F Mk 1 began to operate their weapons. To the dismay of service pilots (and to the horror of the Hawker and Rolls-Royce teams) the Avon suffered violent engine surging when the guns were fired, due to the ingestion of the Aden's exhaust gases. Although the Sapphire engine had coped almost faultlessly, the Avon engine's compressor was rather more temperamental, and it was quickly established that gun-firing engine surges could happen in a variety of situations, sometimes without any obvious warning or reason. At altitudes above 25,000 feet and speeds in excess of 400mph, engine surging seemed almost inevitable, and with the Hunter's airframe and power plant effectively 'fixed' by this stage, Hawker had a potential disaster on its hands. As a short-term measure, a system was developed that automatically reduced fuel flow to the engine whenever the guns were fired, and this provided a solution to the problem, although it was a less than ideal fix when one considers that the Aden guns produced a hefty recoil force when they were fired, and reducing fuel to the engine at the same instant was hardly a good idea. Likewise, firing the guns produced a very significant nose-down pitch, which did nothing to enhance the Hunter's ability as a stable weapons platform. As Neville Duke recalled:

'The gun-firing trials with the Sapphire Hunter went extremely well, and things looked good when the first production Hunter Mk 1 flew on May 16. The modified WB188 made its first flight as the reheat-equipped Hunter F.3 on July 7. All seemed well with the Hunter programme until we ran into the gun-firing results achieved at Boscombe Down with an early production Avon-powered Hunter Mk 1. At first engine surging occurred at altitude, but later (and after we had practised setting up the right combination of flight factors) this undesirable phenomenon could be obtained under a variety of conditions. Surge had not shown up at all during the gun-firing trials with the Sapphire Hunter, the difference lying in the variable swirl-vane system on the Avon engine. Fuel dipping (briefly reducing fuel flow as the guns were fired) eliminated the problem in due course, but the cure was a long and trying process coming at a politically awkward time, just when the Hunter was going into service with No 43 Squadron.'

In fact, the gun problems were never completely solved until later developments of the Avon engine were made available, which were able to handle the gun gas ingestion without undue difficulty, and the Hunter Mk 1 was doomed to remain permanently saddled with an engine that was incompatible with the weapons system of the aircraft it powered.

It was also with some degree of disappointment and resignation that the feared problems with ejected gun cartridges and links began to come true. The Beaufighter gun trials had predicted such a problem but there had been no early evidence to suggest that the Hunter would be affected, and it was assumed that the Aden's heavy cartridges would simply fall away from the aircraft. But as service experience began to grow and more firing exercises were conducted, aircraft were suffering damage and it was clear that something had to be done about it. The simple solution (and the one that was adopted) was to create a blister housing that caught the links as they were ejected, leaving the heavier cartridges to fall clear. This system worked well, but resulted in a pair of rather ungainly bulges appearing under the Hunter's fuselage, which quickly became known by service pilots as 'Sabrinas' in honour of an appropriately well-endowed television star of that era. Like the air brake saga, the ammunition link collectors also gave the impression of being hasty 'add-ons', which in essence they were, but like the air brake they became a familiar part of subsequent Hunter airframes (albeit with some exceptions that will be described later).

Not the best of photographs, but a nostalgic early illustration of one of No 74 Squadron's first Hunters, complete with 'Sabrinas' but devoid of underwing stores.

Intensive gun firing trials resulted in significant damage to the fuselage and air brake area. Subsequent lengthening of the case chute eventually solved the problem. *Hawker*

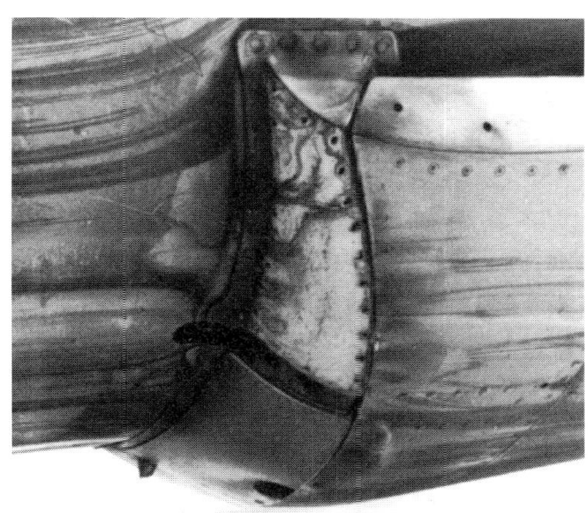

During intensive gun firing trials XE588 almost shot itself down after a 30mm shell ricocheted off the sea surface and entered the wing leading edge. *Hawker*

In addition to the link collection fix, a great deal of attention was also devoted to the vibration caused by the Aden's operation. The Aden gun pack was an impressive piece of kit and it was hardly surprising that, when it was fired, the airframe suffered a great deal of localised punishment, and the simultaneous operation of all four guns quickly led to the development of structural cracks in the forward fuselage. Stainless steel reinforcement plates were added in the appropriate areas and aircraft were restricted to operating individual pairs of guns (rather than all four) until the reinforcements were embodied into the production line schedule. Further modifications included the fitting of an air scoop under the gun pack, which opened when the guns were fired, ventilating the nose section and enabling gun gases to be vented. However, despite these problems, the Aden gun pack was an excellent system that worked very well, and was only troubled by relatively minor teething problems that could only be expected from what was something of a quantum leap in aircraft firepower.

It is fair to say that although the Hunter quickly demonstrated plenty of potential, the Mk 1 was by no means an ideal aircraft for the day-fighter role, and although its deficiencies could be corrected, the production of F.1s was already too advanced to make any fundamental changes to the aircraft, and it was accepted that, with the air brake problem having been solved, the aircraft would enter regular RAF service without any further significant changes.

The first operational squadron to re-equip with the Hunter was No 43 at Leuchars, and four aircraft were delivered to the squadron during July 1954, with more deliveries continuing until October, when the squadron began operational flying. Although aware of the Hunter F.1's faults, the squadron's initial experience with the aircraft was encouraging, even though a number of serious gun-related flame-out incidents occurred. Despite this, 43 Squadron soon set up its own four-aircraft display team, 'The Fighting Cocks', and continued to operate the aircraft within the limits that had been imposed upon it, pending arrival of later versions.

Meanwhile, at West Raynham the CFE was also becoming accustomed to the Hunter F.1's attributes – both good and bad. Apart from the engine surging problem, endurance was the main deficiency, and many pilots got into some very serious situations because of the aircraft's very short legs. One of the best-known incidents took place on 8 February 1956 when six students (accompanied by two instructors) from the Day Fighter School took off from West Raynham to begin a standard four-versus-four air combat exercise at 45,000 feet. Weather conditions were poor but the Met forecast suggested that things

No 43 Squadron had the distinction of being the RAF's first operational Hunter squadron. Although early experience with the Hunter was less than satisfactory, the squadron was often called-upon to 'showcase' the new fighter, as illustrated by this early publicity shot.

Hunter F.1 WT594 pictured on a pre-delivery flight prior to joining No 43 Squadron. *Hawker*

Hunter F.1 fighters from No 43 Squadron, based at Leuchars, sharing ramp space at Wattisham with F.2s from No 257 Squadron in February 1955.

would improve, but despite this the Air Fighting Development School (one of the CFE's component units) had cancelled flying for the day. The combat exercise progressed well and, because of the Hunter's fuel capacity, the ensuing dogfight was relatively brief and the Hunters began their return to base.

As they returned to the overhead at West Raynham at 20,000 feet it became clear that the weather had actually deteriorated and the airfield was now sitting under a 400-foot cloud base with fog restricting visibility to 1,000 yards. With only twenty minutes of fuel remaining, the decision was made to divert to Marham and the Hunters descended to 2,000 feet, separated by thirty-second intervals between pairs. Unfortunately, Marham's air traffic controllers were unfamiliar with the new fighter (especially fuel-starved ones), and to make matters worse the fog had also begun to roll over Marham too, so when the first pair popped out under the low cloud they almost immediately flew straight into the fog. The first aircraft's pilot (Red One) initiated an overshoot while his wingman (Red Two, who had lost sight of the leader) pressed on and successfully landed after a flight of just forty-two minutes. His leader proceeded to fly three more circuits before finally catching sight of the runway and landing, even though his aircraft ran out of fuel shortly after rolling off the runway.

The next pair fared less well and Yellow Three (who could only snatch brief glimpses of the ground) elected to climb away and eject. Meanwhile Yellow Four opted to continue his approach but crashed into a field just a couple of miles from Marham. Yellow Two, with just twelve gallons of fuel remaining, decided to abort his approach and climb away before ejecting. Yellow One, having made a low approach, was forced to climb away after spotting trees in his path, but after making a circuit of the airfield at 150 feet he settled onto the approach again only to run out of fuel, but landed straight ahead with a dead engine. Red Three abandoned his approach and climbed to 4,000 feet where he ejected, while Red Four abandoned his approach at 600 feet and climbed away until the aircraft flamed out at 2,500 feet, at which point he ejected. In just forty-five minutes a total of six Hunters had been destroyed (and one pilot killed) for no reason other than a shortage of fuel.

With some 150 Hunter F Mk 1s having been completed by the end of 1954, more units were able to take delivery of the new aircraft, and No 222 Squadron (another Leuchars-based unit) re-equipped in December of that year. No 54 Squadron at Odiham followed in February 1955, to be joined by co-located 247 Squadron in July. Because of the Hunter F.1's various problems, it was decided that Germany-based squadrons would not receive the aircraft, as there was good reason to keep the Hunter F.1s within economical range of UK-based Maintenance Units and the various manufacturer sites. However, aircraft were delivered to the Empire Test Pilot's School, the Fighter Weapons School at Leconfield and the Operational Conversion Units at Chivenor (229 OCU) and Pembrey (233 OCU) so that a steady through-put of new Hunter pilots could be maintained while gradually converting more experienced pilots from the Meteor.

Chapter 3

Into service: Marks 1 to 5

Although the introduction of the Hunter F Mk 1 into RAF service was a fairly complicated and sometimes frustrating affair for both Fighter Command and Hawker, the Hunter F Mk 2 enjoyed a slightly smoother transition into operational status. Having been produced with a Sapphire engine, the F.2 did not suffer from the same gun firing problems that dogged the F.1, and with the air brake difficulties having been largely solved, the Hunter F.2 was certainly a more successful and 'user-friendly' machine, even though it still had the same serious lack of range and endurance.

The F Mk 2 proved itself to be slightly faster (698mph at sea level as compared with the F.1's 693mph) and it also had a slightly better rate of climb. With hindsight, it is clear that the Sapphire engine should have been selected as the primary power plant choice for the Hunter, as it was a reliable and less-expensive engine than the Avon and did not suffer from gun-induced engine surges. Exactly why Sydney Camm maintained a very strong preference for the Avon engine remains unclear, especially when Armstrong-Siddeley (manufacturer of the Sapphire) was part of the same Hawker Siddeley group. It is possible that Camm's preference for a Rolls-Royce engine was purely a symptom of his traditional approach (having worked with Rolls-Royce engines for many years), but it is also worth remembering that when both the Avon and Sapphire were under development, there was no indication that the Avon's production would be so protracted, nor was there any hint that it would suffer from flame-out difficulties.

In total just forty-five Hunter F Mk 2s were produced at Coventry (the first being WN888, which took to the air on 14 October), and they followed the same path into RAF service as the F Mk 1, being delivered to Maintenance Units prior to being assigned to their operational bases. The first deliveries were made to No 257 Squadron at Wattisham in September 1954, and the co-located No 263 re-equipped in February 1955. Other aircraft joined miscellaneous units such as the CFE in much the same way as the F.1 had done some months previously.

Both the F.1 and F.2 were effectively regarded as being the same 'basic' early-production Hunter, which, although far from perfect, was a suitable aircraft with which Fighter Command could begin to build experience with a modern, transonic high-performance interceptor. Not surprisingly, Fighter Command's pilots flew their new Hunters intensively. In January 1956 No 43 Squadron embarked upon firing trials with the Hunter's Aden cannon, and although their pilots were restricted to gun firing in a

WN888, the first production Hunter F Mk 2 on display at Baginton. Although outwardly identical to the F.1, the Mk 2 was fitted with an AS Sapphire engine. *Hawker*

narrow band between 20,000 and 25,000 feet, the restriction was later relaxed so that firings could take place from sea level up to 25,000 feet at speeds up to 550 knots. The intensive trials were temporarily suspended while modifications to the guns were made (the fuel-dipping system as previously described), but eventually the flights resumed and continued until Fighter Command terminated the trials because of the heavy toll on aircraft serviceability – many Hunters had been flown so hard that they required plenty of attention from the Leuchars engineers.

As mentioned previously, 43 Squadron also formed its own display team, as did No 54 Squadron. One of No 54's display pilots managed to take his enthusiasm for the Hunter just a little too far, and managed to distort the canopy of his aircraft after flying too close to the jet exhaust of the team leader's aircraft. Other mishaps were quite common, as might be expected when a new (and comparatively fast and manoeuvrable) aircraft enters service. At the CFE, one aircraft (an F.1) collided with a Hunter F.2 during an air combat exercise, with one of the aircraft making a belly landing while the other landed more conventionally, albeit without a tail cone, starboard tailplane and elevator. This incident certainly said a great deal about the Hunter's rugged construction. Another CFE pilot managed to land safely after having flown his Hunter straight through a tree while reading his map. Some of the accidents were less serious, however, not least a commonly reported sequencing problem with the aircraft's undercarriage, resulting in a number of door retraction failures.

Not surprisingly, more accidents occurred at the Operational Conversion Units where student pilots were slowly becoming acquainted with the new fighter. One student inadvertently made a wheels-up landing at Chivenor, causing his aircraft to bounce violently back into the air, at which stage he elected to climb away and make a more conventional landing with the wheels extended. Even more exciting was the accidental (but successful) ejection of a pilot at 41,000 feet during a negative-g manoeuvre in an air combat exercise.

The OCU (229) at Chivenor received its first Hunters during April 1955, beginning a long and happy association with the aircraft that lasted for two decades. Students arrived at Chivenor having completed basic and advanced flying training and continued their flying on the aircraft with which they were already familiar – the Vampire T.11. However, having already learned how to handle the Vampire in pure flying terms, the OCU was tasked with the responsibility of teaching students how to use an aircraft as a weapons system, and for this purpose the OCU had a fleet of Vampire T.11 and single-seat Vampire FB.5s at its disposal, together with Mosquito TT Mk 35 target tugs to provide banner-towing facilities for the students. Of course, making the transition from the Vampire to the Hunter was quite a leap as the two aircraft were from distinctly different generations. The sudden increase in speed and manoeuvrability was certainly quite significant (the rather pedestrian flying attributes of the Vampire contrasted starkly with the sprightly performance of the Hunter), but the most important skill that the OCU students had to learn was to understand and manage the Hunter's low fuel reserves. Managing the aircraft's fuel state was vital, and a great deal of attention was given to this aspect of OCU instruction; indeed, most training sorties even required Chivenor's Air Traffic Control to radio fuel state reminders to each student as they flew their sorties.

Such was the demand for new Hunter pilots that a second OCU was established to take some of the workload from Chivenor. No 233 OCU was formed at Pembrey in 1956, equipped (like 229 OCU) with Hunter F Mk 1s, Vampire T Mk 11s and FB Mk 5s, together with a handful of Meteor F Mk 8 target tugs (which replaced the aged Mosquito).

Hunter F Mk 4 aircraft also joined the Central Flying School, with headquarters at Little Rissington but operating from nearby Kemble. Alasdair Liddle was one of a handful of CFS pilots who gained some first-hand experience of the Hunter F.4, as he recalls:

> 'Like so many youths in the early 'fifties, I fell in love with the Hunter at first sight. In fact, I joined the Royal Air Force in October 1958 to fly that beautiful aeroplane. Approaching "Wings" at Oakington a year and a half later, I had to complete a list

RAF Chivenor's flight line shortly after the arrival of 229 OCU's first Hunters. Initially, the OCU operated a mix of Hunter F.1 and Vampire (single- and twin-seat) jets. *Mark Gauntlett*

of preferences: I wrote, "111 Squadron", "43 Squadron" and another Hunter outfit, which also appealed. Shortly afterwards I was ordered into the Flight Commander's office, to be told that I was a fool and that I should have nominated Fighters, Light Bombers and CFS. However, he added, it did not matter as I was going to CFS anyway! Not what I wanted, but, he said, I could choose my posting after the instructional tour – and this did in fact lead to a splendid tour in Aden with No 43 (Fighter) Squadron, there to see and use the Hunter in its operational element. As almost all my course seemed to be going to Hastings, Shackletons and Beverleys, I acquiesced with my "Hobson's choice"!

Thus on 9 May 1960 I joined No 203 Course at "Hell on the Hill", as some called Little Rissington, although I actually enjoyed both the place and the course. Basically I was doing the same as I had been doing for the previous year and a half on the Piston Provost and Vampire, albeit now in the other seat – so maybe I knew no better. The big improvement was that I was now being treated more like an adult and an officer, rather than just another dim student! The great attraction of the course was the opportunity to fly the Hunter, or the Canberra – both lovely aircraft that I wanted to fly. When the "type flying" started, the Hunter was naturally my top priority, so I opted for that.

"No," I was told. "You'll have to fly in the Canberra, not the Hunter."

"Why?" I asked.

"First tourists fly in the Canberra," was the curt reply.

"Why?" I repeated.

"Not enough first pilot jet hours."

"What then is the minimum number of hours for the Hunter?"

"70 hours," was (if I recollect correctly) the answer.

So I rushed off to Shawbury at every opportunity and shot about a million QGH/GCAs, for the training of new controllers. Eventually I announced at Rissington that I now had the requisite number of hours and that I should like to fly the Hunter. To my immense disappointment, I was yet again told that I could not. "Why?" Answer: "Policy – first tourist!" End of conversation!

To spite myself, I declined to fly the Canberra. In some ways I regret that decision, so someone please ask me now!

Late in the evening of the final dining-out night, my boss, Squadron Leader Peter Hicks, asked me if I had enjoyed the course. I answered, honestly, that I had, but added that I was very disappointed that I had been unable to fly the Hunter. I think that there was a non-committal reply – but it was late at night!

On the following morning, the Commandant, while presenting me with my QFI certificate, mentioned that he had heard that I regretted not being allowed to fly the Hunter. I agreed. Air Commodore Whitworth then said, "Well, get a bloody move on. Pete Frame is in a Vampire T.11, waiting to take you to Kemble, which shuts today at 1300 hours!"

I cannot remember how I answered. I suspect that I spluttered my thanks as I exited his office – supersonically! Ever since that remarkable order, I have hoped, fervently, that I had expressed my gratitude properly; if not, excitement must excuse any lack of decorum. Within milliseconds I was leaping into the T.11, the Goblin already being wound up by Pete Frame, as I continued strapping in as we taxied out. I really cannot remember the short trip to Kemble. Etched forever in my mind though was the thrill of climbing up and into this gorgeous swept-wing Hunter F Mk 4.

After finally reaching Kemble, I was soon strapping in. "Sharky" Hastings was on the ladder, as he fixed an emergency checklist to my thigh and asked if I knew how to start it. "No!" I apologised with a grin. Sharky leaned down and pressed something in front of the "stick". "Bang!" went a cartridge, and the Avon 121 engine slowly wound up, the ladder was whipped away in a trice … and he was gone.

"Now what the hell do I do next?" I wondered briefly. After a simple taxi round the perimeter track, in my excitement I would scarcely notice much at all like, say, that hard high-pressure nose wheel tyre and oleo right beneath me, before calling for take-off, lining up and then catapulting down that centre-line of Runway 27. Needless to say, I twitched my way into the blue sky, as the Hunter's powered ailerons were considerably lighter than the Vampire's manual ones. There then followed 30 minutes of sheer bliss, rolling and looping in that 8/8ths clear, mainly Gloucestershire sky. What joy! All too soon, as fuel and curfew dictated, the short flight ended in a surprisingly smooth landing with those high-pressure tyres and that lovely wide track. This was, by far, the most exciting and memorable flight in my whole fifty-year career with 20,000 hours on a wide variety of aircraft types. Days later, I arrived at Cranwell as a brand new instructor, allegedly the youngest QFI at that time.'

Meanwhile, as the RAF's Hunter force slowly built up, other early production Hunters were retained by the manufacturer to conduct a variety of trials, some of which would have a direct bearing on the Hunter's future and some that would influence other programmes. One trial that is often overlooked in accounts of this period was the 'area rule' experiment applied to Hunter WT571. The theory was that airframe drag could be reduced (thereby improving endurance) by revising the contours of the Hunter's fuselage so that the aircraft presented a fairly constant frontal cross-section to the airflow as it moved across the airframe. In essence, this meant widening the rear fuselage so that it matched the cross-section area of the wings immediately in front of it. The result was a rather unusual Hunter with a fat rear fuselage, but the experiment worked and the aircraft did produce a slight increase in speed. The modification was not adopted for production Hunters, however, as the speed increase was fairly small and it created airflow separation around the tail that would have required further aerodynamic modifications to that area. However, the concept was sound and 'area ruling' was applied to many subsequent designs (the Buccaneer's 'Coke bottle' fuselage being a good example).

In addition to a completely revised flap design (and associated air ducting) WT571 was also fitted with a nose-mounted instrumentation boom. *Hawker*

Hunter F.1 WT571 was modified to investigate the properties of 'area rule', which resulted in a widened 'Coke bottle' rear fuselage. Speed increases were only small and the modification created significant airflow separation around the tail area, which would have required more design and development work, so the project was eventually abandoned. *Hawker*

Another Hunter (WT656) was fitted with 'blown flaps', which featured a compressed-air system drawing air from the engine and forcing it over the Hunter's landing flaps to increase lift, thereby reducing the necessary landing speed. Again, the trials were successful but production Hunters were never fitted with the system, simply because it was judged to be unnecessary.

Hunter F.1 WT656 was employed by the RAE to investigate the effectiveness of 'blown flaps', which circulated engine bleed air over the flap surface, thereby increasing lift. The trials were successful but the concept was not incorporated into any production aircraft. *Hawker*

But even as these early Hunters continued experimental work and others entered operational RAF service, attention had already turned towards developments of the Hunter airframe and power plant, which would transform the aircraft into a truly effective fighting machine. Most importantly, Hawker addressed the fuel capacity problem by installing eight small bag-type fuel tanks in the Hunter's wing design, as had been fitted to the one-off Hunter Mk 3, increasing fuel capacity from 337 to 414 imperial gallons. Likewise, after having successfully completed flight trials of an external drop tank, provision was made for two inboard wing pylons that could carry 100-gallon fuel tanks, and later aircraft would be equipped with a modified fuel system (often referred to as 'Mod 228'), which would enable an additional pair of drop tanks to be carried on outboard pylons, providing the Hunter with a fuel capacity of 814 gallons. Rolls-Royce had virtually solved the Avon's surge problems by this stage and the new Avon 115 was judged to be a much more reliable and almost surge-free power plant, and with this new engine, a better fuel capacity, a full 'flying tail' (replacing the interim version used previously) and the gun ammunition link collectors, a completely new Hunter variant was created – the F Mk 4.

The increase in East-West tension and the Korean War had produced a dramatic effect on international politics. The US Congress, traditionally an advocate of isolationalism, voted to introduce the Offshore Procurement Bill, which enabled American funds to be allocated to European defence programmes, thereby expediting the re-armament of Europe's defences. The Hunter was one such beneficiary of this programme, and a total of 367

Hunter F Mk 4s were ordered for the RAF, paid for by American tax dollars. Production of the aircraft was split between Kingston and Blackpool and commenced with the 114th F Mk 1 (WT701), which was the first to be completed to F Mk 4 standard; the first flew on 20 October 1954. Rather perversely, most of the early production F.4s did not go directly to operational squadrons and were immediately placed in storage (many such aircraft being ferried to Kemble and other RAF stations following completion), as they had emerged from the production lines with F.1-standard Avon 113 engines, and in order to simplify the F.4's introduction into RAF service it was decided that it would be more practical to retain these early F.4s in storage until they could be fitted with the later Avon 115 engine (late-production F.4s were fitted with the Avon 121). Likewise, even though the Hunter now had provision for the carriage of drop tanks, these were virtually never carried on the F.4 in RAF service, the improved internal fuel capacity being sufficient for typical training sorties.

Such was the pace of Hunter production that the first F.4 took to the air before the production of F Mk 1 airframes had even been completed. It is also interesting to note that the first F Mk 5 (WN954) took to the air the day before the first flight of the F Mk 4. In the same way that the F Mk 4 was a direct development of the F Mk 1, the F Mk 5 was an equivalent Sapphire 101-engined derivative of the F Mk 2, with the same improvements in fuel capacity. Evaluation and pre-service trials with the F.4 and F.5 went well although they were marred by the tragic loss of WT707 on 25 January 1955. Test pilot Frank Murphy was flying the aircraft over the South Coast when he was forced to make an emergency wheels-up landing at RNAS Ford. Although the initial landing was successful (and Murphy emerged from the incident unscathed), the aircraft skidded on wet grass and careered through a nearby caravan park, killing three people in the process.

On a more positive note, the first Hunter F.4s entered RAF service three months later when Nos 98 and 118 Squadrons, both located at Jever in Germany, each received twelve aircraft. Many service personnel thought that the re-equipment of RAF Germany squadrons was long overdue, but the decision not to introduce the still-unproven Hunter at an earlier stage was probably a sensible one. Even so, the later F Mk 4 aircraft suffered their own minor problems and were delivered in varying states of modification, so much so that aircraft often had hastily scribbled notes applied to their instrument panels to remind pilots of the particular modification state of particular aircraft.

Hunter F Mk 4s were also delivered to the AFDS and DFLS at West Raynham, although most of their flying continued on the F.1 and F.2, and the later-standard Hunter was 'tagged on' to their courses to provide relevant experience for new flight and squadron commanders before they were assigned to their own squadrons. Further deliveries took place in May 1955 when No 14 Squadron at Oldenburg re-equipped, followed by No 247 Squadron at Odiham, this unit gratefully ridding itself of the earlier F Mk 1 Hunters. A second Oldenburg Squadron (No 26) replaced its Sabres with Hunters in June, while back in England No 111 Squadron also received its first Hunters at North Weald, finally saying goodbye to the Meteor F Mk 8. Back at Odiham, No 54 Squadron received a fleet of F.4s to replace its Mk 1 Hunters, and a handful of Hunter Mk 4s were

The Hunters of No 118 Squadron, one of Germany's 2ATAF units, carried particularly eye-catching unit markings on either side of the fuselage roundels and, in smaller form, on either side of the unit badge on the nose.

also delivered to the OCU at Chivenor, together with the RAF Flying College at Manby and Central Flying School at Little Rissington. No 20 Squadron at Oldenburg then re-equipped, followed by No 222 Squadron at Leuchars.

Over the next five months the flow of completed Hunter F Mk 4s continued and more squadrons received the new aircraft, these being 67 Squadron at Bruggen, 93 Squadron at Jever, 43 Squadron at Leuchars, 92 and 66 Squadrons at Linton-on-Ouse, 130 and 112 Squadrons at Bruggen, and 234 Squadron at Geilenkirchen. Still more Hunter F.4s were delivered to No 72 Squadron during April 1957 at Horsham St Faith (now better known as Norwich Airport), and were also destined to re-equip No 245 Squadron, but this unit was transferred to Stradishall prior to receiving Hunters. The move was made necessary by the sheer volume of Hunter activity at Horsham St Faith where space (both on the ground and in the air) had to be shared between operational flying and the activities of a Hunter modification unit that had been set up there to handle work that was required subsequent to factory completion, combined with growing unrest from local residents in response to the noise from the ever-increasing numbers of Hunters at the base. No 74 Squadron exchanged its Meteors for Hunter F.4s at Horsham St Faith in March 1957.

Prior to this date, and despite some typically wartime-esque attempts to keep bad publicity out of the media, the Hunter's shortcomings had been raised in the House of Commons. In the 1956 Defence Debate, the Minister of Supply, Reginald Maudling, stated the following, in response to questions from MPs:

> 'The aircraft itself is an excellent one. That is agreed by the pilots who fly it. The difficulties which now remain are those with the formidable installation of four 30mm cannon which were introduced at the same time as the aircraft. I do not quarrel with that decision, but it was a big decision to take to introduce simultaneously a new aircraft of advanced design and a new armament of very advanced design... The difficulties which were first apparent were the troubles relating to engine surge when the guns were fired, which occurred in certain marks of the Hunter, and also occurred at certain heights... That problem has been solved. The necessary modifications to the engines are being installed retrospectively. In fact, I should have thought that most of them would by now have been done...'

Of the Hunter gun-firing limitations, Maudling said:

> '...it was not one that stopped the firing of the guns, but one that limited the amount that the guns could be fired. It had been found that the "enormous kick" of the big battery of cannon began to do some damage to small components in the nose. Modifications had been devised to strengthen the parts and deal with the trouble. Obviously, in the meantime, in peacetime, it is foolish to go on firing guns to an extent which damages aircraft. That is why limitations have been placed on the duration of firing of the guns. However, if it were necessary to take part in operations tomorrow before the modifications were installed, the guns could certainly be fired. The problem would then be that after a certain number of sorties we should have an increased maintenance requirement because of the replacement of the parts damaged by the firing of the guns. Whatever Right Hon and Hon Members opposite may think about the Hunter, it is interesting that people abroad, who have to pay money for the machine, seem to like it quite well. Apart from off-shore orders, nearly 200 Hunters have been exported to overseas customers.'

The Hunter's export story will be explored in later chapters, but this Government statement illustrates a great deal of support for the fledgling fighter. It was all the more surprising therefore that, just a year later, the Government delivered a savage blow that destroyed a huge chunk of the service's fighting capability, and had profound effects on the future of the Hunter, some negative, but which ultimately laid the foundations of the Hunter's well-earned place in aerospace history.

A quartet of 93 Squadron Hunters in formation over Germany. F.4 WV318 was subsequently converted into a twin-seat T.7 and continues to fly in civilian hands with Delta Jets at Kemble.

Many of 93 Squadron's Hunters were delivered to Germany without ammunition link collectors, but these were retrofitted as they became available.

All photographs on this page courtesy of www.RAFjever.org

Fighter Command's new Hunter squadrons regularly assembled their own formation display teams, including 2TAF in Germany following the arrival of the Hunter F.4. No 93 Squadron demonstrated their Hunters at a number of venues around Europe.

XF319 earned a place in RAF folklore after receiving a distinctly tongue-in-cheek question-mark in place of the usual aircraft code letter while serving with 112 Squadron. Following retirement from 2ATAF, the aircraft became a ground instructional airframe.

In much the same way that RAF Germany had been omitted from initial Hunter re-equipment plans, it was decided that the Sapphire-powered Hunter F Mk 5 would also be confined to UK-based units, so that the additional logistics and expense of supporting a different power plant could be avoided. Consequently it was No 263 Squadron at Wattisham that received the first Mk 5 aircraft in March 1955, followed by No 56 Squadron at Waterbeach in May, this squadron having finally abandoned its short and very unhappy association with the Supermarine Swift. No 1 Squadron re-equipped in June, as did No 41 Squadron at Biggin Hill. In July No 257 Squadron exchanged its Mk 2 aircraft for new Mk 5s, and towards the end of the year No 34 Squadron received Hunter Mk 5s at Tangmere, production of the 105 Hunter Mk 5s (all built at Coventry) ending in August 1955. By the end of 1956, RAF Fighter Command was successfully integrating the Hunter into its Order of Battle; indeed, two Hunter squadrons (Nos 1 and 34) had been deployed to Cyprus in support of Operation 'Musketeer', the Anglo-French occupation of the Suez Canal zone. The Hunters were employed in support of fighter-bomber sorties, but with their limited endurance (and lack of opposition from Egyptian forces) they were mostly confined to maintaining an alert posture from their temporary bases in Cyprus (No 111 Squadron Hunters were also later deployed to fly patrols over Malta), and were not called upon to fire any shots in anger.

By this stage Fighter Command had sixteen operational Hunter squadrons, these being Nos 43 and 222 at Leuchars, 66 and 92 at Linton-on-Ouse, 19 at Church Fenton, 56 and 53 at Waterbeach, 65 at Duxford, 257 and 263 at Wattisham, 111 at North Weald, 41 at Biggin Hill, 54 and 257 at Odiham, and 1 and 34 at Tangmere. This might suggest that Fighter Command had finally established a degree of stability and commonality, but events in the following year threw the Command's plans into chaos.

The infamous 1957 Defence Review was one of the most ill-conceived and destructive political acts of the post-war period, and had a far-reaching and permanent effect on the RAF's future. Based largely on the aftermath of the Suez crisis (which had demonstrated a woeful lack of resources and ability), the Review was far-reaching, and produced some political statements that would still seem relevant fifty years later:

An excellent action image of a 56 Squadron Hunter getting airborne from Waterbeach. Worthy of note are the unit markings (red/white checks) applied to the wingtip, and the streak of engine oil along the fuselage.

> 'A defence plan, if it is to be effective and economical, must be based on a clear understanding of the military responsibilities to be discharged... The aim must be to provide well-equipped forces sufficient to carry out these duties, while making no greater demands than are absolutely necessary upon manpower, money and other national resources... Experience has shown that the rapid progress of scientific development and fluctuations in the international situation make it difficult to foresee future military requirements with any certainty, and that consequently a good deal of flexibility must be maintained. Nevertheless, an attempt must be made to establish a broad framework within which long-term planning can proceed... The new defence plan set out in this paper involves the biggest change in military policy ever made in normal times... The Government are confident that this defence plan, while helping to relieve the strain upon the economy, will produce compact all-regular forces of the highest quality, armed and organised on the most up-to-date lines.'

Unfortunately, the Review contained some fundamental and fatal flaws, particularly the assumption that 'fighter aircraft will in due course be replaced by a ground-to-air guided missile system' – a foolish conclusion that destroyed much of Britain's future fighter development at a stroke.

The Hunter force was one of many casualties of the Government's new policy. By 1957 it had been accepted that the early marks of Hunter would eventually be passed on to Royal Auxiliary Air Force units to replace their aged Meteors and Vampires; indeed, many RAuxF pilots had already been making familiarisation flights with co-located Hunter squadrons in anticipation of this event. The decision to disband all of the RAuxF fighter units was a bitter blow that nobody had expected. Fighter Command was obliged to dispose of many Hunter squadrons that had barely declared themselves operational, and over in Europe RAFG had to do likewise with nine of the thirteen Hunter squadrons, and these were gone within a year of the Review.

By this stage the Hunter had completely replaced the 'interim' F-86 Sabre in Germany and had also re-equipped some of the Venom fighter-bomber squadrons. Formed into Wings comprising two or three component squadrons, each RAFG Hunter base maintained a two-aircraft Battle Flight (the duty being rotated through each Wing) with the designated aircraft prepared to scramble in immediate response to an order, with the rest of the duty squadron maintained at fifteen-minute readiness. Bruggen and Geilenkirchen Hunters were charged with the defence of the Ruhr, while Jever and Oldenburg Hunters were responsible for the defence of different sectors of the North German Plain. In a wartime scenario, the Hunters would also have been required to escort 2ATAF Canberras, possibly carrying defensive 'chaff' in their air brakes and flaps.

The Sabre had enjoyed a brief but successful operational career in Germany, but even at the time that they were ordered (some 430 aircraft being supplied to the RAF under the terms of the Mutual Defence Aid Programme), the aircraft was intended to be nothing more than a 'stop-gap' to give the RAF a capable fighter until the Hunter could be brought into service. Although the three former RAFG Venom squadrons were more than keen to get their hands on new Hunters (the Venom was very much a 'first generation' jet and had also begun to suffer from serious fatigue problems), the Sabre squadrons were less enthusiastic; the Hunter F Mk 4 was perceived as offering little improvement over the aircraft with which they were already equipped. Hunters powered by the Avon 115 were still restricted to firing their weapons below 25,000 feet – hardly ideal for an aircraft tasked with the interception of Soviet bombers that could be flying at twice that altitude – and until the Hunters received Avon 121 engines this restriction was a very unwelcome burden when the Sabre was already able to operate without any such difficulties. Likewise, the Sabre's 'all-flying tail' (the movement of the aircraft's elevators being linked to the tailplane) was much more effective than the Hunter's 'interim' tail design, which only linked the movement of the tailplane with its trim tab, and this gave the Sabre a clear edge in terms of longitudinal control.

With a much better endurance (even without drop tanks) and a similar top speed, it was perhaps understandable that the RAFG pilots did not share any eagerness to convert to the Hunter. However, from an operational point of view, even the relatively unsophisticated Hunter Mk 4 was a superior aircraft to the Sabre and was much better suited to fighter operations in Germany. With a cartridge start system, the Hunter could be airborne much more rapidly than the Sabre, which relied on an external power source. Moreover, once in a combat situation the Hunter's awesome firepower made the Sabre's 0.5-inch machine-guns seem almost archaic. Likewise, the new Hunter was marginally faster and could certainly go supersonic much more easily, needing only a forty-degree descent to achieve Mach.1.0 whereas the Sabre (which had to be flown 'clean') required a rather more brutal vertical dive to achieve the same speed.

It is also important to emphasise that the Hunter was also a much stronger airframe, routinely cleared to 7.5g and tested to 11.25g (one pilot even managed to pull a brutal 14g manoeuvre from which both pilot and aircraft survived intact), and was capable of taking much rougher handling than the Sabre. In later years, when some Peruvian pilots transitioned from exported Hunters onto Sabres, it was interesting to note how many Sabres were overstressed (one aircraft reportedly being written off) simply because the pilots had wrongly assumed that the Sabre would absorb the same amount of punishment. To use a typically British phrase, the Hunter really was as tough as old boots!

Once the Hunter F.4 began to arrive in Germany, the aircraft settled into service with relative ease and quickly earned the respect and enthusiasm of the former Sabre and Venom pilots who were assigned to it. The only significant disruption was caused by aircraft being temporarily withdrawn from service for delivery to the Command Maintenance Unit at Buckeburg to enable engineers to make a variety of post-manufacture modifications (such as extending the cartridge ejection chutes for the Aden cannon, to avoid damage to the fuselage, and the addition of the link-collecting 'Sabrinas'). However, most of the MU activity was fairly uncomplicated in nature, so the Hunters were not away from their assigned squadrons for long.

Day-to-day flying activities generally comprised high-altitude interception practice under the control of GCI radar stations. Because the Hunter was always light on fuel

Hunter F.4 XF310 performed flight trials with Fairey's Fireflash AAM, although the project was not pursued beyond the trials stage, the Air Ministry being more interested in the DH Firestreak, which was ultimately carried by the Lightning. XF310 appeared at a Farnborough show minus national insignia, in order to avoid any impression that the Fireflash fit had any governmental support. *Hawker*

P.1109B XF378 (an F Mk 6) was fitted with AI Mk 20 radar to undertake Firestreak AAM trials during 1957. Although the system was not adopted for the Hunter, it later became an integral part of the Lightning's weapons system. *Hawker*

F Mk 6 WW594 (designated by Hawker as the P.1109A) was also fitted with AI Mk 20 radar as part of the Firestreak trials programme. *Hawker*

(especially so without the addition of drop tanks), it was capable of reaching 50,000 feet without difficulty and could remain there until the limited fuel load was used up. Alan Pollock was a pilot with 26 Squadron at Oldenburg and recalls his experience with the Hunter Mk 4 and its limitations:

> 'Engine surges were only occasionally experienced, but one always flew with diligence to avoid both the rpm and the thrust vanishing as one's JPT [Jet Pipe Temperature] began to soar. Only by diving away with wings level could the situation be cured, which would mean the other aircraft having to support an F.4 that could be found quite quickly a mile or two below. Surges were usually caused by a high angle of attack (as in an emergency break), coupled with an over-zealous throttle handling of the 100-series engine's misadjustment. This tendency at altitude was prone to happen to the less experienced guys for more than one reason. Firstly, they were inevitably flying the outer aircraft, which needed to be the most agile if a battle break was called towards you, and you certainly had to get your own bank, turn and power applied smartly for technique, position and flight safety collision avoidance. Secondly, the more experienced on the squadron tended to have the choice of aircraft and some aircraft were more prone to problems and known to be 'surgers'. If someone was diving to sort out his surge problem, inevitably the formation's vector flexibility was curtailed, seemingly for a variable time ranging from just thirty seconds to much more.'

Even with the remaining restrictions, the Hunter F.4 proved itself to be a successful aircraft and a great improvement over the aircraft it replaced, particularly in the case of the Meteor and Venom.

By this stage plans to introduce an even more capable variant of the Hunter were already under way as the F.4 settled into service, and there was certainly never any intention of retaining the Hunter F.4 and F.5 for any longer than strictly necessary. Indeed (as mentioned previously), the aircraft would ultimately have been transferred to the RAuxF had the 1957 Defence Review not intervened. The fact that the RAuxF fighter squadrons suddenly no longer existed effectively meant that the Hunter F.4 and F.5 would quickly be redundant. The savage cuts that took place from 1957 onwards meant that having peaked at a strength of eighteen operational squadrons, Fighter Command would now be left with just eight, while in Germany nine of the thirteen squadrons would go. A smaller force would remain in Germany, tasked primarily with policing duties, while in the UK the Hunter force would be largely responsible for defending the new V-Force nuclear bombers, even though it was already accepted that the bombers would possess performance figures that were somewhat greater than those of the fighters that would be protecting them. However, a much more advanced interceptor was already being developed (the Lightning) and, unlike many other programmes, this fighter had survived the 1957 Review and was still expected to enter RAF service. But in the meantime, until the Lightning became available, it would be the Hunter that remained at the forefront of the RAF's fighting capability. The Hunter F.4 and F.5, good though they were, still suffered from deficiencies that rendered them less than ideal for the front line, but at long last the Hunter was about to mature into a much more capable version – the classic Mark 6.

Chapter 4

Mark 6: the 'definitive' Hunter

When the truly supersonic P.1083 was cancelled, work on the prototype aircraft (WN470) was already well advanced, and although the re-designed wings were no longer applicable to any other development programme, the basic fuselage was new and unused, and Hawker opted to use it as the basis of a new Hunter variant that could be fitted with the latest version of the Avon engine, which Rolls-Royce was still doggedly developing. The Avon had persistently remained inferior to the Sapphire engine, which, as even Rolls-Royce admitted, had 'a greatly superior flow/pressure characteristic' and delivered more thrust. It was fortunate for Rolls-Royce that information on engine developments was exchanged fairly freely between itself and Armstrong-Siddeley, which provided detailed information on the design of the Sapphire's compressor stage. This enabled Rolls-Royce to develop the Avon so that it took advantage of the Sapphire's first four compressor stages, while incorporating a further eleven stages that were re-designed, maintaining a straight-through flow into the combustion chamber. The result was the first of the 200-series engines, which was first tested in November 1951, and quickly demonstrated that an output of 10,500lb thrust would be possible.

Hawker assembled a new Hunter based on WN470's redundant fuselage and, with a new RA.14 engine fitted, the new aircraft (XF833) became the prototype P.1099 – the Hunter F Mk 6. The new Hunter variant also had a greater fuel capacity, with the small centre fuselage tanks having been replaced by a pair of bag tanks adjacent to the jet pipe, containing 25 imperial gallons, together with a 200-gallon fuselage tank and 140-gallon bag tanks in the wings. Additionally, with 'Mod 228' wings, the aircraft could carry a total of four 100-gallon drop tanks on the inboard and outboard pylons, and could also be fitted with a pair of huge 230-gallon drop tanks on the inboard pylons, although the use of these tanks would foul the operation of the landing flaps, and could only be carried for long-distance ferry flights. Even so, the new tanks certainly gave the Hunter a much-needed boost in range capability, as demonstrated in October 1958 when a Hawker test pilot made a non-stop flight from Dunsfold to El Adem in Libya, covering a distance of 1,588 miles.

Another very significant improvement over earlier Hunter variants was the fleet-wide introduction of a fully functional powered elevator system, which linked tailplane and elevator controls to produce much better longitudinal handling. This was arranged so that, after two degrees of elevator movement, micro-switches were contacted to change the tailplane incidence at the rate of 0.7 degrees per second. This had the effect of an 'all-flying' tail, but small adjustments remained free from tailplane follow-up, making fine control during formation flying very much easier.

The Hunter Mk 6 also disposed of the unreliable cartridge starter system and employed an Avpin (Isopropyl Nitrate) starter system (later used on the Lightning). This new system worked well, but with hindsight it is clear that it also had a fair share of disadvantages, not least the unavailability of replenishment at airfields that simply did not handle the relatively unusual Avpin. Consequently, if a Hunter was forced to land at an ill-equipped airfield, it was likely to remain there for some time. Avpin was also very unstable and prone to explosions, many of which caused catastrophic damage to airframe components, and many Hunter pilots (and more than a few members of the hard-working groundcrews) claimed that it might have been wise to have stuck with the older cartridge system, troublesome as it sometimes was.

The prototype F.6 first flew on 23 January 1954 – ironically, this was before the air-brake-dogged Hunter F.1 had even been accepted for RAF service, and some nine months before the first production F Mk 4 first flew. XF833 was transferred to Boscombe Down to conduct flight observation trials and it was at this stage that, with much disappointment, the new Avon was found to be no more trouble-free than its predecessors. In less than two weeks the aircraft was force-landed following engine failure and the aircraft was transported by road back to Dunsfold where it was fitted with a new engine. Five days after the aircraft resumed flying the new engine also failed, and Rolls-Royce was left with the embarrassing headache of establishing what the latest in a long line of problems was. Investigations revealed that fatigue fractures were causing compressor blade failures and this resulted in more modifications, including a new steel stator and rear rotor blades, as well as a revised fuel system, de-rating the thrust from 10,500lb to 10,000lb.

Flight trials resumed at Boscombe Down in July 1954, and at long last things began to run smoothly and the handling trials were completed successfully. Having learned the lessons of the F.1's protracted development, Hawker set aside seven Mk 1 airframes to be used as 'interim Mk 6' pre-production aircraft, and the first of these (WW592) took to the air on 25 March 1955. Flight trials revealed that in certain conditions, particularly at speed and altitude when g was applied, the aircraft had a tendency to pitch up, and the problem was investigated by A&AEE pilots, resulting in various modifications such as wing fences, all of which proved ineffective. The final solution was to extend the wing's outer leading edge in order to reduce its thickness/chord ratio and move forward the aerodynamic centre of pressure. The result was a 'dog tooth' extension of 9 square feet, which became a standard fit on all Hunter Mk 6s and subsequent variants, although the first aircraft to emerge from the production line were completed before the modification had been finalised and had to be re-fitted retrospectively (a few late-production Hunter F Mk 4s were also retrofitted).

The ever-present engine surging problem was still far from being completely resolved at this stage, and the aircraft still suffered from a pitch-down tendency when the guns were fired. Hawker was now determined to eradicate these deficiencies once and for all. A series of intense firing trials was commenced, and more than 20,000 rounds were fired off at high speed, the complete ammunition load being exhausted each time. The aircraft was then returned to Dunsfold, re-armed and immediately sent off to repeat the process. The gun pack was then removed and inspected, and, after a few small fixes, was re-installed and the whole firing process repeated. The key to solving both the surge and pitch problems completely was to design a suitable gun blast deflector around the gun muzzle, and after a variety of designs had been fitted and flown, a final modification was found that successfully diverted blast away from the aircraft at 90 degrees and in the process provided 150lb of upward thrust from each gun as it fired. The result was the familiar small protuberances that subsequently appeared on many (but not all) single-seat Hunters' gun muzzles, which finally put to rest the seemingly endless saga of problems surrounding the use of the mighty Aden cannons.

Flight testing of the F.6 then continued without any major difficulties and proved to be a much smoother process than had been experienced with the F Mk 1. There were still occasional 'exciting moments', however, as test pilot David Lockspeiser recalls:

> 'The cockpit canopy of the Hunter is operated by an electric motor, and the operating switch is mounted on the port cockpit wall. To open from closed the switch is selected aft; there is a five-second delay while the pressurisation system is automatically switched off, also ensuring that the canopy seal is deflated. The canopy seal is rather like the inner tube of a bicycle tyre and is attached to the fuselage cockpit periphery; when pressurised it inflates, making firm contact with the canopy. Immediately underneath the switch is a small slightly rounded shelf that protrudes about three-sixteenths of an inch further into the cockpit than the switch. This small shelf was introduced as a modification and fitted to all single-seat Hunters following an incident that occurred during a production test flight on an early Mk 6.
>
> The first item on this test schedule was a fixed full-throttle climb, timed from brakes off to 50,000 feet, during which readings of engine and aircraft systems parameters were recorded at specific altitudes. For comparative purposes the flight

RAE Bedford's F6 XG290, illustrating the distinctive gun muzzle deflectors fitted to some (but by no means all) Hunters. *RAE*

F.6 WW598 was converted to P.1109A standard and fitted with AI Mk 20 radar for missile trials. The aircraft subsequently joined RAE Bedford's Aero Flight (wearing a dark blue colour scheme as illustrated) before being assigned to the TSR2 development programme (appropriately painted all-white), during which it flew low-altitude gust response investigations and canopy visibility trails, involving the fitting of a variety of canopy blanking sheets. The aircraft was then transferred to Llanbedr where it received blue paint trim (and a Welsh dragon emblem at a later stage) before being withdrawn in 1974. It was then refurbished and sold to Lebanon as an FGA Mk 70A (L-280). *via Dave Fleming*

path was kept as straight as possible, so taking off from Dunsfold toward the west a gentle turn was made onto a more southerly heading to avoid the South Coast civil traffic and Lyme Bay danger areas. This climb was completed over the Cherbourg peninsula, levelling out ten or eleven minutes or so later, in trim, level at 50,000 feet and .85M, throttling back to an intermediate power setting.

Almost immediately a strong pressure from under my bum forced me hard up into my harness; my immediate thought was that the dinghy, which is stored in the seat pan of the ejector seat, had for some inexplicable reason decided to inflate (in the RAF this had been an anticipated problem, and while on a Meteor squadron we had dinghy stabbers made from steel tube, three-eighths of an inch in diameter, drilled with holes all around and fitted with a sharp conical tip). Grabbing the dinghy stabber from my flying overalls I started stabbing away at the seat, taking great care to avoid that highly valued part of my anatomy for whom hope eternally springs! It soon became apparent that it was only the air trapped during packing that was expanding as the upward pressure remained constant, so I replaced the stabber. Almost simultaneously the cockpit completely filled with condensation, through which I saw the bright red cabin altitude warning light come on.

On the oxygen control panel I selected the emergency toggle switch "on" and this slightly increased the oxygen pressure, the system automatically going to 100% when passing 30,000 feet, but I moved the switch from "normal" to 100 % for good measure and immediately pushed the tab down on my mask. Dr Gabb of the Institute of Aviation Medicine had designed a frame with a toggle joint that, when operated, dramatically increased the pressure of the mask onto the pilot's face, thus obviating any escape of oxygen. The next thing that happened was that the canopy started to motor open. No conscious thought that I can recall remains in my memory except that this was not the time or place to dilly-dally, and with both hands I pulled the control column as fast and as hard as I could onto the forward stop. I was as determined as I could, summoning my whole mind and body, to avoid becoming unconscious, and, focusing on the artificial horizon – the only thing I could see, accurately described by the doctors as tunnel vision, all peripheral vision having gone – I kept the wings level and kept repeating to myself, "You have a baby son – you must remain conscious."

From 50,000 feet in a shallow dive the Hunter easily slips past Mach 1 and in a steep dive it will reach Mach 1.3. At around 20-25,000 feet full consciousness returned; decreasing speed and levelling out, everything seemed perfectly normal, except that the canopy was open. A check around the cockpit revealed that the canopy switch was selected to open, and when selected to close it worked normally. It was rather baffling, so, at this safer altitude, the logical thing to do was to put the aircraft into the same configuration as it had been when the problem occurred, in trim full throttle and reduced power to an intermediate setting, and after five seconds the canopy started to motor open. The cause then became apparent – there is a loom of cables that comes from the outboard side of the throttle, controlling the throttle twist grip, manual gun sight range control, the R/T press-to-transmit button and the speed brake control switch. This loom fouled (but not always) the canopy selector switch when brushing past from the fully forward position.

All now being understood, after completing the schedule I returned to Dunsfold. On the ground, after shutting down the engine and repeating the sequence several times, it was revealed that the loom would strike the switch about three out of five times. Hence the introduction of the little shelf. Other than the Almighty, to whom I again express my deepest gratitude, there were no observers to this performance, so I can only claim that it is on the cards that, unofficially, I hold two world records for an "open cockpit" aircraft – speed Mach 1.3, and height 50,000 feet.'

The first standard production F.6 aircraft were XE526 and XE527, which were delivered to Nos 5 and 19 Maintenance Units respectively on 11 January 1956. Their arrival was at a time when both MUs were already stretched to capacity with the delivery of F Mk 4s, so the F.6s languished in storage until October of that year, when No 19 Squadron at Church Fenton took delivery of them as the first squadron to operate what was often referred to as the 'definitive' Hunter variant. Later that month No 66 Squadron at Linton-on-Ouse also began converting onto the Hunter F.6, and No 43 Squadron at Leuchars began to exchange its F.4s for the new version in November. Likewise, No 65 Squadron at Duxford took delivery of yet more aircraft, together with No 111 Squadron at North Weald and 63 Squadron at Waterbeach.

This unidentified F Mk 6 wears the markings of No 63 Squadron and appears to have overshot the runway. The underwing pylon stencilling reveals that the aircraft was based at Brawdy when this photograph was taken.

Shortly after No 111 Squadron began flying the F.6 it was earmarked as Fighter Command's official display team and quickly assembled a polished display sequence utilising all sixteen of the unit's aircraft. Eventually named the 'Black Arrows', the team's aircraft received a variety of trial paint schemes in anticipation of their public debut and eventually an all-black design was chosen, and this became the team's standard paint finish throughout its four-year tenure. Ultimately the team's name lived on, being part of the title now adopted by the current Royal Air Force Aerobatic Team (the 'Red' element being derived from the CFS display team, the 'Red Pelicans'). The 'Black Arrows' became famous throughout the United Kingdom and beyond, thrilling crowds with their top-class formation flying (the display often being split into two smaller elements to maintain visual interest in front of the audience). During the 1958 SBAC Show at Farnborough, the team assembled a magnificent twenty-two-aircraft formation (which included six Hunter F6s borrowed from other units) and proceeded to fly the whole formation through two complete (and perfectly executed) loops. To this day, this feat has never been matched and probably never will be – fifty years later, even the very idea of getting twenty-two Typhoons airborne together at the same time sounds like a technical and logistical nightmare.

No 111 Squadron's tenure as Fighter Command's display team ended in 1960, after which No 92 Squadron assumed the role, having operated for a year as the reserve team known as the 'Falcons'. After having moved east to Leconfield, the squadron painted its sixteen aircraft in an attractive blue colour scheme (with white trim) and became the 'Blue Diamonds', flying two successful display seasons before moving to Germany to re-equip with the new Lightning interceptor.

The introduction of the F Mk 6 into RAF service enabled squadrons to dispose of virtually all of the preceding Hunter variants, together with their various vices and restrictions. It marked the end of the Sapphire-engined Hunter completely, even though

A nostalgic view of the flight line at Wattisham with No 111 Squadron's famous 'Black Arrows' in the foreground, and Javelins just visible in the distance.

The awe-inspiring sight of twenty-two Hunters looping in formation during the 1958 SBAC Show week. No 111 Squadron's 'Black Arrows' performed this spectacular feat, using additional aircraft borrowed from other units. The record-breaking achievement has never been equalled.

This is a well-known photograph, but one that perfectly illustrates the perfection of formation flying by the 'Black Arrows'. Five Hunter Mk 6 aircraft are in line-abreast formation – note that at this stage the aircraft had yet to receive wing leading edge extensions.

This stunning pilot's-eye view of the 'Black Arrows' team illustrates that No 111 Squadron's Hunters retained their underwing pylons. The squadron continued to fly operational training sorties while maintaining its formation flying expertise. The gun muzzle deflectors and link collectors are clearly visible on the lead aircraft.

The 'Black Arrows' team pictured during training over the Norfolk coast.

in performance terms the F Mk 5 was not greatly inferior to the F.6 and could outperform it in some situations, such as a flat-out climb to 45,000 feet, in which it would beat the F.6 by nearly half a minute. Certainly there was no obvious performance deficiency that hastened the F.5's withdrawal – it was the logistical aspect of operating aircraft with completely different engines that ultimately sealed the Sapphire's fate, even though the engine had been proved to be just as good as the Avon.

In Germany the F.6 was delivered to Nos 4 and 93 Squadrons at Jever in February and March 1957, with Nos 14 and 20 Squadrons at Oldenburg re-equipping six months later (the units then moving to Ahlhorn and eventually to Gutersloh). They were joined by No 26 Squadron in June 1958 when it was realised that the drastic reduction in Hunter units within 2ATAF had become so severe that the RAF could no longer fulfil its obligations to NATO, prompting the formation of the additional Hunter squadron. It was almost comical to note that this additional squadron was composed of Hunters drawn from other RAFG squadrons, which meant that the total number of available aircraft remained much the same.

As with deliveries of the F Mk 2, the first F.6s to reach Germany arrived in various states of modification and inevitably had to be routed through the MU for completion to a common standard. Most aircraft arrived without the 'dog tooth' wing leading edge extension and also lacked the gun blast deflectors and extended cartridge ejector chutes. Unlike the earlier F.4, the increased thrust of the F.6, combined with a slight reduction in internal fuel capacity, meant that the aircraft had somewhat shorter 'legs' than its predecessor. This meant that external tanks would now be a necessity rather than an option. Unfortunately, production of 100-gallon tanks lagged behind production and delivery of the F.6, and many months passed before significant numbers of tanks became available. In the meantime the Hunter pilots took advantage of the 'clean' F6's outstanding performance and regularly enjoyed some distinctly one-sided combat exercises with other RAF and NATO units, but always with a very careful eye on the Hunter's fuel gauge.

A good example of the exuberant fashion in which pilots flew their new Hunter F.6s is described by Group Captain A. J. 'Bugs' Bendell OBE AFC:

> 'Since watching that excellent practitioner, Ken Goodwin, at Jever I had been itching to have a go as a solo display pilot, so I volunteered to take on the commitment. The boss finally agreed. The Hunter was a splendid display aircraft, powerful, robust and highly manoeuvrable, even at low speed. Manoeuvrability was essential for display flying – after all, it made for an extremely dull show if the aircraft covered three counties and disappeared from view every time the pilot turned about through 180 degrees!
>
> After practicing at medium altitude, I was given clearance to perform over the airfield. One of the advantages of solo acrobatics, as opposed to being a member of a formation team, was that I could practice my routine at the end of any standard training sortie. The boss, Sqn Ldr Peter Bairstow AFC, witnessed every practice, as did the station executive officers and the other squadron pilots. It was not just because they were interested; it was a known fact that anyone who witnessed an accident could not he called upon to serve on the subsequent board of inquiry. My first display in front of the public was at Colwyn Bay on 29 August 1959.
>
> For convenience, I usually flew my display with the external 100-gallon wing tanks fitted, though I made sure the tanks were empty before throwing the aircraft around. The loops, rolls, wing-overs and such like were easy to perform in the Hunter. Stalling was not permitted, which precluded flick rolls and true stall turns, but it was possible to cheat a bit and perform an impressive high vertical wing-over. A senior staff officer from Group Headquarters once tried to tell me that the half loop I was using immediately after take-off to reposition in front of the crowd was dangerous. He insisted that the aircraft would stall at the

top of the loop. I knew that was unlikely. Under zero-g conditions, an aircraft would neither stall nor spin. Left to its own devices, the Hunter would drift over the top of a loop quite safely, at a remarkably low airspeed. I suggested to the staff officer that, to lose control at that point, the pilot would have to be extremely ham-fisted, which seemed to upset him. Next day the boss called me into his office. Apparently the self same staff officer had once lost control while looping a Hunter and had had to eject. I rested my case. The boss accepted that the manoeuvre was safe and I did not have to change my display sequence.

I included an inverted loop and an outside break in my show, mainly for the benefit of regular aircrew, who could appreciate the finer points of that kind of flying. The uncomfortable moment was when the aircraft first inverted and one fell away from the seat – there was always some slack in the harness, even when the groundcrew had helped me tighten the lap straps. All manner of doubts flashed through one's mind at that moment. Would the straps hold? Would I still be able to reach the controls, or would I finish up sitting in the cockpit canopy? Minus 3½g was considered to be the practical human limit, which may well have been true. I had regularly experienced the pink flood of vision, signalling the onset of red-out – the opposite of blackout. I'd thought no more of it until after a display in Cyprus, when an RAF doctor had confirmed that my eyes were bleeding internally – proving, if nothing else, that the bleary-eyed, bloodshot appearance of some negative-g display specialists was not solely due to the demon drink.

Inverted flight could spring other unpleasant surprises. On one occasion, due to a fault in the aircraft's fuel system (the recuperator reservoir pressurised the supply of fuel for a limited degree of inverted flight), the engine flamed out as I pushed up at 600 feet for an inverted loop. Everything went very quiet. Fortunately the engine responded immediately after I rolled upright and pressed the relight button, but from then on I made a point of checking each aircraft at a safe altitude before using it for a display. At 360 knots, which was the basic airspeed I used for my show, the Hunter's pitch control – that was the aircraft's response to fore and aft stick movements – could be improved by judicious use of flap. At full power, with ten degrees of flap selected, the Hunter could be hauled around in a seven-g turn – without losing airspeed – which was impressive. But it was essential to raise all flaps before attempting any rolling manoeuvre, otherwise the aircraft tended to "dish" out. On one occasion I was distracted during a Derry turn and mistakenly rolled under with the flap still down. The aircraft dished, losing a hundred feet or so of height and, more by luck than good judgement, I just managed to avoid the control tower. At low altitude there was little margin for error.

Display flying was tightly controlled in the RAF, but the display pilot's self-discipline and his ability to analyse his own performance were equally important. After all, the pilot was the man on the spot and he had to take appropriate action before the situation got out of hand. It was fine to appear as though one was having to pull out all the stops to recover from a vertical dive – indeed, that added to the thrill of the show – but it was foolish ever to allow oneself to be trapped into that position. Then again, a fractional loss of altitude during a hesitation roll might not be critical on the day – it might even go unnoticed by the crowd – but at low level any unintentional loss of height was potentially dangerous.

I thoroughly enjoyed display flying in the Hunter, though I must confess it was physically exhausting. It was the element of risk – a sort of living on the edge – that probably appealed to me. Many of the squadron's groundcrew also enjoyed being involved. There was no shortage of volunteers to prepare my aircraft and one particularly amiable giant made sure he was always on hand to assist with strapping me in.'

The 1957 Defence White Paper had a drastic effect on Hunter F Mk 6 production, which was terminated with immediate effect. Cancelled aircraft were spread between the Blackpool factory (where fifty were to have been built) and Kingston, where 100 aircraft would have been built. This left a total of 119 manufactured by Armstrong Whitworth at Baginton and some 296 aircraft from Kingston, although twenty-three of these aircraft were not delivered to the RAF and were sold to India. Further new-build batches comprised 128 aircraft for India and 100 for Switzerland. This might have been a devastating time for Hawker had it not been for two key developments: first, interest in the Hunter from potential overseas customers was growing (and would eventually result in orders that surpassed even the most optimistic predictions), and second, Hawker was already making strenuous efforts to expand the Hunter's capabilities in terms of the weaponry it could carry. These factors ultimately served to save the Hunter (and Hawker) from disaster. Rather than being a major set-back, the 1957 Review became a seminal moment in the Hunter's history, which prompted Hawker to intensify its efforts to develop the Hunter's versatility. For the RAF, the drastic cuts in F.6 production were certainly unexpected and unwelcome, but the embryonic Lightning had survived the Government's axe and, with the promise of a supersonic interceptor now a reality, the Hunter (at least as a fighter) was rapidly becoming regarded as yet another 'interim' aircraft that would soon be replaced by the Lightning and Javelin. Of course, as is often the case with optimistic predictions, the arrival of the Lightning was delayed quite significantly and the Hunter remained active at the forefront of Fighter Command well into the early 1960s. Thankfully, the trouble-free, fast and manoeuvrable Hunter F Mk 6 was more than adequate for the task.

Commentators have often remarked that it seems odd how the Hunter had already progressed to production status before any attention was given to the possibility of designing a dedicated twin-seat trainer derivative of the aircraft. In fact, although this might seem almost bizarre some fifty years later, at the time it was certainly not standard practice to develop trainer versions of emerging fighter or fighter-bomber designs. For example, all of the classic piston-engine types such as the Spitfire, Hurricane, Typhoon and Tempest had been operated by the RAF throughout the Second World War without any need for dedicated trainer versions, and likewise both the Venom and Sabre had been operated in RAF service only in single-seat form. The new Hunter was not perceived as being in any way different from the types that had preceded it, therefore it was accepted that training for the type would be conducted on the Meteor and Vampire, which were considered more than adequate for students to learn the basics of twin- or single-engine jet flying.

It was, in fact, overseas interest in the Hunter that first drew Hawker's attention to the possible need for a trainer version, and as this concept began to emerge it was only then that the RAF showed some interest in the idea. However, even at this stage the RAF only regarded the Hunter trainer as being a potential useful addition to single-seat operations, and had no interest in the aircraft as an advanced trainer in its own right. Eventually, however, it was accepted that if a twin-seat variant of the Hunter could be developed, it would improve the RAF's ability to train new pilots, especially when the early Hunter F.1 – complete with its various problems and restrictions – was demonstrating that conversion onto the Hunter from the Meteor and Vampire was not as simple as had been envisaged.

The most fundamental issue that occupied the minds of both the Air Ministry and Hawker was the choice between producing a tandem-seat layout (the instructor sitting behind the student) or a side-by-side arrangement. There was no clear indication which would be the better option and the final choice was essentially one of personal preferences. Sydney Camm undoubtedly advocated the tandem layout, believing that it would enable the clean lines of the Hunter to be largely retained (and Camm always believed that if a design looked right, then it would fly right), but some service chiefs were convinced that side-by-side seating would be better, especially for applied flying training such as weapon delivery and instrument flying. Likewise, experience with the Meteor and Vampire had encouraged the belief that, although the Meteor's tandem seating encouraged students to develop a single-seat 'go it alone' mentality, the rather more cosy seating in the Vampire allowed instructors to be much more 'hands on' in terms of demonstrating and explaining techniques to students. Consequently, the side-by-side arrangement won the day and in 1954 Specification T.157D was issued for the manufacture of two prototype P.1101 trainers.

Initial progress on the design of the aircraft was slow, as both attention and resources were largely directed towards production of the Hunter F Mk 1, which was suffering from delays in achieving CA release status. The basic design concept was built around a standard Hunter airframe but with a redesigned twin-seat cockpit section, which would simply be attached forward of the Hunter's transport joint (where the rear and centre fuselage section was divided from the forward section), just behind the air intakes. This new nose section precluded the installation of the Aden gun pack, but it was assumed that the trainer would not require such armament; however, provision was made in the design process for potential single and twin Aden arrangements.

As the design progressed towards the mock-up stage, the new trainer exhibited some rather quirky features. The side-by-side seating concept had been adopted quite literally to such an extent that even the single-seater's canopy layout had been retained, with two single-seat sections simply being joined together in the new design to create an unusual 'double bubble' effect. Even more bizarre was the initial proposal (subsequently changed) to locate the instructor's throttle quadrant on the right-hand side of the cockpit. Wind tunnel tests quickly demonstrated that the canopy arrangement was unsatisfactory and it was swiftly redesigned and smoothed out to produce a more conventional smooth contour.

When XJ615 (the prototype T Mk 7) began flight trials (after first flying on 8 July 1955) it was quickly established that even the redesigned hood arrangement was still far from perfect and was responsible for creating severe buffeting, directional snaking and a considerable amount of noise inside the cockpit as airflow broke away behind the widened cockpit section, the contours of which contracted rapidly before rejoining the standard Hunter fuselage at the transport joint. A variety of trials were embarked upon using different types of vortex generators fitted in different positions around the windscreen arch, all designed to solve the buffet problem. Further changes to the canopy shape were also attempted and, in order to save on both time and cost, the prototype's canopy was replaced by a temporary metal substitute that could be reconfigured as necessary until a satisfactory design shape was fixed. After a great deal of experimentation, it was eventually accepted that the whole area would have to be redesigned again from scratch, and it was Cliff Bore – a Hawker project engineer – who applied the principles of 'area rule' to the fuselage, resulting in the Hunter T.7's distinctive bulged fairing aft of the cockpit, which extends beyond the transport joint before smoothly rejoining the standard fuselage contours ahead of the fin leading edge. The redesigned fuselage might have been expected to create all manner of drag difficulties, but thanks to the application of the 'area rule' principle the twin-seat Hunters proved to be capable of virtually matching the single-seater's performance.

T.7 WV383 returns to its home base at Farnborough while assigned to the RAE Avionics Flight during the early 1970s. The colour scheme is Light Aircraft Grey with Roundel Blue trim. *Paul Tomlin*

An excellent 'plan' view photograph of the Hunter T.7, illustrating the contours of the twin-seat nose and the fuselage spine that conformed to the 'area rule' aerodynamic principle.

T.7 XL565 was temporarily assigned to the RAF Institute of Aviation Medicine at Farnborough and arrived from storage still wearing its FRADU paint scheme. The aircraft eventually went to the USA but crashed on its first flight after a lengthy restoration. *Terry Senior*

T.7 WV318 was used as a procedures and continuation trainer with No 15 Squadron at Laarbruch. The aircraft is now registered as G-FFOX and flies from Kemble with Delta Jets in 'Black Arrows' colours. *Paul Tomlin*

WV318 also spent some time assigned to No 5 Squadron as a 'hack' aircraft while the unit was working up on Lightnings. *Ray Deacon*

Many people also firmly believed that the redesigned fuselage created a Hunter with an even more elegant outline than its single-seat predecessor. It was also considered prudent to address the twin-seat Hunter's airfield performance (particularly with export customers in mind) and a drag parachute was designed to improve the aircraft's landing distance. The parachute installation was a fairly simple process thanks to the layout of the Hunter's fuselage and tail, which provided a very handy space to accommodate the packed tail chute in a fairing built above the tail exhaust pipe, under the fin trailing edge.

Rather perversely, the Air Ministry decided that the new twin-seat Hunter should be based on the conversion of the standard Hunter F.4 airframe. It was accepted that, by the time the first trainers entered RAF service, the Hunter F.4 would have been almost completely replaced by the Hunter F.6, therefore it would have been logical to design the trainer around the more powerful F.6 in order to match handling performance and to avoid the complication (and cost) of maintaining Hunters with different engines. However, the Air Ministry took the view that as so many Hunter F.4s would become surplus to the RAF's requirements, it made sense to use them as the basis for the trainer conversion; indeed, the ability to easily convert the Hunter F.4 into a trainer had been written in to the production specification from the outset. Of course, the decision to convert F.4s made good sense in the short term, but the complications and cost of operating two very different versions of the Hunter for so many years suggests that the choice was rather short-sighted. The irony of the decision was that only six Hunters were converted from F.4 airframes for the RAF, the rest being new-build aircraft that could have been based on the F.6.

An unusual image of WW659 during overhaul at Kemble. The aircraft was built as an F Mk 4 and later converted to GA Mk 11 standard (as illustrated). It was refurbished by HAS in 1971 and re-emerged as a Swiss Mk 58A. *Ray Deacon*

Hawker (being rather less short-sighted than the Air Ministry) proceeded with the production of a second P.1101 trainer that was based on the later F.6, complete with a 10,000lb-thrust engine, and the aircraft (XJ627) first flew on 17 November 1956. Unlike the RAF's trainers, XJ627 was also equipped with a pair of Aden cannons, whereas the Air Ministry opted to arm the T.7 with just one (as a trainer variant, it was assumed that the aircraft did not need to carry the 'punch' of the Mk 6's four cannons). XJ627 performed well during flight trials (apart from a high-speed rudder 'buzz' that was rectified by the installation of a small spoiler unit on the rudder), but despite considerable efforts being made by Hawker, Air Ministry interest in a 'large Avon' trainer was not forthcoming, especially when the Hunter's future was effectively sealed by the infamous 1957 Defence Review. However, the creation of a more powerful trainer variant was undoubtedly a wise decision for Hawker and provided the know-how for the production of many export versions. XJ627 was transferred to Martin Baker, where it flew many high-speed and low-level ejection seat firing flights before being handed back to Hawker for refurbishment and re-sale to Chile.

The first Hunter T.7s entered RAF service in May 1958 with 229 OCU at Chivenor while further examples (usually just one aircraft but sometimes two) were assigned to operational squadrons to act as continuation trainers. Despite the Hunter T.7's less powerful engine, the aircraft was just as manoeuvrable as its single-seat counterpart, and in terms of overall flying performance it was not significantly different, and could never be described as being deficient in any way. Instructors liked the aircraft's rugged construction and vice-free performance, and also appreciated the excellent forward view and the ability to closely monitor their students. Naturally (for much the same reason) students were less enthusiastic and preferred the better visibility and the 'go it alone' feeling of the single-seater. When the RAF's Vampire advanced trainers were replaced by the Gnat, many commentators were surprised that the Hunter was not chosen for the job, as almost everyone believed that the Hunter would have been a much better aircraft for the job. Ultimately, Hunter trainers did join No 4FTS at Valley but only to supplement the Gnats rather than replace them – students were often assigned to Hunters if they were physically just too big to fit inside the diminutive Gnat's cramped cockpit. Certainly, when given the choice between the Hunter and the Gnat, the advanced flying training students inevitably wanted to fly the Hunter.

XL616 was a new-build T Mk 7 that served with a number of RAF units prior to joining the RAE, where it received a grey colour scheme with red trim. It later returned to No 208 Squadron when the unit's Buccaneers were temporarily grounded, and was subsequently sold to a civilian buyer in Sweden. *Paul Tomlin*

Hunter T Mk 7 XL596 was assigned to No 4 Squadron (at which stage it was painted high-speed silver with dayglow orange nose, tail and wingtips as illustrated), prior to joining No 4FTS. It crashed on approach to RAF Shawbury during November 1973. *Richard Caruana*

A pilot's-eye view of a mass formation of Gnats and Hunters from No 4FTS at Valley.

Although the Hunter F Mk 6 was regarded as being the definitive version of the breed, it was the development of the FGA Mk 9 that ensured that the Hunter would remain active as a significant part of the RAF's armoury for many years. Hawker had been aware of the Hunter's potential as a ground-attack aircraft from the very beginning of the aircraft's design and development and had already devoted a significant amount of time and resources to an exploration of the F6's capabilities in this role. However, it was the decision to replace the Venom in 1958 that finally kick-started the creation of the FGA.9. The Central Fighter Establishment embarked upon a series of trials in Aden to select the most suitable aircraft to replace the Venom, and Hawker sent a pair of F.6s (XK150 and XK151, both standard F.6s fitted with tail chutes) to take part in the trials. The competition was almost a foregone conclusion, with only the Jet Provost and Gnat being selected as alternative candidates for the role. It was hardly surprising that the Hunter was quickly recognised as being the ideal aircraft, and Hawker was instructed to begin work on the new Hunter variant, designed for the ground-attack role.

With Mk 6s already in production, the new FGA Mk 9 airframes were created from the existing Mk 6 design by simply incorporating the new modification standards into the production line. An initial order for forty aircraft was placed in 1958, followed by a further order for thirty-six 'interim' aircraft (sometimes referred to as F.6As), which would be delivered with the F.6's Avon 203 engine before being brought up to full FGA.9 standard complete with a new Avon 207. Other modifications introduced on the FGA.9 included full 'tropicalisation' (cockpit ventilation and refrigeration, necessary for operations in the Middle East), and the introduction of 230-imperial-gallon underwing tanks as standard fit, necessitating the fitment of a bracing strut between the wing and the tank's outboard surface, to enable them to withstand the rigours of combat manoeuvres. In order to enable the aircraft's flaps to be operated when the tanks were carried, a small section of flap surface was removed to allow the flaps to drop inboard of the tank's tail. The heavier landing weight of the two huge external tanks (and the 'hot and high' conditions in which the aircraft would probably be operating) prompted Hawker to introduce a braking parachute, which was fitted over the tail cone in a similar fashion to the T.7, although the parachute diameter was slightly larger. Perversely, having spent so much time developing gun blast deflectors for the F.6 (and yet more time on a similar arrangement for the T.7), the muzzle modifications were judged to be unnecessary for the FGA.9's low-level role and were omitted.

The resulting aircraft was a sturdy, manoeuvrable, fast and versatile machine, and it became the 'standard' model from which many of the Hunter's expert versions were created. It could have been even more versatile had further development been pursued by Hawker. If the FGA.9 had any flaw, it was the arrangement of its underwing hard points, which, had they been relocated, could have given the aircraft a better offensive capability. A centre-line pylon under the fuselage would have enabled the aircraft to carry at least a couple of 1,000lb bombs, whereas the FGA.9's inboard pylons were too close to the undercarriage to allow twin stores carriers to be fitted (although a single bomb could be carried). Likewise, the outer pylons were too far outboard and aft to allow bombs to be carried safely (they would create a dangerous pitch change on release, as was demonstrated during Hawker's trials). Fuel capacity could also have been improved, but, as is often the case with RAF combat aircraft, there was never any appetite to allocate more resources to the Hunter than was strictly necessary; at many stages of the Hunter's career it was inevitably regarded as yet another 'interim' aircraft that would be superseded by more capable machines – an attitude that prevented many British aircraft from realising their full potential.

While Fighter Command Hunter squadrons eventually re-equipped with the Lightning, other units (with the exception of those in Germany) re-equipped with the Hunter FGA.9. Although the Hunter had already seen some operational service in 1956 during the Suez Crisis, the FGA.9 was to see much more during its career. The Kuwait crisis in 1961 (mirroring events of decades later) saw the country being declared part of Iraq, and this promoted the British Government to redeploy aircraft to the region, with Hunters moving from their bases at Khormaksar and Eastleigh to Muharraq (Bahrain), before moving directly into Kuwait shortly thereafter. The situation remained tense for some time, but the Hunters were not called into action. However, the MEAF Hunter Wing at Khormaksar was eventually brought into action when the Aden Protectorate became unstable during the early 'sixties. Hunters were regularly called upon to suppress terrorism from local tribesmen, although the typically ridiculous rules of engagement often meant that the Hunters were more of a nuisance than a deterrent. Strikes on rebel villages were preceded by leaflet drops warning the locals that an attack was imminent, and almost inevitably they simply retired to the safety of nearby hills to watch the Hunters in action, before returning to their villages after the aircraft had gone.

No 43 Squadron (which had already been deployed from Leuchars to Cyprus) joined the resident Hunter squadrons (8 and 208) at Khormaksar in March 1963, effectively restoring the shortfall caused by the need to maintain a continual detachment at Muharraq, because of the Kuwait-Iraq situation. Following Yemen's coup, incursions into the area began and the Khormaksar Wing was kept busy defending the local air space, while suppressing the tribesmen and maintaining a presence in Bahrain. Although some action was taken against Yemen (including a retaliatory strike performed by eight Hunters), attention was largely devoted to the Radfan, an area of mountainous territory where rebel strongholds were difficult to reach. In support of British forces in the area, the Khormaksar Wing flew more than 600 operational sorties during May and June 1964.

The Middle East situation was never settled and in many respects it became worse when the British Government announced its withdrawal, which encouraged the warring factions to intensify their activities in an attempt to occupy the resultant vacuum. But despite the politics of the situation, it was fortunate that the RAF had an ideal aircraft to handle the diverse tasks assigned to it in the Middle East. The Hunter FGA.9 proved itself to be a rugged machine that did not suffer from the harsh conditions in which it operated. Much of the day-to-day servicing was conducted outdoors, but despite the heat, the sand and the logistics of maintaining supplies a very long way from home, the Hunter performed admirably. It was undoubtedly this demonstration of the Hunter's capabilities that generated a great deal of overseas interest in the aircraft.

If the Hunter's activities in the Middle East were not enough to demonstrate its reliability, the RAF's presence in the Far East was an even more impressive illustration of how adaptable the aircraft was. Having replaced the Venom in the Middle East, the Hunter was an obvious choice to replace yet more Venoms based in Hong Kong and Singapore. Consequently, No 20 Squadron re-equipped with the Hunter at Tengah, and 28 Squadron did likewise at Kai Tak. Once again Hunters were used operationally in the area, following an armed revolt in Brunei backed by Indonesia, which embarked upon incursions into Sarawak and Sabah, prompting Britain to set up an Air Defence Identification Zone (ADIZ) over the area, policed by 20 Squadron's Hunters. The squadron became heavily tasked with strikes against Indonesian forces to protect the ADIZ, and to escort RAF transports bringing troops and supplies into the area.

Although no evidence has appeared to support the story, it was reported that a Hunter pilot achieved what may have been the Hunter's only air-to-air 'kill' in RAF service, when he forced a MiG-17 into a steep dive from which it was unable to recover. Because the incident is believed to have taken place on the wrong side of the border, no details have ever been forthcoming.

As was the case with the Middle East Hunters, the FGA.9 performed without difficulty in the heat and humidity of Malaysia and proved yet again that, in addition to being an impressive fighter, the Hunter was equally capable of tackling the ground-attack role with ease. However, there was yet one more significant role that the Hunter undertook, that of photographic reconnaissance.

Towards the end of the 1950s the RAF's reconnaissance assets were many and varied. A flight of Meteor FR.9s was based at Khormaksar, assigned to PR duties, while in the Far East Meteor PR.10s were assigned to the same role. Back in Europe, the Swift provided the RAF with a fast and effective reconnaissance platform (ironically, the Hunter's direct competitor was ultimately fit for no other purpose), but as it gradually moved towards the end of its service life, it was hoped that the mixed fleet of reconnaissance types would be replaced by the high-flying, fast and manoeuvrable Canberra PR.7, which itself would eventually be replaced by the even more capable Canberra PR.9. Unfortunately, the cost of such a programme meant that the Canberra would only be able to be part of the solution, and with significant numbers of Hunter F.6s now being withdrawn from fighter squadrons, it was decided (in 1958) to convert forty Hunters into fighter-reconnaissance aircraft.

Hawker had already given the concept a great deal of attention, having fitted a five-camera pack into the nose of a Hunter F.4 (WT780) some time previously. The Air Ministry opted for a less ambitious three-camera option and this was trial-fitted to F.6 XF429 during 1959, this aircraft becoming the prototype Hunter FR Mk 10. The conversions that followed brought the airframes up to FGA.9 standard, but in place of the redundant ranging radar in the nose a new camera pack was installed, covered by a new nose cone into which glazings and a forward-looking 'eyelid' were incorporated. Armour plate was fitted under the cockpit floor (to compensate for the removed radar) and a voice recorder was fitted to enable pilots to give 'visreps' (verbal accounts of target sightings) when required. Nos 2 and 4 Squadrons in Germany re-equipped with the FR.10, and No 8 Squadron's reconnaissance flight in Aden did likewise. The Hunter FR.10 was – at least officially – replaced by the Phantom FGR.2 by the end of the 'sixties, but the aircraft did not completely disappear from the RAF's inventory for many more years.

Although the Hunter is inevitably associated with the Royal Air Force, the Fleet Air Arm also showed some considerable interest, which culminated in an order for single- and twin-seat aircraft to be used primarily for weapons training – in effect a naval equivalent of 229 OCU's activities. Ten aircraft from the RAF order for new-build T.7 trainers were diverted to the FAA (together with another ten that were diverted to the Netherlands – which explains why the RAF's fleet was surprisingly small) and were re-designated as T Mk 8 variants, although they remained identical to the RAF machines save for the addition of an arrestor hook under the rear fuselage. The single-seat aircraft (of which forty were ordered) were taken from surplus stocks of former RAF F.4 airframes, those in

Hunter FR Mk 10 XF432 wears the markings of No 2 Squadron. Note the opened tail parachute housing.
Ray Deacon

FR Mk 10 XE625 is also pictured in No 2 Squadron's markings during a visit to RAF Luqa, Malta. The aircraft was delivered to Chile in 1971 as FGA Mk 71 J-729.

After service with Nos 2 and 4 Squadrons in Germany, F Mk 6 XJ637 joined 229 OCU at Chivenor (pictured here in the markings of No 79 Squadron). The aircraft then moved to 1TWU at Brawdy and crashed in Wales during 1979. *Paul Tomlin*

One of twenty Hunter T Mk 7 twin-seaters purchased by the Netherlands, N-302 was later sold to Denmark and finally returned to the UK as a non-flying exhibit. *via Alf Blume*

T.7 N-307 served with both the Netherlands and Danish Air Forces before being re-purchased by HAS in 1976. It subsequently became G-BOOM and operated on the UK show circuit for some time in an all-red paint scheme. It then became part of the RJAF Historic Flight based at Mafraq AB.

the best condition and with the lowest number of flying hours being selected. At Kingston they were completely rebuilt, the Aden guns removed, Tacan navigation equipment installed and an arrestor hook fitted. Brake chutes were not fitted, but the wings were modified to incorporate the F.6's 'dog tooth' leading edge extension.

Subsequent to the initial order for ten T8s, a further thirty-one twin-seat aircraft were converted from F.4s, and although the Admiralty wanted to order more single-seat aircraft, Treasury pressure forced the FAA to accept an additional batch of unmodified F.4s, which were simply refurbished and transferred from RAF stocks. Unusually, although only one designation was applied to the new aircraft – Hunter GA Mk 11 – there were in fact two standards of aircraft, with sixteen from the fleet being fitted with the four-pylon 'Mod 228' wing. Additionally, a handful of GA.11s were also fitted with the FR.10's camera nose (sometimes referred to as the Hunter PR Mk 11), although in all other respects they were identical to other GA.11s.

Three Hunter twin-seaters were converted to T Mk 8M standard, fitted with Blue Fox radar and instrumentation and used for Sea Harrier conversion training. XL602 was assigned to trials work and, after retirement, was sold to Classic Jets at Exeter, where it is being refurbished to airworthy standard. *Andrew Lawson*

A handful of Hunter GA Mk 11 aircraft were fitted with practice bomb carrier pylons under the forward fuselage. Pictured at Dunsfold, XE685 (the trials aircraft for the pylon fit) carries a full load of four bombs under the fuselage, with a further pair under each wing. *Ray Deacon*

Hunter PR Mk 11 XF977 in action over a South Coast weapons range, releasing rocket projectiles. *FAA*

XE685 was originally an F.4 serving with 98 and 93 Squadrons in Germany. After service with the FAA it was sold on the civilian market and has been passed to a variety of owners. It remains airworthy, based at Exeter. *Ray Deacon*

Tony Ogilvy was one of many Naval pilots who had considerable first-hand experience of both the GA.11 and T.8:

'I commenced flying training in 1965 at Linton-on Ouse with a great course of like-minded chums – people who remain real friends even if we do not see too much of each other these days. I did not do too well in my initial training on the Jet Provost Mk 3 and only struggled through thanks to people like Pete Jarvis (Flight Commander) and, specifically, Bruce Chappell (Instructor). I tended to lag behind the aircraft and despite careful pre-planning I was too slow to correct incipient errors, which became magnified as a result. Also, I did not enjoy the actual environment with the instructor literally breathing down your neck.! However, Pete became a good friend in our later years together on the Buccaneer, despite using minor physical violence on me at least twice during the JP course when he could take no more!

After this initiation came advanced training at Brawdy on the Hunter T.8 – my first introduction to this great aircraft. Although I liked the twin-seater very much, I did not enjoy the course and once again I scraped through thanks in the main to Dickie Wren, the Senior Pilot. Much as with the JP, things were difficult but everything happened faster with a greater chance to mess things up, particularly at low level. Instrument flying I found difficult, and again the instructor proximity was something I disliked. I was drinking in the last-chance saloon, and I knew it. I was also by then wondering if this flying caper was really for me, having had a grim time getting to the last stage of what was in the late 1960s a very short all-jet training en route to the front-line-embarked squadrons of Buccaneer, Scimitar or Sea Vixen aircraft.

After completing advanced training I moved 100 yards down the pan at Brawdy to No 738 Naval Air Squadron for final training prior to front-line conversion on the single-seat Hunter GA.11. Suddenly the lights came on. I finally got it! From the very first moment that I lifted off in that wonderful little fighter I knew that I had been right all along in trying so hard to get into naval fixed-wing aviation. My elation was a combination of freedom from the instructor, plus the wonderful and, in my experience, unique view from the bubble canopy, particularly in the forward sector. The handling and flight control balance was perfect. It looked beautiful and it flew that way. Nothing, but nothing, will ever compare with the feeling of total freedom and fun that I had in those early days in the GA.11. The course was a complete revelation to me; it all started to gel together as did the realisation that the previous work was a very necessary grind to get ab initios like me into a safe(ish) condition before the Queen's money was spent in spades on the front line. Thanks to people like John Middleton – if you are out there, thanks – I made a decent job of the course and progressed to the Buccaneer, which I flew from 1967 to 1978 with two excellent interludes back on the Hunter GA.11, first as a Tactical Instructor back on 738 NAS for a tour in 1968, followed immediately by the Air Warfare Instructors course at Lossiemouth.

Life with 738 Naval Air Squadron as a Tactical Instructor was probably the most enjoyable flying of my career. The setting in South Wales was unbeatable. I had the magnificent Hunter GA.11, Wrens, Newgale beach and elastic licensing laws in the pubs around Brawdy – the perfect combination for a 22-year-old Sub-Lieutenant. Aside from the day job, the then Commanding Officer, Nigel Grier-Rees and his Senior Pilot Bill Peppe, formed, for one hilarious season, the "Diamonds" RN aerobatic team. "Diamonds" was the official name, not "Rough Diamonds" as sometimes stated, although I admit we were a little rough at times. We numbered nine men in total, and we practised either before or after the normal day's work of instruction, and performed during the season at air shows throughout the country.

For the formation flying I was stuck out on the wing where the 7,575lb of shove from the "little" Avon engine (as opposed to the 10k in the RAF Hunters) made

A beautiful image of four FRADU Hunter GA.11 aircraft in a lazy loop over the English Channel. The four-ship 'Blue Herons' were the last of many British Hunter display teams, delighting crowds across the UK. *FAA*

life difficult in some of the rolling stuff. We filmed the practices and they looked awful; we were not aware that in a number of line-astern manoeuvres, for example, the aircraft had to step up and fly a very fine echelon to make it appear, to the ground observer, that we were in level line astern. Enter the Red Arrows, who gave us many trade secrets on how to position and stack up in certain formations to look good on the ground (although it still looked anything but, in the air). I flew behind Mike Todd, a great character from the Sea Vixen world. As a solo performer to keep the crowd amused in the blank spots the illustrious Pete Shepherd filled in with typical gusto. His forte was a continuous rapid aileron roll, which, by rotation number seven, had a look of real danger about it.

Carmarthen Air Show that year was notable for the uncharacteristic and never-to-be-repeated cry from the boss of "Pull for your lives!", as nine noses buried in the vertical following a much slower and lower than usual loop. We did just that, and most of us yanked a "g" or so over the Hunter's limit, but everyone, including the fine citizens of Carmarthen, survived. We recovered somewhat earlier and more quietly than usual to Brawdy that day, where I managed to touch down ever so gently in the grass undershoot. Could I get away with it – no chance… "Diamond Nine Tower, you landed short … hey, wings, you may have missed the numbers but did find the grass" etc.

On another occasion I got so close to Mike Todd in a loop that I pushed my nose into his jet pipe, which caused his engine to momentarily stall. We fell out of that loop as well! Mike was not amused, and I blistered all the paint from my nose cone. I think I just let the stick take its own position in forward travel and eased out of the ensuing dive, avoiding the aircraft nearest to me when I got back full control in all planes.

Our hugely likeable CO also encouraged an unusual weekend activity, namely taking a Hunter away for the weekend. This topped up our flying hours, and provided useful turn-round experience at unfamiliar bases. The only stipulation given was to stay in the UK. This form of visiting impressed the girls no end!

Everyone who has flown this wonderful aircraft has their own unique memories. Most aviators would agree that if an aircraft looks right it will fly right. The single-seat Hunter is still one of the most beautiful aircraft ever designed, and it did fly as beautifully as it looked. It was a wonderful high-performance single jet with impeccable handling qualities, and the ability to forgive crass handling like few others. I honestly think that with a few hours of ground school, a struggle into the weird parachute harness and some time in the cockpit, I could still get airborne (VMC) in a GA.11 and return in one piece, even to this day. I could not say that about the Buccaneer, and certainly not the Sea Harrier! Like your first love, the single-seat Hunter is never forgotten.'

Chapter 5

From the Central Flying School to Aden

RAF Hunters saw service in a variety of roles and an even wider variety of locations. Some air and ground crews were assigned to the Hunter during tours of duty both in the UK and overseas. Ray Deacon was one such individual who, after training as an Air Wireless Mechanic at Cosford, was assigned to work on first-line servicing of the Vampire T Mk 11s on 2 Squadron, Central Flying School, based at RAF Little Rissington in Gloucestershire:

> 'In addition to handling our twin-boomers, No 2 Squadron was responsible for turning round CFS Hunters when they visited Little Rissington from their operating base at Kemble. Unlike most Hunter units, the CFS Hunter Mk 4s and 7s were not equipped with underwing tanks, enabling the flowing contours of this most exquisite aircraft to be fully appreciated. No matter from what angle a Hunter is viewed, on the ground or in the air, it must surely rank as the most beautifully proportioned jet fighter ever built. Likewise, the sound of its Avon engine on the taxi-way, on take-off or at full bore, was always sufficient to make one stop and admire this aeronautical masterpiece. Thus began my initiation on an aircraft that soon gained my respect and admiration.
>
> To facilitate the installation of Category 2 lighting, the main runway at "Rissy" was closed in January 1962. The Jet Provosts and all piston-engine aircraft continued to operate from the other two runways while the Vampire, Meteor and Canberra squadrons moved to RAF Kemble for a six-month detachment so that flying on the fast jets could continue. There we shared a D-site hangar with the Hunters of the CFS Type Flight. Each day our flying and technical personnel were transported to and from Kemble in the "luxury" of RAF Bedford coaches. At that time No 5 MU's storage hangars held large stocks of Hunters of various Marks, including some F.6s still in their "Black Arrows" livery, while others were undergoing refurbishment on the main site for re-issue to service units. In April it was all change for me as I was posted to Aden where a much more intimate relationship with the Hunter awaited me.

Hunter F Mk 4 XF944 was first assigned to No 234 Squadron (part of 229 OCU) at Chivenor before being allocated to the CFS at Little Rissington (as illustrated), but detached to nearby Kemble. It was later refurbished and sold to Switzerland as a Mk 58A. *Ray Deacon*

Since its formation in 1912, the Central Flying School mission has been to turn out qualified flying instructors [QFIs] for the Royal Air Force, Royal Navy, Army and many overseas air forces. By the mid-1950s students would normally have flown around 100 hours on their CFS courses, equally divided between instruction on basic and advanced aircraft types, by night as well as day. Renowned for its reputation as – at 730 feet – the highest operational airfield in the RAF, Little Rissington was the main operating base for the Central Flying School from 1946 until the airfield's closure in 1976.

For QFIs undertaking the advanced courses, the final phase comprised a week's familiarisation on some of the new types then entering RAF service. On 7 November 1955 the first of three Hunter F Mk 4s (WV319, 325 and 331) flew in to Little Rissington to form the nucleus of a Hunter Flight that formed on the following day. It soon became apparent, however, that the somewhat short main runway (1,600 yards in total) was the reason for a number of landings terminating with burned-out brakes and tyres. With this and a seriously congested circuit to contend with, the Commandant and his staff deemed it wiser to relocate the Hunters elsewhere, thus they became a detached unit at the more suitable location at RAF Kemble, twenty-plus miles to the south of Little Rissington.

One of a handful of F Mk 4 aircraft that served with the Central Flying School at Kemble, XF973 first served with No 66 Squadron. After RAF service the aircraft was converted to Mk 58A standard and sold to Switzerland. *Ray Deacon*

Kemble was home to No 5 Maintenance Unit, which was already a major Hunter MU, and its civilian staff were assigned to look after the flight's everyday needs. A conversion course was prepared for students and, commencing on 12 November, four instructors were converted every two days. Two months later the first incident occurred, when the pilot of Hunter F.4 WV319 was unable to stop his aircraft after touching down and overran the runway before crashing into a Cotswold stone wall bordering the airfield. He escaped without injury.'

Having completed tours as a Flight Commander both on Sabres in Germany and on Balliols and Vampires at RAF Cranwell, Flt Lt Peter Hicks embarked on his first posting to CFS as a 'Trapper' on the Advanced Examining Wing in 1956:

'As has been mentioned already, a Type Flight was established at Kemble to give CFS instructors the opportunity to be familiar with current first-line aircraft. In 1956, and for years after, this effectively meant the Hunter F Mk 4. It was a fine aircraft, though I would have preferred a Sabre with an Avon engine and a Martin Baker seat (but not the Mk 2H seat, as you will see!).

On 24 August 1956 I was happily detailed for a day on the Hunter at Kemble, having left my wife and three children in the old cottage we were renting in Chedworth, and I drove to Kemble in our first car, a 1932 Austin 7, bought for £35. It was a bright day but with the sky obscured by 8/8 alto cumulus at about 10,000 feet. My intention on the first trip of the day was to climb to 35,000 feet heading out to the Bristol Channel, make a sonic bang (still a novelty fifty years ago) followed by stalling, aerobatics, perhaps a spin, and a return to Kemble. It was not to be. The medium cloud presented no problem as far as Air Traffic Control was concerned. In those days there were very few regulations compared with those needed in the much more crowded airspace we now have. If you wanted to climb through cloud you did so; there was thought to be no requirement in uncontrolled airspace, nor were there allocated facilities to be monitored, for collision avoidance. However, an accident was about to occur that would help to provide ammunition to bring about change!

I took off heading west and commenced to climb at the recommended 430 knots. Almost immediately after clearing the thin cloud there was an indescribable impact. All that followed is very clear and occupied no more than about 30 seconds, I would say. I thought to myself that it must have been another aircraft. I was tumbling and this was where one should eject. I couldn't find the canopy ejection handle so I went straight up to the seat ejection handle and pulled. In the wind outside (400-500mph initially – very rough) it was so bad that I decided that that was that! I noticed that my right leg was trailing behind at an impossible angle but felt nothing. Then objectivity failed – the safety harness box was rotating, which caused me a moment's panic (very illogical). Looking up produced some happiness – the parachute canopy. But there was a missing segment. There was also something else very obvious to me – the seat was attached to the apex of the chute and was dangling over the side above my head.

I decided I should climb up the rigging and try to release it otherwise it might have hit me on landing. I raised my right hand to start climbing and at that point the delayed shock of a very severe bang on the head finally took effect – and the next thing I remember was regaining consciousness in Gloucester Hospital (before being subsequently transferred to RAF Hospital Wroughton). The "bone dome" ended up in the Black Museum of the Institute of Aviation at Farnborough entitled "The wearer lived". A good design, I have to say, and well produced. Thanks to that, and above all to the Martin Baker seat, which functioned in spite of considerable damage, the tale can be told.

The cause of the accident was a collision between myself in Hunter F Mk 4 XF 980 (barely a month after it had been allocated to the CFS) and Javelin FAW Mk 4 XA644, which was being flown by a Gloster test pilot with an RAF navigator on loan to the company. The Javelin must have broken in two laterally for the navigator saw the floor opening beneath his feet. Apparently part of the wreckage must have operated his ejection seat and he was safely ejected. The pilot ejected but sadly his parachute was damaged in contact with wreckage and he did not survive. The Hunter broke in two somewhere between the wing and the tailplane and crashed near farm buildings at Wotton-under-Edge. Part of the Javelin smashed the Hunter's canopy (and my "bone dome") and part, probably the pitot head, made a three-inch hole in the side of the seat about ten inches behind my left ear. More significantly it cut the drogue-withdrawal line. The results were potentially unpleasant, for the drogue remained throughout in the seat, unable to develop half a second after ejection to slow and stabilise the seat and, at 10,000 feet, to release the wearer from the seat, and finally to pull the ripcord and open the chute. Since none of this happened except the automatic release of the seat harness, I was lucky that only one leg went up over the thigh guard and broke at the hip (there being no leg restraints on the Mk 2 seat). And I was even luckier that, since the drogue could not open the main parachute, I fell out of the seat so that the pull of the seat against me opened the chute, and it developed without wrapping round either the seat or me. Finally, the Board of Inquiry found that in view of the closing speeds of the two aircraft and the very short time available between possible visual contact and impact, no blame could be allocated to either pilot.'

The following extract was taken from the report produced at the conclusion of the inquiry into the accident:

'On 28 August at Little Rissington we examined the Mark 2H Ejection Seat, Serial No 55917, which was used by Flt Lt P Hicks, the pilot of the Hunter who fired himself out immediately after the collision. On this seat the drogue gun had fired but the drogues were not extracted from their container and the reason for this non-extraction was that the drogue withdrawal line had been severed, having been cut by impact with aircraft structure. The blow which had been received on the port side of the headrest was by impact with aircraft structure and not by impact with the ground. Crystallised Perspex and Perspex dust had accumulated in crevices in the top region of the port seat beam, particularly in the drogue gun top mounting bracket, and this was indicative of a shattering blow on the aircraft hood before ejection. The drogue line top stowage clip on the port seat beam was broken, also by structure impact damage, and this broken clip acted as a cutter, cutting the withdrawal line completely through about three inches from its attachment to the drogue gun piston. Also, the "neoprene"-sleeved automatic line was badly cut at the same position, evidently by the same means. This damage to the top port side of the headrest area indicated that the blow was sustained by impact with aircraft structure and not by the seat penetrating through the Perspex aircraft hood, and it seemed from the marks that the impacting object came from a forward port direction. It was noted that severe damage to the pilot's protective flying helmet had occurred in a coincidental area and not on the top of the helmet, which would be expected from contact with a closed Perspex hood.

The main time-release mechanism had correctly operated and opened the harness thus freeing the pilot from the seat. The pilot then separated from the seat, and, as the automatic line was still connected to the automatic rip cord, this separation opened the parachute. Evidently, the parachute developed, with the ejection seat still fastened to the parachute apex, as it was when we examined it. The pilot's descent would have been made in this condition. This would have a two-fold deleterious effect on the descent. First, the drop speed would be increased by the additional weight of the ejection seat, and, second, the ejection seat would continually obstruct the full development of the parachute. This second factor was substantiated by black dye marks on the parachute canopy, made by the ejection seat rubbing on it during the descent. The only other damage to the parachute is that No 14 panel ripcord had its "V" strengthening gusset torn away from the periphery hem. As the stabilising drogues had not developed on this seat and because of the absence of leg restraining harness, the leg injury to the pilot is not surprising. [Leg restraining harnesses were retrospectively fitted on the Mark 2H seat.] From the examination of the ejection equipment in this accident and a study of the location and distribution of aircraft wreckage, it seems that the collision occurred at about 10,000 feet and that the intercepting speed of the two aircraft, at the time of collision, was over 1,000mph.'

In summarising his findings Wg Cdr Jewell stated that the safe escape of these two airmen, in the exceptionally difficult environmental circumstances, was noted to be remarkable. A replacement Hunter, F.4 XF999, arrived on 9 September and Peter Hicks was able to return to flying duties on 25 February 1957.

In July 1957 a dedicated Type Squadron was formed at Little Rissington, comprising separate Type Flights of Canberra, Meteor and Vampire trainers. At the same time, the renamed Hunter Type Flight at Kemble was absorbed within the Type Squadron and its aircraft establishment was doubled following the arrival of three more F.4s. During the late 'fifties, when the RAF first sought a replacement for the Vampire T.11 in the advanced training role, Hawker was invited to fly a Hunter two-seater into Little Rissington for the Commandant to try out. Flown by Hawker Test Pilot Bill Bedford, the aircraft arrived in typical Bedford style with a high-speed roar across the airfield and a landing so short that he was able to turn off the main runway at the intersection.

Intent on 'selling' the aircraft to CFS staff, Hawker had sent a trainer equipped with nose-wheel braking to demonstrate the aircraft's capability to operate out of small airfields. While his aircraft was being refuelled, Bedford and the Commandant discussed the merits of the Hunter as a Vampire replacement over a cup of coffee before going off together for a short demonstration flight. After a couple of touch-and-goes, they landed and returned to the flight line, Bedford stressing the Hunter's short landing run. Air Commodore Whitworth responded by stressing Bill's experience of test flying against that of trainee instructors.

In 1960 the first of two Hunter T Mk 7s destined for the CFS moved from 5 MU storage and across Kemble airfield to the Type Flight's hangar. WV318, which had recently been converted from a single-seat F.4 fighter, remained with the flight for three years before seeing subsequent service with many other units. It was one of the final four Hunters to operate with the RAF when retired in 1994. Purchased by Delta Jets two years later, it was returned to airworthy status at Kemble and is currently the company's flagship aircraft, nearly fifty years after its CFS debut. Once a second T.7, XL611, had joined the flight, two of the unit's six F.4s were withdrawn. At the end of 1960 no fewer than fifty-two Flying Training Command pilots, sixty CFS Staff and forty-nine CFS students passed through the hands of Type Flight's three instructors, all ex-Hunter squadron pilots who had completed a tour of instruction at a Flying Training School, a creditable achievement and testimony to the high serviceability rates that could be maintained by the civilian staff of 5 MU.

Having been a B2 category instructor on the Hunter T Mk 8 since its introduction at RNAS Lossiemouth, Lt Michael Doust decided that, together with one of his colleagues, Arthur Stewart, it was time for them to go south to the CFS and be examined for their A2 category:

> 'Over the next six weeks or so we sat down together in Arthur's married quarters dining room studying the RAF Instructional Manual, and everything to do with the Hunter T.8 jet trainer. By the time we arrived at CFS we knew every rivet, nut, bolt and piece of piping in the aircraft. In addition, we had 764 Squadron's home-made instructional manual for the T.8 (there was at that time no RAF Instructional Manual for the Hunter trainer). This manual was to stand us in good stead, because the CFS examiners could not believe that we had gone to such lengths to produce it. They were so intrigued with it that we left a copy with them, and it would eventually become the standard RAF manual after some modifications and re-writing.
>
> We arrived at RAF Kemble where the school kept its Hunters. Arthur went for his flight test first, while I stayed on the ground being thoroughly quizzed on all aspects of flight, aeronautics and weather. When Arthur had finished his flight, I was briefed and up I went and was back on the ground in time for lunch. We both passed the flying examination, and the examiners were extremely pleased at our standard of airmanship and handling ability in an aircraft that was not used too much for advanced flying training. When lunch was over, they quizzed both Arthur and me even further about our academic flying knowledge, until 18:00. We both thought that the examiners were not going to stop, then suddenly they stood up, said, "Thank you, well done both of you, you have passed with flying colours!"
>
> We looked at each other, with great big smiles all over our faces, and shook hands with each other and the examiners – we had done it! As it was now so late, and the airfield personnel had secured for the day, we stayed overnight at the CFS

Officers' Mess and celebrated somewhat. Early the next day we were on our way back to Lossiemouth, where congratulations all round were waiting for us. Arthur and I were chuffed to the core – we felt like two little suns glowing with success!'

More recollections of life with Hunters at Little Rissington are provided by the former OC of the CFS Hunter Type Flight, Flt Lt T. A. 'Sharky' Hastings:

'The Hunter Flight of Central Flying School gives pilots air experience of an operational type of aircraft without the need of a lengthy simulator or instruction course. The method by which hundreds of pilots in Flying Training Command completed a Hunter conversion course at CFS is described in this article.

The Flight, still part of CFS Type Squadron, was re-designated No 3 Squadron (under Sqn Ldr D. G. T. Franklin), and had three instructors on the staff, all ex-Hunter squadron pilots who had done a tour of instruction at an FTS. The aim of the Hunter Flight was incorporated in the directive to No 3 Squadron of Central Flying School: "No 3 Squadron at CFS is responsible for giving students flying experience of representative service aircraft of an advanced type, so that they may appreciate better the problem their own students will have to face." Under the terms of a more recent AMO, CFS was also responsible for the re-categorisation and standardisation of Hunter QFIs and pilots. At the time, Flying Training Command used the Vampire, Meteor, Jet Provost, Piston Provost, Chipmunk and Varsity as training aircraft. More modern swept-wing aircraft differed from all of these aircraft in such aspects as stalling and transonic characteristics, take-off and landing techniques, control, trim, brake, oxygen and anti-g systems. With the introduction into the services of such aircraft as the Lightning, Victor, Comet, Sea Vixen and Buccaneer, it became of paramount importance to acquaint FTC instructors with the characteristics of swept-wing aircraft. Thus they would be better equipped to discuss knowledgeably with their students all the important differences of swept-wing flying.

The comparatively vice-less Hunter was ideally suited for swept-wing familiarisation and the aircraft in this role provided one of the major tasks of Hunter Flight. The other major commitment of the Flight was the provision of support for the Advanced Standardisation Flight of CFS, which re-categorised and standardised Hunter QFIs and pilots. Aircraft were provided for their continuation flying and the development and evaluation of new techniques. Hunter Flight instructors helped in this work and often completed standardisation teams when visits were made.

Pilots in Flying Training Command who came to the Hunter Flight for flying could be divided into three categories. First, there were those pilots passing through on CFS instructor courses. They did a conversion course of three sorties, making a total of two hours on type. Second, Flying Instructors from the RAF College Cranwell, and the other Flying Training Schools, came to the Hunter Flight for a one-week course, which consisted of nine sorties, giving a total of six hours flying. Three sorties were for conversion, three for consolidation in general handling, and three for formation flying. The third category included the instructors on the staff of CFS who came to Kemble at frequent intervals to keep in current practice on the Hunter.

The aircrew bus left Little Rissington at 08:30, but by this time the driver had collected the official mail for the Flight and any safety equipment that had been in for inspection. The journey to Kemble takes about an hour, and this time was often used for a quiz on the aircraft systems and emergencies. In the meantime the groundcrew, who started work at 07:30, were completing their daily inspection on the aircraft. On arrival at Kemble a Met briefing was held, pilots were briefed for their appropriate exercises and the first detail was airborne at 10:00hrs. After the Met briefing one instructor started the three-hour technical briefing if there were pilots requiring it. The day's programme was planned for three to four aircraft each flying three forty-minute sorties in the morning and two in the afternoon.

This was fitted into the airfield opening times, which, being a Maintenance Unit, were 09:00-12:30 and 13:30-16:00 from Monday to Friday. During this time the QFIs were briefing for exercises, supervising starts, leading formations, debriefing and in the tower as duty QFI. The staff QFIs obtained their flying from air tests, weather checks, leading formations, night flying and SCT. After flying had finished, the inevitable "paperwork" was completed and the bus left for Little Rissington again, arriving there between 17:00 and 17:30. The groundcrew remained at Kemble to after-flight the aircraft, then "clocked-off" at 16:30.

All pilots who flew with the Hunter Flight attended a lecture, which lasted about three hours, covering the systems and emergencies. The student instructors were briefed by courses, while visiting instructors were briefed on the first morning of their attachment. After the lecture a comprehensive questionnaire was given to the pilot to complete. Before flight all pilots received a further briefing on the handling of the aircraft and local procedures, and they had time in the cockpit to familiarise themselves with its layout.

The three familiarisation flights were:

Familiarisation 1 – Take-off; climb to 35,000 feet; effect of controls and turning; descent to 27,000 feet; incipient stall; CDTC; circuits.

Familiarisation 2 – Climb to 42,000 feet; sonic dive under radar control; return to Kemble for circuits.

Familiarisation 3 – Climb to 10,000 feet; practice manual flying; aerobatics; GCA.

The first exercise was usually given dual in the Mk 7, but pilots who had recently flown the Hunter and who were attending for re-familiarisation would often be sent off solo in the Mk 4 at once. All pilots carried a knee pad covering all checks, vital actions, limitations and emergencies. In addition, a card outlining emergency procedures was fitted to the aircraft in place of the gunsight. All the starting, after-start checks and vital actions were closely supervised by a Hunter QFI for the first two flights, and a Hunter Flight QFI was on duty in the control tower for the first solo. All pilots were debriefed after each flight. After three familiarisation sorties, the initial conversion or re-familiarisation was complete. This, therefore, constituted the CFS student instructor's Hunter flying syllabus. Instructors from the FTSs and CFS Staff used these exercises as a basis for their Hunter flying and continued with aerobatics, practice forced landings, manual landings and formation flying.

Although the aircrew travelled the sixty-mile return journey from Little Rissington daily by bus, the servicing of the aircraft was carried out by civilians who were part of No 5 MU Kemble. The Foreman of Trades, Mr J. Phillips, had twelve men to maintain the five aircraft. One aircraft was usually undergoing minor servicing and four aircraft were available for flying. The groundcrew all lived locally and had been associated with the MU for many years, most of them having been with the Hunter Flight since it was formed, and they had become an efficient servicing team. They knew their aircraft and were extremely conscientious – as a result the serviceability of the flight was very high. No 5 MU also provided Hunter Flight with Air Traffic Control, crash facilities, fuel, and midday meals at the Officers' Mess. The co-operation between Kemble and CFS had always been very close, and the Station Commander, Wg Cdr P. A. S. Thompson, and the Senior Technical Officer, Sqn Ldr W. W. Suckling, were responsible for first-class backing to all the Hunter Flight operations.

It will no doubt be of interest to readers to know that No 5 MU Kemble had the function of preparing many types of aircraft for Royal Air Force Units and foreign Air Forces, and was also used as a relief landing ground by CFS jet aircraft. Far from being a quiet MU, there were up to 2,000 aircraft movements each month.

By the end of 1961 the CFS Hunter Flight had successfully completed hundreds of initial conversions and re-familiarisations. In spite of the fact that the flying programme had to be arranged to fit the airfield opening times and that most of the flying consisted of forty-minute conversion sorties with a necessarily restrictive weather minima, the Flight had logged a total of more than 6,000 hours since November 1955. During 1961 alone, 204 pilots were converted, all of whom had a minimum of three trips and some considerably more. While achieving this, the Hunter Flight flew 1,260 hours, which was also the best-ever total, some twenty-three per cent above the previous highest in 1959. There were no incidents of any description in 1961, for example, with the exception of a heavy landing in a Mk 7. The CFS Hunter Flight can therefore look back with some pride on its achievements over the years.'

A change in policy brought about the demise of the CFS Hunter Type Flight at Kemble on 28 February 1963, the swept-wing commitment being transferred to the Gnat Squadron. The unit's two remaining Hunter F.4s, XF944 and XF973, were the last of the Mark in regular service with the RAF and were placed in long-term storage with 5 MU. Some ten years later both were converted to F.58A standard for the Swiss Air Force, while mention has already been made of T.7 WV318, which is still flying out of Kemble, albeit under the auspices of Delta Jets.

It was during the CFS detachment to Kemble in January 1962 that, while watching a pair of CFS Hunters take-off, Ray Deacon was called into the 'Chiefy's' office and, for various reasons, had a fair idea why he had been summoned:

'Chiefy informed me that my overseas posting had come through and yes, it was to the dreaded Aden! My heart sank. Having listened to countless horror stories from airmen who had experienced the place, the prospect of two years on a desert posting filled me with apprehension. Being a red-head who could not sit out even under a British sun without burning to a frazzle, I couldn't imagine how on earth I would survive in Aden's heat. No point in dwelling on it though, I thought – just make the most of my final months at Little Rissington.

To add insult to injury it transpired that I would not even be flown to my new posting, as most people were, but I was to join an "elite" group of thirty or so other airmen on HM Troopship *Nevassa* on her last voyage to the Far East. So, early in April 1962, we slipped away from Southampton Docks and began our cruise through the Mediterranean and Suez Canal, arriving in Aden some ten days later. On disembarkation at Steamer Point, those destined for Khormaksar were taken by RAF transport to the airfield and assigned transit accommodation to await their unit allocations. Assigned to No 8 Squadron, I moved into the aptly named Hunter block, which was one of three newly constructed buildings and comprised four-man, air-conditioned rooms. With hindsight I need not have worried about the sun and the heat as, after an initial month or so of discomfort and gradual loss of my 'moonie' image, I got quite used to it. "Don't forget to take your salt tablets" was regular advice, but having

The CO of No 8 Squadron applied special markings to his aircraft; here the unit's colours and badge are proudly emblazoned on the nose-wheel door. *Ray Deacon*

dumped my lunch over the sand outside the mess after my first couple, I gave them up as a bad job!

My whole tour in Aden was spent with No 8 Squadron, a Hunter ground-attack unit based at RAF Khormaksar. Located at the eastern extremity of the airfield right next to the civilian airport, the squadron operated fourteen FGA Mk 9 Hunters assigned to fighter and ground-attack duties, four FR Mk 10s for fighter and photo-reconnaissance missions, and a couple of two-seat Mk 7 trainers. The other end of the pan was allocated to the second Hunter squadron, No 208, equipped with twelve Mk 9s and a solitary trainer. The squadrons were tasked with supporting the multifarious British and Arab Army units policing not only in the Aden Protectorate, but the whole Middle East Command area from Kenya, 1,000 miles to the south, to Bahrain Island, some 1,300 miles to the north. The heavy responsibility was shared between the two Hunter units.

Looking rather battle-worn, T.7 XL597 is pictured at Khormaksar in the markings of No 208 Squadron. It subsequently served with Nos 4 and 20 Squadrons in Germany (as a continuation trainer, both units operating Harriers by that stage) before crashing in Suffolk during May 1980. *Ray Deacon*

FGA 9 XE530 at rest on the Khormaksar flight line, the under-fuselage air brake unusually left in the extended position. *Ray Deacon*

Hunter F.6 XG209, complete with DFLS titling on the nose and the unit's distinctive yellow spine and tail paint scheme, is pictured during a detachment to Khormaksar. *Ray Deacon*

Tranquillity in the region was a rare phenomenon! At the time of my arrival in April 1962 there was relative calm within the Command, with much of the Hunter flying being mainly devoted to training and familiarisation of new pilots in the area of operations. Back in the previous July, however, things could hardly have been more different. As General Kassim of Iraq aligned his forces along the borders of Kuwait in readiness for invasion and the capture of the lucrative oil fields, the British Government ordered No 8 Squadron to fly up to Kuwait and prepare to defend the tiny kingdom from Iraqi attack. No 208 Squadron, which was stationed at RAF Eastleigh in Kenya at that time, was also ordered up to Bahrain, some 250 miles from Kuwait, to support 8 Squadron should an attack occur. A number of Canberra units were despatched from European bases to others on Cyprus and RAF Sharjah, and a third Hunter squadron – No 43 – moved from Leuchars to Cyprus, leaving the errant General in no doubt as to the British intent should he dare to cross the border. In addition, troop numbers in the Gulf area were rapidly increased with deployments from Europe and other Middle East bases.

In the event, the General backed down and the situation soon returned to normal – at least as normal as normal is for that area. Concerned that a total withdrawal of British forces from the Gulf would tempt Iraq into making a quick strike, the decision was taken to maintain a squadron of Hunters at Bahrain on permanent stand-by together with a pair of photo-reconnaissance Canberra PR Mk 9s from No 13 Squadron on Cyprus. The former was achieved by rotating detachments of 8 and 208 Squadron aircraft from Khormaksar to Bahrain on a two-monthly basis, and this was standard routine by the time I arrived. Every other month Hunters from one of the squadrons would depart from Khormaksar, the groundcrews flying up in a Beverley, Argosy or even a Britannia if one happened to be passing through, for their two-month stint at Bahrain, passing the returning squadron on the way up.

At one stage during the RAF's presence in the Middle East, the unit markings of 208 and 43 Squadrons were merged as part of the Khormaksar Hunter Wing. XJ632 was later sold to Singapore as an FGA Mk 74.

Although blessed with permanent sunny skies, the climate in the Gulf was different from that in Aden – it was hotter and much more humid in the summer, yet cool enough to warrant a change into standard working blue [uniform] in winter. During these eight-week detachments, the opportunity would be taken for the Hunter squadrons to spend two weeks down at Sharjah on the air-to-ground range at Jeb-a-Jib.

When I arrived, because of the intense heat of the afternoon the normal working day at Khormaksar was scheduled from 07:00 to 13:00 followed by lunch and a two-hour siesta. This could change at any moment, and often did when Hunter support was called for from up-country Army units. Assigned to the aircraft line as one of a handful of Air Wireless Mechanics, my job was to ensure that the radio equipment on each Hunter was in good working order before and after each sortie. Inspecting the aerials required an interesting technique that required one to clamber onto the wing, and it did not take long to master it, if blistering of the knees on the hot drop-tank surfaces was to be avoided. The knack was to stand back some ten paces from the front of the wing then run at it and, in

No 208 Squadron's XK151 poses for the camera with her ventral air brake extended, clearly illustrating the distinctive 'saw tooth' wing leading edge extension fitted to later variants of the aircraft.

an instant, bounce one knee on the drop tank while hauling yourself up onto the wing with one hand on the "saw-tooth" leading edge. Everyone had their own technique and the activity made for good spectator sport. The metal surfaces were so hot during one Bahraini summer that it was possible to fry an egg on a Hunter wing, one intrepid pilot taking twenty minutes to prove the point.

One of the toughest jobs as a radio mechanic was working in the cramped, oven-like space of the radio bay, which was so small that you could not turn your body round. With temperatures in the bay reaching 140 degrees Fahrenheit, with much of the heat being reflected from the concrete surface of the pan, it was a very uncomfortable place to work. Changing one of the two 30lb VHF sets was not in itself much of a problem as it could be lowered on your shoulder as you crouched down and out from under the bay, but it was a different story once the 56lb Collins UHF units superseded them. Changing one of those was a real test of strength and control, the sweat literally pouring down, but it no doubt contributed to our keeping fit.

The radio and radar guys may have had it easier than some trades, in fact; the armourers, engine and airframe guys in particular had to handle jobs that demanded that they spend much longer in physical contact with the aircraft than we did. The energy of the armourers in particular deserves a special mention, for it was they who had to change the gun packs and load rockets with great gusto when under operational conditions and when they were on air-to-ground range practice. Working in teams, each would try to beat the local informal record for changing a gun pack, which, if I recall correctly, was a little over eight minutes. I thought I would help once and found lowering the empty pack onto the trolley quite painless, but winching a fully armed pack up to the locking points was a totally different ball game. Nearing the top I ran out of energy and received some stick for making the two armourers stop and await my recovery.

From the first day of training, the importance of flight safety, both for pilots and groundcrew, was drummed into you. Irrespective of whether you flew in or worked on prop or jet, trainer or fighter, the safety of aircraft and personnel was foremost in one's mind. On front-line squadrons such as ours, where Hunters were regularly armed with live weapons, additional precautions were necessary and you had to keep your wits about you. The start-up procedure on a Hunter was straightforward enough but an alert mind and an awareness of what was going on around you was essential. One airman was assigned to each Hunter and he would check the area round his aircraft for clearance of ground equipment and that a ladder was attached to the aircraft as the pilot proceeded with his external pre-flight checks. With the pilot seated in the cockpit, the airmen aligned the parachute and ejector seat straps over his shoulders and, when requested, handed him his helmet and "bone dome", before plugging the radio connector into its socket. Finally, and again on request, the ejector seat safety pin was removed, shown to the pilot, and stowed in a pocket in the side of the seat. On climbing down, the ladder was removed and positioned in a safe place on the pan. A short wait then ensued as the pilot carried out his internal pre-flight checks. When ready to start, the pilot gave a quick twirl of his forefinger and, on checking everything was clear behind, the airman reciprocated.

On the press of the start button the automated engine start sequence began. The highly volatile Avpin fuel that was used to start the FGA.9 and FR.10 habitually ignited the acrid exhaust gases as they exited from a duct under the fuselage, and as this had been the cause of several starter bay fires, the door was left open until the starting cycle finished. An asbestos glove was used to pat out flames issuing from the duct. The door was then closed and the airman stood ready to remove the nose-wheel chock before walking to a suitable point to marshal the aircraft out.

FGA 9 XJ683 served in both the Middle East and Far East theatres before joining 1TWU at Brawdy. After withdrawal from RAF service the aircraft was sold to Zimbabwe. *Ray Deacon*

Some time during the early period of Hunter operations in Aden, the cannon on an armed Hunter parked on the pan fired off several rounds across the Khormaksar runway, hitting the wall of the FRA barracks on the opposite side of the airfield. Fortunately, there were no aircraft taking off or landing at the time and no one was hurt in the barracks, but the outcome could have had much more serious implications. On investigation, the reality soon dawned that no one was working on or near the aircraft at the time. A swift investigation discovered that excessive temperature had caused an electrical short in the gun firing circuit. A microswitch located in the port undercarriage bay broke the circuit between the invertors that provided the power to fire the gun; the undercarriage was down, so in theory the guns should not have been able to fire with the aircraft on the ground. A double security measure was therefore introduced whereby a safety plug – known as the Master Armament Safety Break (MASB) – was inserted into the same gun firing circuit. Access to the plug was via a small panel under the port wing directly behind the engine air intake. It was the duty of an assigned armourer to stand by the perimeter track and disconnect the plugs on aircraft as they taxied in, and reconnect them again on the perimeter track when they taxied out. An order was issued prohibiting personnel from walking in front of armed aircraft, which were identifiable by a red-painted ammunition box being placed in front of the nose and attached via a rope to the top of the nose-wheel door.

One related incident in particular remains very vivid in the memory of my former colleague, SAC Instrument Mechanic Mike Bennett, perhaps because he only had two weeks left to finish his tour when the incident occurred on 2 June 1963. A pilot, whose name escapes him, offered him a trip in a Hunter T.7 as a treat, but he turned it down, being so near to going home. So off flew the pilot in a single-seat FGA.9 instead, unaware that the aforementioned MASB had not been connected properly. He was, however, very soon aware that he could not raise the undercarriage and would have to fly around at low level and at just 150 knots for an hour and a half to burn off his full fuel load. Needless to say he was steaming when he eventually taxied in and all the groundcrew were assembled in the line hut as the proverbial sparks flew. On examining the F700 and discovering that a certain short, blond armourer had carried out the Before Flight inspection, the culprit was duly admonished in no uncertain terms in front of us all. Not a pretty sight!

Despite all the precautions to avoid situations like that one, accidents still occurred, and on one occasion a Sergeant armourer was walking behind the line of Hunters without concentrating and stepped directly into the jet blast of a Hunter that was about to taxi out. The force tossed him across the pan, which in itself did not cause any injury, but the side of his face quickly began to redden and blister. He was rushed to sick quarters for treatment and, on return to work a few days later, the toasted half of his face was jet black, earning him the nickname of "Black and White Minstrel" after the popular TV show of the era. It was a good example of how the Hunter always had to be treated with respect.

Life wasn't all hard work, though, and after an afternoon's kip and with the heat of the day relenting, it was off to haggle with shopkeepers in The Crescent, take a dip in the pool or one of the shark-net-protected beaches, or have a game of tennis, football or rugby if you were feeling rather more energetic, before winding down with a can or more of "Slops" (Allsops lager) in the Camel Club. Life was fairly relaxed.

All that changed in September 1962 when, following repeated strafing attacks on villages close to the Yemen border by Yemeni Air Force MiG-15s, the decision was taken for the Hunter squadrons to fly dawn-to-dusk patrols along the Yemen border to deter such incursions. With the quieter times confined to history and with one of the two Hunter squadrons up in Bahrain, it was left to the squadron at Khormaksar to maintain the border vigil. Pilots and groundcrews were split into shifts, one shift starting at 06:00 and finishing at 13:00 with the second overlapping from 12:00 to 19:00. The routine was for two Hunters to take-off at dawn and fly north-east up the coast towards Yemen, then turn west to begin their lengthy patrol along the mountainous, indistinct border before returning to Khormaksar from the north-west. Further pairs would take-off at hourly intervals, thus ensuring that there was at least one pair in the vicinity of the border. Fully armed and equipped with long-range underwing fuel tanks, a round trip would take approximately one and a half hours. Additional Hunters were maintained on armed stand-by, ready for despatch at a moment's notice to support Army units operating up-country. Needless to say, there was no longer any time to get bored!

Pressure on both air and ground crews during this period was intense, the high workload taking its toll on both man and machine. Aircraft unserviceability became a critical issue and, with only half the number of pilots and airmen on site at any one time, relief appeared with the detachment of eight Hunters from the UK, No 1 Squadron arriving at Khormaksar in late October 1962. A new work pattern evolved whereby No 8 Squadron started work at midday and continued through

Wearing the fuselage and wing tip black and white checks of No 43 Squadron, Hunter FGA 9 XE655 clearly illustrates the strut fitted to the aircraft's external fuel tank, enabling it to carry the store during manoeuvring. *Ray Deacon*

to dusk on day one, then from 06:00 the following morning until 13:00 on day two. No 1 Squadron started work at midday on day two and became the operational squadron for the next 24 hours. And thus it continued, alternating day by day. The system worked very well but after a month in the desert, and with the crews' suntans suitably honed, No 1 Squadron departed for home, leaving No 8 Squadron to revert to the split-shift system.

In January 1963 No 54 Squadron brought its Hunters out to Aden for a month's detachment, bringing short-term relief for 208 Squadron, with No 8 having moved up to Bahrain on its two-month detachment. It was not until March that a final solution was arrived at – the permanent transfer of No 43 Squadron's Hunters from Nicosia, Cyprus, to Khormaksar. Once the squadron had settled in, the rotating detachments to Bahrain took place every four months instead of two. The arrival of 43 Squadron's Hunters led to the formation of the Khormaksar Tactical Wing. In addition to the three Hunter units, the wing comprised Shackletons from No 37 Squadron, Belvedere helicopters of No 26 Squadron and an element of Sycamore helicopters from the Search and Rescue Flight. The four No 8 Squadron Hunter

FR.10s and both its T.7s, plus a T.7 each from Nos 43 and 208 Squadrons, were reallocated to a reformed No 1417 Flight, together with 8 Squadron's five photo-recce pilots and fifteen groundcrew. With forty-six Hunters under his command, Wg Cdr John Jennings had formidable firepower at his disposal, and the years up until our withdrawal in 1967 would see it increasingly called into action.

Up in Bahrain, the NAAFI was the hub of social life for the airmen and many an evening was spent sampling (canned) Tiger beer. On one particular detachment, in December 1963, the Khormaksar (8 Sqn) and Cyprus (13 Sqn) contingents held a joint Christmas Eve party, at the end of which everyone was unquestionably drunk. Having decided to awaken the sleepy camp, we made our way round the billeted area giving a rendition of raunchy songs and carols and, on reaching the station's fifty-foot-high water tower, some bright spark suggested we climb it and continue our serenade from the top. So off we went, one by one, up the steep, narrow ladder. It was not long before a couple of snoops [military policemen] appeared below demanding we come down or face the consequences. A few choice words in response had the desired effect and they soon disappeared out of sight. On walking by that tower a day or so later, it suddenly dawned on us how lucky we had been that no one had fallen off the three-foot-wide ledge encircling it – and it had no safety rails. So life in Aden wasn't always a serious business.

In February 1964 I achieved my long-held dream of flying in a Hunter T.7. With two months of my two-year tour remaining and not a single airman having experienced a flight in a T.7 during my time in Aden, SAC "Taff" John and I took advantage of a unit bash in one of the officers' married quarters to ask our Commanding Officer, Sqn Ldr Tam Syme, if it was possible for airmen to have a trip in a T.7. He said he saw no reason why not and that he would bear it in mind. Not long after, a policy of offering airmen a flight in the right-hand seat, two months before their tours were due to expire, was implemented, so thank you Tam Syme – we very much appreciated it! In January 1964 "Taff" became the first 8 Squadron airman to savour the delights of a trip in a Hunter, and my turn came early the following month.

With Flying Officer Sid Bottom at the controls and myself in the right-hand seat, XL613 (coded "Z") took to the clear blue skies in the late afternoon of 4 February. Keeping low, we flew north-east along the sandy coast for a while, passing a customs post and village on the way, before climbing to 10,000 feet or so for some aerobatics. Sid showed me how to loop and barrel roll before letting me have a go. I'd been taught how to do both on Vampire T.11s by instructors while based with the CFS, so I had some idea of what was required. My efforts were not as successful as those in the T.11, but nevertheless it was fun and I certainly enjoyed having a go. Sid then suggested that we do some "wadi-bashing", which sounded exciting, so off we headed for the mountainous region towards the north of the Protectorate. Some of the peaks there rise to more than 7,000 feet, which was a fact that I found surprising. My driver clearly knew the region like the back of his hand and we hugged the sandy valley floors, flying at around 500 knots, the steeply sloping escarpments rising high above on either side. The roar of the Avon must have echoed long after we had gone. I had been quietly hoping for a more sedate ride so as to take photographs, and on enquiring as to why we were flying so fast, exciting as it was, Sid calmly elucidated that it was to avoid being shot at. A plausible riposte, I thought, and I settled back to watch in awe as we weaved in and out of the high-sided wadis.

One startling sight that sticks in my mind was a narrow, U-shaped rock formation jutting out from the walled sides as it appeared through the windscreen, narrowing the gap through which we would fly. Once through it, the valley floor dropped a thousand feet or more. Down we went to continue our thrilling high-speed ride through the mountains. Then suddenly we were out in the open, flying low and fast over a flat landscape endowed with a generous covering of green vegetation, as we headed south and home to Khormaksar. On nearing Khormaksar I asked Sid if we could climb up and fly "round the volcano" in order to take some shots of Aden and Shamsan, and he kindly obliged. So, after one of the most exhilarating experiences of my life, we touched down and I flopped forward in my harness as the brake parachute deployed and the wheel brakes were applied. Thank you, Sid!

Shortly before the end of my tour, Yemeni aircraft flew a number of raids across the border, and in one of them a frontier guard post and a village near Beihan were attacked with rockets and bombs by an armed helicopter and two Migs. Immediate retaliation was ordered and at 03:30 on the morning of 28 March 1964 we were rudely awoken by the dulcet tones of the duty Sergeant, ordering us to report to Squadron HQ immediately. No shaving, no washing – just get dressed and board the transport waiting outside. Fortunately, the rare sobriety of the previous evening meant that hangovers were rare and, apart from lack of sleep, most of us had clear heads. As we were driven away, the realisation soon dawned that we had never been called out in the middle of the night before, so something special was happening. On arrival at the pan, several pilots could be seen talking to each other as they moved between the various offices. No explanation was given to us however, until "Chiefy" called us together and told us to prepare seven FGA.9s, two in reserve, and two 1417 Flight FR.10s, one in reserve, for an important operation. No 43 Squadron personnel had also been called in to prepare five FGA.9s (one as reserve) for a joint operation.

With pre-flight checks completed, those of us that were free assisted the riggers and armourers with the removal of the outer drop-tank pylons and the fitting of additional rocket rails, a lengthy and fiddly task on the poorly lit pan. An hour or so later, the rocket trolleys were towed along the back of the line and, while some helped with the mounting and connection of these high-explosive projectiles, others stuffed warning leaflets in the wing-flaps of the Mk 10s and two Mk 9s.

At this stage an outline of the tactics employed on the more concentrated air-to-ground strikes by the Aden Hunters may be useful. Two Hunters would depart some fifteen minutes ahead of the main force and, while flying fast and low over the target, would lower their flaps to release warning leaflets onto the target. The lead aircraft of the pair, usually an FGA.9, would be closely pursued by an FR.10 taking photographs of the leaflets being released as evidence of a warning. By the time the inhabitants below had read the instructions, they would have thirteen minutes or so to vacate the area, or face certain oblivion. The main force, usually comprising four to eight FGA.9s, would then carry out the attack with 30mm HE cannon and rocket fire, while the leaflet-carrying aircraft maintained top cover. Once the attack was over, the FR.10 would make a final sweep to photograph the damage inflicted on the target.

As dawn broke, the first two pilots walked out to the line hut and signed the F700s for their particular aircraft – one FR.10 and one FGA.9, both fitted with HE gun packs but no rockets. Once strapped in they started up and taxied out for take-off. I was assigned to start one of the eight main force FGA.9s, this particular aircraft being flown by a young-looking Flt Lt Martin Johnson. He was very excited and eager to complete his pre-flight checks, almost running up the ladder and jumping into the cockpit. I handed him his parachute and seat straps, waited until he was safely belted in, then removed the ejection seat safety pin and stowed it away before removing the ladder. At a given signal from Wg Cdr John Jennings in the lead aircraft, eight pilots hit their starting tits and seven Avpin starters burst into life, their Avons lighting up within a few seconds. Why only seven? Sod's law – if something can go wrong it will, and it did, and the starter on my aircraft had failed. Having applied a couple of hefty thumps on the starter panel relay boxes with the handle of my trusty screwdriver in the vain hope that it might fire at the second attempt, I gave the signal to try once more, but again the starter failed. In an instant, the straps were flying over Martin's shoulders and he was standing on the seat ready for the ladder to be re-attached. He climbed out and dashed to a spare aircraft, with me in hot pursuit. As he clambered in, the other Hunters began to taxi out and, between the two of us, he was belted-up in record time. Safety pin out, ladder away, a twirl of the forefinger and a welcoming blast from the Avpin starter, closely followed by the healthy roar of the Avon. It could hardly have reached idling speed before he throttled it up and raced out onto the taxiway.

By this time the others were accelerating down the runway on their take-off runs, while he must have reached a similar speed on his way to the threshold. Round the corner and on to the runway with the Avon seemingly at full power throughout, off he went in pursuit of the rest of the Hunter pilots who were rapidly disappearing into the distant haze. Did he catch them? Of course he did, and the attack was executed precisely as planned. Photographs taken by the FR.10 before and after the raid revealed that a stone-built fort was virtually destroyed, as was an anti-aircraft gun and a number of vehicles. So good were the photographs that the warning leaflets could be clearly seen lying in and around the fort. And the attack had the desired effect. Incursions by Yemeni aircraft ceased thereafter – at least for the time being.

All in all, with ten months on detachment in Bahrain, three weeks in Kenya to cover a rebellion by the country's army, and a month on leave, my two-year tour literally flew by, and by April 1964 my time was up. The camaraderie and family-like spirit on the squadron had made it a much more bearable experience than anticipated and, having survived the sun, heat, sand and flies, I could now look forward to the sanity of life back in the UK. Or so I thought! After two years in a faraway land where the locals spoke foreign, the beer was tolerable and the climate unbearable, my posting came through – this time RAF Valley on Anglesey, where the locals spoke a foreign language, there was no beer on Sundays and the climate was iffy, to say the least. What had I done wrong?'

Chapter 6

Unscheduled exits

On Friday 6 July 1979 a single Hunter made an unexpected appearance in a Cornish village. The local residents (and countless holiday-makers) had a miraculous escape from possible disaster when the fighter ploughed into the village just yards away from a loaded petrol tanker. The pilot, who was on a training flight, had experienced control difficulties some miles out to sea and was forced to eject, leaving his Hunter XG197 to crash into the ocean. Instead of careering into the Atlantic, however, the Hunter remained airborne and veered towards land at 250mph. It was something of a miracle that the jet eventually ended its flight wedged in a twelve-inch gap between Klymiarven guesthouse and King Arthur's Terrace, damaging just three cars, a greenhouse and a swimming pool in the process. The guesthouse owners, Jim White and John and Muriel Jewell, said they felt lucky to be alive, as did local resident Ivan Irons, who was up a ladder painting the first house to be hit, in Atlantic Road. Garage-owner Jack Cann was preparing to receive 1,500 gallons of petrol when pieces of the Hunter landed only ten yards away. As locals commented, 'If the jet had exploded the whole village would have gone up.' If the Hunter's path had not been stopped by the houses it would have ploughed down the High Street, killing many people.

A grounded Hunter: FGA 9 XE643 was serving with Nos 8 and 43 Squadrons when it was abandoned at Mombassa in Kenya following an aborted take-off. The airframe was recovered by HAS but scrapped, the nose section being preserved at RAF Aldergrove.

In this amazing and (thankfully) unique incident, no one was seriously injured and even the pilot survived his ordeal after being picked up by a Port Isaac fishing boat, the *Lady Jane*. At the time the plane crashed, David Flower was on the telephone speaking to his friend Fred Whiting, who at that time owned the launderette in Tintagel. He remembers that

> '...we were speaking about a TV repair when he suddenly said, "I've got to hang up the phone – a plane is coming straight at the house!" In fact it ended up a few yards away from his house as it veered across a field before crashing. I quickly gathered up a few movie films, grabbed my camera and dashed down to the scene. I shot quite a lot of film at the time and later showed it at the Social Hall in Tintagel at a special thanksgiving night held by the Parish Council. It really was an incredible escape from a major disaster in the village and as the accident took place at the height of the tourist season the fact there was no loss of life added to the miracle.'

The pilot of XG197 was Alick Nicholson, and his personal account of that fateful day makes fascinating reading:

> 'Nearly thirty years have passed since the accident, yet I still see comment on the incident, some of it on the internet – much of it inaccurate, too – and I am aware of continuing interest in the event from many quarters, and so I believe that it is probably time to set the record straight. This, therefore, is my own account of what happened that day.
>
> In 1979 I was being employed as an instructor at the Tactical Weapons Unit at RAF Brawdy, in South West Wales, with No 79 Squadron. In addition to being a Qualified Weapons Instructor [QWI], for many years I had also been a Qualified Flying Instructor [QFI] and Instrument Rating Examiner [IRE], and at the time I was regularly flying both the Hunter and the Hawk (the aircraft that was then gradually replacing the Hunter in RAF service). My primary role was training future QWIs, which was one of the main tasks assigned to 79 Squadron. As an occasional task I was also one of a small number who acted as Unit Test Pilot, and in this role I regularly flew both types of aircraft on various test schedules after major servicing or engine changes.
>
> I believe that I understood the Hunter and all its characteristics very well. The Hawk had only recently entered service and we were just beginning to use it as a tactical and weapons platform. It was a nice enough aircraft, but came nowhere near the Hunter in terms of the sheer joy of pure flying and available firepower. However, all the pilots were aware that it put the Hunter well into the shade in terms of reliability and safety. Along with all the other staff pilots at Brawdy, I had recently come off a front-line aircraft where it would have been inconceivable that reliability and safety were not always excellent. In my case I had flown the Phantom FGR.2 with both 6 and 31 Squadrons, and had developed a justified trust in that fine aircraft.
>
> The Hunter remains one of the greatest aircraft ever built. It was a sensory aircraft full of emotive sounds and vibrations. You could hear and feel the engine. There were haunting smells, especially when firing the guns. The handling qualities of the Hunter were both unusual and magical – an interesting balance between a short-stroke aileron and a long-stroke elevator that gave superb lateral responsiveness coupled with stable and very progressive pitch control. Quirky systems too – a fuel system that had a sense of humour, flaps that almost spoke to you in combat, and a longitudinal control system that could kill you at a high Mach number, and that had in fact killed more Hunter pilots than any other single cause, even though the MoD had been advised of a simple solution to the problem in the late 1950s but had chosen to do nothing about it. Four mighty 30mm Aden canons were internally mounted, and always with you whatever other weapon load was carried, supported by a gyro gunsight based firmly on an ancient Second World War design. It was a gunsight, it has to be said, that required very high levels of imagination, skill and respect to comprehend how to get good air-to-ground results from a device originally designed for air-to-air use. The aircraft was also a potent force. It was fast and electronically silent. The gunsight radar had long been removed. There was no radar altimeter or any other equipment that could be

electronically detected. With a speed of 610 knots in level flight at sea level, the Hunter was not matched by many other aircraft at the time.

Despite the magnificence of the aircraft, though, its day had long gone. It was unreliable – its failure rate was three times that of the Hawk. It was expensive to maintain too, and each repair cost four times the average cost of a Hawk repair. It was undoubtedly considered unsafe by many of us, but it was the Hunter – simply one of the most respected aircraft ever to see service, and the aircraft that every pilot wanted to fly.

My story is one of mismanagement and obfuscation at the highest levels, often by past colleagues – a story of incredible support from the least expected quarters, of downright untruths by the "great and the good", of lack of proper forward planning, and of disregard for the well-being of the pilots who flew the aircraft. And so it was that I came to be flying Hunter XG197, a Mk 6A single-seater, on 6 July 1979. The aircraft had flown less than four hours since the engine had been changed, and I had flown all those hours, including the engine air test. These flying hours had included some aggressive low-level flying in Scotland with some of my students at the end of their QWI course, and the engine had been operated at full stretch without any problems. The night before its final flight the ejection seat had been replaced because the one installed was time-expired. The newness of the seat harness was to have consequences a few hours later.

A close-up view of an RDAF pilot in the cockpit of his Hunter, illustrating the typical pilot head gear and part of the ejection seat assembly. *RDAF*

It had been a hectic week. My wife ran a small dairy farm, and I mucked in when I could. Hay-making was well behind schedule and being away in Scotland had not helped the harvest. The evening of 5 July was magnificent. Warm and friendly, it was one of those memorable times when neighbours really co-operate, when everything seems worthwhile, and it was after 11pm before we brought the final load of hay into the yard.

The following day, the 6th, started much earlier than usual. A telephone call told me that an exercise had been called and this required all aircrew and aircraft on-state as soon as possible. I was assigned the lead of a four-aircraft formation, call sign Bronze, for a low-level sortie mainly over Devon, involving simulated attacks on various targets. Fighter opposition was to be provided by Hawk aircraft, also coming from Brawdy. A key part of the exercise was the preparation of the aircraft to an operational state, and this should have involved loading all four guns in the Hunters with live (high-explosive) ammunition. In my aircraft, armament difficulties led to two of the guns being loaded with HE, and the other two with inert practice rounds. All four guns were primed with a round in each breech, as is normal, and the electrical firing circuits were disconnected. The guns were therefore incapable of being fired. It remains a great blessing that the crash wreckage at Tintagel did not burn, as I believe it would have been quite possible for the guns could have cooked off live rounds if they had been sufficiently heated, and the Aden's gas return system could have cycled further HE rounds into the breech. This was actually a bigger danger than the petrol tanker that was close to the crash site, which got so much publicity, and Tintagel was luckier than anyone realised. A year later the MoD issued a press release in the form of a Parliamentary Question and Answer, which stated that only practice ammunition had been loaded and that the guns themselves were not loaded. This was to be one of many convenient inaccuracies. Another interesting fact, which was never made public, is that two of the HE rounds were never even recovered. So, if you are reading this and you have a couple of 30mm shells on your mantelpiece as a souvenir, please hand them in to a safe place!

The flight did not start well. Two of my Hunter formation went unserviceable before take-off, though one was fixed and joined up when we were over Devon. On the run-in to the second target we were intercepted by a pair of Hawks. We countered them, attacked the target, then re-engaged the Hawks. As I turned in behind one of the Hawks, at a range of about a mile, I coasted out about five miles north of Tintagel and began to set myself up for an easy burst of well-tracked gunsight film. Once over the sea I descended to very low level and accelerated to around 600 knots to close on the Hawk that was maintaining about 450 knots ahead of me. As I closed to 800 yards I gently closed the throttle, aiming to run through an ideal firing bracket between 400 and 300 yards. With speed dropping towards 525 knots I reapplied power to stabilise my closure rate.

Immediately the engine surged with three very loud thuds that shook the whole aircraft. I had experienced engine surges before, but this was different. A glance at the engine instruments showed a large mis-match between the rpm and temperature, and the throttle position. The target Hawk was now perfectly positioned in the gunsight so I recorded a short burst on the gunsight film for old time's sake, jettisoned the external 230-gallon fuel tanks using the Emergency Jettison, started a gentle climbing turn towards RAF St Mawgan, and switched the radio to Emergency. With the throttle left fully open the engine rpm continued to decay as I converted speed to height. I topped out at 3,500 feet with a speed of 180 knots, which is the best glide speed for the Hunter. A clean Hunter at 180 knots will glide two nautical miles per 1,000 feet, so I estimated that I had a range of seven miles and a remaining flight time of a little over two minutes. I knew that St Mawgan was my nearest suitable airfield and that it was at least twenty miles away.

At exactly the same time as my initial engine problem, Bronze 3 in my formation came up on 243.0 MHz – the International Distress Frequency – with a PAN call, having had a failure of both of his aircraft's fuel booster pumps. He also wanted to divert to St Mawgan, and Drayton Centre, the emergency controllers, took control of him. I was about three miles north-east of Bronze 3 when I made a Mayday call to announce my engine failure. Confusion then developed. Two aircraft were now in the same piece of sky, with almost identical call signs, and they were both wanting to go to St Mawgan. To Drayton it must have seemed that the PAN call had simply been upgraded to Mayday and therefore I simply didn't exist as far as they were concerned. There was an obvious link between a fuel problem and an engine problem, so the confusion was very understandable, but in my cockpit there was little time or enthusiasm for sorting out the confusion, as I still had the problem with my engine.'

Ian Hartley was the pilot flying Bronze 3, and he, too, recalls the event:

'Basically, we had been tasked with three- and four-ship formations, two to attack simulated targets in Wales and the third in Devon, with 234 and 63 Squadrons in their Hawks to practice their Low Level Air Defence role against us. I was number two of the latter four-ship formation with Alick as leader. During the planning we decided to indulge in a bit of a gamesmanship – our low-level routing to North Devon from Brawdy would have to be via Worms Head on the Gower peninsula, and it was obvious that the first Hawks would be waiting for us there. We decided therefore to transit to North Devon at high level, then to descend rapidly at the Devon coast to attack our first target at 540 knots (well above the top speed of the Hawk at low level).

We subsequently went out to our aircraft but both mine and number four's were reluctant to start, so Alick and number three left on their own. Five minutes later I finally managed to fire mine up and scrambled airborne to try and catch up with the others, leaving number four behind (he never did get going). I caught up with number three just off the first target, but could not see Alick because of the very hazy visibility. Almost immediately we were bounced by some Hawks, which we managed to evade, but as I recovered from my evasive manoeuvres I found a cockpit full of warning lights showing that I had lost AC power and, most importantly, my fuel booster pumps. I climbed up and put out a PAN call on the emergency frequency and started heading for St Mawgan.

It was exactly at this point that Alick put out his call. My call sign was Tartan Seven Four and his Tartan Seven Five. It was no surprise therefore that London Mil ATCC immediately assumed that my problem had simply got worse. The problem was made slightly worse by Alick still being at low level and not being able to hear London's calls. The confusion was quickly overcome and I was vectored towards Alick as he tried to deal with his worsening situation. Although I was talking to him I did not get to him before he ejected so I did not see his aircraft at all, but I did see him in his parachute as he went into the sea where he was quickly picked up by a small inshore trawler.'

In XG197 the rpm had by now decayed to 6,100 (note that the Hunter did not have the normal rpm percentage gauge). Alick continues his story:

'I knew that the aircraft would just maintain level flight with 6,000rpm, so getting to St Mawgan began to seem like a real possibility, if only I could keep the rpm above 6,000. But slowly the rpm decayed. Throttle movements had no effect. The aircraft was very quiet, but with the engine running at low rpm I still had full hydraulics for the controls. In the haze I could see the coast about three miles to my left as I headed south. They were a peaceful few moments. What had caused the engine to surge? What had caused the loss of throttle control? What options did I have for increasing power, even by just a small amount? There was clearly nothing wrong with the Avon engine itself or the airframe. It had to be the low-pressure side of the fuel system. Deselecting the top temperature control had no effect. Then I remembered that the engine relight system provides an eighteen-second burst of fuel when the engine relight is operated, even when the engine is already running. I was now at 5,950rpm, and operating the relight immediately gave me 6,050rpm. So the answer to my problems was simple – I would go home on the relight system, relying on a quick burst of power every twenty seconds or so.

Unfortunately, my theory failed. The first burst of power was also the last, and the rpm continued to decay towards 5,000. I was now down to normal flight idle and, holding speed at 180 knots, my rate of descent was picking up. I descended still further to 1,500 feet. Land was now coming close on my port side. I was already below the prudent ejection height for my ejection seat, which was set at 1,250 feet.

Drayton had finally worked out that they were dealing with two aircraft, but so many other well-wishers had by now come on to the frequency that I was unable to take much part in the radio chat. By this time I was down to just 1,000 feet. The Chivenor rescue helicopter was already on frequency and getting airborne, which was comforting. I really was out of options, and down to 800 feet above the sea. The surface of the sea was now very clear but land seemed to be too close. With Tintagel Head on the nose, I gently rolled into a right turn. I knew I should have abandoned the aircraft by now. Down to 600 feet, I continued the turn until Tintagel Head is about thirty to forty degrees left of the nose. Just 400 feet now and I was very conscious that I had rather stupidly run myself out of height. Probably caught out by the rapid increase in the rate of descent as the rpm died, coupled with a reluctance to admit that such a fine aircraft really was going to give up on me, I'd descended to just 300 feet. I stretched the glide to 160 knots to briefly kill the descent rate, trimmed a little nose-up so that I could let go of the controls without any risk of a nose-down movement, then I fixed myself squarely in the seat, grasped the bottom ejection handle with both hands and pulled as hard as I could, at a height of 300 feet above the sea.

Strapping in on the FRADU flight line at Yeovilton: this image shows the Hunter's bulky ejection seats and also illustrates the twin-seater's windscreen wiper assembly, which, although effective, appeared rather out of place on a high-performance jet. *FRADU*

Instantly the canopy jettison gun fired. The canopy was immediately gone. All the muck and dust of ages was sucked into my space. My map rapidly followed the canopy out of the aircraft. I knew the seat would fire a second after the canopy went, but for some reason it didn't. There was a loud ticking noise in my left ear – the sound of the Time Delay Unit. I instinctively started to turn towards this loud noise just as a sickening, crushing explosion struck pain into the middle of my back, followed by an even louder crack as my neck was whipped like a rope, throwing my head into my knees as the seat fired after a very, very long one-second delay. Suddenly every tumbling and rotating sense was saturated and I was aware of a comforting air blast. The stabilising drogue had fired and the seat was separating from me and my parachute. Total peace and silence. I was flying feet first over the sea with my aircraft just a little in front and about 100 feet below me. It seemed to be maintaining height and turning very slightly left. The only noise was from my flying suit flapping in the wind.

I was yanked abruptly away from this peaceful scene as the parachute snapped open. Pre-landing drills. Get rid of oxygen mask, to avoid breathing in sea water. Inflate life jacket. Lower dinghy pack by releasing the clips on each side of the harness. Prepare for impact with the water. Rotate and squeeze the harness release box on touchdown. Every action had been practiced many times, and some done thousands of times, but never with only one arm functioning. Somehow, my right shoulder would not function. My height above the water was too low to do much about anything. The oxygen mask was removed and the life jacket was inflated just before touchdown. Reaching the dinghy pack release clips was impossible and the release box would not move. Next thing was a rush of sea water full of bubbles. I was aware of rising to the surface but I was unable to break through into fresh air.

I was floating on my back, head down, in a very stable position supported by my life jacket and by the dinghy pack still firmly attached to my backside. However much I struggled with the release box with one hand, it simply refused to open. The new straps just installed on the ejection seat assembly had pulled so tight, and had so little flexibility, that nothing moved at all. I was very conscious of a growing pain in my back and lack of response from my legs. I let go of the release box, relaxed, and finally floated free from the harness. I broke the surface just in time. Houdini used this escape trick by design; mine was involuntary! I inflated the life raft, hauled myself on board and deployed the "Sarbe" beacon for the helicopter to find me. My aircraft was nowhere to be seen, and I regretted not actually seeing it hit the water. I felt heavy, immobile, cold and in pain.

Within a few minutes a row of faces appeared looking down at me. It was the *Lady Jane* from Boscastle. Expert hands lifted me onto the deck, guided by one of the senior designers of the ejection seat who fortuitously happened to be a tourist fisherman on board! The thud of rotors overhead announced the arrival of the helicopter. Quickly winched on board, we collected a doctor at St Mawgan and flew on to the landing pad at RN Hospital Stonehouse in Plymouth. Journalists were already at the landing site with all the usual questions. How? Why? It was in the examination room that I heard the shocking news that my aircraft had crashed on Tintagel. To make it worse, there were rumours of casualties, but no details. My own problems seemed less important at that moment. So began a long recovery period, with expert attention at various military hospitals, but the effects of multiple and serious crush fractures to vertebrae and whiplash injuries to the neck do not ever go away. One gradually adapts to a different degree of facility, and unexpected opportunities and challenges open up.

So what made my aircraft turn away from the open area of sea that I had so carefully selected? Why was XG197 able to make landfall at all? First, I confess that I was closer to the coast than I had intended. I was trying to get to St Mawgan and I knew that every mile was a life-saving mile, so initially the shortest route was more important than the location. In the haze above about 1,000 feet, and with my attention on other things, I had not taken early enough action to steer away from the coast. But the real problem was that Hunter XG197 was simply one of the best airframes. It was straight. Most Hunters had become bananas in their old age and they would generally roll or yaw at low speed and high angles of attack. My Hunter did not, and in air combat it was a delight because it could be controlled at very low speeds.

When I ejected, the reduction in weight as the seat and I left would have inevitably caused the nose to rise. I had also left it trimmed slightly nose up. It is possible that the engine rpm may have picked up slightly. I had jettisoned all wing stores, so the airframe was clean and the aircraft was light. Very little good performance data existed for a Hunter in that state, but everyone knew that the Hunter flew exceptionally well when clean. I believe that as the nose came up and the speed bled away, the aircraft gracefully slipped into a gently climbing left turn, probably at about 140 knots, after which it turned left through about sixty to seventy degrees, dropped its nose as the speed fell away below about 125 knots, then slowly gained some speed and rolled back to nearly wings level. A "normal" bent Hunter would probably have departed from steady flight during that slow-speed turn and would never have made landfall.

XG197 hit the ground from a shallow descent in a gentle left turn with the engine still running, and the investigation proved that the rpm at impact was less than 6,100. The air brake and outer section of the left wing were torn off on impact. There was then sufficient energy left for the aircraft to bounce and tear its way a little over 100 yards up to the houses, shedding other parts as it did so. The fuselage came to rest between two houses and the nose section, including the cockpit and gun pack, broke away and came to rest in the main street. So what caused this event that touched several lives and totally changed my own way of life?

A jet engine has a Fuel Control Unit [FCU] that controls the flow of fuel to the engine, mainly depending on throttle position, height and speed. An aircraft engine must be able to cope with extremes of conditions, for example from very high air speed at low rpm to very low air speed at high rpm, at all heights up to 45,000 feet. Two of the vital parameters that are used to control the FCU are engine

Final resting place: WT809 served with 66 Squadron and Linton-on-Ouse Station Flight before being converted to GA.11 standard. It crashed near Ilchester during June 1988. *FRADU*

intake pressure (P1) and engine compressor pressure (P2). These two values are taken from small bleeds and are fed to the acceleration control unit section of the FCU via small-diameter pipes. The two bleed pipes cross in a P1/P2 metering orifice, and the highest pressure determines which side of a diaphragm is pressurised; this in turn controls the initial position of the FCU. The pipework has to be leak-free and clean, and where joints are required, dry paper gaskets are used.

The engine in my Hunter was a newly reconditioned unit. When it was ground run at the manufacturer's factory, the P2 bleed from the compressor leaked. The gasket had to be replaced. However, access to this joint is very restricted, and this was made worse by the engine now being mounted in a cradle for ground running. A technique for holding any paper gasket in place while the flange bolts are positioned is to tack it into place with a little smear of grease. But the use of grease was specifically forbidden on P1 and P2 pipes. The faulty air bleed was successfully fixed, and the engine was re-issued to the RAF. Three hours and forty minutes of flight time later, a small blob of grease, the size of a match head and now hardened by heat, broke free from the P2 gasket when I applied power at 525 knots behind the Hawk, and lodged in the P1/P2 metering orifice of the acceleration control unit in the FCU. The FCU immediately sensed that the compressor pressure (P2) had dropped to zero, and responded as though I was instantly parked on the ground with the engine shut down, despite being at over 500 knots.

That was the moment when I thought the engine had surged, and I was right in thinking it was an unusual surge – in fact, the fuel flow had gone unstable as the blob of grease moved into the P1/P2 orifice. It then became impossible to make the engine accelerate, simply because the engine compressor was sensed as being shut down. When the FCU was removed from the wreckage it was one of the few parts that was undamaged. It was bench-tested and gave exactly the fuel flow rate that I had reported. It was then stripped down and the same type of grease that had blocked the orifice was also found on the P2 paper gasket. With the blob of grease removed, the FCU was fully serviceable. Analysis showed that it was an unusual type of silicon grease that is only used on the Rolls-Royce Dart engine, and the Dart was also overhauled at the same factory that looked after Avon engines. No explanation was ever offered as to how this grease came to be present on the P2 gasket in my aircraft, and nobody admitted any involvement.

In his accident report the RAF Inspector of Flight Safety, an Air Commodore, commented that, "Any deviation from laid-down procedures or established safe practices can cause results out of all proportion to the original action." He was absolutely correct. He concluded, "On this occasion it was merely costly: it could have been tragic." Absolutely no mention or concern, though, about the pilot's injuries. So began a litany of cunning and deceit by many very senior Government, MoD and Civil Service officials, all intent upon manipulating the truth and avoiding any real concern for the pilots under their charge. Some years later, the even more despicable besmirching of the reputation of the pilots who were killed in the Chinook crash on the Mull of Kintyre was to resonate to the same methods shown in the handling of my accident. Many questions could be asked. In the ten years between 1971 and 1980, total Hunter flight time was 202,486 hours. In that time there were twenty-six ejections, of which six were fatal. So technically speaking, there was an ejection every 7,788 hours and a death every 33,748 hours. We each flew about 250 hours per year in the Hunter, so Brawdy instructors had a 1 in 31 chance of ejecting in any year, and a 1 in 135 chance of being killed every year. If these were the odds on winning the lottery, you would bet your pension plan! No aircraft that is so unreliable should have been left in service, even if it was the magnificent Hunter. Which Air Marshal and Minister was responsible for forward planning? I bet they didn't gamble their pension plans.

If the ejection rate was so high, then what about the seat? Surely the Hunter would have the best seat possible? Front-line aircraft had superb rocket-propelled ejection seats. The seat in the Hunter was the oldest type left in RAF service, yet it was installed in the aircraft most likely to use it. The complete sequence was slow to operate and the firing mechanism was a spine-breaking gun. Six months before my accident I had presented Strike Command Engineering and the Directorate of Flight Safety with a well-researched case for refurbishing the Hunter with a more modern rocket seat. The rocket seat has a hugely increased safe operating envelope, and the likelihood of injury to the pilot is small. I was officially told that no other seat could be fitted into the cockpit. This was a lie. I had personally installed several rocket seat variants into a cockpit in our ground training facility to prove that they could be installed on existing cockpit mountings, and Martin Baker Ltd, the world-leading ejection seat manufacturer, agreed that the installation of the Mk 10(L) rocket seat would be appropriate and simple.

The official line was then changed to one where the refurbishment could not be justified because of insufficient remaining in-service life. Interestingly, a report in July 1980 by the Institute of Aviation Medicine, following the ejection of Hunter XG261, commented that, "Hunter escape capability has not improved since 1957 but the role of the aircraft has changed considerably and there is a very real need to bring the escape system up to an acceptable standard." Deaf ears won the day, and unnecessary injuries and deaths would continue. As a result of the accident I lost my medical category, and with that went my flying career, much of my pay, all my prospects in the RAF, and a great deal of my mobility. There then followed several years of quite unnecessary personal harassment by the RAF, when my well-being was over-ridden to protect the status quo and one of the main RAF contractors. In simple terms, I had been injured on duty, and the system did everything possible to avoid responsibility, because to have done otherwise would have been to have acknowledged a perceived danger to the system. The Data Protection Act of 1998, which gives the right of access to all documents, has now allowed me to see the extent to which I was tossed to the wind to protect the system. However, too much time and pain has passed to rake over old wounds, and all involved will have long retired, but I am concerned that nothing has changed. Events continue to show that subsequent generations of management at the core of our armed services remain unable to recognise the need for unfailing fairness, honesty and respect for service men and women under their command. And finally, I acknowledge and regret the disruption to the people of Tintagel, when I inadvertently came your way.'

It is probably no surprise that even after the Hunter's operational career has come to an end, occasional mishaps still occur, affecting some of the many Hunters that remain active around the world, in civilian hands. Craig Penrice was one individual who was unfortunate to have been forced to abandon a Hunter as recently as 2003, as he recalls:

'On the afternoon of Sunday 1 June 2003 I had reason to part company with Hunter Mk 6A XF516 (G-BVVC) rather more rapidly and prematurely than I had otherwise intended to. Previously I had often said that a good pilot is defined as one having the same number of take-offs in his log book as landings, but that was something I said before my ejection from Lightning XS921 on 19 September 1985. Now that I have two more take-offs than landings in my log book, I can add some more wisdom to the tale – having ejected once does not make you exempt from further intervention by the fickle finger of fate.

The Hunter flight in question was supposed to be the end of a fine weekend's flying displays at the Portrush Air Show in Northern Ireland. The jet had been operating out of Blackpool (where many Hunters had been built) to assist in the overall logistics. Following the display the intention on this day was to fly directly back to Exeter where the jet was normally based before driving back to my home near Warton in Lancashire. The aircraft had been recently troubled with some electrical problems, but these were suitably mitigated it was thought. Nothing in this mitigation, however, could cope with the fact that at 25,000 feet over the middle of Wales the engine suddenly flamed out. Prior to the flame-out the electrical system had failed completely and the battery had become exhausted, which left me without communications. In anticipation of this happening, ATC had been warned that I might lose the radio and as a result I was following my flight planned route with the intention of a landing at Exeter without radio and using visual signals only.

Out of the blue there was a slight shudder felt through the airframe, not too dissimilar from the normal IGV [Intake Guide Vane] chatter familiar to all "big engine" Hunter operators. I thought this to be somewhat strange as this shudder was not at the normal rpm where one would expect it; indeed, the engine was in essentially steady state cruise conditions. It took some time before the realisation dawned that the engine had actually stopped. I guess I had always imagined that when the engine stops in a single-engine aircraft it will all go very quiet (like it does on the ground after shut-down). However, the noise level in the cockpit was essentially unaltered. This was due to the fact that the engine was windmilling satisfactorily and the surrounding noise was actually from the pressurisation system and external wind rush.

Anyway, on acceptance of the fact that the engine had quit and I was now in a flying brick, the following events unfolded. First off I set course for the general direction of nearby Llanbedr in the hope of a dead-stick attempt onto the runway there. It quickly became apparent that I had insufficient altitude to glide that far. I discounted jettison of the external tanks primarily because I could see the banner headlines if they landed in someone's back garden – or worse. I really didn't want to jeopardise future ex-military jet operations following an outcry about my tanks landing somewhere inappropriate. The same concern was true when it came to planning my exit from the jet – something that was slowly beginning to dawn on me as inevitable.

My previous ejection from the Lightning had been a bit of a rushed affair; a control restriction had robbed me of the ability to control my destiny and in a very short time (less than ten seconds) I went from flying along happily to going for the ejection handle. Since that experience I have said many times that I was thankful that I didn't have a long time to think about things, but here I was now with a long glide running up to an inevitable ejection. There was very little thought involved with attempts to get the engine going again. I had no electrical power therefore there was no way to get the flame lit again – it's kind of hard to get a jump start for a plummeting brick! My attentions were then focused on the best course of action for me, the jet and the unsuspecting population below.

Having determined that Llanbedr was out of the equation, I set about making for the coastline in an effort to jump and dump into Cardigan Bay, and it looked like this would be possible. I elected to stay with the jet below the recommended height for a premeditated ejection of 9,000 feet for the reasons of potential third party damage I have already mentioned. As the glide progressed it became clear that it was going to be a very close-run thing, but I had the consolation that there was an estuary running along my flight path, just prior to the coast. My personal preference has always been the water landing rather than smacking into the ground, or trees, or buildings. We get dropped into the water twice a year for practice, but we don't get the same exposure to the risky ground landing. The time during the glide was taken up with tightening straps and going through in my head all the drills, posturing and a lot of swearing. Why was this happening to me – again? I think I was really annoyed at the fact that I had a lot of time to think about it, but the inevitable was fast becoming a reality.

In the end the final moment came a bit sooner than I expected. I quite suddenly spotted a village coming into view from behind a mountain spur. The village was Borth and it was clear to me that my ejection point at or near the coast was going to leave a pilotless aircraft to make up its mind as to its final resting place in close proximity to the village. As I was currently over the estuary and could see some sparse marshland ahead, that was my subliminal message to go for it. I pulled the seat pan handle. I had descended from 25,000 to 2,000 feet during the preceding five or six minutes.

In comparison to my previous experience, where I have no recollection from the point of pulling the handle until the time I was sitting in the rescue helicopter, this one is in crystal-clear Fuji Colour. There was a huge, massive explosion and the most enormous force acting on my behind. There was a pain in my back like I had been hit by a plank of wood. I watched the cockpit disappear from around me and I watched from above as the jet flew on without me. I saw it pitch up steeply and I saw it start to wing over at the apex of its short climb, and that was the last I saw of it. During this time I was aware of the various seat mechanisms operating, and being jerked around like a rag doll as the drogue and then the main chute left me hanging. This was accompanied with a feeling that my legs had swollen to enormous size and the pain in my back was now excruciating. I did what I could manage of my parachute descent drills and managed to reason that as I could still wiggle my toes the pain in my back could not be that serious. I was aware that the water below me was rushing up at quite an alarming rate, and it was at about this time that I realised I was still attached to the seat – automatic seat separation had not occurred. I was able to release the appropriate harness and the seat fell away – I could not have been at more than 300 feet when this happened. I will return to this fact a bit later.

Just prior to hitting the water I took a deep breath, closed my eyes and placed my hands in a position to release the parachute harness on water entry. I next recall being desperate to release my breath, but not being aware that I had floated back to the surface. I opened my eyes to find that I was in all of eight inches of water. I had hit the water as the tide was out and all my best intentions of a soft landing were gone in an instant. I was dragged for a short distance before I released the parachute harness. It was a lovely warm afternoon. The pain in my back was awful, but if I laid still it was bearable. I could not move my legs, but I could once again wiggle my toes, so I could therefore believe it was not a broken back – but it was.

I was able to remove my helmet and, after some time, managed to get myself into a position that was vaguely comfortable. I was now lying on a sandbank with the water receding. The sun was shining, my back was broken and I was about a mile from the nearest shoreline. I got out my Oakleys and even got out my mobile phone. I had the intention of calling my wife to let her know I had banged out (again) but that I was alive. Luckily, perhaps, the phone was soaked and did not work. I could do nothing but wait to be rescued. I could here sirens from various directions around the shoreline, and I had no worry that I would be rescued. I had no place to go and no way of getting there.

It was about forty-five minutes before the first person arrived on scene. A man out walking had seen the jet and my ejection and had waded, swam and walked from the shore over to me. We established that he could do nothing on his own but he did have some drinking water, which was gratefully received. Next on the scene was an off-duty policeman who had called the emergency services before he too had made his way to me. Shortly thereafter I could hear the sounds of a boat repeatedly getting grounded on the sandbanks as the estuary continued to drain – this was the RNLI from Borth making their way to me. Very soon the sky was filled with choppers and the RAF Valley and Chivenor aircraft arrived on scene closely followed by the North Wales Police helicopter. Quite soon there was a fairly large group of people on the sandbank to keep me company and get me to hospital.

As you would expect, the professionalism of this bunch and in particular the winchman was exceptional, and I was soon on a back board and winched into the helicopter and on my way. Without going into vast reams of medical details about the days, weeks and months that followed, the force of the ejection resulted in a burst fracture of one of my vertebrae. The fragments of bone embedded themselves into my spinal chord. The result of this was to effectively paralyse me from the waist down. The spine was fixed by the introduction of yet more metalwork in the form of a supporting cage around the burst vertebrae. I have been able to regain the use of my legs but am still devoid of feeling and function below the waist. My days of flying bang-seat-equipped aircraft are over, but I guess it was time to grow up and find a proper job. As those of us who fly know all too well, it could have been worse. I'm still here to tell the tale.

Returning to the fact that I found myself still attached to the seat at a late stage of the descent and the apparent failure to achieve man and seat separation, this was investigated. It appears to have been a result of my tightening of the lap straps possibly a little over-zealously. The mechanism worked as advertised and released the lugs in the seat pan, but the tension from the straps acting at ninety degrees to the release direction has resulted in a geometric lock being set up. This was only released when I was able to activate the QRF and the seat dropped away. The AAIB said that this accident was a salient reminder to those operating old aircraft that any snags that are being carried must be carefully considered and that my injuries may have been lessened if I had elected to use the face screen handle instead of the seat pan handle. This, they suppose, would have given me a better posture. An interesting observation, as most modern seats have only a seat pan handle. I presume the move away from "bang" seats to "rocket" seats has resulted in a transition from good posture afforded by the face screen handle against the more rapid access of the seat pan handle. With gentler (relative term only) rides from rocket seats, the importance of posture has diminished in favour of more rapid egress. I never even contemplated the face screen handle as an option. My routine training had always been to use the seat pan handle.

I have many people to be thankful to in getting me from there to here. The rescue services, the surgeons, nurses and physiotherapists who helped to patch me together physically. My friends and workmates who helped keep my spirits up. But most of all my wife and family, to whom once again I am for ever indebted – without their love and support this would have been far, far worse than it already was.'

The analysis of the accident was undertaken by the Air Accident Investigation Branch, and its report was as follows:

'Total electric failure

The aircraft had a recent history of electrical generation problems and the risk of double generator failure had been realised the day before but probably without completely exhausting battery power. For the accident flight, provided single generator failure was the only fault, the remaining generator should have been sufficient to supply all this aircraft's electrical needs and charge the batteries. Replacement batteries had been fitted before the flight and external power was available to support pre-start preparations so the aircraft should have had well-charged batteries and one functioning generator when it taxied. The pilot did not notice the exact moment when the second generator failed in flight. Since the primary indication would have been an additional red warning light down near his knees, with no immediate loss of any electrical services, it is not surprising that he did not notice it immediately. At first the only secondary indication of double generator failure would have been the changeover of a single "doll's-eye" magnetic indicator from black to white. If at the time he was concentrating on instrument flying in cloud or carrying out his display beneath it, he was unlikely to notice that indication either. However, until he took action to shed non-essential electrical loads, particularly the booster pumps, the load on the batteries would have depleted them after about ten minutes. The time to climb out to FL220 would have been about three to four minutes so the failure of the radios followed by some flight instruments during the climb suggests that the second generator failed during the display. Given the pressures, skills and concentration required to carry out a display, especially in restrictive weather conditions, it is not surprising that the pilot did not notice either warning until he had completed his display. As the batteries discharged, any warning lights would have dimmed and once the batteries were depleted there would have been no warning lights and no internal fuel gauging. It seems likely that battery power expired during the climb to high altitude or shortly afterwards so the load shedding carried out in the cruise would have had little beneficial effect.

Landing options

Once at altitude with few flight instruments, no internal fuel gauging, and no communications, the pilot was committed to landing at an airfield where good visual flight conditions prevailed, and preferably one that could be reached without penetrating controlled airspace. The extension of the Hunter's wing flaps creates a significant change in pitch trim so, depending on the tailplane position when the batteries expired, the inability to trim might mean that the pilot would be forced to land his aircraft "flapless" and with no retardation from the electrically released brake parachute. Consequently, he would have needed at least 1,500 metres of hard surface to be confident of stopping on the runway. These constraints had to be set against the Aircrew Manual's advice to "land as soon as possible". The aircraft was above cloud so the pilot's decision to continue in accordance with his flight plan towards central Wales, where the weather was much improved, was sensible. Photographs of the aircraft departing Blackpool showed that the weather there was fine and cloudless less than an hour earlier so landing there might have been practicable. Also, since the pilot could see Lake Bala from FL180, a landing at nearby RAF Valley or at Llanbedr were potential options. However, the pilot decided that since he had plenty of fuel, and the appropriate technical expertise to rectify the aircraft was at Exeter, he would continue at altitude and land there.

Engine flame-out

The engine ran down at some time between throttling back to descend and advancing the throttle to level off at FL180. The immediate action of trying to relight the engine was bound to fail because there was no electricity to power the igniters. There were no mechanical failures evident within the engine, which continued to "windmill", and plenty of fuel on board the aircraft, so it seems highly probable that the flame-out occurred through temporary fuel starvation.

Fuel starvation

With no booster pumps, fuel should still transfer from the front tanks to the engine through air pressure. The electrically motored fuel sequence valves would have been at the "wing" transfer position when electrical power was lost so fuel from the drop tanks should still have transferred to the wing tanks and from them to the front tanks under the action of air pressure. There had been no reported problems with fuel transfer in recent flights and the pilot noticed that both drop tanks were about one-third full over Wales (they have mechanical quantity indicators). Consequently, there is no reason to assume that the electrical failure was responsible for a fuel transfer failure. However, if fuel transfer from one or both "side" groups of wing and drop tanks into the associated front tank stopped, the engine would have flamed out when one of the front tanks emptied. There would have been no amber light warning of low fuel contents in the front tank because of the total electrical failure. Each front tank's capacity is 790lb and it would have been supplying only half the engine's demand. If fuel transfer ceased when the electrical system failed, a maximum of 200lb of fuel from each tank would have been used in completing the climb and the remainder would have been consumed at about 18lb per minute (half the engine consumption rate from each tank). This would have been sufficient for thirty-two minutes cruising at FL200. With a ground speed of 380 knot (6.3nm/min), flame-out through fuel exhaustion would not have happened until the aircraft was more than 200nm from Portrush. The distance flown from Portrush to 15nm east of Bala Lake was about 185nm, so this explanation for the flame-out is plausible but unlikely.

Throttle handling

With no booster pump pressure and hence no recuperators, throttle handling must be gentle and maximum power is restricted because air pressure transfer alone cannot meet the engine's fuel requirements at full throttle, but the engine should still run if it is managed sensitively. Closing the throttle reduces the fuel demand so this action is probably less critical than opening the throttle, which must be done slowly so that an uninterrupted fuel supply to the engine is maintained. The pilot stated that he opened the throttle slowly but this was the moment that he noticed the engine failed to respond. Consequently, throttle handling seems a likely explanation for the flame-out but it may be that the engine was unusually sensitive to throttle advances under the prevailing conditions.

Forced landing

At a speed of 210 knots IAS the Hunter glides about 2nm for every 1,000 feet of height and, when downwind, the pilot needs at least 2,500 feet height abeam the touchdown point in order to land near the runway threshold. The distance from a position 15nm east of Bala Lake to Llanbedr is about 35nm, so from 18,000 feet the aircraft may have had just enough potential energy to reach Llanbedr, but probably not enough to land because the runway was aligned at right angles to the aircraft's approach track. Also, with fuel in the drop tanks the aircraft was still fairly heavy and the pilot opted to glide at 250 knots IAS, thus reducing the aircraft's glide ratio slightly but improving the chance of a relight, had there been any electrical power to achieve it, but there was none. For similar reasons, the drop tanks would probably not have separated if he had tried to jettison them because they are released by explosive ejector units initiated by an electrical pulse. An accumulation of these factors probably deprived the pilot of any realistic chance of reaching Llanbedr, but it was well worth the attempt. When it was obvious to him that he could not achieve a forced landing, he sensibly and considerately decided to head for open water and eject, thereby minimising the risk to other people and property.

Ejection sequence

The pilot prepared for the ejection but chose to use the seat pan handle rather than the face screen handle. He may have done so instinctively because he was more familiar with seats that have just a seat pan handle. Whether his injuries would have been less severe had he used the face screen handle as recommended in the Aircrew Manual is impossible to determine, but on this type of seat a better ejection posture is achieved if the face screen firing handle is used. Unlike modern ejection seats, which have rocket assistance, the range of ejection seats fitted to the Hunter had only the telescopic ejection gun with which to achieve ejection velocity. These seats are, therefore, of necessity somewhat harsher in operation than "rocket seats". The rate of acceleration has to be greater, so spinal injuries are more likely. Another factor that influences the probability of spinal injuries is the PSP upon which the pilot sits. The Mark 2H and 3H seats have a "soft" PSP whereas modern seats generally have a more rigid fibreglass "shell" covering, containing the pack, which is contoured and forms the base of the seat. This rigid shell transfers the ejection impulse to the pilot's body more swiftly than the soft pack and so minimises the peak acceleration imparted into the pilot's spine. The pilot saw his abandoned aircraft pitch up. This is not surprising given the likely change in centre of gravity position due to the lost weight of the seat and its pilot, and their relatively long moment arm from the centre of gravity. Moreover, the aircraft was out of trim in the nose-up sense when he ejected and so likely to behave erratically after he ejected for both reasons. The ejection seat was recovered but not examined by the AAIB. It was, however, examined by a knowledgeable person. Apparently, all the automatic sequencing had operated and there were no obvious pre-impact defects. Pilot separation from the seat should have happened automatically but the most likely explanation for his retention was a stuck torso restraint harness. The pilot is attached to his parachute by one harness and held in the seat by a totally separate harness, which should have been released by a cable connecting its fastener to the "barostatic" time release unit. The unit must have operated to deploy the pilot's parachute so it is likely that the cable did not move the fastener sufficiently to release the harness. However, the pilot had only to operate the seat harness release mechanism with one hand in order to free himself from the seat. The pilot's spinal injuries may have been aggravated by his heavy landing on the sandbank. Numerous people had seen him descending on his parachute and had reported the event to the emergency services, resulting in the scramble of the RAF rescue helicopter. Bystanders also made their way through the shallow water to assist the injured pilot. Had he descended into deep water or further out to sea, the outcome might have been different.

Conclusions

The accident was preceded by total electrical failure but caused by engine flame-out. The engine probably flamed out because its fuel supply was delivered to the engine at abnormally low pressure and was temporarily interrupted when the throttle setting was changed. Without robust electrical power the engine could not be relit and the pilot was unable to reach a suitable airfield for a forced landing so he ejected. He received spinal injuries during the ejection and had to separate himself from the seat when the seat harness remained fastened after his parachute had inflated. His spinal injuries might have been less severe had he used the face screen firing handle and not landed in very shallow water.

Recommendations

There were no formal safety recommendations arising from this accident but it does serve to remind operators of ex-military jet aircraft that safety is compromised if recurrent faults are allowed to persist. Moreover, pilots of such aircraft must bear in mind the likely consequences of accepting pre-take-off defects before undertaking a demanding flight.'

Chapter 7

A 'third life'

Following the withdrawal of British forces from the Middle East and Far East, the Hunter ground-attack squadrons were rendered largely redundant and the Hunter was officially withdrawn from front-line RAF service in 1969. The RAF's attention was turning towards the next generation of Hawker-designed ground-attack aircraft in the shape of the legendary Harrier, and while further developments of the Hunter were considered, Hawker's resources (and interest) was now firmly behind the new Hawk trainer, and many senior members of the Hawker Siddeley empire were unwilling to divert attention or finances towards the Hunter, even though the aircraft could have had its capabilities stretched still further.

While some export customers requested modifications to the Hunter, the Hawker management were less than willing to sanction such concepts, and when Singapore opted to install centre-line hard points on its Hunters, it was Lockheed (or at least former Hawker staff within that company) that handled the task. Likewise, when Switzerland and Oman opted to fit Sidewinder AAMs to their Hunters, the modifications were done locally. The aircraft could have been developed into a second-generation Hunter with expanded gun magazines, a stretched fuselage (carrying more fuel), new navigation equipment and Sidewinders (such a proposal was put forward by the CFE), but despite the fact that such a design would probably have been more effective than the slowly emerging Harrier, Hawker Siddeley was firmly behind the revolutionary new vertical-take-off aircraft.

In Royal Air Force service the Hunter was gradually withdrawn from operational squadrons to take up second-line duties, and although this move might have appeared to mark the beginning of the end for the Hunter's RAF career, it was in fact the beginning of what was effectively a 'third life', which took the aircraft from the late 1960s through into the 1980s. No 4FTS at Valley received a fleet of Hunters, which were assigned to the advanced flying training role, augmenting the diminutive Gnat (indeed, many RAF personnel felt that the Hunter should have been used to equip the entire FTS in preference to the Gnat), but always operating as a distinctly separate unit, situated on the far side of the airfield. Officially, the 4FTS Hunters were assigned to the training of overseas students, but a considerable number of RAF pilots were unable to physically fit into the Gnat's cramped cockpit, and these students were automatically posted onto Hunters, together with a few other individuals who (by luck or by design) managed to get a much-sought-after seat on the Hunter course. The twin-seat Hunter T.7 was used for basic handling and conversion training onto the type, but much of the 4FTS course was conducted in the single-seat Hunter F.6, which, in stripped-down form minus various items of equipment (and the removal of the Aden cannon), was a hugely popular machine, being both fast and manoeuvrable, often referred to locally as the 'GT6'.

Possibly the most colourful line-up of RAF Hunters ever seen, this is the 4FTS line at Valley shortly after the unit's Hunter fleet had been repainted with the high-visibility training colours in the mid-to-late-1970s. Four lightweight 'GT6' Hunters are accompanied by a quarter of T.7 trainers.

A former Jever and Sylt Station Flight aircraft, XL623's last assignment was to No 4FTS at Valley (as illustrated). The aircraft crashed near Blaenau Ffestiniog during May 1971. *Paul Tomlin*

Prior to being repainted in high-visibility training colours, the Hunters assigned to No 4FTS wore standard RAF camouflage with white trim on the fuselage spine and tail, complete with unit codes superimposed on a white disk.

Meanwhile, in 1974 Hunter FGA.9s (and some T.7s) were delivered to Wittering to form two new squadrons (45 and 58), which were assigned to the ground-attack role, acting as temporary units for newly qualified students who would ultimately be assigned to the new Jaguar. Until substantial Jaguars could be brought into squadron service, the Wittering squadrons provided a means of maintaining pilot skills and qualification, while also providing the RAF with an additional ground-attack capability. When the Jaguar started to settle into squadron service, Nos 45 and 58 Squadrons were scheduled to disband and their Hunters re-assigned to tactical training. However, in order to effectively keep the TWU concept in business, these two squadrons were disbanded rather sooner than originally planned. Wittering had also played host to a small number of Hunter T.7s and FGA.9s, which formed the basis of what would become the Harrier Operational Conversion Unit, providing the unit with additional aircraft while deliveries of the new Harrier slowly trickled through. Hunters also remained active with the RAF's Buccaneer squadrons, with a handful of twin-seat aircraft being assigned to the conversion of new Buccaneer pilots. As the Buccaneer was never produced in a dual-control version, a small number of Hunters were equipped with an Integrated Flight Instrument System (IFIS) to enable instructors to train new Buccaneer pilots in the Hunter, prior to making their first flight in the RAF's new bomber, and it was ultimately in this role that the Hunter finally ended its RAF career many years later.

Pictured at Wittering, XF419 carries the markings of No 58 Squadron. Unusually, the aircraft is fitted with a modified nose containing a sideways-looking camera. It is believed that this equipment fit was created in response to a demand for a covert anti-terrorist reconnaissance capability. *Ray Deacon*

Hunter FGA 9 XF430 was one of only a handful of single-seat Hunters assigned to the Harrier Conversion Team (title applied on the nose), as new-build Harriers were slowly delivered to Wittering. Later renamed the Harrier Conversion Unit, the HCT later became the well-known No 233 OCU. *Paul Tomlin*

Ever since the Hunter had entered RAF service, RAF Chivenor had been regarded as its 'home', responsible for the training of huge numbers of British and overseas students. When the Hunter was withdrawn from front-line service, 229 OCU's role began to change and, instead of converting students onto the Hunter prior to being assigned to a front-line Hunter squadron, it began to widen its scope by preparing students for the Harrier, Phantom, Lightning, Buccaneer and Jaguar. As former 79 Sqn Flight Commander Ian Hartley explains:

> 'Throughout the operational life of the Hunter its OCU was No 229 and all those destined for Hunter squadrons passed through its portals. With the arrival of the Lightning and its very high pilot workload, it was deemed necessary to insert an extra training element between AFTS at Valley and 226 (Lightning) OCU, so all prospective Lightning pilots left Valley to do a "pre-Lightning" course at the Hunter OCU and I was one that did this in 1968. We did the same course as our Hunter brethren initially but did not do the low-level elements nor any of the air-ground weaponry but instead had increased air combat, IF and night-flying phases (I see from my logbook that the course was just under seventy hours). When the Buccaneer, Jaguar, Phantom and Harrier arrived in service it was decided that their pilots, too, should do this pre-OCU training, so by the early '70s, with the demise of the West Raynham and RAFG Hunter squadrons, 229 OCU had ceased to be a Hunter OCU and was now a pre-OCU for these five types. When the unit moved to Brawdy the opportunity was taken to rename it to reflect its new role.'

The first Hunter T.7 trainers delivered to No 229 OCU at Chivenor were finished in high-speed silver paint, with yellow 'trainer bands' applied to the wings and fuselage. The yellow trim was subsequently changed to 'stick-on' dayglow orange patches (as illustrated) and the silver finish was eventually replaced by Light Aircraft Grey. *Ray Deacon*

XE606 served with 229 OCU at Chivenor and moved to Brawdy to continue service with No 1TWU, assigned to No 79 Squadron. More recently the aircraft has been placed on static display at Cottesmore, adorned with No 4 Squadron's markings. *Terry Senior*

An unusual image of a 229 OCU T.7 on the flight line at Chivenor, being photographed following a bird strike. The battered nose cone has been removed, revealing the ranging radar equipment fitted to the nose bulkhead. *Ian Hartley*

229 OCU earned quite a reputation for large formation demonstrations, as illustrated by this early-1970s image. Chivenor's annual air show usually included a mass take-off and formation fly-past performed by OCU Hunters – an event still vividly remembered by those fortunate enough to have seen (and heard) the spectacle.

This shift in emphasis was reflected in the change of unit title when Chivenor finally closed in 1974 and the Hunters moved to Brawdy to form the Tactical Weapons Unit. Generations of Hunter pilots were saddened to witness the closure of Chivenor, tucked away in a picturesque corner of North Devon, where so many young students had enjoyed countless happy hours learning to fly the Hunter. The move was promoted by a mix of financial and political considerations prompted by the Fleet Air Arm's withdrawal from Brawdy, where the local economy was suddenly left without its major source of income and employment. Chivenor's domestic accommodation had always been rather rudimentary (often in the shape of wooden huts, which many pilots regarded as part of Chivenor's 'charm') and, with more modern accommodation available at Brawdy, easier access to the Welsh low-flying areas, and a rather longer runway (three runways in fact, one being 1,500 feet longer than Chivenor's, which had often been a problem for new students), the move seemed to make perfect sense. Thus, it was decided that Brawdy's association with the Hunter would continue and having said goodbye to the last FAA Hunter GA.11's and T.8's, the RAF's Hunters moved-in to take their place.

Pilot's-eye view of a pair of No 1 TWU's Hunters pictured from a T.7 during a training sortie.

Unfortunately, having been driven mostly by political considerations, it was no surprise that the move to Brawdy proved initially to be rather less than successful. Most significantly, Brawdy's weather factor was significantly worse than that of Chivenor, and foul conditions (particularly 'horizontal' fog) often hampered training for long periods. Likewise the number of servicing groundcrew was reduced with the move so that aircraft serviceability – particularly that of the T.7 – dropped dramatically. There were huge problems with the medium/upper-level training airspace out to the south-west, not the least of which was the avoidance of the new Concorde route that went straight through the middle of it, Then, to add to these problems, Brawdy's runways started to break up with the increased use and they had to be completely re-ballasted and re-surfaced. All this quickly led to the TWU getting further and further behind its training schedules, causing a huge bottleneck in the fast-jet training system.

FGA 9 XG225 was one of three aircraft that received an eye-catching paint scheme as part of a high-visibility trial conducted at Brawdy. This particular aircraft received bright yellow trim and is seen during a visit to Binbrook. *Terry Senior*

It was as a direct result of this problem that Nos 45 and 58 Squadrons were prematurely disbanded and their aircraft given to Brawdy together with their groundcrew, most of whom were only detached to Brawdy, which did little for their morale. Ian Hartley continues his description of the TWU set-up:

> 'The three Chivenor squadrons – 63, 234 and 79 – took their roles with them to Brawdy. Nos 63 and 234 were each responsible for the training of the Valley output prior to their posting to the Buccaneer, Harrier, Jaguar, Lightning and Phantom OCUs. No 79 Squadron was effectively the post-grad unit and was responsible for the training of new instructors, refresher courses for those who were going back to the front line after ground tours, weapons courses and instrument rating examiners' courses. Those ab initio students going through 63 or 234 all did the same course, which consisted of the following phases: Convex (Conversion Exercises), Instrument Flying, Formation, Tactics (air combat), Weapons (air-ground gunnery, SNEB rocketry, and dive bombing), Air Cine followed by Air-to-Air gunnery, low-level navigation and finally the SAP (simulated attack phase). The course was about eighty hours. Refreshers on 79 Squadron only did those phases that were appropriate to their subsequent OCU. Ab initios did not get their postings until towards the end of their course as it was necessary to assess their ability and aptitude for the various types, whereas the refresher students already knew their next destination before joining the course.
>
> As instructors we would assess students on their workload capacity, aptitude for a role and their initiative. The first decision was whether they were suitable for single-seat, then whether they were better suited to the low-level or fighter role. There was a definite pecking order on ability: Lightning came first closely, followed by Harrier, then Jaguar followed by Buccaneer and lastly the Phantom (the Buccaneer was more difficult to handle than the F4 although the role was marginally easier).

At Brawdy were the three former Chivenor squadrons, Nos 63, 234 and 79. F.6 XK149 was a former AFDS aircraft and, as can be seen by the pristine camouflage finish, it was used as a solo display mount during service with No 63 Squadron. *Michael Freer*

A familiar backdrop for any former 229 OCU pilot: a Hunter FGA 9 from No 63 Squadron descends over Bideford Bay, with Chivenor just coming into view below.

As can be seen, No 234 Squadron's F.6 XG152 was one of relatively few RAF Hunters to have gun muzzle deflectors fitted. *Ray Deacon*

An unusual pilot's-eye view of a pair of 1TWU Hunters on a training sortie. The nearest aircraft carries No 79 Squadron's markings, while the other wears those of No 234 Squadron.

The pilots that served as instructors at Brawdy look back on it with tremendous fondness despite the tribulations. It was fantastic fun with a great bunch of guys. One of the huge attractions was that it was the only unit in the RAF where Harrier "mates", Lightning, Jaguar, F4 and Buccaneer pilots all served together. Needless to say the banter could be testing at times!'

Indeed, Brawdy was a unique base within the RAF, with a mixture of personnel coming from a wide background of RAF types, combined with many young and very enthusiastic students and a fleet of rather old but much-loved aircraft. The venerable Hunters were not the only 'historic' aircraft at the base; a Meteor T.7 and a Meteor F.8 were brought back into RAF service, as Ian Hartley explains:

Hunter FGA 9 XJ687 proudly wears the distinctive markings of No 79 Squadron, another of the component units of No 229 OCU at Chivenor.

'On 229 OCU at Chivenor, the target [banner]-towing was done primarily by a small number of Meteors, later supplemented by a small number of 79 Squadron FGA.9s that were modified with the target-towing attachment and release mechanism under their rear fuselages. When the move to Brawdy took place the Meteors had already been withdrawn from service, so the task was done exclusively by Hunter FGA.9s. However, the shortage of airframes at Brawdy quickly proved to be critical and, back at HQ No 11 Group, in the OCU's office that I shared with John Allison (later to be ACM Sir John Allison, C-in-C Strike Command), we persuaded Strike Command to allow the re-introduction of a Meteor T.7 and F.8. For some time HQSTC strenuously denied that any Meteors were left in existence, but I had already visited colleagues at 5MU RAF Kemble to check on the ones that I knew to be there (I had been on a holding posting at Kemble some years earlier, so was familiar with the storage sites.) So it was that F8 VZ467 'Winston' and T7 WA669 'Clementine' arrived at Brawdy to assist with the target-towing task, which they continued to do until the early 1980s. Eventually some Hawks were TT-modified and took over the role with the demise of the Hunter. F6/F6As never towed targets.

This beautiful in-flight shot shows a pair of classic jets, both in service with No 79 Squadron during 1972. Meteors were used by the OCU as target tugs, providing Hunter students with an 'in-house' training facility. *via Glenn Sands*

This pre-show publicity photograph shows a trio of OCU Hunters led by a twin-seater, flanked by target tug Meteors. Of particular interest is the white paint applied to the spines of the single-seat aircraft, as a visual aid for various training exercises.
via Glenn Sands

> Shortly after the move to Brawdy, and the simultaneous closing of Chivenor, it quickly became apparent that diversion availability was a real problem, so nearby St David's became a "crash" diversion manned by a couple of fire tenders. Air Traffic was not a problem as the airfield was only two miles away and easily visible from the tower. I cannot recall it ever being used regularly but I do recall it being used on a trial basis as a detachment for target-towing. Setting up a banner and attaching it to a tug was very expensive on runway time so moving it away to St David's, where there was already some fire crew, had its attractions. It was done as a trial but was short-lived as it was expensive on manpower, which was already in short supply – the ageing Hunter was not the most serviceable of beasts.'

The component phases of the TWU course have already been listed, and as a former instructor with No 79 Squadron, Ian Hartley had plenty of first-hand experience of Brawdy's Hunter operations:

> 'On 79 Squadron, those destined for the air defence role (Lightning and Phantom) did not do the Air-to-Ground/bombing/SNEB and did only a short low-level navigation phase. Those going to Harrier/Jaguar/Buccaneer did not do the Cine/Air-to-Air sorties and, of these three, only Harrier students did Tactics. The Convex phase comprised three dual, three solo and two dual Instrument Flying sorties, the second of which was the Instrument Rating Formation sortie, comprising one dual and three solo. The first two sorties were at medium level and covered close, battle and fighting battle formations, sortie three was a pair at low-level covering battle and fighting battle, and sortie four was a four-ship repeating the content of sortie three. Note, however, that only students on 79 Squadron flew four-ships. Low Level Navigation comprised one dual and seven solos. Sortie one was flown dual as a singleton, sortie two was a chased solo, sortie three was another dual, this time with a couple of simulated ground targets, sortie four was a chased solo version of three, sorties five and six were flown as battle pairs with a bounce aircraft, while seven and eight were four-ships with bounce.
>
> The Cine sorties were one dual and three solos covering simple cine (a non-manoeuvring target) and cine weave (a target following a pre-set manoeuvre pattern). Air-to-Air sorties comprised of one dual and three solos, the air-to-air

As one of three Hunter FGA Mk 9 aircraft included in a high-visibility trial at Brawdy, this particular aircraft received dayglow orange wing and tail trim. The trial did not result in any long-term change from the standard camouflage colours applied to all TWU Hunters. *via Glenn Sands*

gunnery against a towed banner. Air-to-Ground sorties were one dual and three solos with the air-to-ground gunnery conducted on Pembrey range. Bombing sorties comprised one dual and three solos employing ten-degree dive bombing at Pembrey. The SNEB sorties comprised one dual and three solos employing ten-degree rocketing at Pembrey. These last two phases would each end with two solo composite sorties where you would do air-to-ground as well as bombing and SNEB profiles. Tactics flying involved one dual and three solo flights. Sorties one to three were a pair one-versus-one, and sortie four was two-versus-two.

Apart from the Convex phase very few of the training sorties were flown as singletons, most flights being flown as pairs and some as four-ship flights. Consequently, someone watching the TWU activities outside the base would usually see aircraft leaving the airfield in pairs. Take-offs were either made in close formation or in a stream. Stream spacing was either ten seconds for visual departures or thirty seconds if departing through cloud. There was a crosswind limit for close-formation take-offs so stream take-offs had then to be the only option. Recoveries were made as close pairs if descending through cloud or via a formation "run and break" if not.

The flying programme was very rigidly managed. Each squadron had its own formation call signs; I remember "Bronze" and "Tartan" being two used by 79 Squadron, and they had their own recovery slots. From what I recall we could put a maximum of eight aircraft into each slot, but that would be reduced when the weather was poor. Although one did not normally have to take-off in a slot it was absolutely de rigueur that you landed in it. Obviously by recovering as a formation you could easily get your eight aircraft on the ground in a matter of minutes.

I instructed on 79 Squadron – the refresher squadron – and our course content differed a little from the two ab initio squadrons on two counts: first, our sorties were generally more complex, reflecting the experience level of our students, and second, our students did not do a common course, doing only those phases that were relevant to their next posting.'

Ian continues:

'For several years a permanent detachment of two FGA.9s was kept at Gibraltar nominally to protect military aircraft flying in and out of the base. Crews would be on cockpit readiness for specific movements and on stand-by in the operations room for others. Outside these times some training sorties were flown but daily reconnaissance sorties were flown over the Russian anchorages off the North African coast. Three pilots manned this detachment and were provided on a rolling detachment by the TWU. A pilot was rotated every week and did a three-week stint. It was the most fantastic fun as everyone on the Rock knew "Red Section" and the social life was, as a consequence, brilliant. Unfortunately all good things come to an end and on 3 August 1978, after some twelve years, the detachment was finally ended and the aircraft were flown home. I remember it well as I was the detachment commander at the time and, with one of the other pilots, had the job of flying the aircraft back to Brawdy. That flight is a story in itself!

Pictured far from home, this Chivenor-based Hunter is captured during a detachment to Gibraltar. No 229 OCU (and later No 1TWU) maintained a presence at Gibraltar, conducting training and reconnaissance missions.

The three TWU squadrons' operational reserve role was the low-level air defence of the United Kingdom. This was practiced two or three times each year. The concept was that the TWU aircraft would be based at the UK's front-line air defence bases at Wattisham, Coningsby, Binbrook and Leuchars. From these bases low-level visual CAPs – combat air patrols – would be mounted. It was of dubious merit as the Hunter's only armament was its four 30mm Aden cannon, and it had no radar; it had therefore to depend on visual sightings, so could only really work in good visibility, the snag being that the enemy could then easily see you and avoid you. The role was taken over by the Hawk, which also had its limitations. Although the Hawk could carry two Sidewinders it was 100 knots slower than the Hunter and desperately fuel-limited – realistically a maximum of forty minutes on task whereas the Hunter could comfortably do over an hour or nearly two hours with four drop tanks under the wings. As for the merits of each Hunter variant, the T.7 was not an operational aircraft and never had been. It was used exclusively

Brawdy's Hunter pilots were unashamedly enthusiastic about the Hunter and seized every opportunity to demonstrate the aircraft's outstanding performance. In this picture an FGA 9 is 'beating up' the flight line shortly before retirement – new Hawk trainers are visible below. *Ivan Pennell*

for those sorties that required dual instruction – generally the first sortie of each phase. They were never used as substitute single-seats especially as they were already in short supply for the dual-training task. While at Chivenor there was little to choose between the merits of the F.6 and FGA.9. In fact, 63 Squadron and 234 Squadron operated F.6s exclusively, and only 79 operated the FGA.9. At Brawdy they all went into the same pot, but now the limited external fuel capacity of the F.6 became significant – it could only carry 100-gallon drop tanks whereas the FGA.9 could carry 230-gallon drop tanks on the inboard pylons.

Brawdy was not so well served with nearby diversions as Chivenor, where we had Brawdy, St Athan, St Mawgan and Yeovilton all within easy reach, whereas at Brawdy we really only had St Athan's 6,000-foot runway. In order to be able to conduct an acceptable-length sortie of around fifty minutes and to land at Brawdy with sufficient diversion fuel for anywhere other than St Athan, the F.6s had to be modified to carry 230-gallon drop tanks. Unfortunately this extra weight then led to stopping problems on a 6,000-foot runway, especially when wet, so a brake chute had to be fitted too. Because the aircraft now had longer endurance, the oxygen capacity had to be increased, and thus was born the F.6A; all the Brawdy F.6s were so modified, apart from two, which I believe were XE656 and XG164.

The Hawk, which ultimately replaced the Hunter at the TWU, was a great little aircraft, but had two big limitations compared with its older brother. Its low-level performance, especially with any external stores – gun or bomb carriers – was quite poor, with a maximum speed of around 450 knots, and a fifty-minute sortie was very tight. The Hunter could easily do 550 knots with external stores and comfortably fly one hour and twenty minutes with the 230-gallon tanks fully filled, although on normal training sorties they were two-thirds full. Its other big limitation was that it was insufficiently testing of the students. It was a much easier aircraft to fly than the Hunter and, moreover, the students had already been flying it for six months prior to their arrival at the TWU. To get over this latter "limitation" the content of each phase of the course had to be modified in order to increase the workload on the students so that they could be assessed properly.'

Ian also remembers some of the Hunter's (and perhaps the TWU's) more quirky characteristics:

'The Hunter was fitted with an MASB – Master Armament Safety Break – which was in the wing root for F.6s and in the cockpit of FGA.9s. When carrying any weapons the MASB was only made live at the runway holding point. Additionally, for SNEB sorties there was a safety plug in the SNEB pod itself. The routine for this was that you would taxi to the holding point, put the parking brake on and signal the waiting armourer to come and make the pod break. As an additional safety precaution the pilot would hold his arms up with his hands outside the cockpit – one of the advantages of the Hunter's sliding hood.

On one occasion one of our foreign nationals had been briefed for his first solo SNEB sortie with all the safety actions. He taxied out, canopy open (we never taxied with the canopy closed – not cool!), halted at the holding point and signalled the armourer. It was a typical Brawdy day with persistent rain and our subject was used to sunnier climes so he decided to close the cockpit. He reached inside, selected the switch and instinctively put his hands back outside the cockpit again. The canopy slid forward and trapped his wrists on the canopy arch. Fortunately for him the canopy motor runs out of power at the end of its travel, so it did not crush his wrists, but did trap them very effectively. His first reaction was to select the canopy switch open, but this was out of reach on the left-hand cockpit wall, so he decided to radio for help. No can do – the transmit switch is on the control column inside the cockpit! He was now helpless and in considerable discomfort. Meanwhile the armourer had returned to the runway caravan to continue his cup of tea with the runway controller. A queue of impatient pilots quickly built up behind him and ATC kept calling him to find out why he was not moving. Of course he could not respond. Eventually the armourer turned round and saw a pair of very limp arms trapped outside the cockpit and summoned help. Needless to say the student got little sympathy, but did collect a huge amount of banter to go with his bruises!'

As described, the TWU training syllabus was a busy and demanding one, and a combination of poor weather and increasingly poor Hunter serviceability led to the TWU being barely able to meet its current task . When, early in 1976, HQ No 11 Group was given new training task figures for the TWU, which took into account the dramatic rise in the number of fast-jet pilots needed with the replacement of the V-Force with the (then) new Tornado, it quickly became apparent that Brawdy could no longer cope. The biggest problem was that the TWU physically could not exact the daily number of aircraft movements over the runway. Moreover there were not enough aircraft, and the Hawk (which was destined to replace the Hunter) was still too far away. Some short-term stop-gaps were consequently implemented, including deploying the target-towing task to nearby St David's as described previously, but another runway was the only solution – the hunt was on for a second TWU!

The essential criteria were access to air-to-ground and air-to-air ranges, good local low flying, existing spare runway capacity, a weather factor better than Brawdy and suitable domestic and operational accommodation. The obvious choice was the Hunter's former home at Chivenor, but senior figures were reluctant to be seen to have effectively made a mistake by closing Chivenor in the first place. So the shortlist came down to Coningsby, West

Raynham and even Akrotiri in Cyprus. At the time Lossiemouth was not an option because of a lack of accommodation. (No 11 Group did look at moving 8 Squadron's Shackletons to Coningsby but that, too, was also felt to be politically unacceptable.) Finally, it was accepted that, despite accommodation difficulties, Lossiemouth provided the best solution, so the TWU was divided into No 1TWU, which remained at Brawdy, and No 2TWU, which formed at Lossiemouth with a fleet of thirty single- and twin-seat Hunters. In April 1981, when new-build Hawks began to be delivered to the RAF, No 2TWU was finally relocated back to the base from which (if good sense had prevailed) its aircraft and personnel would never have left in the first place – Chivenor. The Hunters were returned to Brawdy and the unit re-formed with new Hawks, the first aircraft having arrived there in August 1980. Sadly, no Hunters returned to Chivenor to renew their long association with the base.

The all-new Hawk trainer (effectively the very last Hawker creation) entered service with No 234 Squadron at Brawdy in May 1978, followed by No 63 Squadron, where the last Hunter course ended in May 1979. This left only No 79 Squadron operating the Hunter, and operations continued into 1984 when the unit finally began to dispose of its aged aircraft, to make way for more new Hawks. A formation of nine Hunters was flown to St Athan in July (the aircraft being placed into storage there), but a handful remained at Brawdy, used for general 'hack' flying, chase and 'bounce' flying and for towing targets, until the end of 1984, when, at long last, the final Hunters left Brawdy for good.

Hunter FGA 9 XF418 was one of three aircraft to receive an unusual paint scheme while taking part in high-visibility trials with No 1TWU at Brawdy. This particular aircraft carried bright red trim and has mistakenly been referred to as a 'bounce' aircraft in many publications. In fact, Hawks were almost inevitably used as 'bounce' aircraft by this stage in the TWU's history. As can be seen, it was also one of a handful of Hunters that received a nose camera modification. *Terry Senior*

Constructed as a Mk 6 and serving with 63 and 66 Squadrons, XE651 was converted to FGA 9 standard and assigned to No 1TWU, carrying the markings of No 234 Squadron as illustrated. The aircraft crashed near Brawdy during May 1977. *Terry Senior*

F.6 XG226 was assigned to No 234 Squadron within No 1TWU at Brawdy by the time this photo was taken. However, the aircraft still retains the 'high-visibility' national insignia and yellow/black tail code presentation as carried by 229 OCU's fleet. *Terry Senior*

Hunter F.6 XE608 is pictured at Binbrook wearing the markings of No 63 Squadron, 1TWU Brawdy. Still retaining a 229 OCU-style fin code, the aircraft has received 'low-visibility' national insignia. After retirement the aircraft was assigned to battle damage repair training at Bruggen. *Terry Senior*

Having served with Nos 66, 43 and 208 Squadrons, XL623 moved to No 1TWU at Brawdy before being assigned to the Gutersloh-based Lightning squadrons as a continuation trainer. It is currently on display (pole-mounted) in Woking, Surrey. *Paul Tomlin*

A former AFDS and DFCS aircraft, T.7 XL593 is pictured with No 1TWU at Brawdy shortly before being lost in an accident during August 1982 when it crashed near Carmarthen following engine failure. *Michael Freer*

Hunter FGA 9 XF511 spent a considerable amount of time stationed in the Middle East before being assigned to No 1TWU at Brawdy. The aircraft also spent some time with No 2TWU at Lossiemouth. *Michael Freer*

Pictured during 1984, FGA 9 XG228 is seen during its final flight in RAF service, en route from Brawdy to St Athan where it was placed in storage prior to disposal.

An impressive but slightly sad sight as five Hunters from No 1TWU complete their final RAF sorties, destination St Athan for storage. As can be seen, external stores have been removed (retained at Brawdy as a spares source).

The new Hawk was certainly a welcome development for both the hard-worked groundcrews and the TWU's syllabus planners, but from a pilot's viewpoint it was received with mixed emotions. It was undoubtedly easier to fly, but that was not necessarily an asset for a unit that needed to test its students thoroughly. The Hawk was unable to compete with the Hunter's range and speed, and even the tandem seating arrangement was not always regarded as being an improvement over the Hunter T.7's side-by-side arrangement, which many instructors preferred. On a more subjective level, many pilots (both instructors and students) regarded the Hawk as an excellent aircraft, but the Hunter was in a class of its own – a special aircraft with all the characteristics, quirks and demands of a 'real' aircraft from the 1950s. More than a few TWU instructors often remarked that if a student expressed a preference for the Hunter, he was likely to be the sort of student who truly relished a challenge and would, therefore, be ideally suited to the demands of an operational squadron.

Old and new: No 1TWU operated both the Hunter and Hawk for some time while the component squadrons gradually took delivery of the new trainer. Although the Hawk quickly proved to be a much more reliable aircraft (and much easier to fly), most pilots still preferred the speed and manoeuvrability of the classic Hunter.

The withdrawal of the last TWU Hunters marked the end of a very significant chapter in the Hunter's history, but it was by no means the end of the Hunter's military activity in the UK. The Buccaneer operational squadrons and the Buccaneer OCU still employed twin-seat Hunters as lead-in trainers, equipped with Buccaneer navigation equipment and instrumentation (including the Buccaneer's Airstream Attitude Detector). They remained an essential part of the Buccaneer force and were destined to remain in service until the final withdrawal of the remaining Buccaneer fleet at Lossiemouth. However, it was also the Buccaneer that indirectly led to the re-introduction of the 'standard' twin and single-seat Hunter back into front-line squadrons during 1980-81, a quarter of a century after having first entered RAF service.

When the Buccaneer force was temporarily grounded, No 216 Squadron received a small number of Hunters for continuation training. Among these was just one single-seater, F.6 XF383. It was the only single-seat aircraft ever to be painted in wrap-around camouflage and the only one to carry 216 Squadron's markings. It was later transferred to 12 Squadron (suitably adorned with that unit's markings) and ultimately became the very last single-seat Hunter in RAF service. Unusually, the starboard insignia on the tail was reversed, with the blue segment ahead of the red. *Chris Ayre*

Following two fatal accidents involving the Buccaneer, the bomber fleet was grounded after serious metal fatigue problems were found in large numbers of the aircraft. Some were relatively undamaged but, when investigated, others were already beyond economical repair. The majority of the remaining Buccaneer force was repairable but a significant amount of time would pass while the aircraft were modified and strengthened, and this would leave five operational squadrons (and an OCU) without any aircraft in which to maintain flying proficiency. The solution to the problem was the Hunter, with a variety of airframes in storage and a relatively large number of aircraft still actively flying on behalf of the FAA with the FRADU at Yeovilton.

Twin-seat T.7s and single-seat F.6As and FGA.9s were quickly brought back up to flying condition and delivered to Honington, Lossiemouth and Laarbruch, while additional T.8 aircraft were 'borrowed' from the FRADU, enabling the Buccaneer squadrons to maintain a flying programme while the Buccaneers were slowly brought back up to a serviceable standard. It was quite a remarkable sight to witness a line of 'front-line' Hunters (albeit assigned to training duties) back in business, particularly at Honington, where a number of aircraft could be seen wearing the markings of No 208 Squadron – a unit that had relinquished the Hunter some ten years previously.

XG225 proudly wears the markings of No 237 OCU during a brief period of duty with the unit, while the Buccaneer fleet was grounded. *Terry Senior*

Although 208 Squadron disposed of its last Hunters when the unit reformed as a Buccaneer squadron, the Hunter returned to the squadron when the Buccaneer fleet was temporarily grounded. XG152 is seen on Honington's flight line. *Terry Senior*

After the Buccaneer's problems had been resolved and the aircraft settled back into squadron service, most of the Hunters were again withdrawn to storage or flown back to Yeovilton as appropriate, although some of the 'temporary' Hunters lingered with the squadrons for quite some time after the Buccaneers had returned. Ultimately, only the specially modified trainers remained, and the last operational single-seat Hunter in RAF service (No 12 Squadron's F.6 XF383) was withdrawn from flying duties at Lossiemouth in 1984. The remaining RAF Hunters soldiered on with the Buccaneer OCU before being passed on directly to the remaining Buccaneer squadrons (12 and 208) when the OCU was disbanded. They remained active until the final Buccaneer squadron (No 208) disbanded on 31 March 1994, when both the Buccaneer and Hunter were simultaneously retired from Royal Air Force service.

With the Royal Air Force's Hunters finally gone, only the Navy's Hunters remained, save for a handful of aircraft assigned to test duties with the RAE at Farnborough and the ETPS and A&AEE at Boscombe Down. Deliveries of Hawks to the FRADU at Yeovilton began in May 1994, shortly before the unit began relocating to Culdrose. The transition to the Hawk was a gradual one and it was not until January 1995 that the last FRADU Hunter was withdrawn at Yeovilton, leaving just four Hunters assigned to military duties, one (WV383) being used as a platform for a series of night-flying infrared trials until 1998, when it was withdrawn, largely because its operator (DERA) could not source sufficient funds for its scheduled overhaul. Two more T.7s remained active with the Empire Test Pilots School, being particularly valuable as the only swept-wing aircraft type routinely cleared for intentional inverted spinning, but one (XL564) was lost in an accident during 1998, leaving just one other T.7 (XL612) and single-seat F.6 XE601 (which had been progressively modified to FGA.9 standard); the latter earned the distinction of being the very last active British military Hunter, retiring on 15 July 1999 with little celebration – a rather sad ending to the conclusion of the Hunter's incredibly long and successful military service. Thankfully, this rather low-key retirement was certainly not the end of the Hunter.

Constructed as an F.4 and first assigned to the RAF Flying College at Manby, WV383 later served in the Far East and in Germany before being allocated to the A&AEE and RAE. *Paul Tomlin*

Chapter 8

Civilian operators

Following the withdrawal of the last Hunters from RAF, FAA and research/test flying duties, Hunters quickly disappeared from the skies over the United Kingdom, with just a few overseas operators still maintaining them in relatively small numbers, with the exception of Switzerland, where the Hunter was still very much in business. Although a variety of former military aircraft could still be seen in the UK appearing at air shows around the country, these machines were mostly unsophisticated piston-engine aircraft that could be maintained, certified and operated relatively inexpensively. Few people imagined that the Hunter would ever re-appear in civilian hands, and it was only the occasional appearance by Switzerland's Patrouille Suisse aerobatic teams that provided British audiences with a rare opportunity to see the magnificent Hunter in the air.

It was therefore a welcome surprise for every Hunter enthusiast when businessman Spencer Flack appeared on the air show scene with G-HUNT, a former Danish Hunter F.4 (E-418), which was restored to flying condition and certified for operations on the British civil aircraft register. G-HUNT became a popular air show performer, painted in an eye-catching red paint scheme with white and blue trim. It was subsequently joined by a twin-seat Hunter (G-BOOM) and both aircraft thrilled crowds across the UK for some time until the aircraft were sold on; the single-seater eventually went to the USA (where it remains grounded, having proved to be too expensive to overhaul), while the twin-seater joined the Royal Jordanian Historic Flight (eventually moving to Jordan).

A former 229 OCU and AFDS aircraft, XL573 is now operated in civilian hands as G-BVGH. Thankfully, her former military paint scheme and markings have been retained. *Jeff Middleton*

This recent photograph shows XL573 still wearing the markings of No 12 Squadron, the unit with which it last operated. In the background is a Tornado GR Mk 4, No 12 Squadron's current equipment. *Jeff Middleton*

The beautifully finished G-HUNT (a former Danish F Mk 4) was the first of many former military Hunters to re-emerged in civilian hands as air show performers. *via David Griffin*

Operated by the RJAF Historic Flight, this Hunter masquerades as a Jordanian machine but is in fact a former Swiss aircraft, as can been seen by the extended gun link collector pods, which were unique to Swiss machines. *Paul Tomlin*

It was not until Switzerland finally disposed of its last Hunters that the aircraft re-appeared in civilian hands, with many aircraft (some with surprisingly low airframe hours) offered for sale. Inevitably, many of the Hunters came to England and so began another stage in the Hunter's story, as a civilian-operated jet. Although the Hunter is, by any standards, a fast and manoeuvrable jet, the simple design and construction combined with relatively unsophisticated systems has enabled many private individuals and groups to restore the aircraft to acceptable certification standards, and while many other 'Cold War' jets stand little chance of being operated safely without the full weight of military servicing and support facilities, the 'no nonsense' Hunter has proved to be the most significant exception.

It was with some irony that some military agencies eventually regretted their decision to dispose of the Hunter; many FAA pilots were certainly never convinced that replacing FRADU's Hunters with Hawks was necessarily a good idea, and the unit's re-equipment was largely prompted by what was (at the time) an apparent surplus of RAF Hawks, created by a reduction in flying training requirements. The opportunity to replace the ageing Hunters with modern trainers was (at least superficially) a reasonable decision, but the Hawk's better serviceability and slightly lower operating costs were outweighed by the type's inferior performance when compared to the Hunter, which could fly much faster and much further. When the RAF's training requirements increased again, No 4FTS at Valley was suddenly faced with a shortage of Hawks, and there was serious consideration given to the possibility of withdrawing Hawks from test flying duties and from FRADU, so that they could be returned to training operations at Valley. It would have been the supreme irony if these aircraft had indeed returned and their duties had been resumed by Hunters.

Undoubtedly, there must have been a great deal of regret that many Hunters that had been assigned to non-flying duties had been sold to civilian buyers. Had they remained in use, they would almost certainly have returned to flight with either FRADU, the MoD(PE), or both. More recently, when additional Hawks were required by various test units, the inability to find surplus airframes led to the Hunter being considered yet again as a suitable replacement, effectively taking on the roles from which it had been withdrawn many years previously. Ultimately, the MoD did not opt to buy back or lease any civilian-owned Hunters, citing various technical reasons why the idea would not have been practical. The truth may well be that the decision to buy back Hunters that had been sold with unnecessary haste might have been just a little too embarrassing to seriously contemplate. In the end a batch of surplus German Alpha Jets was purchased for operations at Boscombe Down and Llanbedr – although the absurdity of having to buy European trainers to supplement and replace British trainers seemed to be no less embarrassing than the concept of buying back Hunters.

The magnificent Hunter F Mk 6 G-KAXF (XF515), fully restored to flying condition in the markings of No 43 Squadron. *Jeff Middleton*

More recently, the Hunter's post-military history has taken yet another twist. While various civilian-owned Hunters continue to appear at air shows across the globe (many having already changed ownership at various times), others have resumed a pseudo-military existence with companies that have realised the potential for providing 'outsourced' training facilities for military agencies faced with ever-decreasing budgets. In the UK, it was management consultant Mat Potulski who re-introduced the Hunter into the world of military operations. During the 1990s he joined company with the late Mark Hanna to investigate the possibility of creating a 'contractorised' fast jet unit using former military jets. They looked at a variety of types including the L-39, F-100 and the A-4 Skyhawk but, as Mat explains:

> '...the Swiss Government's decision to retire its Hunter fleet offered us the opportunity we'd been looking for. They were all low-hours airframes maintained in superb flying condition, so we purchased twelve single-seat Mk 58s and obtained hangarage at Scampton.'

Unfortunately, the project stalled when Mark Hanna died following the crash of a vintage aircraft he had been flying in Spain. It was not until 1999 that Potulski decided to pursue their original business plan and create Hawker Hunter Aviation, which was registered in 2000. Years later and millions of pounds the poorer, Potulski has now created his own 'private air force' comprising fourteen Hunters (two being twin-seaters), a Sukhoi Su-22 and a Buccaneer. With a team of eight former RAF maintenance personnel and a group of retained former military pilots, HHA succeeded in gaining AvP67 status – the vital approval that endorses HHA's maintenance and operating standards as being equivalent to military levels, enabling the company to bid for MoD taskings.

During 2007, to the astonishment of many observers, two of HHA's former Swiss Hunters re-appeared in British military markings and serials, and embarked upon a series of trails with FRA, which used them as radar targets for the new Sampson radar being installed in the Royal Navy's new Type 45 destroyers. Mat explains that HHA

> '...beat all other competition on the grounds of overall performance, cost effectiveness and flexibility. Under a serviced contract arrangement, a number of Hunters are being tasked as agile fast jet targets, operating from Bournemouth. This is the first ever contract where taskings of this nature have been outsourced to a dedicated fast jet private contractor. The Hunter is a perfect choice. It offers viceless handling, superlative subsonic performance, excellent reliability and a good internal and external payload.'

With more than a little hint of irony, HHA has also brought the Hunter back to Boscombe Down, where one of its T.8 trainers is employed by the Empire Test Pilots School as part of their training syllabus. ETPS students fly the Hunter to show students specific parts of the flight envelope that other aircraft in their fleet simply cannot replicate. Potulski explains that HHA

> '...is not a flying circus. The Hunter allows military and defence industry customers to significantly cut the costs associated with fast jet taskings. These operations are by definition expensive, and with pressure on budgets, and new equipment costs spiralling, there is increasing interest in the use of a low-cost "legacy" platform for support and training. We are able to replicate the majority of the profiles once performed by air forces but now increasingly outsourced to the private sector. That ranges from all aspects of aerial combat training, trials support, high-speed photo chase and radar calibration.'

Already, HHA's Hunters have completed more than 500 incident free sorties, and so far not a single sortie has been lost due to technical unserviceability. As Mat says, this is 'testament to the quality of our engineers and also a reflection of the simplicity and reliability of the aircraft.' With perhaps another ten years of so before the Hunters reach the end of their airframe lives, HHA's aircraft may well become a familiar part of military operations in the UK. Now that FRADU's Hawks are approaching their scheduled retirement, it would be another great irony if HHA eventually assumes FRADU's training tasks, using the Hunter once again. That prospect is by no means unlikely.

Across the Atlantic in Canada, the Hunter has also embarked on a similar career in the 'contractorised' world of military operations. Northern Lights Combat Air Support offer similar facilities as those being provided by HHA in the UK. Flight Operations Manager Steve Legassick explains:

> 'We are Canada's first, best and most experienced provider of superior airborne tactical services, operating incomparable Hunters flown by highly skilled fighter pilots. Currently contracted to Canadian (ICATS), American and other NATO air, ground and naval forces, we supply first-rate electronic warfare, target-towing, air-to-air gunnery, aggressor fighter, attack simulation, air intercept and forward air controller training around the world.'

Truly a fantastic sight: the Northern Lights Hunter fleet ready to embark on another combat air support mission. *NLCAS*

Northern Lights Hunters returning from a simulation mission in company with an ATAC F-21. 'The jets perform very well and have blown away the Canadian, US and French forces in particular, who want more and more and love the cost-effectiveness and massive productivity, since our planes almost never break, and by using us they free up their own resources, pilots, crews and flying hours, and, most importantly, it saves a lot of money'. *NLCAS*

With more than 4,000 hours flown since 2002, Northern Lights has become a recognised name in the world of customised combat training. Steve continues:

> 'Our high-performance Hunters are more cost-effective and fly faster, further, higher and longer than Alpha Jets, L-39s, L-59s or A-4s, while configured with more training equipment, for unrivalled realism and speed. Our Hunters are flown by former NATO fighter pilots with experience in F-14, F-15, F-16, CF-18, F/A-18, Mirage and Rafale aircraft. Northern Lights is capable of safely and properly simulating opposing forces in virtually any capacity. Truly able to attain speeds of up to 600 knots, even with external fuel tanks, and able to pull 7.5 gs, the Hunter's performance, size and manoeuvrability is ideal in the hands of an experienced fighter pilot with the knowledge and skill to conduct valuable and realistic air defence and air combat training. Depending on the required configuration, on-station training time can be from 1.5 to 3 hours. The Hunter has several interesting capabilities regarding air-to-air training: ACMI, threat emitter, jammer, chaff, flare, radar warning receiver and typically the endurance to take-off before and land after the training fighters, providing the maximum training value possible. As aggressors, the Hunters can simulate strikers or air-to-air fighters, or anything else you wish. We can provide a cooperative training aid for basic fighter pilot training in air interception, basic fighter manoeuvring or as any air combat opponent. In more advanced air combat training, we can execute two v twos, four v fours, or more. In large force exercises such as Red Flag, Maple Flag or Fighter Weapons Instructor courses, we can augment the opposing force to increase the training quality. For both pilots and air weapon controllers conducting NORAD training, the Hunters can accurately supply interception targets, flying at high or low speeds at high or low altitudes, and in any weather,

for flights lasting up to 3 hours. Since all our efforts are dedicated to the client's training, we accommodate whatever requests or profiles are required to attain the needs of the fighters and controllers. Military aircraft and personnel never have to sacrifice their own training and they can conduct fighter exercises with the confidence that highly experienced and properly trained professional fighter pilots are providing the simulated enemy.'

Back in the UK, the Hunter has also become a very familiar sight in the skies over Devon, as John Sparks, Chief Engineer at Hunter Flying Ltd, explains:

'We are the largest civilian operator of the Hawker Hunter in the world, and a CAA-approved operation The organisation has a wealth of knowledge and a reputation that has been built up over a number of years. This in turn has brought us to the attention of Hunter owners who have purchased aircraft and require them to be restored back to airworthy condition, or want their aircraft dismantled and transported. All the aircraft in the care of Hunter Flying are maintained and operated for their individual owners. The fleet contains a wide variety of different variants, from the twin-seat T.7s and T.8s through to single-seat aircraft, such as the Navy's GA.11 and PR.11. Also in our care is the well-known former Boscombe Down FGA.9 XE601. Many of these aircraft are maintained in airworthy condition and they appear at air shows around the UK. Hunter Flying Ltd also rebuilds aircraft to airworthy condition for other operators, such as Hawker Hunter Aviation and the Canadian operator Northern Lights, with whom we have been closely involved. Northern Lights has benefited greatly from Hunter Flying expertise in overhauling and operating these aircraft and this has resulted in a number of joint projects, which included converting a single-seat Hunter to a twin-seat configuration. Over the years, Hunter Flying has celebrated a number of anniversaries connected with the Hunter, including the 50th anniversary of the Hunter speed record and the 50th anniversary of the Hunter entering RAF service, as well as the 50th anniversary of XE601's first flight. Likewise, we were proud to be closely connected with Hawker Hunter test pilot Neville Duke, who attended many Hunter events held here at Exeter.'

Similar operations also take place at Kemble, where so many Hunters first embarked upon their military careers, being stored there prior to onward allocations to operational units. Delta Jets also has CAA-approved authorisation to restore, maintain and fly Hunters (and other jet types), and currently has three Hunter T.7s under its management, all of which regularly make hugely popular appearances at air shows both in the UK and Europe. Beyond Europe, in the sunny skies of South Africa, yet more Hunters remain active both as air show performers and as 'thrill rides' for those who are lucky enough (and rich enough) to be able to take a joyride in Camm's classic fighter. Getting the Hunters to South Africa to begin their new lives in civilian hands was by no means as straightforward as might be imagined. The former Chief Test Pilot of Classic Jets, Keith Hartley, recalls the story of one such ferry flight:

'Early in 1995 Mike Beachyhead, a South African businessman and private pilot then running an air freight company in Cape Town, bought two Hunters, T.8 XL598 and F.6A XE653, and a Canberra T.17, WJ607, at a Sotheby's auction of surplus RAF aircraft. He was then faced with the problem of collecting his purchases and getting them to Cape Town. After enquiries, he found his way to the Classic Jet Aircraft Company at Exeter in Devon. Classic Jets, owned and run by Chief Engineer Barry Pover and myself as Chief Pilot, owned and operated a variety of Hunters and had an established and approved ex-military aircraft maintenance organisation. Mike asked Barry to collect the aircraft, take them to Exeter, do the servicing and modifications to get them civil-permitted and then, subject to the necessary planning and costs, deliver them to Cape Town. That was a fair amount of work. The T.8 and F.6A were prepared sufficiently for the issue of a ferry flight permit and taken to Exeter for the serious work. The RAF delivered the Canberra to Exeter, where it needed some additional preparation before delivery, but that's another story.

The F.6A was at RAF Scampton, where it had resided for the previous twelve years as a ground runner, used for training line groundcrew in marshalling skills; this meant it was regularly run and in generally good condition – well, that's what we thought at first look. Externally, it looked terrific; it had been freshly painted in pristine gloss black and original 1960s Treble One squadron markings for a recent squadron anniversary. Internally it wasn't as good, but having been regularly run it was in much better condition than some aircraft that had come on the market after being in long-term disuse and storage.

A rare and unforgettable sight in 2008 as a pair of HHA Hunters taxi out for take-off at RAF St.Mawgan, ready to embark on a training flight in support of Royal Navy operations. *HHA*

Hawker Hunter Aviation's magnificent twin-seater XF995 pictured during a sortie from Boscombe Down. *HHA*

Resplendent in an unusual variation on the RAE's 'raspberry ripple' paint scheme is Hunter F Mk 6 XE601, the last active British military Hunter, now flying in the hands of Classic Aviation/Hunter Flying Ltd, at Exeter. *Jeff Middleton*

These views of the Classic Jets/Hunter Flying Ltd hangar at Exeter were taken in February 2007. Every inch of hangar space is crammed full of Hunters, all airworthy or under restoration to flying condition. *via Jeff Middleton*

XE601 spent her entire career with the A&AEE, often performing unspecified duties on behalf of Porton Down. The aircraft became a familiar sight at RAF airfields during exercises, when it often flew simulated chemical attack profiles, spraying water from specially designed underwing tanks. Although modified to FGA 9 standard, the aircraft was never redesignated. *Jeff Middleton*

The main problem area was the fuel system. Having done the work necessary to get the aircraft a CAA Permit to Fly, the air tests showed problems with fuel transfers. We concluded that although it had been frequently ground run, the taxi training environment only involved putting a limited amount of fuel into the aircraft, which meant the wings and rear fuselage tanks were little used and rarely, if ever, filled. Most of the problems came from these areas, thanks to sticking float switches and leaks. After a fair bit of work, the problems were resolved, the air test completed and the Permit issued. We then put the 230-gallon drop tanks on and had another burst of fuel system fixing to get them working right. Other systems, the engine and aircraft handling, were fine.

Eventually the ferry shakedowns were completed satisfactorily on all three aircraft for delivery. Meanwhile, I'd been beavering away – with lots of assistance from Flight Ops at BAe Warton – on route planning and getting the necessary diplomatic clearances and overflight approvals for the delivery. That's a long, slow and tedious business. Civil aviation bureaucracies are not speedy at the best of times, and when you're trying to deal with some in the middle of darkest Africa about a set of strange ex-military aeroplanes it can, and did, take months to sort out. The delivery route was, necessarily, a compromise. The Hunter T.8 defined the maximum range for each leg (the Canberra would have flown for ever in comparison, but it was going to carry all the support equipment and thus needed to stay together) – this was about 900nm, allowing for some contingency fuel to play with for diversions and ATC-induced delays. Having studied the available and suitable airfields along the route, the plan was Exeter-Genoa-Brindisi-Iraklion-Luxor-Addis Ababa-Djibouti-Nairobi-Lilongwe-Johannesburg-Cape Town.

The difficult bit was between Luxor and Nairobi; Luxor-Khartoum was short, but Khartoum-Nairobi was too tight for the Hunter T.8 with slightly less fuel. Djibouti-Nairobi was much better, but then tight from Luxor to Djibouti. On paper, it was possible to cut the corner and go direct from Luxor to Djibouti; that would have saved the expected hassles of an arrival in Addis (7,600 feet high and surrounded by serious mountains), but would involve arriving at Djibouti with precious little reserves. Fine if the weather remained OK and, more important, ATC was accommodating, but none of us had been there before and my yellow streak was healthily wide – like it or not, Addis had to be planned.

At last we were ready. The Canberra was loaded to the gunnels with tools, essential spares, wheels, brakes and ground equipment for all three aircraft, not to mention a small arsenal of starter cartridges for the Canberra and T.8 Hunter, and several gallons of Avpin starter fuel for the Hunter F.6A. Bravely, we Hunter pilots decided we'd move a healthy distance away from the Canberra if we came across any decent thunderstorms and lightning along the route... With excellent weather forecast for the first day's flying, we launched from Exeter bright and early on the morning of 19 April 1995. The Canberra took off first and orbited the field while the Hunter pair got airborne, then we joined up and set sail southwards across the Channel. We settled into the cruise – I can't remember now, but it would have been around flight level 330 or 370 – in glorious weather over northern France. That was the good news: the bad news was that my right 230-gallon drop tank wasn't transferring at anything like the correct rate.

I re-did the arithmetic and concluded that I wasn't going to make Genoa with what I could rely on getting. One option was to divert somewhere else – say, in southern France – refuel, then leapfrog to join the others in Brindisi. While that would get me to Brindisi, the jet would need fixing before going from the comfort and plentiful airfields of Europe into the sparse and difficult lands of Africa – and that would likely result in a significant delay of all three aircraft for an unknown time and cost. Nope, I bade the others farewell and headed back to Exeter. They had their own set of adventures (a euphemism for occasional excitement and trepidation, lots of frustration with bureaucracies, and a great sense of final achievement) before arriving safely in Cape Town.

Meanwhile, I put the F.6A back to bed in the Exeter hangar, and waited for Barry Pover, the trusty Hunter engineering expert, to return from the back seat of the Canberra via Cape Town. At least I was able to act as route monitor, handling difficult phone calls from Barry in strange places, and able to sip gin and tonic while hearing how hot, dusty and difficult life was with the rest of the team. On Barry's return he once again attacked the fuel system with gusto – two and a bit months later, we had a serviceable aircraft that satisfactorily completed all the ferry shakedown tests, including long high-altitude transits. I replanned the route and got all the clearances and approvals ready for the solo delivery. There was some excitement at doing the job on my own – a real challenge that not many have done in recent years – but I knew from many other deliveries I'd done with BAe that it was going to be hard work doing the aircraft servicing as well as taking on ATC, customs and all the other bureaucracies en route. This was only twelve years ago, yet it's easy to forget how much we now take for granted, things like effortless international mobile phone roaming, and being able to get cash anywhere by sticking plastic into a hole in the wall. I went armed with a large bucket of cash in US dollars, a bigger wad of traveller's cheques, and a detailed list of contacts along the route to assist with communications. I also had Shell and BP fuel carnets to help with paying for the fuel along the route.

Around 7.00am on 14 August, I launched into a brilliant, cloudless summer's morning and again turned left to cross the Channel. France was equally clear and we – the jet and I – cruised effortlessly down to the Alps, which were cloud free in all their beauty. One of the nice things about ferrying the Hunter is that it likes high altitude: the engine's efficient up there, and the lovely aerodynamics make for a fast cruise – 0.85 Mach for good range, 0.90 Mach if you've fuel to spare. It's a nice fast way to eat distance. Without the Hunter T.8 I was trying to make best use of the F.6A's range and make Brindisi direct from Exeter in one go. However, there was a headwind – unusual, since the prevailing winds are usually westerly – and that was adversely affecting range. I watched the arithmetic as I flew down Italy's Adriatic coast and decided that expected fuel at Brindisi was less than I'd like, so I diverted to Bari.

Bari was good: they seemed pleased to see me and were very helpful, in particular by getting the flight plan into the creaking European ATC system and sending me speedy fuel. I turned the jet, paid the bills and set off post haste for Iraklion in Crete. I arrived there after another easy cruise. It was peak holiday time in Crete and the airport was packed with transiting holidaymakers and lots of charter jets to ferry them back and forth. As I taxied in there were several admiring comments from the airliner crews and ATC on the stunning gloss black paint job. I waved nonchalantly at them and the crowds on the terminal building balconies as I parked right in front of the terminal. I asked for the fuel, slipped on the dark shade sunglasses and tried to look really cool as I checked the oils and connected the fuel bowser when it arrived.

That's when the God of Aviation decided he'd had enough of me posing. The fuel poured nicely into the jet, filling the front fuselage tanks first. But instead of then filling the other tanks, it went straight into the wrong bit of the fuel system and straight out through the vent mast, at 30psi through a big wide pipe. The bowser driver didn't help – he just shouted very loudly and ran around waving his arms instead of shutting off the bowser. I lost Joe Cool in an instant, leaped up on top of the jet and started stamping on the wing roots behind the cockpit, in the hope of shaking free what I assumed to be a sticking high-level float valve that hadn't closed as it should. Meanwhile, an increasingly large pool of fuel was flowing inexorably towards the Airbus A320 parked right next to me and all the calm bonhomie of two minutes ago had long disappeared.

I eventually got the bowser shut off and the fire trucks rushed up and sprayed the fuel away. Quite reasonably, no one – myself included – was willing to have another go at refuelling and I had nowhere near enough to go anywhere. After some discussion with the airfield manager I started the Hunter and moved it to any empty area well away from real aircraft, over near the military dispersal, while I tried to work out what was wrong. With some help from the handling agent, I went through some diagnostics but quickly concluded that this was beyond my skill set, so it was back to the agent to phone Barry at Exeter. Barry collected his kit and some likely spares and tried to work out the quickest travel plan to Iraklion. Meanwhile, I went and found a local hotel and retired for some excellent Greek food and dreadful retsina.

The following morning Barry arrived, having somehow blagged his way onto a passing charter jet, and we started dismantling the Hunter's fuel system. The first area of access I came to know well – under the fuselage, behind the nose wheel electrical bay and forward of the engine and gearbox bay is a panel with several hundred bolts. Of course, that involves lying underneath it and, when it comes off, it happily drops any fuel trapped in the panel onto you. It was high summer, in the 30s, and the sun was baking, so my immaculate poser dayglow orange flying suit rapidly developed the characteristic smell of sweaty Avtur fuel. We took off pipes – more fuel deodorant – and swapped valves, twiddled with electrical thingies and eventually decided it was fit for another attempt at refuel. The bowser driver returned and this time stayed very close to his bowser with hands poised for instant disconnection. That slowed us down a bit, because most Hunters leak occasional drips from the vent mast during re-fuelling, but he assumed it was the start of the second Great Flood and kept stopping. But we got it all in OK and Barry was satisfied it was working properly. We then re-assembled all the missing bits and Barry did up several hundred bolts again while I filed a flight plan for Luxor. I then got airborne while Barry retired to the hotel for a prolonged clean-up and some decent overnight sleep.

The leg to Luxor was painless. As usual the Egyptian desert was yellow and brown and in startling contrast to the blue Mediterranean. Far away to the east I could see the equally startling green of the Nile valley, gradually closing towards me as I headed for Asyut before looping away to the East again and snaking back to me at Luxor. Luxor was doing a typical Luxor in August – clear blue sky, nil wind and very, very hot. ATC said it was 42°C when I arrived, but the sun-baked concrete made it a lot hotter than that after shutdown while I was servicing the aircraft. I refuelled the jet, checked the oils and topped up the Avpin tank. The big-engined

Hunter (Avon 200 series) uses an Avpin starter to get it going. Avpin – isopropyl nitrate – is a self-oxygenating fuel that is nasty stuff: it's corrosive to paint and skin and very volatile. You can't get it anywhere en route so I had to carry it with me. I had several one-gallon petrol cans of the stuff, padded out with sponge rubber, stored in the sabrinas – the spent links containers for the Aden guns, seen as the little bulges below the cockpit. The Avpin tank for the starter is on top of the fuselage on the spine behind the cockpit. It was so hot up there on top of the black fuselage that I couldn't kneel down to do the job, and my boots were melting and leaving black smears on the wing roots. I poured in the Avpin but I reckon a sizeable proportion was evaporating before it got inside the tank. I tried checking the level with the dipstick, but as soon as it got clear of the liquid inside the tank – I couldn't see the level in the dazzling sunlight – it instantly evaporated. In the end I just kept pouring and made sure that none came over the side. I don't think the UK Health & Safety Executive would have approved of the process.

I'd been unable to get any oxygen in Luxor, which was a nuisance. I needed to go at high level between Khartoum and Nairobi, to get the required range, so I opted to go to Khartoum (only about 570nm) at flight level 220, which gave me a cabin altitude of less than 12,000 feet and meant that I could leave the oxygen off and save it for the next leg, on the assumption that I couldn't get any in Khartoum either. I did the revised fuel planning, which showed no problems. Work done, I retired to the Winter Palace Hotel, a very genteel establishment full of Howard Carter and Tutankhamun memorabilia. I dived into the shower – fully clothed – and gave my flying suit a serious wash, before taking it off and repeating the exercise on me. That done, time for the patio, the gin and tonic and the Nile sunset.

The following morning I was up bright and early, jet prepped and packed, and on my way. The next leg was direct to Khartoum. With the slight additional range of the Hunter F.6A compared with the T.8, the Luxor-Khartoum-Nairobi option was better than the originally planned Luxor-Addis-Djibouti-Nairobi route. The down side was that Khartoum would be the worst place along the route to go u/s: difficult to get to, probably more politically sensitive and bureaucratic than many places, and no one I'd spoken to had any experience of it in aviation terms. The route was easy, following the Nile past Aswan to Khartoum, where the White and Blue Niles meet. The only problem was that I seemed to be using a bit more fuel than I'd expected, notwithstanding the lower cruising altitude and the very hot temperatures.

Khartoum ATC seemed surprised when I arrived on frequency, and I was faced with repeated questions about where I was from and what was my aircraft type. On arrival overhead there was a bit of faffing about delaying my landing clearance, but after a few minutes of sightseeing I landed. My parking area was swarming with people, half of them in military uniforms, the other half civilians, typical airport workers. They watched with interest as I shut down, climbed out and started checking round the aeroplane for signs of fuel leaking – I was concerned over the fuel usage on that leg. Some of them wanted to drag me off somewhere but I politely declined while I did my checks. I was also waiting for the agent to arrive: I'd arranged handling agents for many of the places along the route, particularly in Africa. Agents cost money – sometimes big money – but they can markedly smooth the way through the local bureaucracy and can save significant turn-round time, important when I was also having to do all the aircraft servicing.

However, the agent didn't arrive and I was being urged to "come along" somewhere, so I secured the aircraft and obediently followed. I was taken to a small private lounge in the airport, where I met a young military man who turned out to be a Sudanese Air Force Mig 17 pilot. He offered me some very welcome orange squash and we sat and chatted about aviation and other polite small talk. He was a very pleasant guy with excellent English, and I'd have really appreciated the dialogue had I not been more keen to get back to the aeroplane, check it out, refuel it and be on my way. After about an hour the agent arrived. He was a very interesting chap, as I came to discover. He was physically large – well over six foot and heavily built – and spoke excellent English. His name was Youssef Gillespie – he was of mixed Scottish and Sudanese heritage (when was General Gordon in Khartoum?) and, having spent a lot of time in the UK, he was very familiar with both cultures, something that proved to be of enormous value to us.

After a rapid Arabic dialogue with my escort, he whisked me off back to the Hunter, telling me as we went that he'd just released me from arrest. Apparently, the civil and military bits of the government hadn't copied my diplomatic clearances to those that needed them. At that time, Sudan and Egypt were having a mild head-banging over their common border and my unannounced arrival from the direction of Egypt was taken to mean that I was an Egyptian Air Force defector coming to seek refuge. That theory had fallen under some scrutiny when they recognised the aircraft as a Hunter (a lot of them had been trained in UK) and that I clearly didn't look very Egyptian. I marvelled at how naïve and innocent I was, arrested in Sudan without even realising it. At the aeroplane there was an ominous drip, drip, drip from the vent mast under the fuselage, worrying because there wasn't that much fuel in the jet and certainly not enough to get to the vent pipes.

After checking what I could – not much – we got the bowser and I refuelled it. It didn't burst into a flood, and actually refuelled fine, but there remained the ominous drip, drip, drip. After briefing the agent, I started her up and got him to check whether the dripping stopped when the fuel system was pressurised. Nope – drip, drip, drip. Shut down again and think, but not for long. It was 1,000nm to Nairobi, over some very inhospitable countryside littered with civil war. I had no idea how much of a leak rate the drip, drip, drip represented. There were no sensible or usable diversion airfields en route that I could use if the fuel system didn't behave. I was going nowhere and my worst fear – u/s in Khartoum – had come to pass.

I told the agent we needed to find hangarage – I had to get the aeroplane out of the baking sun. I'd guessed – rightly as it transpired – that the blistering heat of Luxor had made something fail and cause the leak, and I didn't want to expose the poor old girl to any more roasting than necessary. He frowned and looked worried, then he said, "Come with me. We'll go and drink tea with several people – while I talk, you look polite and smile a lot."

And that we did: the Sudanese officials were polite, hospitable and friendly, and after several cups of hot sweet Arabic tea I was told we had hangar space. We had no tractors or a towbar that would fit a Hunter, so I started it up and taxied it into the hangar, stop-cocking the engine just before I rolled through the hangar door. Just get it inside, I thought – I'll worry about how to get it out later. I came to a halt next to a tired-looking Fokker F27 that looked normal apart from having bent propeller blade tips, which made me look closer and notice the nicely rippled and torn bottom fuselage. Ah, another wheels-up landing…

I arranged some drip trays to catch the fuel, and we set off for Youssef's office, which, blessedly, had one of the very few telephones and faxes with international access. Barry Pover got the unwanted call from Khartoum. After some discussion, I went back to the airfield to check a few things on his instructions, then returned and reported back. Clearly, there was nothing more to be done that day, so the agent took me to the Hilton Hotel and I went to bed while Barry pondered. The following morning I went back to the airfield and arranged more drip trays and some willing labour to move them when they filled up. The labour was very happy to do that – my full aircraft would fuel cooking fires all over Khartoum for the next fortnight!

Talking again with Barry, it was clear that there was nothing I could do to fix it, that it would take significant time to arrange his travel to Khartoum with tools and spares, and that both he and I were needed back at the day job for a bit. So I ensured the aeroplane was secured with the help of Youssef, found a flight in the general direction of Europe, and flew back to the UK.

A fortnight later Barry and I boarded a Sudan Air flight from Heathrow back to Khartoum, Barry bringing along a significant excess baggage load of tools and spares, including a fuel recuperator, which he had deduced was the guilty culprit. The recuperator is a small tank split into two halves by a seal: on one side is pressurised nitrogen, on the other fuel pushed in by fuel pump pressure in normal flight. If the aircraft turns inverted and the fuel pump gets uncovered, the recuperator nitrogen pressure forces the fuel back out to the engine and keeps it going – it's good for fifteen seconds on inverted flight at full power. In this case, the seal had split, almost certainly as a result of age and disuse (it probably wasn't ever pressurised and would never have been needed when the aircraft was run for training marshallers), together with the baking hot temperatures at Luxor. The fuel was then able to track through to the pressurisation pipes and eventually through the vent pipes to fall out of the drain, and would continue to do so even when the aircraft was on the ground engine off, because of gravity pressure on the full tanks.

Back in the hangar the aircraft was filthy, covered in dust, sand and bird droppings – serious bird droppings. Once again we took off the fuel panel – by now requiring thousands of bolts – and dismantled the fuel system. Ah, the joy of another Avtur shower! It took us a couple of days, watched by attentive locals who were incredibly friendly and helpful to a humbling degree. Here was a country where the average annual income was then US $1.80 per person, yet they would always come and offer us half of their lunch. As soon as Barry looked around for something, they'd leap to help, and not in an annoying and intrusive way, but proactively. Inevitably, there were some bits that Barry didn't have, but since a lot of the Sudanese aircraft engineering personnel had been trained in the UK with the RAF – they all recognised the Hunter – they were familiar with UK fittings and bits, so they'd take Barry off to some dark and dusty store where Barry would find a treasure trove of British couplings, seals and widgets. They were excellent, friendly people and allowed what could have been a very demanding and difficult exercise to be completed in very short order.

They were unfailing tolerant too. One morning while we were sloshing away at leaking fuel pipes the hangar filled with chairs and a lectern, filled soon after by a large gathering. A man in Arab dress then began to make a long and energetic speech, ignoring the two pasty Westerners dropping spanners and swearing at skinned knuckles. Some time in the middle of this one of the locals came along and chatted quietly to us, explaining that this was a political rally and the speech was a Muslim extremist harangue against the children of Satan and other non-believers. Barry and I obligingly kept our heads down and avoided attention, but when it finished the speaker and assembled audience waved a friendly goodbye as they left, or came over to find out how it was going. I can't say I ever felt threatened by any of it.

The jet was fixed, I ground ran it and everything worked as it should – no drip, drip, drip. We washed the aeroplane to get rid of several hundred pounds of guano, and I went round and cleaned all the gear microswitches I could find – everything was covered in a fine dust and I thought we'd risk trouble with the undercarriage retractions if I didn't. We'd finished it all by sunset, put it to bed (yes, we'd found a towbar and tractor by then) and went to the hotel for a good night's sleep before a very early start.

Just after dawn I was in the aircraft, ready to go. The diplomatic clearances had all been re-jigged and the flight plan was filed. I started up and the first thing I noticed was that the electrical inverters weren't working. They're gadgets driven by the batteries or the engine DC generator that produce AC power to drive the artificial horizon, the compass and the engine oil pressure gauge. There are two of them for redundancy; on the ground run the night before they were fine. Pause for reflection, while Barry looked quizzical outside. Normal training and behaviour would make me shut down, diagnose and fix it. However, this was Khartoum. The inverters were down behind the ejection seat, which meant canopy off and seat out to get access to them – all work that was going to be difficult in Khartoum and hard to get any required spares, which was not a positive prospect and one that could keep us there for another fortnight. On the plus side, the weather was fine there and at Nairobi, allowing for the usual afternoon thunderstorm risk, but at least there are several good diversion options within easy reach. I didn't need an artificial horizon, I'd have liked to have a decent compass, but I had a swung and reliable stand-by and two GPSs that did my serious navigation. Decision: go for Nairobi.

I taxied. After take-off I pressed the undercarriage up button and it wouldn't move. Easy diagnosis: the dreaded dust hadn't been cleaned off from the relevant places. OK, use the over-ride and I'll worry about it at Nairobi. I disappeared from sight as Barry packed his tools and got ready to fly out (but only as far as Egypt – he wasn't going to go further until he was sure I was somewhere near Cape Town!). As I roared up the climb I became aware that the cabin was pretty cool – no, it was actually very cold. I twiddled with the temperature control knobs, normal and manual, without success. Again, I guess more fine dust had somehow got into the temperature controller and seized it in the fully cold setting. This was potentially a real problem: I was dressed for the Tropics in just flying suit and t-shirt, and spending two and a half hours at up to 41,000 feet with the aircon stuck fully cold wasn't just going to be uncomfortable, it was potentially life-threatening. This was getting very frustrating: the little jet was still stuck in a mindset that played on the ground, but didn't want to fly.

At that stage I had a real John Cleese/Fawlty Towers moment: I gave the aircraft a very good telling-off. If I could have ripped up a handy bush and thrashed it with it, I would have. At the back of my mind I resolved to continue the climb while continuing to work the control knobs: if I couldn't get any warmth out of it, I'd have to turn round and go back to Khartoum, but that's not what I told the aircraft. I made it very clear that she and I were going to Nairobi, that I was fed up with her tantrums, that she'd better realise who was in control here and accept that that spot on the far horizon in front of her nose was where we were going.

Just as I finished my rant, out of the corner of my eye I caught sight of the left fuel gauge suddenly dropping to zero. There was one of those little frozen moments in time while I stared at it, then I looked out at the drop tanks (which have little float gauges on them so you can see they're transferring, which they both were) then back at the inert gauge and all the other properly behaving fuel system controls and indications.

"Right, that's it, I'm not pandering to your sulking any more – we're going to Nairobi even if I have to drag you there by your wingtips!" were my final words, and I stared fixedly out of the windscreen, with occasional peeps at the sky above to see if the GPS satellite constellation was now falling to earth.

The mighty Avon and everything else continuing to work properly, and we headed for the stratosphere. After fifteen minutes of juggling cabin heat controls I managed to get adequate temperature and then left them well alone. The GPSs continued to point unerringly at Nairobi, and occasional peeps out of the window confirmed the drop tanks were transferring normally; every now and again the left fuel gauge would spring back to normal, showing that it was reasonably matched with the right gauge, then go back to sleep again, as if to mock me. We droned on (if zipping at 0.85 Mach constitutes droning) across the empty southern Sudan, with just occasional chat with passing airliners on the Africa chat frequency, who obligingly gave my updated Nairobi weather in passing, along with incredulous smart remarks about why was I there.

I closed in on Nairobi. The good news is that there were no thunderstorms. The bad news was that there was a solid stratocumulus overcast, which fortunately didn't have a low cloudbase. Encouragingly good cross-check of the GPSs and the radio navaids, and it's off through the cloud for a limited panel (turn and slip only) approach pattern to Nairobi International. Long roll out after touchdown – you really appreciate the difference between TAS and IAS when landing at those altitudes! – and taxi to the pre-arranged parking spot. Refuel the aircraft as soon as possible – bit of hassle getting the right bowser, and the inevitable need to "assist" the transaction with some US dollars – and turn the aircraft round ready for the next leg (I wanted to get through to Jo'burg that day if possible).

But when I go to file the flight plan I run into the not unexpected African bureaucracy: I'm told I do not have diplomatic clearance to overfly or land in Kenya, so I am not allowed to leave. I wave my dip clearance numbers at them, and they say they are civilian clearances but I am clearly a military aeroplane. I show them the correspondence from the military and civil authorities that says I only need a civil dip clearance for this aircraft – military is not required. I see no opportunity to "assist" this discussion, and the result is a stand-off. I am informed that the matter is now referred to the President's office – yes, the President of the country, not the local bowling club – and I must wait for his decision. After five hours permission is given, but it's now too late for me to progress on my next leg since I don't want to arrive at Lilongwe in the dark. So I return to the agent, arrange overnight parking – I had to move the jet elsewhere – book a hotel, put the aeroplane to bed, and finally take the agent's clapped-out bus into town.

I'm up at 02.30 in the morning, to get everything ready for a dawn departure – my aim is to complete the remaining three legs in one day if possible, so it needs an early start. After a huge fight with the agent, who wanted me to buy him two new company cars under the guise of the knackered bus ride 'taxi' fare, I'm at my aeroplane, free of bureaucrats, crooks and the corrupt. Everything's stowed, I've done all the checks I can, I'm strapped in and waiting for dawn. At the first hint of lightening in the eastern horizon I call for start. The inverters still don't work, nor does the left fuel gauge, but I'm cool about them now. The weather's fine apart from the prevailing stratocu overcast. I line up on the enormous runway and go, the instrument lights glowing strongly in what is still effectively night. I'm tired but feel free, glad to be on my way again. I clean up the gear and flap and turn right onto a south-westerly heading to follow the departure plate before I punch into the cloud, watching the turn and slip with concentration.

Shortly I pop out of the cloud top, and it's one of those moments that's indelibly printed on my brain. As far as the eye can see there's the bubbly top of the cloud sheet, painted a rich salmon pink by the sun that has just cracked over the cloud horizon. But what captures my total attention, huge and perfectly conical in my left quarter windscreen, is Mt Kilimanjaro, the peak still above me and the top of the cone bathed in reflected pink snow. It's one of those moments you pray for a "pause" button: it was so incredibly beautiful that you just want to freeze it still and study every little detail of the marvel. I fumbled for my camera and took a shot to capture it. It's a dreadful picture because the auto-focus grabbed a scratch on the canopy so it's all out of focus, but I don't care: the colour's there and the shape, and even now it still evokes in me the same sense of wonder and beauty. In that instant, there's the reason I fly aeroplanes – because it gives me moments like that, and I realise how incredibly lucky and privileged I am.

We head south for Malawi and Lilongwe. The cloud soon gives way to scattered cloud and the African savannah rolls on for seemingly ever. The airways are quiet and the little Hunter marches effortlessly southwards: the same things are broken – inverters, temperature control that takes some talking to to get a survivable temperature, left fuel gauge AWOL a lot more than present – but everything else is fine. Lilongwe is quiet, very friendly and helpful: the bowser arrives before I've shut down and I chat with the driver and his large smile under a very pleasant morning African sun. ATC is helpful, the fees very reasonable and no need to "assist" with anything. They ask for a flypast on departure and, after a quick check of the fuel calcs for the leg, I agree. I hope I was able to give them some pleasure from seeing the beautiful black Hunter in its proper environment, in exchange for their natural friendliness and easy help.

Settled back in the cruise for Jo'burg, I suddenly became aware of changes in the aeroplane. All the usual little resonances, squeaks and rattles that I've grown used to have dropped away, and there's a detectable sense of the aircraft waking up. It's as if it's taken the last fifteen hours of flying to remember that this is what it used to do and, actually, it was rather good fun and it would like to do some more. It sounds fanciful to those who've not seen something similar, but I could feel the little jet loosening up and beginning to hum like the thoroughbred that generations of Hunter pilots know and love. The airways traffic begins to thicken up as we head into Zimbabwe. The Hunter's going well now: I set 6,900rpm, which is the magic figure for flying at range speed in a Hunter, pretty much at any height. This is giving me 0.9 Mach indicated, pretty good for a 45-year-old lady, and, checking the fuel figures, she's doing about five per cent better in fuel flow than the Performance Manual suggests.

Listening to ATC, I become aware of a South African Airways Boeing 737 about eighty miles ahead of me on the same route into Jo'burg. Fuel's running fine, weather's good, so I decide to make a race of it. Another 100rpm gives me 0.94 Mach and we're cooking along nicely. I ask for a change in flight level up to 43,000 feet to see if there are any better winds that could help – no benefit, but no loss either, so we stay there. Some time comfortably before the South Africa/Zimbabwe airspace border I overtake the B737 like he's standing still, but my cunning plan to beat him into Jo'burg is foiled when ATC insist on letting him descend first, so I slow down and take the leisurely way around the Jo'burg terminal area, courtesy of Jo'burg approach radar. The owner has fixed for me to get refuelling at Safair's site on the eastern side of the airfield, so I taxi around what seems most of the Transvaal to get there. I get a warm reception, coffee and a sandwich and plenty of help sorting out the aircraft and the ATC and customs admin, great hospitality. I'd have stayed there longer, but the end was in sight and I was keen to go.

A slick departure courtesy of Jo'burg ATC sees me back in the cruise over the emptiness of central South Africa, all strong red desert and occasional winding rivers and farmland. The airways are busy, but the Hunter's now in her element and taking every opportunity to impress with speed and impeccable behaviour – just as well, as the previous few days of unrelenting solo pressure are making me feel tired now I'm back in a good European air traffic operating environment. Over the distant horizon ahead of me appear the little white clouds that hang around the hills east of Cape Town, and I know I've only a couple of hundred miles to go. I call Cape Town approach, and receive a warm and natural "welcome to Cape Town" response in return from the brisk and businesslike controller. My mind takes another step towards relaxation.

Then he says, in his strong South African accent, "Do you have the procedures for a Whisky 2 Alpha arrival?"

I sigh, thinking I'd much rather do a visual approach, but dutifully check my charts and reply in the affirmative. "Roger, continue for Whisky 2 Alpha arrival."

A few more minutes pass and he says, "Do you have the procedures for a Whisky 2 Bravo arrival?"

I sigh more heavily, thinking I'll be really glad when this is done, and reply in the affirmative. "Roger, clear for Whisky 2 Bravo arrival."

Another few minutes pass and I can see Table Bay and the Cape, and he then says, "Alternatively, you could turn right five degrees to position on finals for a run and break: when visual, clear descend, there are no speed or height restrictions."

I thank him for understanding Hunter pilots. The arrival was a joy: on the blue note, fast and low, zipping down a line of Boeing 747 tails and past the tower before snapping into the left break to downwind, the most g I'd pulled in more than two weeks' flying. The owner led the welcome committee by the hangar, and ran up the ladder to greet me as soon as I'd shut down. That done, it took me a surprising time to get out after that: we'd come a very long way together, me and this little girl, a hard journey by anybody's standards. I'd left Nairobi very early that morning, we'd flown for 2,100 nautical miles, landed twice for refreshments, yet it was still not even 2pm in Cape Town.

I'd arrived very tired but with a huge sense of achievement. She'd arrived rejuvenated for flying and arguably rescued from the scrap heap. I just sat quietly in the cockpit for a while, meditating on the beauty of flying and how it was possible for inanimate lumps of steel, aluminium and rubber to have a personality that shared my joy and enthusiasm for what we did together. What a pleasure!'

Chapter 9

Fond memories

The Hunter earned its reputation as a reliable, rugged and versatile aircraft over many years of Royal Air Force service, and countless hours of flying (sometimes in combat conditions) with overseas air arms around the world. Although the aircraft will be remembered by many people for different reasons, few will have anything negative to say about Hawker's classic fighter. Having served with the Royal Air Force over an astonishingly long time span, the Hunter's presence within the RAF spanned the 'Cold War' years between the withdrawal of the first-generation jets (the Meteor and Vampire) through to the more technologically advanced era of the Lightning, remaining very much in business right through to the modern era of the Tornado. Having remained in front-line service through this period, the Hunter performed an outstanding job, but was never called to participate in any serious warfare. Although it did fly many operational missions in the Middle East and Far East, these were never on the same scale as more recent RAF operations in places like Iraq or Afghanistan. Consequently, the Hunter was never able to capture many media headlines although, if the circumstances had arisen where the Hunter's abilities had been needed, few would doubt that the aircraft would have been more than able to do everything expected of it, and more.

Surprisingly, one of the few headline-grabbing events that did involve the Hunter was one that had nothing to do with operational activities. It was, quite unashamedly, the result of one man's attempt to claim some publicity for the men and women of the Royal Air Force. In 1968 Allan Pollock was a serving pilot with No 1 Squadron at West Raynham, flying the Hunter FGA.9. With a considerable amount of experience under his belt (having served in the Middle East and with No 4FTS as an instructor), Alan had become increasingly frustrated at the lack of official enthusiasm to mark what would be the RAF's Fiftieth Anniversary on 1 April. Like many of his colleagues, he was saddened to see how such an important milestone in the RAF's history was going to be ignored, save for a few parades and the Royal Review at Abingdon, which was not scheduled to take place until later in the year. Alan's account is a fascinating insight into a truly unique event:

This unusual vertical view was taken during an exercise, and shows a mixed formation of RDAF Hunters and RAF Lightnings, followed by a Danish Starfighter. The picture illustrates the comparative sizes of three very significant NATO fighters, as well as the old and new RDAF camouflage schemes on both single- and twin-seat Hunters. *RDAF*

Marking an earlier anniversary, T.7 XL583 from the IAM is on display with the Hunter 25th Anniversary line-up at Greenham Common. The colour scheme is overall white with red trim. *Terry Senior*

'At West Raynham, as senior operational flight commander on the RAF's senior and the world's oldest military air squadron – No 1 (Fighter) Squadron – I felt that a particular responsibility lay with us to honour the occasion of the Fiftieth Anniversary in a suitable fashion. For months previously I had frequently badgered my station, wing and squadron commanders to obtain Command approval for our own Command, Group Wing or Squadron to make an aerial contribution. Flypasts make excellent training events, and since nothing else was going to happen from an airborne point of view (and we were scheduled on normal Forward Air Control Strike training on Salisbury Plain and 38 Group would not release us from this) I prevailed on the Boss that we should carry out celebration Anniversary leaflet raids, which would be both good training value for the pilots and a morale-raiser. We prepared our leaflets and our Chiefy had a Hunter specially prepared ready to fly immediately after the scheduled parade on 1 April.

Wearing the markings of No 1 Squadron during a visit to RAF Luqa, Malta, FGA Mk 9 XG228 later joined No 229 OCU at Chivenor before moving to 79 Squadron at Brawdy. *Richard Caruana*

F.9 XE595 is pictured at Luqa, also in the markings of No 1 Squadron. This aircraft took part in bombing attacks on the *Torrey Canyon* tanker off Cornwall. *Richard Caruana*

After the parade I changed rapidly into my flying suit from my No 1 uniform, and we "bombed up" the aircraft with a celebration war load of Anniversary leaflets in the flaps, with a special ceremonial total of fifty-seven leaflets (one for each full year of our own squadron's service) and an adequate supply of four-and-a-half-inch GP ("Government Property") toilet rolls stowed in the air brake. Normally one was always supposed to take-off with forty degrees of flaps down in the Mk 9 at heavier weights, but one could not put flaps down until "bombing" with them or the leaflets would be lost straightaway – they were hydraulically compressed into the space with the flaps fully up.

My Hunter was first up using a flapless take-off technique and, despite a strong gusty crosswind, by getting down below the side of the hangar I was fortunate to make a good drop above 54 Squadron's pilots outside their crew room at West Raynham. The station was then given a beat-up from the south, ending with pull up into a vertical roll. One of our best pilots and weaponeers, Flt Lt "Wally" Walton, was selected to head for Wattisham, another proud fighter base and home for part of the Lightning force. An ex-Lightning OCU pilot, Ken Becker, was suitably chosen for the paper delivery at Coltishall. Successful drops were also made by our ex-Vietnam F-105 pilot and USAF exchange officer, Capt Pete Albrecht, at Chivenor, and by Fg Off Barry Horton in bad weather at Valley. Both Chivenor and Valley responded well to the drops with congratulatory telephone calls. However, both Wattisham (with an AOC's inspection next day!) and Coltishall rang up and blasted our Station Commander, complaining of dangerous flying, bad example, untidy debris and poor airmanship from No 1 Squadron – a happy, successful and operationally very combat-ready unit!

I saw red, particularly with the attitude from an "operational" station like Wattisham, because I knew full well and at first hand Wally Walton's competence as a Hunter driver and live armament deliverer. Throughout the day we had only one reprisal raid, flown by one pair of Lightnings from Wattisham who at least showed some spunk by returning the compliment, even if rather inaccurately, over our wing's dispersal. Apart from the parade as at other stations, there indeed had been no special celebration for the airmen, that vital body of skills and service sacrifice that had made the Air Force great – no Anniversary dance, no party, no half day off. The head seemed to have forgotten the importance of the heart!

Later in the week I carried out a sample survey of twenty airmen at two stations, none of whom knew there was anything special about this week! Virtually nothing was in the press and I was particularly acid with the Government and perhaps less forgiving of the apparent apathy of much of our RAF to its great past. However, RAF Tangmere, doyenne of the UK's fighter stations, had laid on an excellent event for Thursday 4 April. The station had asked us to return to our pre-war home after first flying over its "Freedom of the City" parade through Chichester. The previous Friday I had developed a heavy cold and, with the impending important visit to Tangmere, I dosed myself with some strong quick-cure anti-histamine drugs bought from a chemist before being prescribed more triominic drugs by the station doctor on the Monday (1 April). These drugs, he said, would make me sleepy, but in fact had the reverse effect, keeping me ultra-alert and sleepless.

Despite the lack of sleep I still felt particularly alert on Thursday (the 4th) and was "box man" for a group of four No 1 Squadron Hunters led by Flt Lt Mike Webb, recently posted from No 8 Squadron, for the formation flight from West Raynham to Tangmere. I was apparently given the deputy leadership as "whipper-in" because I had been given my head on 1 April and in any case was to lead the whole squadron of Hunters out to El Adem in Libya later in the month.

En route we performed over Brighton's air display, later arriving on time to the second over Tangmere's parade for a small impromptu work-up with some difficult square box work over Arundel Castle. We had been told to expect visitors and the press on landing so, as was our standard operational procedure, we hid our "bone domes" on the cockpit floor and used our relaxed "boater on" traditional routine for the last part of our taxi in! Tangmere hospitality was excellent, with a cocktail party and dance attended by the Duke of Norfolk, the then No 38 Group AOC, Air Vice-Marshal 'Mickey' Martin CB DSC* DFC*, AFC of 617 Squadron of Möhne Dam fame, the Mayor of Chichester and Group Captain Pater Latham AFC, whom I had met at a 26 Squadron South African reunion and at Raynham, who was now Gp Capt Ops No 38 Group.

In a rather supercharged state and after so little sleep on the toxic drugs I had been prescribed, I filled the odd available ear about subjects dear to my heart, in particular the fighting effectiveness of the RAF and the poor celebrations during this special year. The party at RAF Tangmere was a good one – the champagne flowed! I had two and a half hours' sleep, awakened by the loud throbbing downwash of a helicopter from Thorney Island flying very close to the Mess, partly on a recovery mission to pick off our Boss's boater, which had somehow found its way to the top of a rather inaccessible Tangmere flagpole! With what turned out later to be suppressed pneumonia emphasised by the anti-histamine drugs I had been given, champagne, high spirits and lack of sleep, my mind had evidently "speeded up", and I was quite unable to go to sleep again. My mind went back over the past three or four days and the Government's decisions of the last few years. Surely I was not alone in feeling strongly that the tremendous contribution of the RAF had been sold rather short on its Fiftieth Anniversary?

With the Anniversary week coming to a close, I concluded that at least one celebration flag wave across the Houses of Parliament, Downing Street, the Ministry of Defence and, most of all, the RAF Memorial that day was quite in order. In our young Air Force we had always been taught that an ounce of timely action and initiative was worth whole weighty volumes of passive, pompous pontification. When I was a youngster, the solitary camouflaged Spitfire had done just this to commemorate both the Battle of Britain and so many unsung heroes in the Air Force's worldwide service. The weather was ideal. When I had decided upon the scheme, it was just the normal routine of how best to achieve target objectives both effectively and safely. Hell – I thought – how can any Government court martial the spirit of its most famous and historically important airfield? Tangmere was flying – I was a mere passenger in a single-seat fighter! London lay astride my straight-line return route to West Raynham. All I needed was to lose contact with the other Hunters immediately I was airborne, somehow slipping off to the north unobserved and without arousing suspicion.

The other three pilots were lukewarm about joining in on yet another ding! My route over London would obviously have to be low level. To reduce the Rolls-Royce Avon's noise and inconvenience to Londoners, I would cut my speed down and drop my flaps once I was over the built-up areas. At idling thrust, the Hunter was a very quiet bird of prey. The lower I was, the less likely I would encounter any other air traffic – not too many aircraft transit throttled back at eighty feet! Although I was obviously going to break the odd regulation to carry out my protest celebration flight, my fifteen years of flying dictated as careful and as professional a sortie as possible. A recent squadron intensive bombing programme in minimum planning, dogfight recovery, and opportunity "off the cuff" low-level navigation stood me in good stead. Without any ruler I sketched the route I was to follow on my quarter-million map, scribbling on rough timing marks for my Hunter's various speeds: 420kts, 360kts, 210kts, 420kts and climb out. Fuel figures were rechecked and headings to steer were put on. Next I stowed away my maps as nonchalantly as possible in my large flying suit pockets. Not being in a car, I didn't have a detailed map of Central London in or under this particular dashboard. However, I had wandered into an airmen's crew room and asked to borrow an AA book for a quick peep. The total inadequacy from a flying point of view of the tiny red, squiggly map of Central London discouraged me greatly, without enlightening me one iota.

Eventually, the OK was received from West Raynham about the Norfolk weather. With an acrid hiss from the Avpin engine starter system and a roar, my Hunter 'H' Hotel XF442, a substitute for my own 'Bravo' 'B' Flight commander's aircraft, started up with the others. At once I felt relaxed and safe in my homely cockpit environment. We checked in on Tangmere's local traffic radio frequency and began taxiing out slowly, the last section of the last Hunters and fighters of the RAF to fly into and out of our nation's historically greatest fighter airfield. I was to take-off as number four in the lead section of four aircraft. A singleton Hunter with our squadron commander, Sqn Ldr Spike Jones, on board was taking off immediately behind me, making five in all, all FGA.9 single-seat fighter ground-attack aircraft.

We turned onto the runway after carrying out our normal vital pre-take-off checks – "bang seat" live, trim, fuel, flaps, instruments, oxygen, hood, harness and hydraulics. As we accelerated, rolling down Tangmere's runway in two pairs and a singleton, with a five-second stream spacing, my main pre-occupation was how best to break away from the rest of the section without attracting anyone's attention. I was aware that a pair of eyes was looking down my jet pipe as we took off to the east. Immediately after take-off and the reassuring, quiet rumble and clonk had signalled the wheels locked up, I watched as the Hunter behind me turned belly-on to me to beat up the airfield as a farewell fighter tribute to such a legendary airfield, symbol of fighter excellence – what a splendidly good and indiscretionary Boss! I knew that I would not now be seen by him!

I eased my throttle back, reverse rolled and slipped gently away from the others towards the north, descending rapidly to low level to melt my Hunter's camouflaged drop tank deep into the dense eiderdown of the Sussex countryside. After an intentional one-and-a-half-minute delay, I then told the formation leader, with speechless code on my R/T button, that I had "lost" visual contact and pretended my radio had failed, using speechless transmissions to avoid any further embarrassing conversations on the R/T. With things to do, I certainly didn't need a dialogue on the cancellation of TSR2 or the rights and wrongs of initiative!

Before taking off on my way north I had realised that my route would take me close to Dunsfold's idyllic, rural aerodrome, the airfield home of Hawker where such a small team had assembled and tested so many fine and famous aircraft for the RAF, including my own much-loved Hunter FGA.9. Aware that my flight would have certain consequences, I had decided to pay them a brief Anniversary visit, and after a careful check for other air traffic I streaked low over Dunsfold. Two brief minutes later I was at my "initial point" over the lakes three miles south of London Airport. As I banked to the right, turning clear and safely below London Airport's approach lane, I saw a stately Boeing 707 glide-sloping its studied, final approach onto Heathrow's westerly runway.

Crossing the green area of Richmond Park, I joined the silvered-brown Thames, following its sharp, sinuous bends down-river in a melody of delightful Fairey and Dowty banked reversals. Over London the weather was still one of those rare perfect, 8/8, Gordon's-crystal-gin-clear days when all the colours shout out brightly. Definitely cerulean, not just azure. There was not a breath of wind, and the blue sky shone cloudlessly. For obvious reasons, at this stage my crumpled quarter-inch map of the route became useless. I had no 50,000 or 100,000 target map on this particular sortie!

With bursts of flap and power, I now concentrated all my attention on flying down the centre of the stream, startled by the varied beauty in the arches of each bridge just seconds apart. I swept round over Wandsworth, Battersea and Chelsea bridges, keeping a special eye open for any helicopters. Straightening up after the next bend, I crossed Vauxhall Bridge and there, a mile or so ahead, was that familiar and splendid silhouette of the Houses of Parliament, viewed within a special sense of Triplex privacy. This was my main target area. Descending to around 200 feet, the lower I got over the London houses, for safety I pulled my

A Swedish pilot at the controls of his Hunter Mk 50 (known as the J-34 in Swedish service). Hanging a 'bone dome' helmet on the aircraft's pitot tube probably was not part of Swedish regulations!

Hunters racing across the countryside at 250 feet were a common sight across Germany and beyond during the 1960s and early 1970s. Here a reconnaissance image taken from a Hunter FR Mk 10 shows a reconnaissance Hunter from No 2 Squadron on a low-level sortie.

speed back to be extremely quiet – part of the thought process was that that the lower you are to some great extent the more safe you would be, and also it is easier to see other aircraft above too, where they are better silhouetted against the sky.

Speed was around 320 knots with two notches of flap going down, and as I banked over, climbing slightly to circle Whitehall and the historic seat of British Government, I had to open my throttle fully to maintain a good tight orbit. I realised that this would put the noise level up considerably, which up to now I had been careful to keep as low as possible to avoid offence or complaint. As I put the Rolls-Royce power on, I decided that this was perhaps what was really necessary at this juncture to wake up our MPs and remind other august figures, sitting chairbound at their ministerial desks below, that we still had a fighting Air Force, one small unit of which was celebrating its Anniversary, despite the dead hand of Government policy and the sickening cut-backs of previous years.

The message to Westminster was received. A debate was interrupted. Later, a four-MP cross-party inter-service motion of support for the tribute was ruled retrospectively out of order. Quashed, it was reportedly deleted from Hansard. Three times I circled, fascinated by the scene below like any Gulliver looking down at Lilliput, and with not dissimilar motives from its author either, exactly as Big Ben struck 12 noon.

Having been abroad for some years I was now conscious of the high Vickers Millbank tower block, which came as a surprise to one who had not been in Central London for some long time. Its bulk spoiled the even radius of my turns. Quite suddenly I felt a little tired, rather annoyed that, for once without a target map, I could not readily pick out No 10. With care I kept well clear of Buckingham Palace, which was plainly visible outside my orbit. Surprisingly there was time to wonder idly in the final turn exactly how many regulations I was breaking at this rather exclusive and delicious moment. After my final orbit I levelled out again over the Thames, dipping my wings past the RAF Memorial and coincidentally the statue of Viscount Trenchard, who had helped to found the RAF in 1918 exactly fifty years earlier, despite tremendous opposition.

With a quick glance at my fuel I decided that, rather than cross the heavily populated area to the north or climb through the airspace used by inbound airliners, the safest plan for all concerned was for me to follow the Thames eastward at low level out into a less populated area, then turn north for home. At one minute past noon, I turned happily down-river, descending through 100 feet. Over Waterloo Bridge I could see the architecturally pleasing Post Office Tower away over to the left, then suddenly ahead a fine view of St Paul's Cathedral, at that time with its latticework curtain of scaffolding. Despite thousands of sorties in Europe, Africa, Asia and the Middle East, I felt now like some voyeur or landscape artist, so intoxicated with the various views to the left and right of me that it was literally just before my wings crossed London Bridge that I looked ahead and suddenly saw Tower Bridge. There, standing like some proud, strong sentinel across the Thames, this famous matronly structure blocked my low-level path to the east.

Until that very instant I'd had absolutely no idea that, of course, Tower Bridge would be there. It was easy enough to fly over it, but the idea of flying between the towers suddenly struck me. I had just ten seconds to grapple with the seductive proposition that few ground-attack pilots of any nationality could have resisted. My brain started racing to reach a decision. Years of fast low-level strike flying made the decision simple. From a trained low-level pilot's point of view it provided an interesting penetration problem. To give added decision time I jinked hastily to line myself up directionally and pre-position down low over the river, with my eager fighter well beneath the tall cranes standing like silent, puzzled spectators on the banks. This manoeuvre would give me a full extra three seconds of decision time to study defloration further.

Down at just fifty feet, I could see that there was considerable road traffic, including a red double-deck bus, slowly lumbering across the famous, double-basculed bridge from north to south. With less than half a mile or so to run, I realised that it would be easy enough to fly through, but what would be the best and safest way? Swiftly I concluded that Sydney Camm's favourite fighter would have to fly as close to the top structure as possible – rather like a reverse skip bombing run with target cues above! At the last split second before I passed at around 300 knots, the steel girders suddenly seemed to explode all about my cockpit, above, below and about my ears, totally engulfing the canopy and one's traditional sense of flying fun! In that microsecond my mind felt quite certain that I had overcooked it and the top span would certainly take my fin off.

Something then happened to me that had only occurred once before, when I had "mushed" after pull-out from an FAC attack with over-sufficient aircraft weight and "g" and insufficient speed, power and thought! Thinking I had hit the ground, but missing Cloud Ninety Nine by a whisker, my heart actually had missed a couple of beats with the shock of expected imminent disaster. After that there was the acute, physical reaction as the heart fires up to full stroke again, just like a fighter's fuel pump, trying hard to catch up. My Hunter flew on, rather unexpectedly finding itself still completely functional and not a finless wonder, and I headed out over Greenwich and Hornchurch, heading towards Clacton.

With little to hide now, I opened up a separate Anniversary sortie call sign on the R/T, Romeo Alfa Foxtrot 01 (which was rather lost on the controllers). Then, with R/T permission from the ground, ceremonially I beat up RAF Wattisham, home of some of the Lightning force and whose Station Commander was decidedly an accessory before this particular crime!

Lakenheath was next on my list, and a Mach .93 dive-brake sonic-boom-letting where No 1's fellow squadron with its parked USAF F-100s was based. Lastly, without really the fuel to do it justice and only after a very careful R/T check on the position of a Victor on its extended final approach lane, I flew across RAF Marham. Thus the flight had correctly covered both halves of the former Bomber and Fighter elements that were just now combining as the new "Strike Command". Sic transit gloria.

With less than 400lb of fuel now remaining, I carried out a rather hurried, inadequate inverted run over the squadron hangars at RAF West Raynham before breaking downwind, punching down the gear and landing, with the brake parachute bobbing about contentedly behind my precious fin. Somewhat pensively I taxied slowly in, expecting, like most other privileged fighter drivers in their time, that a formal reception committee might already be there, as on one or three previous occasions in my flying career.

Only the normally cheerful airman was there to greet me onto my chock. Cutting the HP cock, the Hunter's Avon shut down with that unique intermittent cacophony of cooling blades and tolerances like some giant desyncopated lawnmower. Feeling rather pleased that I had a few more minutes before the thunder of the gods was unleashed, I asked the lad if he'd mind getting my overnight bag out of the radio bay, something I invariably did for myself. While he proceeded with this task, I had the opportunity of doing what I wanted to but thought I might not get any chance to. Professionally, I felt thoroughly ashamed of the scribbled marks and tatty appearance of the quarter-million map on which I had navigated up from Tangmere without problem. Certainly I did not want this dog-eared relic produced as an exhibit before some bleary Group Captain at my, or should I say Tangmere's, court martial. So I casually walked into Wing Ops, borrowed some matches and burned the map outside, out of sight. Next I went to ground in the MT section to chat up the switchboard girl and put through private calls to my wife, my wife's mother and my parents. I told them what had happened and why my Anniversary sortie was flown, that I would be under close arrest for at least a couple of days and on no account were they to say anything to the press if badgered by them. The station telephone operator somewhat apologetically said there was a small delay as they'd just had two "Lightning" priority calls, quite outside her long experience.

"The balloon's possibly gone up," I said dryly, without too much revelation to her.

Then I went back to the squadron, climbing upstairs to tell first Boss Jones, who was with our OC Flying Wg Cdr Ron Wood, and later the Station Commander, Gp Capt Basil Lock, what they knew already. By nature used to fast-moving situations, they took it all coolly in their stride. Spike Jones seemed strangely relieved that Whitehall wasn't littered with leaflets! At the end it was left to me to suggest that perhaps I ought to be placed under arrest! The section leader and section downstairs meanwhile were in a slight panic – all five aircraft had taken off, verbally authorised but certainly not signed up on any authorisation sheet. Begged to follow noble traditions and "cook the books straight", and mildly amused at their level of fear, I obliged by signing up! Next I went up to the Mess with the Wing Commander Admin, another splendid officer also called Wood, after the Station MO had examined me, finding me absolutely normal.

After such little sleep, I was restless. Exhausted but unable to sleep, at 1.30pm I asked the doc for something to help me sleep. This was refused and, as was apparently normal service procedure on occasions like this, I had to wait a further seven hours for them to get hold of an RAF psychiatrist. He eventually arrived at 8.30pm and, after an hour or so, pronounced me quite definitely of sure mind, but in a rather tired and cold-ridden body. After a day and a half I became rather fractious at being held under close arrest for obviously "political" reasons and asked to be released to open arrest. My deaf AOC of last month was now even deafer. With my spirits high I then started playing up, locking up my officer guard. However, the press made it uncomfortable and I remained cooped up inside in a state of siege. To take the heat off I was taken to Nocton Hall.

Always an athletic climber and in high spirits – fourteen years earlier I had regularly climbed with, but below, fellow course member Chris Bonington – I climbed out and onto a roof unnoticed and wickedly watched with some glee the intriguing denouement below. I was transferred to Wroughton where the suppressed pneumonia following my terribly heavy cold broke out. A combination of drugs and disillusion now slowly sapped my spirit.

A month later a Summary of Evidence took place, which lasted five days. With the squadron in North Africa and almost incommunicado, my morale was shaken. Two weeks or so later, despite the strong recommendation from Wroughton that I was quite fit enough to stand Court Martial, for political expediency the Under Secretary of State for Air pronounced in a press release that I would not be tried by Court Martial, but would be invalided out of the Air Force.

Having dispensed some Air Force justice myself, I was astounded. I was very angry and repeatedly asked for what I knew was my statutory right, to see my AOC-in-C. This opportunity was denied me. I was told that if I did not accept the invaliding out, with the inducement of my small pension, my services could be dispensed with under a certain Queen's Regulation, without any formal disciplinary action or comeback. With tremendous resolve I fought back in a sinister game, the bureaucratic dice of which were increasingly loaded against me for reasons that were only to become clear fairly recently, when the Hugh Cudlipp and May 1968 'Whitehall plot' against the PM was revealed. The bizarre circumstances included a later arrest; the clear statement of my true position and the drug situation at the Summary of Evidence in front of four witnesses and in writing; the cancelled Court Martial, which I had openly sought and had been declared fully fit for by the specialists; the copy of my Summary being forcibly taken back, then later "destroyed" so it was never found again; myself and the country being told of the outcome of a Medical Board in the press by the Secretary of State for Air before even a board had been convened; the refusal of any help in a ten-year cover-up; the complete and absolute change of Medical Board findings before and after discharge; the tax position switch after a threat to expose the British Government in Zurich; the persistent withholding of any right of appeal until 1982 when my case was finally vindicated; and the total refusal of my AOC-in-C to allow me an officer's sole right in this sort of situation – the right to see him!

Following some friendly counsel, I realised then that I had to accept Hobson's choice and decided that my next move should be after I was actually receiving my pension. Then I would petition against what I knew was an unjust decision, taken to avoid any public enquiry into the Government's other arbitrary decision to cancel the Anniversary Flypast. I received more than a hundred letters from the general public and all strata of present and former members of the RAF, as well as members of the other two service arms I admired most, the Royal Navy and the Special Air Service – both of which I had operated with – and they bore witness to the fact that a lot of other people were not satisfied either.

Subsequently, after a painfully difficult career "re-light" in 1968/69 and having been labelled like an intellectual leper, my languages and aggressive fighter training gave me a perfect career base for success in exporting. Tomorrow is a new day!'

Mike McEvoy was one of countless young RAF pilots who was fortunate to have flown the Hunter in its heyday. As he recalls, his affection for the Hunter has stayed with him throughout his life, eventually encouraging him to track down his 'personal' aircraft many years later, on the other side of the globe:

'I must have first seen the prototype Hunter WB188 at the SBAC show at Farnborough in 1951, but my abiding memory is from the following year when the DH.110 of John Derry and Tony Richards had broken up in mid-air after planting a sonic boom on the airfield, with the loss of several in the crowd as well as the two crew. There was a period of silence while the emergency services did what was necessary, and when it was announced that the flying would start again with Neville Duke and the Hawker P.1067, many of us on the flat near the crowd line drifted, probably only half consciously, under the wing of the Blackburn Universal Freighter. Duke would be "breaking the sound barrier", a phrase that still carried an air of slightly fraught mystery to the general public; the dramatic British film with that title – although it featured the aircraft of Hawker's traditional rival Supermarine – had been released only six weeks previously. Other than the anticipated boom there were of course no other effects, apart from audible relief, and the 110 break-up turned out to be due to something else altogether. Nevertheless, Duke's flight carried an undoubted aura of heroism and even derring-do in an age when these were still admirable qualities.

Two Farnboroughs later I'd almost finished my square-bashing at Kirton-in-Lindsay preparatory to going to Canada for flying training, and the P.1067, now with the suitably predatory name of Hunter, was in service with 43 and 222 Squadrons at Leuchars. No 43's black and white checks had featured in all the aviation press, and it was obviously part of a future that nearly all our course aspired to (there were the odd one or two that saw their future on transports, but they were the old men of twenty-five, or even older!).

Returning to the UK at the end of 1955, we were taught to fly, and navigate, in English weather on sunny Teesside where the Middlesbrough smog gave an added challenge to finding the airfield while still downwind. We awaited our postings, and mine was to 229 OCU at Chivenor.

For the first half of the No 15 Hunter Course we were still flying Vampires, notionally at least learning battle formation tactics and how to fire the guns, though we never quite got to the gun-firing. Meanwhile we studied the Hunter ground school syllabus, and surreptitiously had "hero pictures" taken next to one of the F.1s on the flight line. There were no two-seaters yet, so our "practical" training was in a "Cockpit procedures trainer", a form of sub-simulator that looked just like the real thing but on which nothing actually moved; the instructor at our shoulder made sure that we knew where everything was and could touch it on demand, and that we were word-perfect on the relevant checks and emergency procedures.

Then, on 28 May, I strapped into WT624, started the Avon, with an instructor still on my shoulder to make sure I did it properly, and taxied out to the far end of runway 09. The procedure was to run the engine up to 7,000rpm on the brakes – which gave as much thrust as they could hold – and, on releasing the brakes, to open the throttle to 7,900. This we were assured would give us a kick in the back the like of which we wouldn't previously have experienced; one of us was baffled, therefore, when he didn't get the expected kick, although his take-off was successful, and he only later realised that he hadn't increased the rpm by the extra 900. Mine was as exhilarating as expected, though like all first solos my wings waggled while I got accustomed to the power-assisted ailerons, and those watching could always tell. We were very grateful for this assistance once we had experienced the "manual reversion" mode, a necessary fall-back if the hydraulics failed, but one that needed considerable physical strength on the stick.

My first flight lasted thirty-five minutes, the length of most of our Hunter sorties at Chivenor. The F.1's tankage, typical of the early mark of almost any British fighter, was very limited, and even with eighty gallons more internal fuel the F.4 would normally only record fifty minutes or so per trip. I once achieved forty-five minutes in an F.1, and once an hour in an F.4, but I suspect I had one eye on the fuel gauge and was twitching a bit by the time I touched down.

Chivenor had one of the two best weather records of any UK airfield – the other being Lossiemouth – and I remember the summer of 1956 as being so hot that it needed the little streams that ran through Saunton Sands to keep the bottles of Carlsberg, a newly discovered taste, suitably cool! As a rite of passage we got to make a sonic boom. On 1 June WT688 and I positioned ourselves at 40,000 feet over Lundy on a westerly heading to ensure that only the Bristol Channel seabirds, and perhaps the occasional fishermen, were disturbed, and put the nose down. The Machmeter needle jumped as briefed, and we pulled out somewhere around 10,000 feet. Two years before this had been the prerogative of test pilots, and it was heady stuff for a twenty-year-old, so just to be sure WT697 and I repeated the exercise four days later.

By mid-1956 most UK-based day-fighter squadrons had been re-equipped, and in Germany the Hunter had replaced all the Sabres and some of the Venoms. My posting was to Bruggen, one of four airfields with the type, to join 67 Squadron, on which most of the pilots were ex-Sabre jockeys. As the junior pilot on the squadron one of my primary functions was coffee-boy (Standard NATO, two sugars and condensed milk), a post that I seemed to fill for most of my time on the unit, all those posted in after being senior to me! Those of us in the Second Tactical Air Force, a title with a proud operational record from a decade before, knew we were in the best place as, after all, the threat would come from the east, and we were the first line of defence – Fighter Command was just there to reinforce us if necessary.

A few memory snapshots from the flying. The Möhne Dam was in our Low Flying Area, with restrictions, I think, of 250 feet and 250 knots, and, apart from making suitably low passes at it, one of my less admirable habits was to lower twenty degrees of flap and the undercarriage and make a "carrier" approach towards the stern of one of the large barges on the waterway before overshooting. One of the great perceived advantages of serving in Germany was the duty-free prices on drinks, particularly in the mess bar, and when we went to Sylt, as did all the 2nd TAF squadrons, for gunnery practice, apart from finding out that I was not a natural shot I remember that detachment as being the only time I ever flew with a hangover, giving myself a serious fright returning to land in low cloud. With 2TAF divided into two Groups, 2 (North) and 83 (South), we had regular joint "war" exercises; the casus belli was that one side or the other had treacherously seized control of "Abyssinia Beach", the stretch of sand at this holiday resort traditionally reserved for nudists.

One of the firm precepts drummed into a young fighter pilot was that a wingman should always stick to his leader, and when Terry Filing had to abort his take-off halfway down the runway, as a faithful number two I braked hard and turned off the end of the runway behind him. As I was taxiing round to take-off again – I forget what the mission was, but I must have had a plan at the time – I didn't pay enough attention to the R/T call, "Hunter passing Station Flight, your brakes are on fire." With the habit of listening only for the appropriate call sign – it's probably no longer contrary to the Official Secrets Act to reveal that at the time I was Monkey 113 – it took a second or perhaps even a third transmission for me to realise that it was my Hunter being referred to, and I had the engine stop-cocked and everything shut down with much quicker reactions. I don't think I waited for a ladder to get me back on the ground.

While we were on the Chivenor course *Aeroplane* magazine published a colour spread of RAF squadron markings, and when Dick Whittingham and I got our postings to Bruggen we looked up 67 and 71, our destinations. Somewhere the allocations had been reversed, but thinking I was going to 71 I made a disparaging comment about the "primary colours" of 67, which came back to haunt me after a couple of months. Of the others, 71 wore black, yellow and white and sometimes the eagle from its Second World War history; 130 "Punjab" Squadron had an elephant's head between its red, white and blue rectangles; and 112 had the sharkmouth made famous on its Desert Air Force P-40s (and somehow they were usually chosen for any publicity photos).

I had the temerity to mention to the Boss that our only identification was a light blue/grey letter on the fin, and could we apply the markings that had been devised for the Sabres? I've never been quite sure if he thought it was a good idea, but in the best service tradition I was given the task, starting with a test on "my" aircraft,

XF317/U. I think my motivation was to make our aircraft instantly recognisable, so that although the markings were applied on the nose beneath the windscreen they were much smaller than those of 71 or 130. Our badge featured a drongo, sometimes a term of abuse among Australians but in fact a small Malayan bird that is fierce in defence of its young and its territory; this referred to the squadron's Second World War service when, equipped with Buffalos, it had been part of the defences of Malaya against the Japanese invader (by the end of the war it had Spitfire VIIIs and had fought back to Burma). I schemed a simplified badge flanked by the red/yellow/blue rectangles devised in F-86 days, and once approved by the Boss application to the other squadron aircraft started, beginning of course with his WV403/J.

We only had about five marked up when the need ceased. As a part of the implementation of the 1957 Defence White Paper, nine Germany-based Hunter squadrons were disbanded in April and May. No 67 was among the first, and I made my last Hunter flight on 16 April as part of the squadron's farewell tour of 2TAF, before returning to the UK for alternative employment. Much the same happened with XF317; it was briefly on the strength of 229 OCU, but not long after was transferred to Halton as an instructional airframe with the maintenance serial 7773M, and I was to be reunited with it twice. The first time it was still at Halton, and from a visit by an aviation group in the late 1960s I have a picture of me, very obviously civilian in hat and beard, sitting in its cockpit with it still wearing my name and 67's markings. The second time was on the far side of the world.

Hawker – or more properly Hawker Siddeley – had found a considerable export market for refurbished Hunters, and was buying up low-houred airframes wherever it could find them. 7773M was taken to the old Armstrong Whitworth plant at Bitteswell, from which it emerged months later as an FR.71A with a 200 series Avon, a 'dog tooth' leading edge and a camera nose, and doubtless many other replacements and improvements. It then joined the Chilean Air Force as FR.71A J-734, serving with 8 then 9 Grupo. It was in their final Hunter flypast, and I'd heard persistent rumours of its survival, confirmed in 2002 in a photo looking slightly worn in some rather long grass and reputedly in the hands of the Chilean Aviation and Space Museum.

Hunter F.4 XF317 enjoyed only a brief flying career with the RAF before being assigned to ground training duties. It was re-purchased by HAS in 1971, converted to FR Mk 71A standard and sold to Chile as J-734. It was retired to the Chilean Air Force Museum, where it is seen here. *Mike McEvoy*

Former Chilean AF Hunters in open storage at Los Cherillos. *via David Griffin*

Researching a trip to New Zealand, I found that it was possible to fly via Santiago, and we gave ourselves four days stopover to find it. Through the good offices of the Museum Director, I was reunited with it on our last day, though at that stage it was still on the books of the Air Force and being looked after with the other three from that final flypast by 9 Grupo, now a helicopter unit. Photos were taken, and in them both of us show our age a bit; indeed, both of us were a little down on one leg. After all these years I can probably confess safely that I wasn't more than competent as a Hunter driver, and didn't stay with it long enough to be seriously considered an operational fighter pilot. But it was a splendid experience at that age – I turned twenty-one on my Hunter squadron while on detachment to Zweibrucken with 434 Squadron RCAF – which I've never forgotten, and while I spent less than a year with the Hunter, it had a, possibly disproportionate, effect on the rest of my life. I remember the exhilaration, and the occasional fright, and for me it remains, both visually and emotionally, the most attractive of aeroplanes. Thank you, Sir Sydney!'

Perhaps the final words should be those of David Lockspeiser, one of Hawker's famous and much-respected test pilots, who was responsible for countless hours of Hunter test flying:

'Inevitably a greater affection develops for an individual aircraft with which one becomes particularly involved, and I can think of three. One was XE588, a Mk 6, which was used for most of the single-seater demonstrations in Switzerland and the associated armament development work at A&AEE. So impressed were the Swiss with the Hunter's ability, at high all-up weights with external stores, to turn in narrow mountain valleys that when we turned up at the airfield at Meiringen one morning we found that they had painted the Swiss mountaineers' badge on the side.

Not all the armament development was carried out at the A&AEE's usual Porton range, however, and the Swiss, in addition to holding trials with their own bombs and rockets, also wanted to have demonstrated the safe release of napalm bombs. This was to be carried out over Lake Payerne and, to satisfy the authorities, napalm itself could not be used because of danger to the fish. Water was unacceptable because of its higher specific gravity, and a reduction in quantity would mean an unpredictable change in the tank's centre of gravity. The authorities came up with an answer – a liquid that had the same specific gravity as napalm. To my chagrin I have to relate that two drop tanks each containing 100 gallons of Liqueur de Poire William were dropped in the lake.

The second aircraft was G-APUX, our two-seater demonstrator, which was used at home and abroad for demonstrating at air shows, flying politicians, potential customers, conversion of customer pilots, and service officers of all ranks. At the time of the Hannover air show this included the Minister of Defence and Goering's former deputy, ex-Luftwaffe Field Marshal Milch, and fighter leader General Adolf Galland. Demonstration routines were of course varied to suit the occasion. The performance and the agility of the aircraft was demonstrated, often including a spin from altitude followed by a supersonic dive during recovery then levelling out to display the deceleration that could be achieved from high speed by throttling back, popping the speed brake, dumping full flap and, trimming nose-up, which was quite impressive. The flap on the Hunter, like the Swift, was designed to be the speed brake, but unlike the Swift it gave a strong nose-down trim change, which is why the existing external speed brake was introduced. Also, strangely, the petal-type brakes that were effective on the F-86 gave problems when tried on the Hunter.

XE601 is about to embark on a chemical attack simulation flight during a NATO exercise at Waddington. The colour scheme is Light Aircraft Grey with red trim, prior to the application of the more familiar 'raspberry ripple' colours. *Terry Senior*

A sight that is unlikely ever to be repeated, even with more modern combat aircraft: two RDAF Hunters perform a mirror-image formation over a densely populated area of Denmark. The Hunter's crisp and responsive controls made it ideal for aerobatics. *RDAF*

One incident I recall at Hannover following a photographic sortie during the week of the air show was when I returned rather low on fuel and thus made a straight-in approach. A problem presented itself when selecting the undercarriage down – the port leg remained up. From our Service Dept reports I was aware that this had happened to an RAF Mk 9. The pilot had opted to jettison the 230-gallon tanks and then do a wheels-up landing; however, the shock to the airframe caused by firing the ejector release units [ERUs] brought the leg down. G-APUX unfortunately was not fitted with ERUs, so I could only induce a shock to the airframe by banging the starboard wheel hard onto the runway. Fortunately this had the desired effect and I completed a tightish circuit and landed; refuelling revealed that there were only twelve gallons remaining. The following day I received a relayed reprimand from our Director at Kingston, to tell me to return with more fuel in future, as though that would have made any difference, but such is life!

Finally there was 510, a Mk 74 B/S, which was the aircraft on which most of the Lockheed-designed modifications were carried out for the Singapore Air Force. It was a comprehensive programme greatly enhancing the combat and reconnaissance capability, with a twin-store pylon carriage centre-line station, Sidewinder and increased multi-store carrying capacity. As it happened I had carried out the first flight on this aircraft at Dunsfold on 30 November 1956 as a Mk 6, XG296, and I also flew it on a centre-line twin-store (bomb) handling flight from Tengah on 30 November 1977, its twenty-first birthday, so to celebrate this event in the aircraft's life I produced at "the flights" a bottle of champagne.

On one occasion, assessing the handling during a five-g dive recovery, an attachment failure occurred and one bomb fell off – fortunately missing my chase pilot and the Indonesian sampans that were cruising around below in the Strait of Malacca. This was found to be due to faulty material.

People sometimes ask which of the many variants I enjoyed flying the most, and the answer is the Mk 6, because it combined the power of the larger 200 series Avon with a more aft centre of gravity. The Mk 11 was also delightful, for although having the smaller engine it was much lighter, not having any guns, while having the same

A 54 Squadron Hunter FGA 9 pictured while visiting RAF Luqa, Malta. Clearly visible are the underwing rocket pylon attachment points and the bracing strut attached to the fuel tank.

Hunter PR.11 XF977 sports an interesting code presentation, which appears to be made up of three 9s, one of which has been flipped over to produce a rather odd-looking 6. Also visible are the fuselage-mounted practice bomb pylons. *Paul Tomlin*

FGA 9 XJ692 provides an interesting view of the aircraft's underwing pylon locations, as the aircraft breaks away from the camera ship. The gun muzzles (which have not been fitted with deflectors) are surrounded by a familiar patch of soot.

> centre of gravity as the Mk 6. With the heaviest fixed gun armament of any aircraft, four 30mm Aden (I think the Me262, which of course preceded the Hunter, was the only other), its genuine multi-role capability, excellent handling characteristics, ease of servicing and unmatched elegance made the Hunter the best in its class.'

There are few combat aircraft that can be credited with so many years of operational service as the venerable Hunter. Emerging from a troubled development process, it proved itself to be both a capable fighter-interceptor and an impressive ground-attack 'mud mover'. Its service life with the RAF and FAA extended many years beyond projected retirement dates, simply because there was no other aircraft better suited to the tasks that the Hunter performed so well. Overseas, Hunters achieved remarkable export success, often emerging as 're-born' Hunters, embarking on a second service life after having completed their duties with the RAF and other air arms. Perhaps most remarkably, more than fifty years on, Hunters are still very much in business, not only as hugely popular solo jet performers at air shows around the world, but as modern threat simulators, flying with as much flair and agility in a modern world of high-tech aerial warfare as they did decades ago. It is fascinating to see that, for some specialised roles, military air arms are slowly accepting that when it comes to finding a replacement for the magnificent Hunter, there is only aeroplane fit for the task – another Hunter!

Chapter 10

Overseas buyers

The first overseas interest in the Hunter came from Australia, where serious consideration had been given to the production of the P.1081 and, subsequently, the P.1067. However, a combination of sales pressure from America, and the troubled development of the Hunter, eventually persuaded Australia to opt for the F-86 Sabre, an aircraft that had proved itself as a capable fighter during the war in Korea, often operating alongside Australia's obsolescent Meteors. European interest was also fairly high, especially after Neville Duke demonstrated WB188 at the Brussels Air Show in 1952. Two American NATO pilots came to Hawker and flew the prototypes and Hawker diverted a significant amount of time (which would otherwise have been devoted to the overstretched development programme) to demonstration and evaluation flying. However, Hawker's efforts were successful, and both Belgium and the Netherlands negotiated arrangements to manufacture the Hunter in their home countries, although funding for the aircraft was ultimately provided by America, from its Offshore Procurement Programme.

Rather more surprising was Sweden's interest in the Hunter. As a non-aligned country, Sweden was expected to take a truly international view of suitable fighter designs and could have realistically opted for various aircraft (including indigenous projects) but, much to the delight of Hawker, the Hunter was chosen. Despite purchasing the relatively unsophisticated Mk 4 (which did not even take advantage of the 'dog tooth' wing leading edge extension), Sweden enjoyed a very satisfactory association with the Hunter, which provided the country with a capable interceptor to defend the 'no man's land' between NATO's northern flank and the Soviet border. Erik Johannson describes his own association with the Hunter in Sweden:

> 'As I had been interested in aviation from my earliest years I applied to join the Royal Swedish Air Force [RSAF] when I was called up for my compulsory military training. As luck had it I was posted to the RSAF base near Stockholm. After the initial military training I was appointed as an Assistant Engineer and that was when I met my first love, the Hawker Hunter. As an Assistant Engineer I got some very basic training on the aircraft and was eventually qualified to do the daily check and turning-around of the aircraft between flights – not a big deal as the Hunters, which we had received only weeks before I started my military service, needed very little attention. Like many aircraft, leaving them unattended for periods of time allowed oil leaks and other snags to develop and we had a saying that the Hunter worked best if it was never allowed to cool down!
>
> The Licensed Engineer I was assigned to was responsible for one particular aircraft, "Adam Blue". Before I finished my year of compulsory training I looked at that aircraft as my own, and when the Squadron CO needed an aircraft he also chose "Adam Blue". I must have done rather well at this as, a week before I was due to finish my service, I was asked by the Squadron Commander if I would be interested in joining the squadron on a permanent basis. More as a joke I replied that, providing I could keep "Adam Blue" as my aircraft, I would accept. This I was promised, and the day after I left I was back starting my three years of apprenticeship.
>
> The two first years were spent as an Engineer Trainee followed by a year of basically classroom training. Provided you had a satisfactory report on your work from the Squadron Commander and passed your exams you were now promoted to Engineer. Having then got your type training you were then promoted to Licensed Engineer and could then be given the full technical responsibility for up

to four aircraft. This included the aircraft, the engine and all systems except certain electronics and the armament. A Trainee Engineer was allowed to do most of the basic work and limited trouble-shooting, but all his work had to be inspected and approved by a Licensed Engineer.

When I say that the Hunters performed exceptionally well, I measure this against my experience with the J29 Flying Barrel, the A32A Lansen, and the J35 Draken. The squadron did all the smaller checks and routine rectifications and repairs. All major checks and major repairs went to a special unit that did nothing else. Apart from working on the ejection seat the only unpleasant job one had to get used to was to inspect the first stages of the engine compressor each morning. We used a ladder to get up, then slid down the air intake to do the job. For a big lad like myself, at 6ft 6in, at standard temperature there was not much elbow room in there. To get out you had to snake backwards and slightly uphill, which was a bit of a job. When we got used to these inspections we simply took a few steps back, ran up to the air intake and dived into it. With this speed we slid all the way in to the compressor very easily. I really enjoyed keeping my "Adam Blue" in showroom condition. It was a pleasure, and "Adam Blue" stayed with me until the day I left the RSAF.

As I mentioned earlier, the Hawker Hunter served us exceptionally well. It was not only the pilot's aircraft but also the engineer's. We experienced very few technical problems as the Hunters we had were equipped with the mufflers on the cannon nozzles, which prevented the engine compressor from stalling when the four Aden cannons were fired, the ammunition link collectors, which prevented the ammunition links from being sucked into the engine air intake, and the later engine starters driven by isopropyl nitrate rather than the dreaded cartridges, one of which took the hand of one of my colleagues. I was never trained on the latter system as none of the Hunters we received had the isopropyl nitrate system. The first Hunters delivered to Sweden went to another base, F18, south of Stockholm, and they arrived with the cartridge starting system. I know that there was a holder for three cartridges and when these were spent they had to be replaced. The cartridges were filled with cordite, which, when ignited, produced a vast amount of gas very rapidly. It really was like an explosion. When replacing the cartridges the mechanic could reach them by sticking his arm in through one of the three access panels under the aircraft. In this particular case the cartridge fired as the mechanic was installing it. This we were told at the routine morning briefing as a word of warning should we ever be asked to service a non-modified Hunter.

The squadron I was attached to moved, shortly after I joined, from a conventional hangar into a hangar blasted deep into the mountain with room for two squadrons. The two tunnels leading down to the underground facilities were S-shaped and both had refuelling facilities. During the war exercises we undertook the Hunters would land after a mission, taxi to the entrance of one of the tunnels, cut the engine, then roll down to the bottom of the underground hangar. There was not much space between the wing tips and the mountain walls but, using the brakes, the pilots skilfully manoeuvred their Hunters to the hangar area where we, the engineers, met them. After a quick inspection we hooked our aircraft to a jeep and pulled them up to the refuelling positions in the other tunnel. After refuelling, and with the pilots still in their cockpits, we reported the aircraft ready and were then on high alert stand-by.

When the starting orders came, the massive tunnel doors slid open and we pulled our aircraft out of the mountain and onto the taxiway. As soon as the aircraft was in the open the pilot started the engine and, after his brief cockpit check, he signalled the jeep away. Pulling a lever in the jeep detached the towbar from the nose wheel and we accelerated away from our Hunter, which turned onto the taxiway. On these exercises the important point was that the aircraft battery was in top trim or the engine starter would fail. It was a shameful moment for any engineer to have to tow his aircraft to the nearest exit and then onto the ramp to make way for the Hunters heading for the runway. We became quite good at this. We were always timed and we spent two and a half minutes from receiving the order to take-off until the last aircraft was airborne.

I suppose that few will disagree with me that the most worrying job the engineers had to do on the Hunter was the removal and installation of the pilot's seat, the Martin Baker ejection seat. I still remember the cold sweat when I faced this job the first time. There were three cartridges to be removed before the seat was safe, and to remove one of them you had to remove the drogue chute gun that, if accidentally fired, would penetrate your chest. This did actually happen in one case in the RSAF. The engineer had installed the seat and, to make sure that it was fitted properly, he bent over it and pulled it upwards. The seat was supposed to be able to move about one-eighth of an inch. This was normal. But the engineer did this test with the wire to the drought chute gun attached to the floor and the small movement was enough to fire the gun. There was no way to confirm that the seat had been fasten properly, as trying to pull the seat upwards automatically fired the drogue chute gun. On the ground, and as soon as the engine had stopped, it was the engineer's duty to make the seat safe. This was done by locking the firing handle with a safety pin. This pin was attached to a large red disc so that it could be easily detected from the ground if the seat was safe or not. Handing over the aircraft to the pilot, the very last thing the engineer did was to remove the safety pin and show it to the pilot, then place it in its parking position.

As a Licensed Engineer you were supposed to know the aircraft build and how all the systems worked. We did some rather unorthodox but safe fixes from time to time, especially when we were away from home base. In one such case the pilot of my "Adam Blue" taxied into a low metal fence when returning from a flight, which caused damage to the main landing gear door. I inspected this and found that it had been holed and moved a little from its proper position. Time was short and we were proud of being able to report all Hunters serviceable at all times. The damage was, in my opinion, not too severe, so a few inches of tape fixed the hole and I trimmed the door back into its proper position. There was no time for a gear swing to verify this so I simply judged that everything was well. It was with a lot of excitement I watched my "Adam Blue" take-off and see the gear retract and lock as it should. My superiors, I noted, looked away from my aircraft when walking past during the rest of the exercise. I did a permanent fix as soon as we were back on base.

Another time the engine of my "Adam Blue" would not start. I asked the pilot to have another go and counted the seconds from the moment the starting cycle began until it failed. I got it to fourteen seconds. This gave me a good idea which relay had failed and I asked the pilot to remain where he was and to give me a few seconds. The suspect relay – and relays in those days were big boxes that could be opened and manipulated – was installed inside one of the three quick-access doors under the belly. I opened the door, removed the cover of the relay and hollered to the pilot to try another start. Again I counted the seconds and, arriving at fourteen, I shoved my screwdriver in between two points in the relay to create the contact. The engine started. The relay cover was refitted at double-quick time, I closed the access door and signalled to the pilot to go. "Adam Blue" took off with a very slight delay.

The good relationship between me and "Adam Blue" was well known on the base. When the Hunters were new and we still did not know all the little things that could go wrong, we had a few incidents when, coming in to land, the two main gears would come out all right but the nose landing gear would fail to extend. This day I watched my "Adam Blue" come in over the base with the main gears out but there was no sign of the nose gear. It couldn't be true! But minutes later my Hunter came in for landing. The pilots had developed a way to land these aircraft with the main gear out only in a very clever way. They landed on the runway, and while they still had speed enough to control the aircraft they steered the stricken craft off the runway and onto the grass, where they slowly dropped the nose to the ground while I stood there with tears in my eyes watching my beloved "Adam Blue" roll down the field with grass and earth spraying around its nose. Because of the skill the pilots showed in these cases the damage to the aircraft was minimal. The nose landing gear door had to be replaced and trimmed in and any other minor damage sorted out, and the next day the aircraft was in the air again.

We later discovered why the nose landing gears failed to extend. The simple answer was that if we allowed the pressure in the nose landing gear shock absorber to go down to the lower scale of the allowed pressure, the nose wheel would end up resting on the landing gear door. With this extra weight on the door the locks would not open and the gear stayed in. We also noted that when the pressure in the shock absorber came down towards the lower end of the allowed pressure limit the tyre, still rotating when being retracted, left a small patch of rubber on the inside of the door. Once that was understood, we never had that kind of failure again.

In order to keep us all really sharp the bases had annual competitions, a mixture of field and track and military skills. F8 was one of five bases taking part in this contest. One particular year one of the tasks we were measured on was how fast we could turn around an aircraft. It was assumed that the aircraft had been in battle and as soon as it landed we had to do the turn-around check, refuel and rearm the aircraft. As I mentioned before, the engineers had nothing really to do with the armament except lending a helping hand if required. The Hunter had a weapons unit holding the four Aden cannons and the ammunition boxes. This unit could be winched down onto a specially designed trolley, the spent one removed and a loaded one rolled in underneath and winched into the aircraft. The most important thing to remember when lowering the weapons unit was that, when lowered a foot or so, someone had to disconnect the "one hundred wire loom", as we called it. The loom did not have one hundred cables, but they were many. Failing to disconnect the loom from the aircraft resulted in it being torn apart from the weight of the weapons unit.

For the competition it was a question of time, and everything went a bit faster than it normally did. I was concentrating on my tasks when I suddenly heard this dull bang. I looked around and spotted this oversize shaving brush hanging between the fuselage and the weapons unit, now lowered a good two feet. Everything is allowed in love and war, I adored my "Adam Blue", and we were at war with the other bases, so I quickly ordered three privates to stand between the aircraft and the judges, and told the armament mechanics to get the new box under the aircraft and hoisted up faster than they had ever done it before. Seconds later we reported the aircraft serviceable and, as luck had it, we were not asked to go airborne again. I still have the medal we received for winning the competition.'

Further exports followed, and Denmark ordered thirty aircraft in 1954. Like Sweden's Hunters, the Danish F Mk 51 was essentially an F.4 airframe without wing leading edge modifications. Denmark's Hunters proved to be both versatile and reliable, but funding shortages resulted in an almost constant shortfall in available spares and maintenance facilities. Consequently, serviceability became a problem and many aircraft spent significant amounts of time on the ground, with components 'robbed' to maintain others. When Denmark retired the remaining aircraft in 1974, Hawker bought back the fleet for potential resale, but the condition of the aircraft proved to be so poor that refurbishment would not have been cost-effective. Consequently, most of the airframes were donated to museum collections with the exception of one aircraft, which underwent an extensive private refurbishment before appearing on the air show circuit as G-HUNT.

Far away from Europe, India also took great interest in the Hunter and became the largest export customer. It sent an evaluation team to England and eventually opted to make an initial purchase of 160 Mk 6 aircraft, giving Hawker a welcome financial boost just at a time when the 1957 Defence Review had slashed RAF orders for the Hunter. India's enthusiasm for the Hunter prompted Hawker to draw initial deliveries from the cancelled RAF production and re-allocate aircraft that had already been completed for the RAF (which had rather fortuitously been placed in storage); India even sent pilots to Dunsfold to undertake acceptance and delivery flights themselves. India was also the first customer for the 'big Avon' twin-seater at a time when a great deal of development work on the design had yet to be completed. Hawker was keen to meet India's demands and proceeded to build a demonstrator in May 1959 using salvaged or damaged parts from derelict Hunters. Registered G-APUX (and known by the nickname 'Gappux'), the centre and rear fuselage, tailplane and elevators came from Belgian F.6 IF-19, the wings, fin and rudder from Belgian F.6 IF-67, and the engine from RAF F.6 XE378. The newly built front fuselage had been produced as a ground display unit for the Paris Air Show, being entirely representative of an Indian T.66. India used Hunters extensively over a long service career that saw many aircraft used operationally, and it was not until 2001 that the last aircraft were finally withdrawn from second-line support duties, which credited India's Hunters with a service record that was as long as the RAF's.

Peru was another customer for the Hunter, as described by Hawker's publicity department:

> 'One of the features of the inauguration ceremony of 14th Fighter Group of the Peruvian Air Force was a formation flypast by three Hawker Hunters. A common occurrence, but with this difference: the aircraft were piloted by Peruvians who had never flown Hunters in formation before. The event was typical of the fast work and initiative of the Peruvian Air Force in bringing their new Hunter 52s into service. With the aid of five engineers from Hawker and a representative of Rolls-Royce Ltd, they assembled sixteen aircraft in the space of twenty weeks. Duncan Simpson, Hawker test pilot, went to Peru to deal with all flying questions and, incidentally, provided all the information on which this article is based. This was the first occasion on which Hunters had been shipped overseas since those delivered to NATO countries were assembled in Britain and flown to their new owners. There were, inevitably, difficulties. The aircraft were new to the Peruvians and the assembly line was a stretch of tarmac, with an open-fronted hangar to house the stores organisation. The crates in which the aircraft had been delivered had to be pressed into service to make benches and shelves for the stores. The policy followed was to open the cases and leave the broken-down units of the aircraft on the wooden formers in their cases until such time as the team was ready to assemble them in their correct sequence. By this method they were able to conserve their limited supply of ground equipment and at the same time carry on assembling the units. This applied more to the rear fuselage than any other unit. It was possible to assemble the fin, rudder, tailplane and elevator to the rear fuselage while it was still on the ground and the only time the rear fuselage trolley was used was when the completed rear fuselage was offered up to its aircraft.
>
> Assembly proceeded on the sixteen aircraft in such a way that it was possible to lay down a production line principle, and one or two men, whatever the operation called for, were able to follow through the same operation on all sixteen aircraft. The Hawker team were gratified to report that no major defects were encountered on any test flight and the majority of aircraft were passed off on the two flights called for under the test flight procedure laid down. But they were even more pleased to know how smoothly Hunters can be assembled under unorthodox conditions, provided the people doing it are as energetic and skilful as the Peruvians. The first six Hunters arrived at Callao in the third week of February 1956 and were taken by

road to Limatambo Airport. A second shipment of four aircraft arrived at the airport on 21 and 22 March. Yet by 15 May four of the Hunters were ready for testing and the remaining six were completely assembled and undergoing final inspections. The first flight from Limatambo took place on 1 April, watched by the Air Minister and senior officers of the Peruvian Air Force. It was a critical moment. Limatambo is essentially a civil airport and the loose gravel surface of the runways was hardly ideal for jet aircraft, although the length of 2,000 yards (2,200m) was adequate. Both runways were flanked on either side with a flourishing cotton crop, and the undershoot and overshoot areas had their share of obstructions, including a ten-foot-high brick wall at one end. But through the excellent co-operation of the civil controlling staff, the programme of flying the Hunters proceeded smoothly.

Work did not stop with the final flight tests of the aircraft; Peruvian crews then had to be trained to fly them. Before training began, six officers of the newly formed 14th Fighter Squadron translated the Hunter manuals into Spanish. The final result, produced in just over two weeks, was a beautifully finished flying manual covering not only pilot's notes but instruction on equipment and powered flying controls. Although Mr Simpson gave lectures to the Peruvian pilots, most of whom spoke English fluently, it was found more satisfactory to run through each lecture in detail with their commanding officer and let him answer questions in the pilots' own language. On 22 May, Peruvian pilots made six flights in Hunters, led by the Group Commander, Col Fernando Paraud, who had previously flown a twenty-minute sortie in a Hunter at Dunsfold. He and his colleagues Major Leon Melgar and Captain Thorndike (the Squadron Commander and senior pilot) very soon completed the conversion course and later flew each aircraft on an official acceptance check.

The only feature of Peru that disappointed the British was the weather. From the beginning of May onwards the sun was obscured for most of the time by a cloud layer extending up to about 3,500 to 4,000 feet (1,050-1,200m). Owing to the temporary absence of radio compasses in the aircraft, instrument let-downs were made by a method worked out in discussion with the senior Peruvian pilots, and which was in fact safer than it sounds. Flying above the clouds, the procedure was to set course from one of the Andes foothills known as Cerra Colorado, the top of which was 8,000 feet (2,400m) above sea level. A standard pattern was then flown and an overhead position at Limatambo was cross-checked by the relative position of three other known mountain tops and from a portable DIF radio set, most professionally operated by one of the Hawker engineers. In practice the DIF set was subject to various inherent failings, whereas the mountains were noted for their reliability. It need hardly be added that to make full use of the latter facility one required a good knowledge of the local Andes! From the overhead position, the pattern was confirmed and contact was made about 2 miles (3km) off the coast. The method proved very accurate owing to the light south-westerly winds prevailing and worked well in conjunction with civil approach procedure.

The last six aircraft had meanwhile arrived at Limatambo on 5 May, and by uninterrupted work were made ready by the 30th, the day on which the inauguration of 14th Fighter Group took place. In the first week of June it was decided to begin training the squadron pilots. The aim was to have them flying the Hunters in a flypast on the traditional National Day celebrations on 29 July. It was certainly a formidable one. Each training sortie was carefully planned in advance, and the pilot received a printed card bearing full information on the conversion flight he was about to undertake. The programme was planned each evening for the following day, and amended according to the weather if necessary. At an early morning briefing the senior pilot explained each day's flying programme and discussed with the pilots under instruction any problems from the previous day. In addition, squadron pilots were chosen at random to explain one flight emergency at each briefing. A "chase" aircraft flown by Col Paraud, Major Melgar or Captain

Thorndike accompanied each pilot on his first and second flights, so that the less experienced aviators could concentrate on flying the aircraft rather than on the "Cerra Colorado" let-down, which was usually needed at the end of the flight.

On 7 June a squadron pilot made his first solo; by mid-July all the pilots had completed their conversion courses and begun formation training. Despite the difficult conditions the entire task of conversion was accomplished at the negligible expense of one tail bumper. A more serious accident occurred during formation flying when the wingtips of two aircraft touched. Both aircraft landed safely, however, with damage confined to the port wing of one and the starboard wing of the other, and by putting both intact wings on to one aircraft it was possible to fly fifteen of the Hunters in the flypast. Although they had only two or three hours' formation flying in their Hunters, the Peruvian pilots led the flypast in a box formation that was first-rate judged by any standards. The flypast passed over the centre of Lima, where the armed forces are reviewed by the President of the Republic in the Campo del Marte. That year's National Day was especially significant as it coincided with the election of the new President, Dr Manuel J. Prado. During the eight years of office of the outgoing President, General Manuel A. Odria, the Peruvian Air Force had been re-equipped with the most up-to-date British and American jet aircraft, and those that took part in the National Day celebrations left no doubt of the capabilities of the Peruvian pilots. The day after the flypast the press contained the text of a telegram of congratulation sent by Air Chief Marshal Sir Dermot Boyle to the Peruvian Air Minister.

By 11 July the last aircraft had been handed over to the Peruvian Air Force. Some two weeks later four of the Hawker engineers bade farewell to Peru, the fifth remaining as the firm's representative. Their achievement was one of which they might well be proud. Sixteen aircraft had been assembled in four and a half months, under conditions that were far from ideal. The work had been done in the open air in anything from tropical heat to cold heavy drizzle. Equal credit is due to the civilian engineers from the Peruvian Air Force arsenal, who were drafted in to assist in the work. The facilities of the arsenal were placed at the disposal of the British engineers for any repairs that were necessary. No job, however small or large, was beyond their capabilities and the standard of workmanship was of the highest order. The commandant, Colonel Jorge Vigil, and the liaison officer, Captain Victor Ortega, always proved able to meet the numerous and often seemingly impossible requests for materials and specialist engineers. The knowledge gained by these technicians during assembly will stand them in good stead in the future. Language difficulties, resulting from the British attempts to learn Spanish and the Peruvians simultaneously learning English, did not present any problems that spontaneous good humour on both sides could not solve. The cooperation between both parties resulted in trouble-free test and service flying, which is a credit alike to the Peruvian civilian and service engineers and to the Hawker and Rolls-Royce team. The Hawker team were greatly encouraged by the way in which the Peruvian pilots flew their Hunters. On his return to this country, Mr Simpson said: "One did not need to know the language to realise their enthusiasm over their new aircraft. To have worked with them was indeed a pleasure, and we look forward to hearing more of them in the future."'

Perhaps the most significant overseas order came from Switzerland, a non-aligned country with the ability to 'pick and choose' aircraft purchased from the international market. The decision to purchase the Hunter (after a long and thorough evaluation process) demonstrated to any doubters that it was a formidable machine with the simplicity of construction, ruggedness, versatility and performance to meet Switzerland's demands for a capable and adaptable aircraft that could undertake both interception and ground-attack roles over a long service life. Hawker certainly recognised the importance of a Swiss purchase and went to great lengths to secure it, sending two Hunters to Switzerland as participants in a demanding fly-off contest against other designs. The company then sent two more aircraft (together with a variety of weaponry) to undertake a month-long service trial with Swiss Air Force units. The Hunter proved to be an ideal aircraft for Swiss operations and remained in front-line service until 1994, having been progressively modernised to carry modern systems and weapons. Even their retirement from Swiss service did not mark the end for many Swiss Hunters, as a surprisingly large number of retired airframes were presented to selected museums and civilian operators, and these have now embarked on new flying careers as 'outsourced' military support aircraft and air show performers.

As the Hunter began to capture an international sales reputation, more interest developed in the Middle East, undoubtedly encouraged by the RAF's operations in the area, which demonstrated more effectively than any sales talk that the Hunter was more than capable of performing well in harsh environments. Iraq, Jordan and Lebanon placed orders and more were to come, even though Hawker was ultimately faced with a potential problem – finding Hunters to sell. When RAF production ended and the main overseas exports had been supplied, production of the Hunter ended, and Hawker turned its attention to other projects, most notably the design that became the legendary Harrier. But it soon became clear that there was a developing overseas market for more Hunter sales, many smaller nations having recognised the aircraft's potential as a relatively inexpensive but potent fighter and ground-attack aircraft.

Most of the RAF's Mk 1 and Mk 2 aircraft had been scrapped as soon as the F.4s had been brought into service, and in retrospect this was an unfortunate development for Hawker as most of these airframes were virtually as-new and could have been refurbished for resale. The F.4 fleet was mostly regenerated for continued RAF and FAA use (many aircraft becoming T.8 trainers or GA.11s), but scrapping of the remaining surplus Hunter airframes was finally halted in 1963. However, it was clear that Hawker would not be able to supply the demands of the export market from the stocks of remaining aircraft in the UK. The only option was to buy back Hunters that were nearing the end of their service with existing customers, particularly Belgium and the Netherlands. The idea seemed logical but it was potentially unworkable without the support of America, as these aircraft had been funded by US dollars and it seemed unlikely that the US would allow Hawker to simply buy them back and resell them to customers who might well have otherwise purchased American aircraft. Thankfully, Hawker had ensured that the sales contracts included an option to buy back airframes and, much to the company's relief, no opposition was forthcoming from America, largely because the anticipated customers for the Hunter were recognised as being countries that already relied upon British aircraft, or who relied on British personnel to fly or maintain them, and they would therefore, have no interest in American equipment.

Investigations at Kingston revealed that the Hunter's fatigue life could be extended significantly by strengthening the fuselage forward transport joint, and it was eventually accepted that the basic Hunter airframe could be economically refurbished and re-sold to new customers 'as new'. The first customer for the former Belgian aircraft was Iraq, followed by India, where significant numbers of Hunters were required as attrition replacements following the country's war with Pakistan. Switzerland, Rhodesia, Kuwait, Jordan, Lebanon, Saudi Arabia, Chile, Singapore, Abu Dhabi, Qatar, Kenya and Oman all eventually became customers for 'second-life' Hunters. The very fact that the majority of these aircraft went on to complete long and successful service careers is testament to the soundness of the Hunter design; indeed, more than a few Hunters were sold on for a second time and of course many have more recently found their way onto the civil market, where they look set to remain airworthy for at least another decade. In many respects, the 'disaster' of 1957, when the RAF's Hunter order was drastically cut, became the turning-point for Hawker, whereby the Hunter was effectively re-born as an international, multi-role warplane. Few aircraft have matched the Hunter's phenomenal success.

Appendix 1

Export sales by country

Abu Dhabi

Seven Hunter FGA.9 aircraft were delivered to Abu Dhabi in 1970, equipping one squadron of the Abu Dhabi Army Air Wing at Sharjah. The aircraft were used operationally, flying air defence patrols during a long-running border dispute with Saudi Arabia. Two reconnaissance-configured Hunters and a pair of twin-seat trainers were also delivered. Used initially in the air defence role, the aircraft later switched to ground-attack operations and were withdrawn in 1983, when some of the surviving airframes were transferred to Somalia.

FGA Mk 76	701 to 707
FR Mk 76A	708 to 710
T Mk 77	711 and 712

Army Air 702 illustrating the unusual colour scheme applied to Abu Dhabi's Hunters, comprising of sand and stone camouflage, medium blue undersides with a bright red cheat line. *via David Griffin*

Belgium

Both Belgium and the Netherlands built Hunters under licence with funding from the US Offshore Procurement Programme. One example (an F Mk 4) was delivered on 16 March 1955 as a pattern aircraft, from which production lines were set up by SABCA and Avions Fairey, although further Belgian Hunters were taken from the Fokker production line. The two nations combined their Hunter training programme, so Belgium had no direct requirement for dual-control trainer versions. Entering service with Belgium in 1956, the Hunter fleet remained active until 1963. Not surprisingly, most of Belgium's aircraft were withdrawn with significant hours left on the airframes and most were eventually refurbished and sold on to other export customers. Belgium's 'Diables Rouges' aerobatic team also operated the Hunter.

F Mk 4	ID-1 to ID-64; ID-101 to ID-148
F Mk 6	IF-1 to IF-144

A quartet of Hunter F Mk 4 aircraft from the 7th Ecole de Chasse (the Hunter OCU), based at Chievres.

F Mk 4 ID-109 was built for the Belgian AF by Fokker in the Netherlands. It was written off during 1960 and scrapped, although this image appears to show an earlier, less serious mishap.

These aircraft, from the second batch of Fokker-built F Mk 4 aircraft, were retrofitted with power-assisted elevators by SABCA in Belgium. The Belgian fleet of Hunters retained its standard RAF-style camouflage until retirement.

Hunter F Mk 6, XE548/H, No 4 Squadron, Waddington, September 1960. Dark Sea Grey/Dark Green/Aluminium finish; serial in white on fuselage sides, repeated below wings in black. Code 'H' on fin in yellow, repeated in black on front nose-wheel door; unit badge, flanked by black/red bars, outlined in yellow, with yellow flashes superimposed.

Upper surface views of Hunter camouflage schemes; On the left is Hunter FGA Mk 71 '702' (see page iv), which compares with a standard RAF single seater in the centre. Note that the camouflage demarcation scheme remains the same on most Hunter aircraft although paint colours varied depending on the country with which the aircraft operated. On the right is an illustration of the Training Command scheme applied to aircraft operated by No 4 FTS.

Hawker Hunter F Mk 4, ID27, 350e Escadrille, Belgian Air Force, 1957. Dark Sea Grey/Dark Green upper surfaces with Silver undersides; national markings in six positions, with those on fuselage thinly outlined in light blue. Serial in white, repeated in black, within a red circle, on nose wheel door, above starboard wing (inboard of roundel) in white and below port wing in black; wingtips red with white triangles.

Hawker Hunter FR Mk 71A, J-735, Grupo Nº8, Chilean Air Force, 1994. Dark Sea Grey/Dark Green/Light Aircraft Gray finish with white serial on fin, the latter repeated above port wing; national markings above starboard wing only (no markings under the wings). Unit badge on forward fuselage.

Hawker Hunter FGA Mk 71, J-702, Grupo Nº9 Panteras Negras, Chiliean Air Force, 1976. Dark Earth (FS.20227)/Sand (FS.23594)/Sky Blue (FS.25466) finish with black code on fuselage sides, the latter repeated above port and under starboard wing. '704' in black on nose wheel door; national marking above starboard and below port wing. Note unit badge on nose repeated on fin.

Hawker Hunter F Mk 4, N-110, No. 322 Squadron, Royal Netherlands Air Force, Biak (Dutch New Guinea), 1962. Dark Green/Dark Grey upper surfaces; RAL-7001 (light grey) undersides. National markings on fuselage sides, above port wing and below starboard wing; codes in white. Red/white/blue 'Papua Barat' flag on fin/rudder.

Hunter F Mk 6, XF383, No 12 Squadron, Lossiemouth, mid-1984. Wrap-around finish of Dark Sea Grey/Dark Green with blue/red roundels and fin flash; serial in black, repeated below wings. Unit badge on nose. NB. Fin flash colours reversed on starboard fin.

Hunter F Mk 6, XF383/14, No 145 Squadron, June 1961. Dark Sea Grey/Dark Green/Aluminum finish with roundels in six positions; serial and code in white (the former repeated in black below wings). Unit badge on both sides of nose; white wingtips.

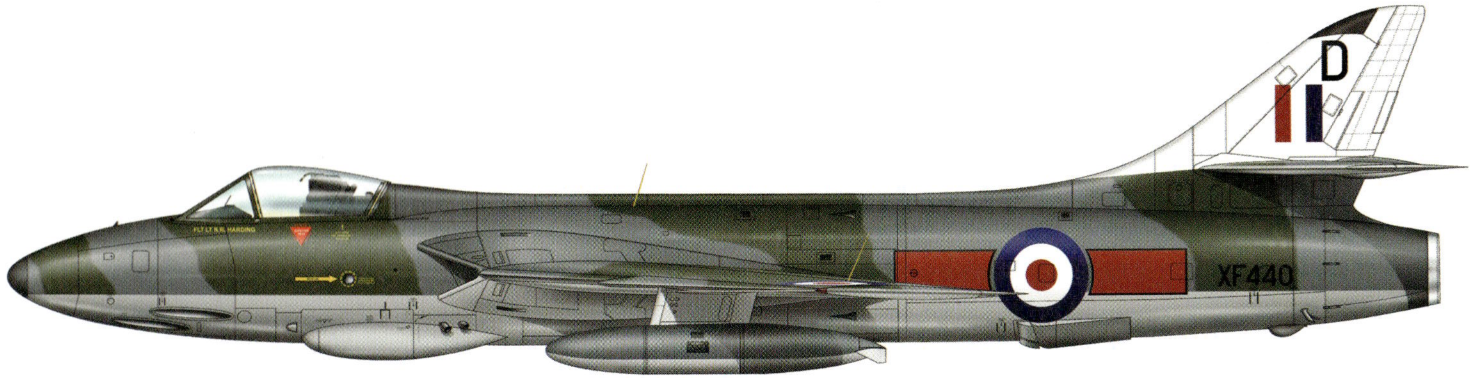

Hunter F Mk 6, XF440/D, No 247 Squadron, 1957. Dark Sea Grey/Dark Green/Aluminum finish with white fin/rudder; black serial, repeated below wings. Red rectangles, bordered in black, flanking fuselage roundel; code on fin in black, repeated on front nose wheel door. Pilot's name (Flt Lt R R Harding) in yellow below windscreen.

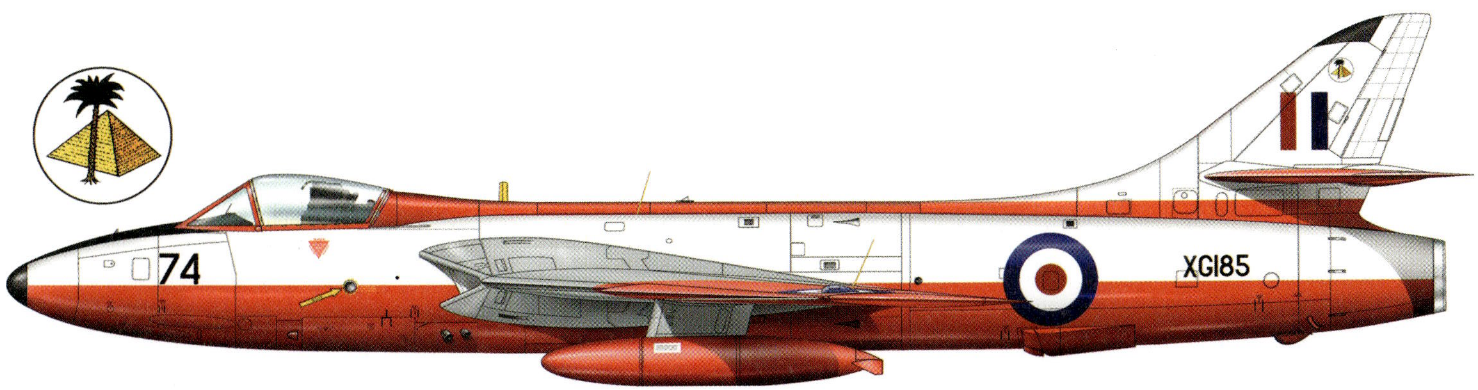

Hunter F Mk 6, XG185/74, No. 4 Flying Training School, RAF Valley. Gloss White top decking to fuselage and sides of fin/rudder; Gloss Signal Red undersides of fuselage, spine, outer wing panels and horizontal tail surfaces. Gloss Light Aircraft Grey engine intake fairings and wing inboard sections; black serial/code and anti-dazzle panel. '74' repeated in black on Light Aircraft Grey nose wheel door.

Hawker Hunter F Mk 56, A476, 'The Banners' Target Towing Flight, Indian Air Force, Kalaikunda, 1999. Light Grey and Dark Green upper surfaces with Silver undersides; national markings in six positions. Serial in black, repeated on nose wheel door (see detail).

Hawker Hunter F Mk 6 (ex-Belgian Air Force), 660, of the Iraqi Air Force, October 1966. Dark Grey/Dark Green upper surfaces and Light Aircraft Grey scheme; red nose, wingtips and tail assembly. '661' in white; triangular national marking in six positions.

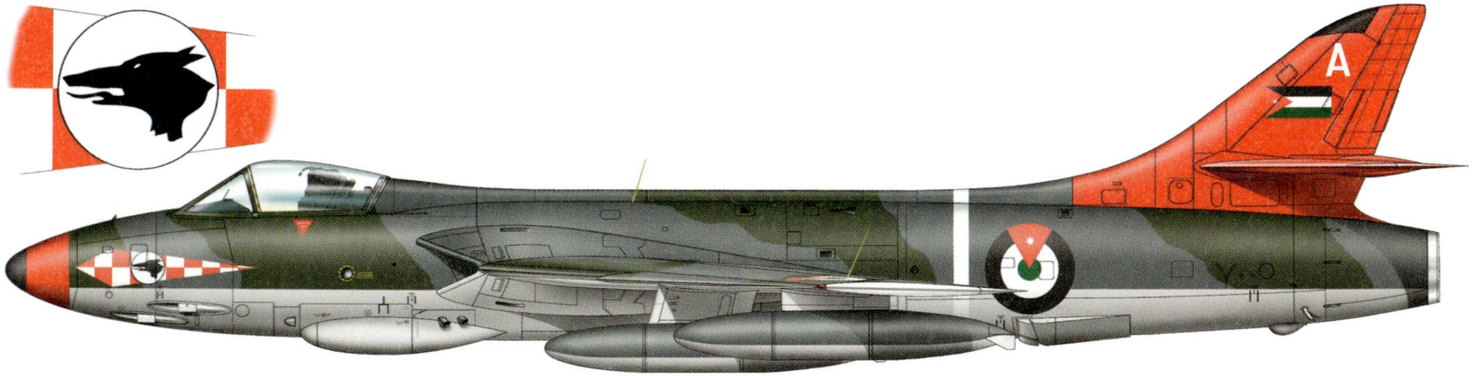

Hawker Hunter F Mk 6, 700/A (ex-XE551), Royal Jordanian Air Force, 1965. Dark Green/Dark Sea Grey upper surfaces and Silver undersides; red nose and tail section. White wing tips and fuselage band; national markings in six positions. Serial in black; 'A' on fin in white, repeated on nose wheel door in black.

Hawker Hunter FGA Mk 80, 806, Kenyan Air Force, 1975. High Gloss Dark Grey/Dark Green upper surfaces with Light Aircraft Grey undersides; national markings on fuselage sides and above wings only. Serial '806' in black, '6' repeated (also in black) on nose wheel door.

Hawker Hunter FGA Mk 73, 847, Sultan of Oman Air Force, 1994. Two-tone wrap-around grey scheme comprising of Dark Sea Grey and Extra Dark Sea Grey; national markings on fin only. Serial in black on rear fuselage.

Hawker Hunter FGA Mk 9, R1821, No 1 Squadron, Rhodesian Air Force, 1972. Dark Earth overall with Dark Green disruptive pattern on upper surfaces; national markings on forward fuselage only. Serial R1821 in black on rear fuselage, '1821' repeated on nose wheel door.

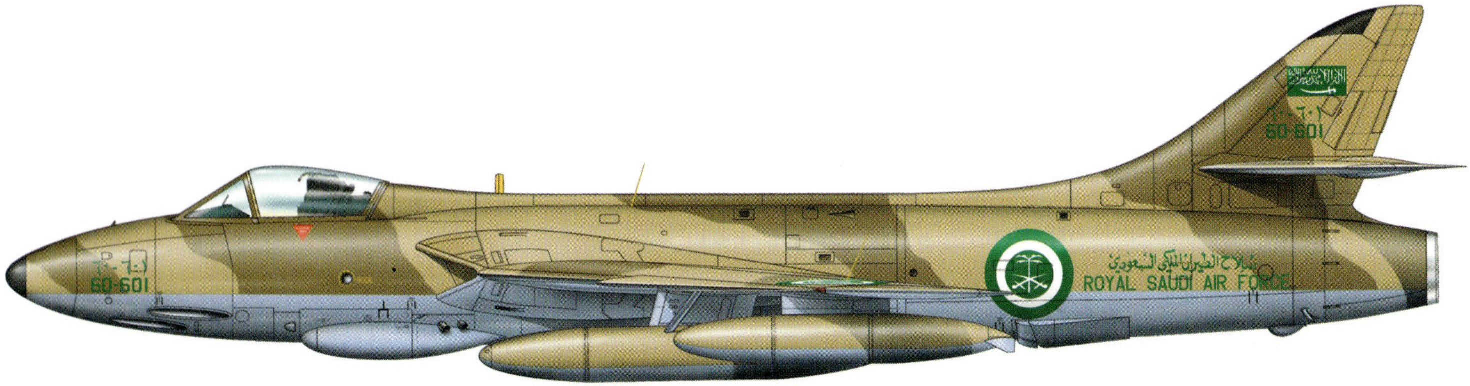

Hawker Hunter F Mk 60, 60-601 (ex-XJ715), No 6 Squadron, Royal Saudi Air Force, 1966. Dark Earth/Middle Stone/Azure Blue glossy scheme; national markings in six positions. Lettering in green.

Hawker Hunter FGA Mk 74B, 524, No 140 Squadron, Singapore Air Defence Command, 1977. Dark Earth/Dark Green/Bronze Green upper surfaces; Pale Grey (FS.36622) undersides. Black/red checks on rudder and flanking unit marking on nose; national markings on fuselage sides only.

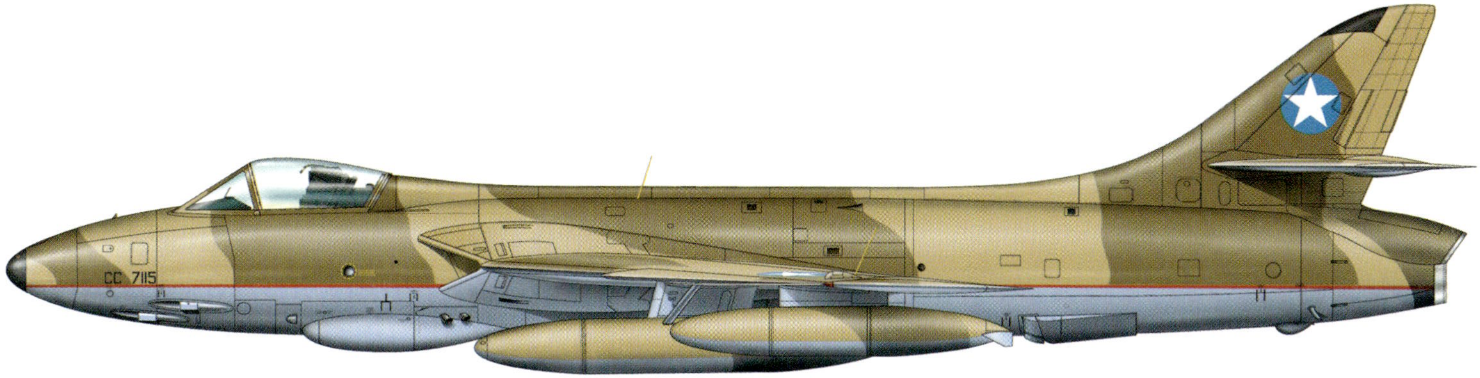

Hawker Hunter FGA Mk 76, CC 7115 (ex-Abu Dhabi aircraft), Somali Air Force. Dark Earth/Mid Stone/Azure Blue scheme with a red line marking between the upper and lower colours on the fuselage; national markings on sides of fin and four wing positions.

Hawker J-34 Hunter, 'M', 2nd Squadron of F.18 Wing, Royal Swedish Air Force, 1960. Dark Olive Green upper surfaces and Light Blue undersides; Silver rear fuselage end and nose cone. Blue 'M' outlined in white on fin repeated in white on blue nose wheel door; Yellow '18' on fuselage sides.

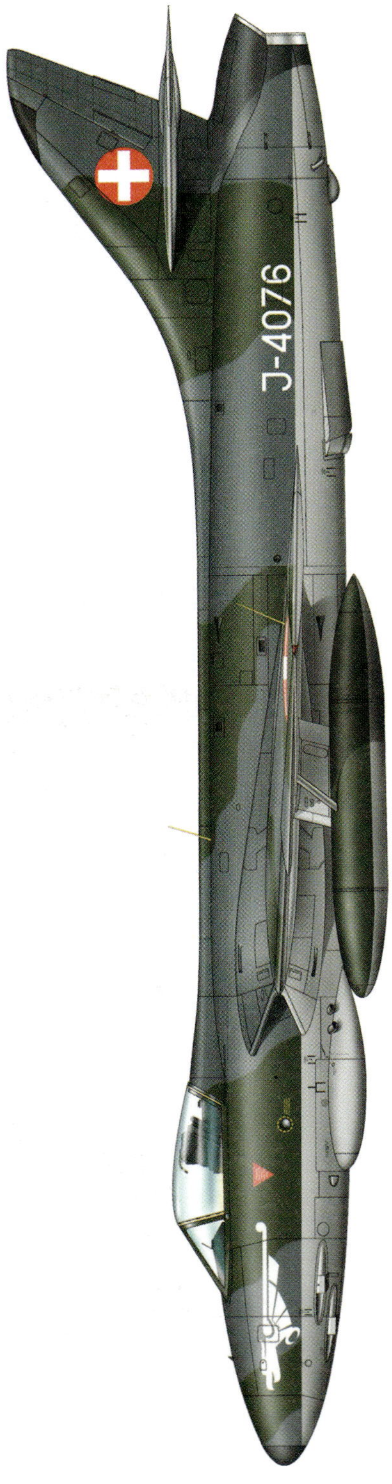

Hawker Hunter F Mk 58, J-4076, Staffel 1, Swiss Air Force. Dark Sea Grey/Dark Green/Light Aircraft Grey finish with national markings in six positions. Serial in white on fuselage sides, with '76' repeated on nose wheel door in black.

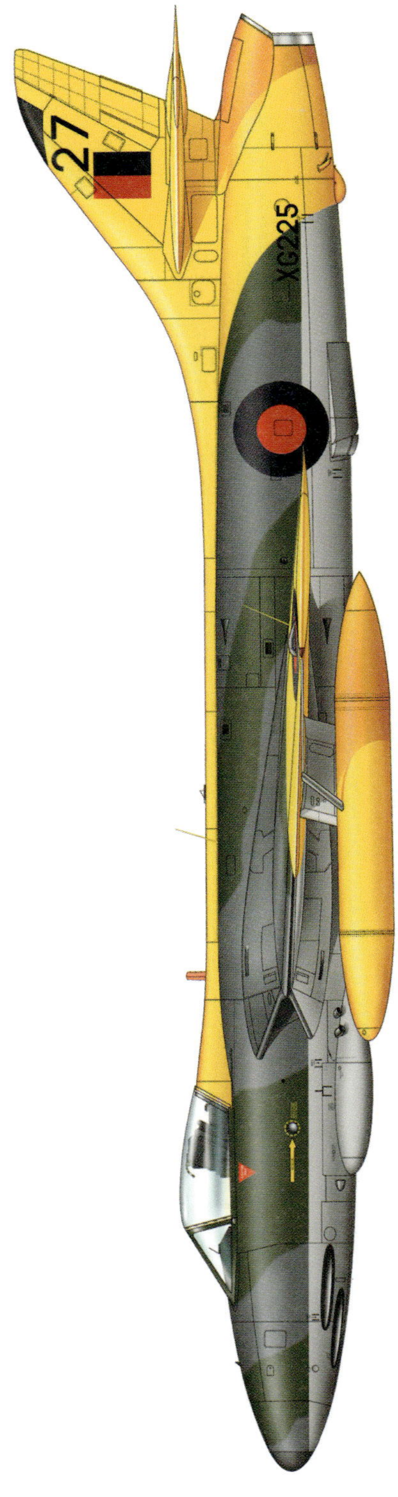

Hawker Hunter FGA Mk 9, XG225, No 1 Tactical Weapons Unit. Dark Sea Grey/Dark Green/Light Aircraft Grey finish with yellow fuselage spine and tail wingtips and fuel tanks; 'B' type roundels in six positions, with black serial and code

Chile

Following the withdrawal of an arrangement to supply F-86F Sabres from the USA, Chile purchased a fleet of Hunters, the first arriving in 1967; eventually a mixed fleet of single-seat, twin-seat and reconnaissance-configured aircraft were supplied. Operational use included attacks on the Presidential Palace (following a coup in 1973). Chile would have purchased more Hunters had their political situation not precluded further sales, and it was not until 1982 that a batch of former RAF Hunters was supplied, effectively in exchange for Chile's vital co-operation during the Falklands conflict. Many of Chile's Hunters were upgraded with modern radar warning receivers, chaff and flare dispensers and IAI Shafrir AAMs. The last of Chile's Hunters were retired in April 1995.

FGA Mk 71	J-700 to J-714; J-724 to J-730; J-731 to J-733; J-737
FR Mk 71A	J-715 to J-717; J-734, J-735 and J-738
T Mk 72	J-718 to J-723 and J-736
FGA Mk 9	J-739 to J-754

Chile operated the Hunter for almost thirty years. This pre-delivery image shows an FGA Mk 71 at Dunsfold wearing standard RAF-style camouflage and Chilean AF markings. *Paul Tomlin*

Hunter F Mk 4 XF317 was converted to F Mk 71A standard and sold to Chile as J-734. It is currently stored at Los Cerrillos Airport in Santiago. *via David Griffin*

Originally constructed as Mk 4 XF317, the aircraft was re-purchased by HAS and modified to FR Mk 10 standard, becoming Chilean AF J-734.

Denmark

Following the delivery of one Hunter F Mk 4, which was used for evaluation (and eventually retained for service use), Denmark received a fleet of Hunter F Mk 51s beginning in January 1956. These were export equivalents of the RAF's F.4 with straight wing leading edges. A pair of twin-seat Mk 53s (also without 'dog tooth' modifications) were also delivered, augmented by a pair of T.7 trainers from the Netherlands in the 1960s. Unfortunately, Denmark failed to fund effective maintenance and spares back-up for the fleet, and by the time the final Hunter unit disbanded in 1974 only ten aircraft remained fully serviceable. Hawker Siddeley did buy back twenty of the fleet but, because of their poor condition, they were not refurbished for re-sale.

F Mk 51	401 to 430
T Mk 53	EP-271 and EP-272
T Mk 7	ET-273 and ET-274

An early image of RDAF Hunters, illustrating the delivery-standard colour scheme that was based on standard RAF camouflage. One notable variation is the natural metal area around the jet pipe. *RDAF*

An excellent view of recently delivered RDAF Hunters at Aalborg. In addition to the two aircraft in the foreground, a further six Hunters can be seen on dispersals in the distance. Aircraft 418 was eventually retired and sold to Spencer Flack, becoming the well-known G-HUNT. *RDAF*

During service with the RDAF, the Hunter fleet was progressively repainted in an overall matt green colour scheme complete with smaller national insignia. Note that the lead aircraft is getting airborne with flaps partially extended. *RDAF*

The switch to dark green camouflage certainly rendered the Danish Hunters less conspicuous, but the matt paint weathered rapidly, resulting in a fleet of rather battle-worn aircraft. This particular example later became the beautifully finished all-red (with white/blue trim) G-HUNT. *RDAF*

A factory-fresh Hunter T Mk 7 at Dunsfold, prior to delivery to the RDAF. As with Denmark's single-seaters, the aircraft has straight wings without the leading edge extension. *RDAF*

Hunter 415 has been abandoned three miles from Skydstrup following an engine flame-out. The aircraft (minus pilot) remained airborne and eventually landed wheels-up on the runway, and was repaired and returned to service within a month! *RDAF*

This image clearly illustrates the 'turned-up' profile of the 'small-bore' engine jet pipe fitted to early Hunters. Hunter 427 now resides at Bruntingthorpe in RAF 'Blue Diamonds' colours, masquerading as XF324. *RDAF*

Hunter WT720 is in fact former Danish F Mk 4 408, which has been on static display at RAF Brawdy, Cranwell and Sealand. *John Laming*

India

India was a major operator of the Hunter and its first aircraft were drawn directly from RAF stocks held at Kemble. These were followed by new-production aircraft, which were F Mk 6 airframes, but modified with gun blast deflectors and brake parachute housings (the earlier aircraft were retrofitted). The first aircraft was delivered to India in 1957. Following significant operational use during India's war with Pakistan, more Hunters were purchased, these being modified FGA Mk 9 aircraft; further twin-seat trainers were also purchased. Front-line use of the Hunter ended during the 1980s but the type was retained in the training role for some time and the very last Indian Hunters remained flying into the new millennium, albeit in small numbers, as target facilities aircraft and continuation trainers.

F Mk 56	BA201 to BA360
FGA Mk 56A	A459 to A494; A936 to A943; A967 to A969; A1010 to A1015
T Mk 66	BS361 to BS376; BS485 to BS490
T Mk 66D	S570 to S581
T Mk 66E	S1389 to S1393

The Indian AF's Target Tug Flight based at Kalaikunda operated a mixed fleet of eight single- and twin-seat Hunters and only recently disposed of its last aircraft. *via Ray Deacon*

Hunter A-491 was one of a batch of thirty-six FGA Mk 56A aircraft delivered to India in 1966 as attrition replacements following India's war with Pakistan. *Richard Caruana*

S-577 is pictured at Luqa during its delivery flight to India. It was one of twelve T Mk 66D aircraft ordered by India in 1968. *Richard Caruana*

India's eye-catching 'Thunderbolts' display team was formed by crews of No 20 'Lightning' Squadron, resplendent in overall blue and white display colours. *IAF*

The 'Thunderbolts' team was formed in 1982 at Hashimara AB in India. Initially equipped with six aircraft, the team later flew a nine-ship display before being disbanded in 1988. *IAF*

After serving with the RNethAF as N-216, this F Mk 6 was returned to HAS and re-emerged as India's FGA Mk 56A A1015.

Former RNethAF N-261 is seen at Kemble prior to delivery to India as S-575. The aircraft was written off in a flying accident during December 1970. *Ray Deacon*

S-571 spent some time at Kemble training Indian AF pilots prior to being ferried to India in 1967. This was one of the last Indian Hunters in service. *Ray Deacon*

Iraq

An initial batch of fifteen F Mk 6 airframes was delivered to Iraq in 1957, these being RAF machines that had been held in reserve. They were financed by the US Offshore Procurement Programme. Further aircraft were delivered between 1964 and 1967 and Hawker's famous twin-seat demonstrator G-APUX was operated on loan for a year before a batch of trainers and reconnaissance-configured Hunters were subsequently supplied. Iraq's Hunters (sometimes flown by Jordanian pilots after many of Jordan's aircraft were destroyed) were flown operationally during the 1967 war with Israel and again during the 1973 Yom Kippur conflict. Although largely replaced as front-line aircraft during the 1970s, Iraq's Hunters remained active in second-line roles into the 1980s, although a precise retirement date remains unclear.

F Mk 6	394 to 409
FGA Mk 59	570 to 587; 628 to 633
FGA Mk 59A	657 to 661; 690 to 702
T Mk 69	567 to 569; 626 and 627
FGA Mk 59B	662 to 665

This factory-fresh Hunter FGA Mk 59B is pictured at Dunsfold prior to delivery to Iraq. Note that in addition to a camera-configured nose, the aircraft also has gun muzzle blast deflectors fitted. *Hawker*

Although of poor quality, this image shows one of just four reconnaissance-configured Hunter FGA Mk 59B aircraft that were operated by Iraq. The nose section is painted red.

This pristine Hunter FGA Mk 59 is pictured at Luqa, Malta, during its delivery ferry flight to Iraq. *Richard Caruana*

Jordan

A batch of twelve Hunters was ordered by Jordan in 1958 followed by three twin-seat trainers. Hawker's demonstrator G-APUX was also operated on loan for a year and a pair of former RAF FR Mk 10 aircraft were transferred from RAF stocks. These were joined by a further twelve FGA Mk 9s, together with eight F Mk 6 airframes that had been modified to FGA Mk 9 standard. Most of Jordan's Hunters were destroyed on the ground during the 1967 war and Jordanian Hunter pilots transferred to Iraq, flying their Hunters extensively during the war. After the conflict, three former Saudi F Mk 6 aircraft and a further four former RAF FGA Mk 9s were delivered to Jordan; finally a batch of refurbished airframes was purchased from Hawker in 1970. The surviving Hunter fleet was withdrawn in 1975, and many airframes were transferred to Oman.

F Mk 6	700 to 711 and 717; 752
FR Mk 10	853 and 712
F Mk 73	801 to 812; 825 and 826; 814 and 832
T Mk 66A	800
T Mk 66B	714, 716 and 718
FGA Mk 73A	828 to 831; 840; 842 to 849
FGA Mk 73B	841; 850 and 851
T Mk 70	835 and 836

A former RAF FGA Mk 9 pictured at Dunsfold after refurbishment, prior to delivery to Jordan. *Paul Tomlin*

Jordan purchased three T Mk 66B trainers, this example proudly wearing the colourful markings of No 1 Squadron, complete with red trim. *Richard Caruana*

Kenya

Established following the country's independence in 1963, the Kenyan Air Force received four former RAF single-seat Hunters and a pair of twin-seat trainers, these being the air force's very first combat aircraft. They remained active until 1981 when the surviving airframes were sold to Zimbabwe.

FGA Mk 80	803 to 806
T Mk 81	801 and 802

Kuwait

Four single-seat Hunters were delivered to Kuwait in 1965, followed by a pair of FGA.9s (from RAF stocks) and five new twin-seat trainers. The single-seaters were replaced in 1976 but the trainer variants continued in service until 1980.

FGA Mk 57	212 to 217
T Mk 67	210 and 211; 218 to 220

Lebanon

Delivered in 1958, Lebanon's first Hunters were six former RAF F Mk 6 aircraft, financed by the US Offshore Procurement Programme. Four more single-seaters were delivered together with three twin-seat trainers, and Lebanon also borrowed Hawker's demonstrator G-APUX for a year. A further batch of six former RAF FGA.9 and FR.10 aircraft were delivered in 1975 and 1977. Lebanese Hunters were used operationally in 1983 when they flew attack missions against Druze positions, operating from a dispersed road site. No precise withdrawal date is know but it is believed that most aircraft were placed in storage during the early 1990s.

FGA Mk 70	L-170 to L-184; L-285
T Mk 66C	L-280 to L-282

Netherlands

The Netherlands (like Belgium) secured an agreement to manufacture Hunters under licence and a pattern aircraft (an F Mk 4) was delivered on 3 March 1965. Production (undertaken by Fokker) comprised a mix of F.4 and F.6 aircraft, and twin-seaters were supplied by Hawker, ten of these being machines originally allocated for the RAF. Although the F.4 served until 1963, the later F.6 and T.7 aircraft remained in use until 1968, when they were replaced by Starfighters.

F Mk 4	N-101 to N-196
F Mk 6	N-201 to N-293
T Mk 7	N-301 to N-320

PH-NLH was originally ordered for the RAF as XM126 but became part of a ten-aircraft purchase for the Netherlands. It was eventually transferred to the NLR (Nationaal Lucht-en Ruintevaartlabortorium, essentially the Netherlands equivalent of the A&AEE/RAE) as a test and trials aircraft. Note the fin-mounted trailing aerial.

The Netherlands built the Hunter under licence. N-189 (323 Squadron) is pictured during an exercise deployment to Cazaux AB in France.

NLR's PH-NLH (formerly N-320) illustrates the modified nose cone (and instrumentation boom) and modern aerial fit on the upper fuselage.

A quartet of No 325 Squadron F Mk 6s, all of which were manufactured by Fokker in the Netherlands. N-271 was subsequently sold to the Indian AF as A1011.

N-108 served only with the RNethAF and, after a period of service at Biak (Netherlands New Guinea), was used by the fire/rescue unit at the RNethAF Electronics & Technical School.

Oman

The Sultan of Oman's Air Force received a mixed fleet of single- and twin-seat Hunters, which were given to the SOAF by Jordan in 1975. During the 1980s SOAF Hunters were modified to carry chaff and flare dispensers and were capable of carrying Sidewinder AAMs and LAU7/A launchers. Hunters were used operationally during the Dhofar War in the 1970s. The surviving Hunter fleet remained in service until August 1993.

FGA Mk 73, 73A and 73B	718; 720; 723 and 724; 816 and 817; 822; 825 to 831; 840 to 854
T Mk 66B	800 to 802
T Mk 67	803 and 804

Glistening in the desert sunshine, this SOAF Hunter F Mk 6 wears an early post-delivery 'high-vis' colour scheme (standard RAF-style camouflage). *via David Griffin*

A rare T Mk 67 pilot's-eye view of a pair of SOAF Hunter F Mk 6s in action over the desert. Sidewinder AAMs are clearly visible on custom-built inboard pylons. *via David Griffin*

Hunter FR Mk 10 853 bakes under the Omani sun in low-visibility tactical camouflage. *via David Griffin*

SOAF Hunter pilots earned a reputation for their spirited flying. Low-level beat-ups were fairly common sights, as illustrated by this ultra-low pass being made by F Mk 73 844 at Thumrait. *Ian Mackay via David Griffin*

Peru

Although orders for the Hunter from overseas customers had already been placed, Peru's purchase of sixteen former RAF airframes in 1956 was the first order to emerge from outside Europe. These F Mk 4 aircraft (with early unmodified wings) were later (in 1960) supplemented by a single example of the dual-control trainer version. Operational flying ended in 1976 but the Hunter remained active in the training role until 1980.

F Mk 52	630 to 645
T Mk 62	681

Hunter FGA Mk 52 No 137 is Peru's former 637, which began life as the RAF's F Mk 4 WT774. It is currently preserved at Las Palmas, Peru. *Sergio de la Puente*

Peruvian Mk 52 Hunters on display at low level, shortly after delivery. The Mk 52 was essentially a standard Mk 4 with an unmodified two-pylon wing.

Qatar

Following Abu Dhabi's purchase of Hunters, Qatar also opted to purchase the aircraft, ordering three single-seaters and a single example of the dual-control variant, all of which were former Netherlands aircraft. They were all in service by 1971 although one aircraft was lost in an accident in 1977. They were withdrawn in 1982.

FGA Mk 78	QA-10; QA-11; QA-12
T Mk 79	QA-13

F Mk 6 N-219 was built by Fokker in the Netherlands. Following RNethAF service, the aircraft was rebuilt as an FGA Mk 78 and delivered to Qatar. QA11 was written off in an accident during 1977 but remained in use as a ground training airframe.

N-316 was originally destined for the RAF as XM122, but after having been diverted to the Netherlands order it served until 1965 when it was re-purchased by HAS. It was refurbished and sold to Qatar as T Mk 79 QA-13. It is believed to be on display within Doha International Airport.

Rhodesia/Zimbabwe

Rhodesia's first twelve Hunters were delivered in 1963. Further aircraft were acquired from Kenya in 1981 (Rhodesia having declared independence from Britain to become Zimbabwe), but three of these aircraft (and six from the original order) were written off following a sabotage attack in 1982. Britain supplied a further five aircraft in 1983. It is believed that the remaining Hunters were withdrawn from service in around 1998/9.

FGA Mk 9	RRAF 116 to RRAF 127; 1805; 8106, 1807; 1088; 1098; 1180; 1811; 8112; 1813; 8116; 1817; 1188; 1198; 1280; 1821; 8122; 1823; 1284; 1258; 1286; 1827
FGA Mk 80	8100; 1801; 1082; 1038
T Mk 81	1084

RRAF Hunters were not reconnaissance-configured aircraft but they were able to undertake photography missions thanks to the creation of a custom-built reconnaissance pack, built into a standard 100-gallon drop tank. The equipment was used to effect during the 'Bush War'.

RRAF Hunter F Mk 6 122 was formerly XG294. It suffered damage following an overshoot at Thornhill and was ultimately destroyed during a sabotage attack in July 1982.

Not a SAAF aircraft but certainly the only Hunter to wear SAAF markings, this unidentified photograph shows an RRAF Hunter that took part in Sidewinder missile trials during the early 1970s in South Africa.

A rare image of a Zimbabwe AF Hunter streaming its brake chute on touchdown. This was one of the final six aircraft operated by Zimbabwe shortly before the type was finally withdrawn in 2002.

Saudi Arabia

Unusually, Saudi Arabia's purchase of Hunters was part of a wider deal that included the Lightning interceptor, Strikemaster and SAM systems. Four refurbished single-seaters and a pair of twin-seaters were supplied to Saudi Arabia to act as advanced pre-Lightning conversion trainers but were also used as interim fighters pending the delivery of the Lightnings. The single-seaters were withdrawn by 1968 but the twin-seaters remained in use until 1974 when they returned to RAF service (XX466 and XX467).

F Mk 60	60/601; 60/602; 60/603; 60/604
T Mk 7	70/616; 70/617

Singapore

Having been re-formed with RAF assistance, it was no surprise that the Republic of Singapore Air Force purchased Hunters as its first combat aircraft. A fleet of twelve FGA Mk 74, four FR Mk 74A and four T Mk 75 aircraft were delivered in 1970/71, followed by 22 FR Mk 74B and five T Mk 75A aircraft. Some of these initially operated out of RAF Chivenor while new pilots were converted onto the type with 229 OCU. The Hunter was phased out of service in 1992.

FGA Mk 74	501 and 502; 505; 507 to 511; 513; 515; 518; 519
FR Mk 74A	503; 506; 512; 517
FR Mk 74B	529; 530; 537 to 539; 541 to 543; 520 to 527; 531; 533 to 535; 545; 546
T Mk 75	500; 504; 514; 516
T Mk 75A	528; 532; 536; 540; 544

G-9-290 (formerly RNethAF N-303) is pictured at Dunsfold in post-restoration primer paint, prior to delivery to Singapore as No 500. *Paul Tomlin*

Somalia

The Somalian Air Corps received six Hunters from Abu Dhabi in 1983, one of which was a dual-control trainer. However, conflicting reports indicate that a total of nine or even eleven Hunters may have been acquired. They were used operationally against Somali rebels but were largely withdrawn from use in 1989, the last attempt to fly any of the aircraft taking place the following year.

FGA Mk 76 and FR Mk 76A	CC-701 to CC-710
T Mk 77	CC-711

Hunter CC-711 was the sole twin-seat T Mk 77 transferred to Somalia from Abu Dhabi. It is believed that Somalia operated a fleet of eight single-seaters in addition to this aircraft.

Formerly XF362 (which served with No 112 Squadron), FGA Mk 76 No CC705 remained in service with the Somali Democratic Republican Air Force until the 1980s after having been transferred from Abu Dhabi.

Sweden

As the very first export customer for the Hunter, in 1954 Sweden ordered 120 F Mk 4 aircraft (designated as the J-34 in Swedish service) with one aircraft being delivered on loan, for evaluation and familiarisation. Delivered without link collectors or wing leading edge extensions, the collectors were retrofitted but the wings remained unmodified. They were, however, modified to carry Sidewinder AAMs. Further modifications included a scheme to fit the fleet with afterburning engines but, despite trials, the idea was not adopted, and the fleet was finally withdrawn in 1966. Unusually, no dual-control versions were ordered.

F Mk 50	34001 to 34120

During 1962 the Swedish AF formed a Hunter aerobatic team comprising aircraft from F18. The lead aircraft of the 'Acro Hunters' formation carries non-standard yellow markings, presumably used as a visual aid for formation-keeping.

Although Swedish Hunters lacked the wing leading edge extensions fitted to later variants, their Hunters were equipped with the 'Mod 228' wing, which enabled additional stores to be carried. Sweden opted to arm the Hunter fleet with Sidewinder missiles, as illustrated.

Hunter Mk 50 34018 appears to have taken the arrestor barrier following a runway overshoot. It was later written off after a crash-landing during June 1959, following engine failure.

Groundcrew fitting a Sidewinder AAM to a Swedish AF Hunter.

This unidentified Swedish Hunter appears to have made a wheels-up landing. Another F18 Hunter is seen taking off in the distance.

Inexplicably, Hawker failed to negotiate a re-sale clause in the Swedish order, so the aircraft were not bought back by HAS upon retirement. Virtually all of the Hunter fleet remained in Sweden and most airframes were scrapped.

The unique reheat system installation devised for Swedish Hunters, but not used operationally.

Switzerland

Following the evaluation of various aircraft types (including the F-86E and Gnat), Switzerland ordered the Hunter after its own indigenous P-16 design was cancelled. Twelve former RAF machines were delivered followed by eighty-eight new-build aircraft, fitted with brake chutes and ballast to allow additional outboard stores carriage. Delivery commenced in 1958 and Hawker's G-APUX twin-seater was also used on a loan basis. A second batch of refurbished Hunters was delivered between 1971 and 1973, and when an evaluation of designs to replace the Venom was ultimately made in favour of the Hunter, yet another batch of refurbished single-seaters was purchased together with eight twin-seaters in 1974/75. Used initially in the interception role (carrying Sidewinders), the Hunters were later assigned to ground-attack duties, armed with CBUs and Maverick missiles, amongst other systems. Switzerland was a major operator of the Hunter, and it was not until 1994 that the last examples were withdrawn from service, most eventually being presented to civilian operators and museums around the world.

F Mk 58	J-4001 to J-4100
F Mk 58A	J-4101 to J-4152
T Mk 68	J-4201 to J-4208

An in-flight view of the Patrouille Suisse, captured en route to a display. Of note is the flap setting (indicating a relatively low air speed) and the gun muzzle deflectors.

Switzerland traditionally exercises a policy of off-base operational capability. Hunters were frequently detached to a variety of sites to maintain this ability, and motorways often became temporary runways.

Appendix 2

Aircraft serial numbers

United Kingdom

Hawker P.1040

VP401 ff 02.09.47	First of three prototypes constructed.
VP413 ff 03.09.48	Second prototype. Navalised version. Used in carrier trials.
VP422 ff 17.10.49	Third prototype. Provision for rocket-assisted take-off gear (RATOG) and attachment points for drop tanks. Rolls-Royce Nene 101 engine.

Hawker P.1052

Specification E.38/46. Swept-wing variant of P.1040. Contract received by Hawker Aircraft Ltd 05.47

VX272 ff 19.11.48	First prototype. Preserved at Fleet Air Arm Museum.
VX279 ff	Second prototype. Preserved at Cosford.

Hawker P. 1067

Designed to Air Ministry Specification F.3/48. Contract placed 25.06.48.

WB188 ff 20.07.51	First prototype. Boscombe Down, Performance & Handling Trials. Fitted with side-mounted air brakes, extra fuel tanks and RR Avon RA7R with reheat for World Air Speed Records 07.09.53. (Mk 3). Air brake trials. GIA No 1 SoTT, Halton 10.53. To Melksham as gate guard. To Abingdon 1961-64, then Colerne until 1975. Transferred to RAF Museum Cosford, 1975-09.92, then loaned to Tangmere Military Aviation Museum Trust.
WB195 ff 05.05.52	Second prototype. Air brake development trials. Initial spinning trials. Armed with full four Aden cannon and radar ranging. GIA 7284M 1 SoTT Halton 03.59. Sold as scrap 12.67.
WB202 ff 30.11.52	Third prototype. AS Sapphire engine. Delivered 09.03.53. Air brake and drop tank trials at Dunsfold. Firestreak installation. SOC 31.12.57. Scrapped at Bedford 1960.

Hawker P.1072

Airframe modified from P.1040 with Armstrong-Siddeley Snarler liquid fuel rocket motor.

VP401 ff 20.11.50	Made six flights, engine explosion on final flight, repaired but not flown again and eventually scrapped.

Hawker P.1081

VX279 ff 19.06.50	Modified airframe from P.1052. Crashed near Lewes in Sussex during test flight, cause unknown. Pilot Sqn Ldr T. S. 'Wimpy' Wade killed. Aircraft landed relatively undamaged. W/O 03.04.51.

Hawker P.1083

Ordered under Contract No 6296. Instruction to proceed received 26.02.52 Draft Specification issued 18.04.52 based on modified airframe from fourth P.1067 prototype.

WN470	RR Avon RA14R and 50º swept wing of 7.5% thickness/chord ratio. Reached advanced stage of manufacture before being cancelled 13.07.53. Rear fuselage and nose used for construction of P.1099.

Hawker P.1099

Ordered under Contract No 6/Aircraft/10032/C.B.5(b). Instruction to Proceed dated 17.11.53.

XF833 ff 22.01.54	Prototype F Mk 6. Flying trials Dunsfold and Boscombe Down. To Miles Aircraft Ltd, Shoreham, for fitting of RR thrust-reversal equipment 05.06.56.

Hawker P. 1101

Ordered under Contract No 11595

XJ615 ff 08.07.55	First prototype twin-seat trainer. RR Avon (113) RA21 engine. Initially with two Aden cannons, subsequently modified for one Aden cannon. Hood Development, Spinning, Gun Firing Trials. To ETPS Farnborough, struck trees on high ground 1.5 miles SE of Haslemere, Surrey. Pilot killed. W/O 24.06.64.
XJ627 ff 17.11.56	Second prototype twin-seat trainer. RR Avon (203) RA28 engine. Boscombe Down 06.12.57. Returned to HAL 04.07.58. To Martin Baker, Chalgrove, for ejection seat trials 19.11.58. Re-painted in blue and white with Martin Baker logo. Starboard side plexiglass canopy panel removed and plexiglass vertical barrier fitted. Re-purchased by HAS, 27.09.68. Refurbished as G-9.296. Converted to Chilean T Mk 72 (721). Delivered via Chester to Chile 19.05.71. Antofagasta, Cerro Moreno AB, Chile.

Hawker Hunter F Mk 1

Schedule of Supply placed with Hawker Aircraft Ltd for 200 F.3/48 aircraft dated 14.03.51 under Requisition Nos 26FX/1/E.23(a)/50 and RD/RAF/46/50.

WT555 ff 16.05.53	Neville Duke 'State Express' in reference to first formation flight. Duke of Edinburgh crest on nose. Handling and performance trials HAL and A&AEE 11.53. Retired 11.53. Transferred to No 5MU Kemble, 28.06.57. Allocated GIA No 7499M. Transferred to No 7 Radio School, Locking 11.57. Transferred to No 2 SoTT Cosford 1967, then to RAF Museum, Cosford. To Vanguard Holdings, Ealing, West London, and displayed on roof of building.
WT556 ff 16.06.53	CA aircraft. RR for engine trials. Used for familiarisation flying at Boscombe Down 7.53.
WT557 ff 17.07.53	CA aircraft. To RRE for radio trials.
WT558 ff 24.08.53	Gun firing trials. Fitted with wing fences. A&AEE Boscombe Down 11.53. No 54 Sqn 'T'. Allocated 7492M.
WT559 ff 02.09.53	CA aircraft. To A&AEE Boscombe Down for hood jettison trials. No 43 Sqn 'A' Leuchars, 1954. Allocated 7493M.
WT560 ff 28.09.53	CA aircraft, Dunsfold. Minor trial installations. To RR for Avon 119. Allocated 7488M.
WT561 ff 23.10.53	CA aircraft, Dunsfold. Minor trial installations.
WT562 ff 29.10.53	CA aircraft, Dunsfold. Trials of one-third-span flaps. Crashed at RAF Thorney Island after engine failure. W/O 06.08.55.
WT563 ff 02.11.53	CA aircraft, Dunsfold. Trials of interim standard flying tail. No 233 OCU 'T'. Rolled into ground after take-off 2 miles north of Pembrey. Pilot killed. W/O 20.06.57.
WT564 ff 28.11.53	CA aircraft, Dunsfold. A&AEE Boscombe Down for interception trials 1954.
WT565 ff 07.12.53	RR for Avon 119 development.
WT566 ff 12.12.53	CA aircraft, Dunsfold, for trials of ventral air brake development. SOC 1979, Catterick 7494M.
WT567 ff 18.12.53	Gun firing trials, A&AEE Boscombe Down. Allocated 7489M.
WT568 ff 16.12.53	CA aircraft, Dunsfold. Trials of wing leading edge extensions. To A&AEE 01.54. Allocated 7490M.
WT569 ff 18.12.53	A&AEE 15.07.54. Trials of full-power ailerons. Tropical trial, Khartoum 09.54. First Hunter to carry drop tanks. GIA 7491M. 2117 Sqn ATC Kenfig Hill, South Wales, 2002.
WT570 ff 22.12.53	CA aircraft, Dunsfold. Trials of full-power ailerons. To A&AEE 27.07.54. 229 OCU. No 2 SoTT Cosford 1956-65 as 7497M. Scrapped 1965.
WT571 ff 08.01.54	Trials of 'area rule' fuselage. To RAE Farnborough 01.07.55.
WT572 ff 05.01.54	HSA company trials.
WT573 ff 05.01.54	CA aircraft, Dunsfold, and A&AEE Boscombe Down. To RR for Avon 119 development.

WT574 ff 22.01.54	RR 12.05.54. Engine failure during sortie from Hucknall. Forced-landing on grass. SOC 31.08.56.
WT575 ff 08.01.54	A&AEE 05.07.54. No 229 OCU 'RS-A'. Crashed at Chivenor, 03.04.58.
WT576 ff 18.01.54	A&AEE 03.06.54. AFDS. No 2 SoTT Cosford 1956-65 as 7498M. Scrapped 1965.
WT577 ff 19.01.54	CFE 05.07.54. AFDS. W/O 20.08.56.
WT578 ff 20.01.54	CFE 05.07.54. AFDS.
WT579 ff 23.01.54	CFE 08.06.54. No 229 OCU. Crashed into sea at Morte Bay, Devon, during aerobatics practice. Pilot killed. W/O 21.01.55.
WT580 ff 25.01.54	No 5MU Kemble 07.07.54. No 43 Sqn 'E' Leuchars 1954. No 229 OCU Chivenor. Swung off runway at Chivenor. W/O 22.03.57.
WT581 ff 03.02.54	No 5MU Kemble 05.08.54. No 222 Sqn 06.12.54. No 43 Sqn 'S' Leuchars 07.12.54. Crashed west of Leuchars after engine flame-out. Pilot killed. W/O 12.07.55.
WT582 ff 08.02.54	No 5MU Kemble 05.07.54. No 43 Sqn 'B' Leuchars 1954.
WT583 ff 10.02.54	First aircraft with Avon 113. To No 5MU Kemble 09.07.54. No 43 Sqn 'D' Leuchars 1954.
WT584 ff 20.02.54	No 5MU Kemble 07.07.54. No 43 Sqn Leuchars 1954. No 229 OCU.
WT585 ff 20.02.54	No 5MU Kemble 07.07.54. No 43 Sqn 'F' Leuchars 1954.
WT586 ff 22.02.54	No 5MU Kemble 09.07.54. No 229 OCU.
WT587 ff 24.02.54	No 5MU Kemble 21.07.54. No 43 Sqn 'C' Leuchars 1954. No 229 OCU Chivenor. Crashed near Lundy Island, pilot unknown. Indian AF. W/O 10.11.56.
WT588 ff 24.02.54	CFE, AFDS. No 43 Sqn 'P' Leuchars 1954.
WT589 ff 15.06.54	No 5MU Kemble 09.07.54. No 43 Sqn Leuchars 1954. Crashed 7 miles E of Leuchars. W/O 16.11.54.
WT590 ff 21.06.54	No 5MU Kemble, 22.07.54. No 43 Sqn Leuchars 1954. St Athan. Scrapped 17.05.57.
WT591 ff 21.06.54	CFE 16.07.54. DFLS. W/O 24.01.56.
WT592 ff 22.06.54	No 54 Sqn 'S' 'Black Knights' team. No 233 OCU. Allocated 7495M.
WT593 ff 01.07.54	No 5MU Kemble 17.09.54. DFLS.
WT594 ff 02.07.54	No 5MU Kemble 17.09.54. No 222 Sqn 07.12.54. No 43 Sqn 'U' Leuchars 21.12.54.
WT595 ff 02.07.54	No 5MU Kemble 27.07.54. No 43 Sqn 'Q' Leuchars 1954.
WT611 ff 08.07.54	Trials with Avon 115. No 5MU Kemble. Crashed 2 miles SW of Kemble, W/O 9.9.54.
WT612 ff 13.07.54	Second aircraft with Avon 115. To No 5MU Kemble 7.9.54. GIA Yatesbury 11.57. GIA 7496 Halton. Gate guard RAF Henlow.
WT613 ff 16.07.54	No 5MU Kemble 05.08.54. No 43 Sqn 'R' Leuchars 1954. DFLS.
WT614 ff I2.07.54	No 5MU Kemble 17.09.54. FWS.
WT615 ff 20.07.54	No 5MU Kemble 21.10.54. No 222 Sqn 'Q' 24.01.55. No 233 OCU 08.12.55.
WT616 ff 20.07.54	CA aircraft. Trials with Avon 115. No 43 Sqn Leuchars 1954. SOC 1980. Allocated 7500M.
WT617 ff 29.07.54	No 5MU Kemble 04.10.54. DFLS.
WT618 ff 28.07.54	No 5MU Kemble 07.09.54. No 43 Sqn 'O' Leuchars 1954. DFLS. Hit tree at West Raynham. W/O 05.06.56.
WT619 ff 28.07.54	No 5MU Kemble 13.09.54. No 43 Sqn Leuchars 1954. No 222 Sqn 'B' 14.12.54. CWP Horsham St Faith 14.12.55. No 233 OCU. GIA 7525M St Athan. Henlow. Museum of Science and Industry, Manchester.
WT620 ff 29.07.54	No 5MU Kemble 22.09.54. No 222 Sqn 'D' 15.12.54. No 233 OCU 14.12.55.
WT621 ff 10.08.54	ETPS 21.09.54.
WT622 ff 08.07.54	No 5MU Kemble 27.07.54. No 43 Sqn 'G' Leuchars 1954. SOC 1979 Catterick.
WT623 ff 15.07.54	No 5MU Kemble 05.08.54. No 43 Sqn 'H' Leuchars 1954. Missing over North Sea during air-to-air firing trials. W/O 19.01.56.

WT624 ff 13.07.54	No 5MU Kemble 12.08.54. No 222 Sqn 'F' 30.12.54. No 229 OCU 25.11.55.
WT625 ff 07.08.54	No 5MU Kemble 20.09.54. No 229 OCU. No 233 OCU.
WT626 ff 28.07.54	No 5MU Kemble 07.09.54. No 229 OCU.
WT627 ff 07.08.54	No 5MU Kemble 04.10.54. DFLS. Crashed on approach to West Raynham. W/O 13.08.56.
WT628 ff 05.08.54	No 5MU Kemble 27.09.54. ETPS '26'. Crashed at Cove near Farnborough due to fuel starvation, pilot killed. W/O 18.07.55.
WT629 ff 06.08.54	No 5MU Kemble 22.09.54. DFLS 'T' West Raynham. Crashed at Great Thornes Farm 2 miles NW of Swaffham. W/O 08.02.56.
WT630 ff 01.09.54	No 5MU Kemble 12.10.54. No 43 Sqn Leuchars 1954. No 222 Sqn 'T' 27.01.55. No 229 OCU 16.11.55. No 43 Sqn No 233 OCU.
WT631 ff 29.07.54	No 5MU Kemble 22.09.54. No 222 Sqn 'A' 07.12.54. No 229 OCU 22.11.55.
WT632 ff 04.09.54	No 5MU, Kemble 27.09.54. Crashed on overshoot at Kemble. Pilot killed. W/O 06.12.55.
WT633 ff 12.08.54	RAE Farnborough 29.09.54 for miscellaneous trials. No 43 Sqn Leuchars 1954.
WT634 ff 10.08.54	No 5MU Kemble 28.09.54. No 222 Sqn 'R' 24.01.55. No 229 OCU 12.12.55. No 233 OCU.
WT635 ff 01.09.54	Delivered 04.10.54 No 233 OCU.
WT636 ff 09.09.54	Delivered 04.11.54 No 233 OCU. No 4 SoTT St Athan as 7514M 22.11.57.
WT637 ff 15.09.54	Delivered 11.10.54 No 43 Sqn Leuchars 1954. No 222 Sqn 'G' later 'F' 30.12.54. To CWP Horsham St Faith 03.01.56. No 229 OCU GIA 7514M 22.11.57. Stradishall fire dump 30.08.67.
WT638 ff 07.09.54	No 222 Sqn 'P' 03.01.55. No 233 OCU 13.01.56.
WT639 ff 03.09.54	Delivered 27.10.54 DFLS 'N' West Raynham. Crashed at Beechamwell after running out of fuel. W/O 08.02.56.
WT640 ff 09.09.54	Delivered 19.10.54 No 54 Sqn No 229 OCU.
WT641 ff 02.09.54	Delivered 11.10.54 No 43 Sqn 'T' Leuchars 07.12.54. No 222 Sqn 12.54. DFLS.
WT642 ff 10.09.54	Delivered 07.10.54 No 43 Sqn Leuchars 1954. No 229 OCU.
WT643 ff 09.09.54	Delivered 15.10.54 No 43 Sqn Leuchars 1954. No 222 Sqn 'H' 03.01.55. No 233 OCU 13.01.56.
WT644 ff 04.09.54	Delivered 07.10.54 No 43 Sqn 'C' Leuchars 1954. Aircraft flew into hill W of Leuchars, pilot killed. W/O 14.08.55.
WT645 ff 16.09.54	Delivered 07.10.54 DFLS.
WT646 ff 16.09.54	Delivered 19.10.54 No 222 Sqn 'U' to 13.01.56. SOC 05.03.57.
WT647 ff 11.09.54	Delivered 17.11.54 No 222 Sqn 'V' 10.02.55. Returned to HAL 04.05.55. No 222 Sqn 'V' 01.11.55. No 233 OCU 13.01.56.
WT648 ff 04.10.54	Delivered 04.11.54 No 222 Sqn DFLS. GIA 7530M St Athan. Scrapped at Hanningfield. Metal cockpit section to Air Defence Collection, Boscombe Down Museum.
WT649 ff 21.09.54	Delivered 12.11.54 No 43 Sqn 'N'. No 222 Sqn.
WT650 ff 17.09.54	Delivered 15.10.54 No 222 Sqn.
WT651 ff 23.09.54	Delivered 21.10.54 No 222 Sqn 'C' Leuchars 16.12.54. No 229 OCU 09.01.56. No 233 OCU. No 229 OCU. GIA 7532M Credenhill. Halton. Newark Air Museum.
WT652 ff 18.09.54	Delivered 12.10.54 DFLS.
WT653 ff 27.09.54	Delivered 09.11.54 No 229 OCU.
WT654 ff 24.09.54	Delivered 03.11.54 No 229 OCU.
WT655 ff 04.10.54	Delivered 24.11.54 No 233 OCU. No 229 OCU. Collided with WT689 during landing break. Pilot killed. W/O 02.09.55.

WT656 ff 14.10.54	Experimental trials of blown flap at HAL and RAE.
WT657 ff 02.10.54	Delivered 10.12.54 No 229 OCU. No 233 OCU.
WT658 ff 06.10.54	Delivered 24.11.54 DFLS. Crashed 3 miles SSW of Wattisham. W/O 23.03.56.
WT659 ff 29.09.54	Delivered 3.11.54 No 54 Sqn 'D' later 'U' Odiham. 'Black Knights' team 1955. No 229 OCU. Crashed in River Taw. Pilot killed. W/O 10.04.56.
WT660 ff 11.10.54	Delivered 17.11.54 DFLS. RAF Carlisle gate guard 7.92. Whitehills Farm, Cullen, Grampian, Scotland 1993. James A. Campbell, Highland Aircraft Preservation Society, Inverness, Scotland 09.09.07, stored in pieces at Glastullich Farm, Tain, Scotland. Wears markings of No 43 Sqn. Intention at later date to reassemble aircraft for display at Highland Aviation Museum.
WT679 ff 28.09.54	Delivered 12.11.54 No 229 OCU. U/c leg shear, aircraft swung into obstruction, damaged beyond repair. W/O 13.01.56.
WT680 ff 06.10.54	Delivered 10.11.54 West Raynham Stn Flt. DFLS. GIA 7533M Weeton 11.57. DTEO Aberporth, 1429 Sqn ATC. To Anglia Hotel, Lincs 26.07.04.
WT681 ff 11.10.54	Delivered 24.11.54 No 54 Sqn 'R' Odiham. No 229 OCU 'RS-T'. CAT5 after bird-strike near Chivenor 22.05.57. GIA 7442M Weeton 20.06.57.
WT682 ff 06.10.54	Delivered 17.11.54 No 229 OCU. No 43 Sqn Leuchars 1954.
WT683 ff 08.10.54	Delivered 24.11.54 DFLS.
WT684 ff 14.10.54	Delivered 24.11.54 DFLS. GIA 7422M Bicester 03.57. South Yorkshire Aviation Museum 10.04.
WT685 ff 09.10.54	Delivered 10.12.54 No 229 OCU.
WT686 ff 14.10.54	Delivered 12.11.54.
WT687 ff 20.10.54	Delivered 15.12.54 No 54 Sqn.
WT688 ff 19.10.54	Delivered 10.12.54 No 229 OCU.
WT689 ff 20.10.54	Delivered 10.12.54 No 229 OCU. Collided with WT655. Pilot killed. W/O 02.09.55.
WT690 ff 13.10.54	Delivered 24.11.54 DFLS.
WT691 ff 21.10.54	Delivered 10.12.54 No 229 OCU. No 233 OCU.
WT692 ff 23.10.54	Delivered 10.12.54 No 43 Sqn 'E' Leuchars 1954. No 54 Sqn 'S'. DFLS.
WT693 ff 21.10.54	Delivered 10.12.54 No 54 Sqn.
WT694 ff 29.10.54	Delivered 02.02.55 No 43 Sqn 'C' Leuchars 1955. No 54 Sqn DFLS. No 229 OCU. GIA 7510M Debden. Newton. Caernarfon Air World 2001.
WT695 ff 08.11.54	Delivered 15.12.54 No 229 OCU. No 233 OCU. Scrapped No 5MU Kemble 02.04.58.
WT696 ff 04.11.54	Delivered 04.02.55 No 54 Sqn 'O'. No 229 OCU Chivenor 'RS-S' 11.55-07.57. GIA 7515M 22.11.57. Scrapped at St Athan.
WT697 ff 01.11.54	Delivered 17.12.54. Returned to HAL for demonstration flights. Scrapped No 5MU Kemble 02.04.58.
WT698 ff 04.11.54	Delivered 10.12.54 No 54 Sqn DFLS.
WT699 ff 03.11.54	Delivered 10.12.54 No 229 OCU.
WT700 ff 10.11.54	Delivered 01.02.55 No 229 OCU 'RS'. Force-landed in River Taw, declared CAT5. Later scrapped. W/O 07.06.57.

Second production batch of twenty-six aircraft built by Hawker Aircraft (Blackpool) Ltd, Contract No SP/6/8435/CB.7a dated 15.08.53.

WW599	Delivered 07.07.54 No 5MU Kemble. No 43 Sqn 'B' Leuchars 29.07.54. First Hunter received by No 43 Sqn. Crashed at South Lambieletham, Fife. W/O 17.04.56.
WW600 ff 07.07.54	Delivered 16.08.1954, No 43 Sqn DFLS. Bird-strike, damaged beyond repair. RAF Fire School, Sutton-on-Hull. W/O 01.11.56.
WW601	Delivered 20.08.54 DFLS 'G' West Raynham. GIA 7538M 22.11.57. Scrapped Halton.
WW602	Delivered 26.08.1954 No 229 OCU 'RS-L'. GIA 7539M 22.11.57. Manston Fire School 1964, later scrapped.

WW603	Delivered 07.10.54 DFLS. Crash-landed at Marham after running out of fuel. W/O 08.02.56. GIA 7413M 23.04.57. SOC to St Athan and scrapped.
WW604	Delivered 14.09.54. No 222 Sqn 'E' 17.12.54. No 233 OCU 09.01.56. GIA 7540M 22.11.57. Scrapped at Halton.
WW605	Delivered 23.09.54. No 233 OCU. Scrapped at Farnborough 1963.
WW606	Delivered 07.10.54 No 222 Sqn 'W' 10.02.55. No 229 OCU Chivenor 'RS-H' 15.11.55. Crashed at Chivenor. W/O 18.09.56.
WW607	Delivered 03.11.54 No 229 OCU. Force-landed west of Chivenor. W/O 11.10.55. GIA 7290M 09.11.55 Manby. Scrapped at Manby 01.03.61.
WW608	Delivered 04.11.54 DFLS. Scrapped 02.04.58.
WW609	Delivered 03.11.54 No 222 Sqn 'S' 27.01.55. No 233 OCU 12.12.55. Force-landed 3 miles NW of Swaffham. GIA Halton 11.60. 7541M 22.11.57. Scrapped at Halton.
WW610	Delivered 04.11.54 No 54 Sqn 'N'. FWS. Scrapped 30.04.58 at No 5MU Kemble.
WW632	Delivered 24.11.54 No 229 OCU. GIA 7516M 22.11.57. Scrapped St Athan 1972.
WW633	Delivered 24.11.54 DFLS 'H' West Raynham. Crashed 3 miles SW of Swaffham when aircraft ran out of fuel. W/O 08.02.56.
WW634	Delivered 24.11.54 No 229 OCU. GIA 7517M 22.11.57. Scrapped St Athan.
WW635	Delivered 25.11.54. Crashed at Great Fransham when aircraft ran out of fuel. Pilot killed. W/O 08.02.56.
WW636	Delivered 03.12.54 No 54 Sqn 'Q' Odiham. No 229 OCU. Scrapped at Kemble 30.04.58.
WW637	Delivered 07.12.54 No 229 OCU. No 233 OCU. GIA 7518M 22.11.57. Scrapped at St Athan.
WW638	Delivered 27.01.55 No 247 Sqn 'J'. DFLS. GIA 7519M 22.11.57. Scrapped St Athan.
WW639	Delivered 28.01.55 DFLS 'X' West Raynham. Crashed 3 miles S of Swaffham. W/O 08.02.56.
WW640	Delivered 17.12.54 No 54 Sqn 'C'. Aircraft caught fire on start-up at Odiham. W/O 02.05.55. GIA 7212M 07.06.55. Scrapped at Kirkham 1957.
WW641	Delivered 08.02.55 No 43 Sqn 'Q' Leuchars 1955. No 54 Sqn 'B'. 'Black Knights' team 1955. DFLS. GIA 7520M 22.11.57. Scrapped St Athan.
WW642	Delivered 02.02.55 AFDS. Scrapped 02.04.56 at No 5MU Kemble.
WW643	Delivered 02.02.55 No 229 OCU. W/O 03.04.57. GIA 7424M 15.15.57. Scrapped at Halton 21.02.61.
WW644	Delivered 08.02.55 No 229 OCU Chivenor 'RS-Y'. GIA 7521M 22.11.57. Scrapped at St Athan 1972.
WW645	Delivered 08.02.55 No 43 Sqn 'S' Leuchars 1954. DFLS. Scrapped 30.04.58. Fuselage at Dunsfold 04.62.

Hawker Hunter F Mk 2

Contract No SP/6/6315/CB.7a. Forty-five aircraft built by Armstrong Whitworth Aircraft Ltd, Coventry.

WN888 ff 14.10.53	Retained by AWA for performance and handling trials. Dunsfold 03.11.53. No 257 Sqn 'B' 27.07.56. No 5MU Kemble 10.12.56. Allocated 7486M.
WN889	CA aircraft. Baginton 18.01.54. Trials with Sapphire AS Sa.12 engine.
WN890	CA aircraft AWA Baginton. Boscombe Down for de-tuner trials 11.62. Nose to Boscombe Down Museum 08.98.
WN891	CA aircraft Baginton. A&AEE. CEPE Canada for winterisation trials. Returned to UK and scrapped 06.01.59.
WN892	CA aircraft. A&AEE Boscombe Down 01.10.54.
WN893	CA aircraft. RAE Farnborough and Bedford 31.08.54.
WN894 ff 26.08.54	CA aircraft. Allocated to trials A&AEE Boscombe Down.

WN895	CFE and AFDS. No 263 Sqn 02.02.55. SOC Catterick 1978.
WN896	No 257 Sqn 'C' 15.10.54. Ground fire at Wattisham. Damaged beyond repair. W/O 20.12.54.
WN897	No 263 Sqn. No 257 Sqn 24.07.56. No 5MU Kemble 10.04.57. Allocated 7513M.
WN898	No 263 Sqn. No 1 Sqn 'N'. No 257 Sqn 'A' Odiham 29.09.54. GIA 7440M Henlow 13.04.57.
WN899	No 263 Sqn. GIA 7542M Henlow.
WN900	No 263 Sqn 'D'. No 257 Sqn 'Z' 16.01.57. No 5MU Kemble 15.04.57.
WN901 ff 12.07.54	Delivered 03.09.54. No 257 Sqn 'B' 29.09.54. To AWA 24.02.55. No 12 SoTT Melksham. GIA 7543M. No 9 SoTT Newton.
WN902	No 257 Sqn 'D' 23.11.54. No 5MU Kemble 26.04.57.
WN903	No 257 Sqn 'X' and 'B' 29.03.55. To AWA 30.08.55. No 257 Sqn 'X' 06.09.55. To AWA 23.05.56. No 263 Sqn 'B'.
WN904 ff 27.07.54	Delivered 13.09.54. No 257 Sqn 'Q' 20.10.54. To Gloster Aircraft 12.01.56. SOC 22.11.57. GIA 7544M. No 12 SoTT, Melksham 1959. No 9 SoTT Newton. To Duxford Museum.
WN905	Crashed East Langton, Leics. W/O 14.08.54.
WN906	Delivered 23.9.54 No 5MU Kemble. No 257 Sqn 'L' 07.09.56. No 5MU Kemble 10.04.57. CFE. GIA No 12 SoTT Melksham 7454M.
WN907	Delivered 8.9.57 No 257 Sqn 'H' 31.01.55. No 5MU Kemble 09.04.57. GIA No 12 SoTT Melksham 1959. Colerne 7416M 08.68. St Athan. Nose preserved at Robertsbridge Aviation Centre, Sussex.
WN908	Delivered 08.09.57 No 5MU Kemble. No 263 Sqn.
WN909	Delivered 27.09.54 No 45MU. No 257 Sqn 'S' 03.12.54. No 58MU Honington 02.06.55. No 257 Sqn 'S' 12.09.55. Crashed near Colne Engaine, Essex. Pilot killed. W/O 17.04.56.
WN910	Trials AWA. Delivered No 41MU. RAF Halton 1964 7487M.
WN911	Delivered 13.09.54 No 5MU Kemble. CFE. AFDS.
WN912	Delivered 23.09.54 No 5MU Kemble. No 257 Sqn 'X' 30.07.56. No 5MU Kemble 10.04.57. No 263 Sqn CFE.
WN913	Delivered 09.10.54 No 5MU Kemble. No 257 Sqn 'U' 17.01.55. To AWA 11.10.56. No 257 Sqn 'U' 22.02.57. To No 5MU Kemble 11.04.57. No 263 Sqn.
WN914	Delivered 28.9.54 No 5MU Kemble. No 257 Sqn 'V' 31.01.55. CAT5 No 54MU 26.09.56. SOC 20.06.57.
WN915	Delivered 21.9.54 No 5MU Kemble. No 263 Sqn 'L' 18.03.55. No 257 Sqn 'T' 01.04.56. No 5MU Kemble 10.04.57. Fire Fighting School, Sutton-on-Hull.
WN916	Delivered 14.10.54 No 45MU. CFE. AFDS.
WN917	Delivered 07.10.54 No 45MU. No 257 Sqn 'E' 03.12.54. CAT4. To AWA 13.08.56. No 257 Sqn 'E' 29.03.57. No 5MU Kemble 09.04.57.
WN918	Delivered 23.09.54 No 5MU Kemble. No 257 Sqn 'W' 31.01.55. No 5MU Kemble 26.10.56. Loaned to No 1 Sqn.
WN919	Delivered 27.09.54 No 45MU. No 257 Sqn 'Q' 08.03.56. No 5MU Kemble 10.04.57. No 263 Sqn 'F'. Loaned to No 1 Sqn.
WN920	Delivered 28.09.54 CFE. No 257 Sqn 'P' 31.08.56. No 5MU Kemble 17.04.57. AFDS.
WN921	Delivered 01.10.54 No 5MU Kemble. CFE. No 263 Sqn 'S'. Crashed into North Sea. Pilot killed. W/O 06.03.56.
WN943	Delivered 22.10.54 No 5MU Kemble. No 257 Sqn 'M' 01.09.55. No 5MU Kemble 11.04.57.
WN944	Delivered 7.10.54 No 45MU. No 263 Sqn 'H' 21.04.55.

WN945	Delivered 21.10.54 No 45MU. CFE 01.12.54. AFDS 'N'. Transferred to Day Fighter Leader School 09.56. Survived mid-air collision and repaired. No 257 Sqn 'M'.
WN946	Delivered 28.10.54 No 5MU Kemble. No 263 Sqn 'N' 04.02.55. CFE.
WN947	Delivered 09.11.54 No 5MU Kemble. No 263 Sqn 'R' 31.05.55. No 257 Sqn 'W' 05.09.56. No 5MU Kemble 24.04.57.
WN948	Delivered 27.09.54 No 45MU. No 263 Sqn. No 257 Sqn 'R' 17.11.54. W/O 11.04.57. To Ouston Fire Dump.
WN949	Delivered 15.10.54 No 5MU Kemble. No 257 Sqn 'M' 30.12.54. CAT4. No 58MU Honington 29.08.55.
WN950	Delivered 01.10.54 No 5MU Kemble. No 257 Sqn 'F' 31.12.54. No 5MU Kemble 11.04.57.
WN951	Delivered 22.10.54 No 5MU Kemble. No 263 Sqn No 257 Sqn 'T' 31.12.54. CAT4. To AWA 03.03.55.
WN952	Delivered 14.10.54 No 5MU Kemble. No 257 Sqn 'G' 17.01.55. No 5MU Kemble 15.04.57.
WN953 ff 19.10.54	Delivered 04.11.54 No 45MU. No 257 Sqn 'C' 04.02.55. No 5MU Kemble 01.05.57.

Hawker Hunter Mk 3

Aircraft modified for September 1953 World Absolute Air Speed Trials.

WB188 ff 20.07.51	First prototype. Tangmere Military Aviation Museum Trust, Chichester, W Sussex.

Hawker Hunter F Mk 4

Eighty-five aircraft built by Hawker Aircraft Ltd, Kingston-upon-Thames. Second part of Contract No SP/6/591O/CB. 7a.

WT701 ff 20.10.54	CA aircraft. Under-wing fuel tank trials. Converted to T Mk 8. No 764 Sqn FAA. Collided with XL599, W/O 23.08.61.
WT702 ff 09.12.54	CA aircraft. Retained by Ministry of Supply and used by HAL and A&AEE for radio trials. Placed in store at Dunsfold. Converted to T Mk 8C 1959. No 703 Sqn FAA. '747/VL' No 759 Sqn FAA Brawdy '806/BY' until 1969. Air Direction Training Unit 1970 '737'. FRADU '877/VL'. Placed in store at Kemble. FRADU '874/VL' until 12.82. Crashed into English Channel 2.5 miles SW of Nab Tower. Pilot killed. W/O 16.12.82.
WT703 ff 00.11.54	CA aircraft. External store trials HAL and A&AEE. Broken up Dunsfold 1964.
WT704 ff 26.11.54	CA aircraft. Broken up 1959.
WT705 ff 12.11.54	CA aircraft. Handling and gunnery trials A&AEE. Collided with banner target during air-to-air gunnery trials. WFU 16.10.57. Broken up 11.64.
WT706 ff 13.12.54	CA aircraft. Handling trials A&AEE Boscombe Down. Converted to T Mk 7 standard by HAL, ff as T Mk 62 15.09.59. Delivered to Peru '681'. Sold to Grace Aire, TX, status unknown.
WT707 ff 13.01.55	Crashed at Ford. W/O 25.01.55.
WT708 ff 15.12.54	Delivered 9.02.55 No 33MU. No 54 Sqn 'F'. FWS. Scrapped No 5MU Kemble 1963.
WT709 ff 15.12.54	Delivered 15.02.55 No 33MU. No 54 Sqn. Crashed Slinfold, Sussex. W/O 03.11.55.
WT710 ff 17.12.54	Delivered 15.02.55 No 33MU. No 111 Sqn 'N'. No 222 Sqn 'Q' 28.11.56. No 5MU Kemble 01.11.57. No 54 Sqn. Broken up HAL 1964.
WT711 ff 10.02.55	Delivered 19.03.55 No 33MU. No 54 Sqn. No 14 Sqn 'A'. Converted to GA Mk 11 11.62. No 738 Sqn FAA Brawdy in 1964 '783/BY' until 1970. Placed in store. FRU Hurn 02.08.71 '837'. Yeovilton 20.10.72 FRADU '837/VL'. Only GA.11 not fitted with Harley light. Shawbury A2731 School of Aircraft Handling Culdrose '833/DD' 18.09.85. Sold 01.94. Air Atlantique 24.11.95 Coventry Airport.
WT712 ff 19.01.55	Delivered 01.03.55 No 33MU. No 54 Sqn. Converted to GA Mk 11 05.63. Collided with WT810. W/O 25.10.65.

WT713 ff 25.01.55	Delivered 02.03.55 No 33MU. No 111 Sqn 'S'. No 222 Sqn 'S' 29.11.56. No 5MU Kemble 01.11.57. Converted to GA Mk 11 02.63. Returned to HAL 1970 as G-9.336. Converted to F Mk 58A 'J-4120'.
WT714 ff 25.01.55	Delivered 29.04.55 Lyneham. No 111 Sqn. No 14 Sqn Oldenburg. Collided with WT807. Pilot killed. W/O 18.08.55.
WT715 ff 25.01.55	Delivered 14.03.55 No 33MU. No 111 Sqn No 229 OCU 'L'. Broken up HAL 1964.
WT716 ff 27.01.55	Delivered 29.03.55 No 5MU Kemble. No 111 Sqn 'D'. No 222 Sqn 'T' 04.12.56. No 5MU Kemble 20.01.58. To HAL as G-9.385 04.72. GIA 7790M. Converted to F Mk 58 'J-4152'.
WT717 ff 25.01.55	Delivered 16.06.55 North Weald Stn Flt. Converted to Peruvian Mk 52 '630'. ff 01.12.55. Delivered Peru 02.56.
WT718 ff 29.01.55	Delivered 14.03.55 No 33MU. No 111 Sqn. No 245 Sqn No 229 OCU 'RS-I'. Converted to GA Mk 11. Lossiemouth. W/O, 31.08.64.
WT719 ff 27.01.55	Delivered 17.03.55 No 5MU Kemble. No 43 Sqn 'E' Leuchars 1954. No 92 Sqn. No 118 Sqn 'L' 20.05.55. CAT4 23.08.55. GIA 7596M 1958. Scrapped 1959.
WT720 ff 27.01.55	Delivered 01.03.55 No 33MU. No 111 Sqn 'Q' North Weald 1955. No 222 Sqn 28.11.56. No 74 Sqn 'B' 17.01.57. Broken up HAL 11.64.
WT721 ff 02.02.55	Delivered 15.02.55 No 33MU. No 54 Sqn. Converted to GA Mk 11. Delivered 21.12.62. A&AEE trials. Converted to PR Mk 11 Lossiemouth 1.66. Crashed near Blair Atholl. W/O 22.09.70.
WT722 ff 04.02.55	Delivered 24.05.55 No 54 Sqn Odiham. No 26 Sqn 'S' 29.06.55. No 5MU Kemble 17.05.57. Conversion to T Mk 8. Delivered 10.04.59. No 703 Sqn FAA '873'. No 764 Sqn Lossiemouth '702/LM'. No 759 Sqn Brawdy '807/BY' until 1969. To ADTU Yeovilton as '742' 1970. Became part of FRADU '873'. Became '878/VL' 1980-82. Retired at Yeovilton 1994 and flown to Shawbury for disposal. Classic Jets at Exeter Airport G-BWGN. Delta Jets Kemble 2006 for storage.
WT723 ff 16.02.55	Delivered 25.03.55 No 5MU Kemble. No 54 Sqn Odiham. No 14 Sqn 'T' Oldenburg. No 229 OCU Chivenor. Converted to GA Mk 11. Delivered 17.08.62 No 764 Sqn FAA Lossiemouth '694/LM'. Short Brothers, re-designated PR Mk 11. FRADU Yeovilton as '866/VL'. Retired 23.05.93. Transferred to Culdrose, School of Aircraft Handling until 1996 as A2616. Classic Jets Exeter by 09.97. G-PRII 21.08.99. Believed to be oldest airworthy Hunter.
WT734 ff 05.02.55	Delivered 09.05.55. Returned HAL. Converted to Peruvian Mk 52 (638) ff 18.01.56. Delivered Peru 03.56.
WT735 ff 12.02.55	Retained by HAL for radio trials. Fitted with American electronic equipment. SOC 08.04.60.
WT736 ff 04.02.55	To RR for engine trials with Avon 121. To Ferranti 04.12.56. Broken up 11.64.
WT737 ff 01.03.55	Delivered 29.03.55 No 5MU Kemble. No 118 Sqn 'N' 28.04.55. No 222 Sqn 17.06.57. No 229 OCU 'B' 29.07.57. Scrapped 11.62.
WT738 ff 17.02.55	Delivered 25.03.55 No 5MU Kemble. No 118 Sqn 'P' Jever 06.05.55. Crashed north of Hamburg after colliding with WT757 of No 118 Sqn. Pilot killed. W/O 20.10.55.
WT739 ff 10.02.55	Delivered. 02.03.55 No 33MU. No 111 Sqn ' R'. Edinburgh to London (Turnhouse to Farnborough) record 03.08.56, average speed 675mph. London to Edinburgh speed record 08.08.56, average speed 717.504mph. No 222 Sqn 11.12.56. FWS 'E' 16.01.57. Broken up 11.62.
WT740 ff 17.02.55	Delivered 29.03.55 No 5MU Kemble. No 54 Sqn. No 247 Sqn 'F'. No 229 OCU. Broken up HAL 11.64.
WT741 ff 01.03.55	Delivered 28.03.55 No 5MU Kemble. No 118 Sqn 'Q' 30.04.55. No 5MU Kemble 26.09.57. Converted to GA Mk 11. Delivered 11.62. Converted to Singapore T Mk 75A '544'. Delivered 10.73. Nose section to Aeroventure, Doncaster.
WT742 ff 28.02.55	Delivered 29.03.55 No 5MU Kemble. No 98 Sqn 'A' 25.04.55. CAT4 09.04.57. Returned to HAL. Broken up 10.57.

WT743 ff 28.02.55	Delivered 29.03.55 No 5MU Kemble. No 118 Sqn 'R' 27.04.57. CAT4 No 71MU Bicester 09.08.57. SOC 01.01.59.
WT744 ff 28.02.55	Delivered 25.03.55 No 5MU Kemble. AFDS. No 247 Sqn 07.11.56. No 5MU 04.01.57. To HAL. Converted to GA Mk 11. Delivered 13.02.63 No 738 Sqn FAA. FRADU '868/VL'. South West Aviation Heritage, Eaglescott, 1998.
WT745 ff 02.03.55	Delivered.14.04.55 No 5MU Kemble. No 14 Sqn. Converted to T Mk 8C. Yeovilton Instrument Rating Flight '947/VL' 12.04.60. No 764 Sqn FAA. To USA as N745WT. Rear fuselage of Swedish J34 34006. Forward fuselage of WT745 and components of ET-271. Crashed during post-restoration test. W/O 17.11.99.
WT746 ff 01.03.55	Delivered 17.03.55 No 5MU Kemble. AFDS. St Athan. GIA 7770M Halton. SOC 12.11.62. Army training airframe at Saighton Camp. Dumfries and Galloway Aviation Museum 1999. Reported to be in poor condition.
WT747 ff 04.03.55	Delivered 25.03.55 No 5MU Kemble. No 98 Sqn 'B' 25.04.55. No 5MU Kemble 12.04.57. Broken up HAL 11.64.
WT748 ff 03.03.55	No 118 Sqn 'S' Fassberg 27.04.55. No 5MU Kemble 17.09.57. Broken up HAL 1962.
WT749 ff 15.03.55	Delivered 01.04.55 No 247 Sqn 'P'. No 229 OCU. Broken up HAL 11.64.
WT750 ff 04.03.55	Delivered 20.04.55 No 247 Sqn No 245 Sqn 'A'. Broken up HAL 11.64.
WT751 ff 09.03.55	Delivered 28.03.55 No 5MU Kemble. No 247 Sqn. No 118 Sqn 'T' 27.04.55. No 5MU Kemble 24.08.57. A&AEE 1962 for jettison trials with 230-gal drop tanks.
WT752 ff 03.03.55	Delivered 01.04.55 No 247 Sqn. No 118 Sqn 'D' 12.05.55. No 5MU Kemble 17.09.57. Broken up HAL 11.64.
WT753 ff 10.03.55	Delivered 01.04.55 No 118 Sqn 'E' 12.05.55 to 19.08.57. No 5MU Kemble 12.09.57. No 247 Sqn. Broken up HAL 1962.
WT754 ff 10.03.55	Delivered 29.03.55 No 5MU Kemble. No 118 Sqn 'Y' 28.04.55. Benson 19.08.57. No 167 Sqn. Caught fire on start-up and destroyed. W/O 05.09.57.
WT755 ff 18.03.55	Delivered 14.04.55 No 14 Sqn. Returned to HAL 1959. Converted to T Mk 8. Lossiemouth. Brawdy 08.64. Returned to HAL 03.73 G-9.417. Converted to Kenyan T Mk 81 (801). Delivered. 24.06.74 Kenya. To Zimbabwe as '8100'.
WT756 ff 11.03.55	Delivered 20.04.55. Converted to Peruvian Mk 52 '639'. ff 17.01.56. Delivered Peru 04.56.
WT757 ff 15.03.55	Delivered 01.04.55. No 118 Sqn 'C' Jever 20.04.55. Collided with WT738 20 miles W of Hamburg. Pilot killed. W/O 20.10.55.
WT758 ff 17.03.55	Delivered 14.04.55 No 5MU Kemble. Converted to Peruvian Mk 52 '633'. ff 15.12.55. Delivered Peru 03.56.
WT759 ff 16.03.55	Delivered 20.05.55. No 111 Sqn 'B'. Returned to HAL 1955. Converted to Peruvian Mk 52 '642'. ff 26.01.56. Delivered Peru 03.56.
WT760 ff 17.03.55	Delivered 14.04.55 No 5MU Kemble. No 118 Sqn 'C' 10.05.55. No 229 OCU 23.04.57. Broken up HAL 11.64.
WT761 ff 28.03.55	Delivered 15.04.55 No 5MU Kemble. No 14 Sqn. No 98 Sqn. Broken up HAL 11.64.
WT762 ff 29.03.55	Delivered 15.04.55 No 247 Sqn. Crashed during display at Farnborough. Pilot killed. W/O 07.07.55.
WT763 ff 25.03.55	Delivered 20.04.55. No 26 Sqn 'A' and 'F' 14.06.55. No 5MU Kemble 08.05.57. Broken up HAL 11.64.
WT764 ff 16.03.55	Delivered 14.04.55 No 5MU Kemble. No 111 Sqn. No 54 Sqn 'G'. No 74 Sqn '1'. No 247 Sqn. Broken up HAL 11.64.
WT765 ff 05.04.55	Delivered 29.04.55. No 111 Sqn 'V'. Returned to HAL 1955. Converted to Peruvian Mk 52 '645'. ff 07.02.56. Delivered Peru 4.56.
WT766 ff 24.03.55	Delivered 02.05.55. Converted to Peruvian Mk 52 '635'. ff 30.12.55. Delivered Peru 03.56.
WT767 ff 12.04.55	Delivered 26.04.55 No 14 Sqn. No 229 OCU 1957. Broken up 1962.
WT768 ff 23.05.55	Delivered 25.04.55 No 118 Sqn 'C'. No 247 Sqn 'G'. Converted to Peruvian Mk 52 '641'. ff 20.01.56. Delivered Peru 04.56.

WT769 ff 30.03.55	Delivered 20.05.55. No 26 Sqn 'B' 17.06.55. No 5MU Kemble 01.05.57. Broken up HAL 11.62.
WT770 ff 24.06.55	To Sweden for evaluation. Became Swedish Mk 50 '34001'.
WT771 ff 14.04.55	Delivered 20.05.55 No 111 Sqn 'C'. No 222 Sqn 'C' Leuchars 04.12.56. No 5MU Kemble 01.11.57. Broken up HAL 11.64.
WT772 ff 30.03.55	Trials of Ram Air Turbine. Trials of extended-span tailplane. CA aircraft HAL. Converted to T Mk 8 by AWA 1959. No 764 Sqn FAA Lossiemouth. No 759 Sqn Brawdy '808/BY'. ADTU Yeovilton as '736' 1970, later '745'. FRADU as '875/VL' 1974. Engine failure on take-off, Charlton Adam, Somerset. W/O 21.09.76.
WT773 ff 19.04.55	Delivered 29.04.55 No 111 Sqn 'U'. Returned to HAL 1955. Converted to Peruvian Mk 52 '644'. ff 13.02.56. Delivered Peru 05.56.
WT774 ff 06.04.55	Delivered 02.05.55. Converted to Peruvian Mk 52 '637'. ff 09.01.56. Delivered Peru 05.56.
WT775 ff 04.04.55	Delivered 25.04.55 No 247 Sqn 'Q'. Damaged in take-off accident at Odiham, W/O 26.03.56. Broken up HAL 1964.
WT776 ff 01.04.55	Delivered 24.05.55. Converted to Peruvian Mk 52 '631'. ff 05.12.55. Delivered Peru 05.56.
WT777 ff 21.04.55	Delivered 02.06.55 No 4 Sqn 'D'. No 98 Sqn 'F' 15.03.57. No 5MU Kemble 15.07.57. Broken up 1962.
WT778 ff 05.04.55	Delivered 25.05.55. No 26 Sqn 'W' 30.06.55. No 5MU Kemble 17.05.57. GIA 7791M 1960. Returned to HAL 03.73. Scrapped HAL 1979.
WT779 ff 19.04.55	Delivered 25.05.55 No 111 Sqn 'E'. Returned to HAL 1955. Converted to Peruvian Mk 52 '643'. ff 27.01.56. Delivered Peru 04.56.
WT780 ff 06.04.55	Retained by HAL. Trials of camera nose, tail hook and drag chute, and ram air turbine. HAL Dunsfold for ground sound suppression trials with engine running muffler pens 1956. Broken up HAL 11.64.
WT795 ff 21.04.55	Delivered 04.05.55 No 247 Sqn 'A'. No 229 OCU. Broken up HAL 09.64.
WT796 ff 22.04.55	Delivered 04.05.55. Converted to Peruvian Mk 52 '640'. ff 18.01.56. Delivered Peru 05.56.
WT797 ff 21.04.55	Delivered 20.05.55 No 14 Sqn. No 111 Sqn GIA 7792M. To HAL 05.72 as G-9.388. Converted to Mk 58A (J4148). Delivered 30.04.74.
WT798 ff 13.04.55	Delivered 22.03.56. A&AEE performance trials. Broken up HAL 11.64.
WT799 ff 26.04.55	Delivered 01.06.55 No 111 Sqn Bruggen. No 4 Sqn Jever. CAT4. Repaired at HAL 24.09.57. Converted to T Mk 8 06.02.59. No 764 Sqn Lossiemouth, later Brawdy, as '802' 08.66. RAE Bedford. No 764 Sqn '689'. No 759 Sqn 24.07.70. FRU Hurn '839'. FRU Yeovilton '948/VL' 12.72. FRADU 1974. FRADU Yeovilton '879/VL' 09.74. Last flew 09.82. Withdrawn to Shawbury. Open store at Exeter. Sunk into lake near Doncaster for Scuba training.
WT800 ff 04.04.55	Delivered 26.04.55 No 54 Sqn. Converted to Peruvian Mk 52 '636'. ff 13.12.55. Delivered Peru 04.56.
WT801 ff 23.03.55	Delivered 01.06.55 No 4 Sqn. Fitted with wing leading edge extension. No 229 OCU. GIA 7789M. To HAL 10.71 G-9.379. Converted to Chilean FGA Mk 71 'J-731'. Delivered Chile 03.09.73.
WT802 ff 24.03.55	Delivered 20.04.55. No 98 Sqn 'P' 12.05.55. No 5MU Kemble 09.07.57. No 247 Sqn. Broken up 1962.
WT803 ff 23.03.55	Delivered 29.04.55. Converted to Peruvian Mk 52 '634'. Delivered Peru 05.56.
WT804 ff 23.03.55	Delivered 14.04.55 No 247 Sqn 'R'. Converted to GA Mk 11. Delivered 09.63 No 738 Squadron. 'Rough Diamonds' team. No 738 Sqn RNAS Brawdy 30.07.69. Transferred to FRU Hurn '831'. To Yeovilton 1972. FRADU '831/VL' 20.10.72. Shawbury 1985. GIA SAH RNAS Culdrose 'A2732' '831/DD' 11.05.95. Transported to Moreton-in-Marsh GIA 'A2646'.

WT805 ff 18.04.55	Delivered 02.05.55. No 234 Sqn 'J' 15.05.56. No 130 Sqn 'X' 10.01.57. No 234 Sqn 'F' 02.05.57. No 5MU Kemble 09.07.57. Converted to GA Mk 11. Delivered 03.63 No 738 Sqn FAA. Crashed at St Bride's Bay. W/O 22.03.67.
WT806 ff 31.03.55	Delivered 25.04.55 No 14 Sqn Oldenburg. Aircraft over-stressed. Returned to HAL. No 5MU Kemble. CFS Type Flt 'A' Little Rissington 01.03.57 to 04.11.57. HSA. Converted to GA Mk 11 '838' 21.07.61. Withdrawn 02.87 to Chivenor. Shawbury 1994. Sold to Northbrook College, Shoreham, as instructional airframe 07.02.
WT807 ff 04.04.55	Delivered 02.05.55 No 14 Sqn. Collided with WT714 of No 14 Sqn, crashed near Bremen. W/O 18.08.55.
WT808 ff 05.05.55	Delivered 24.05.55. No 43 Sqn Leuchars. No 111 Sqn 'G'. No 222 Sqn 'G' 04.12.56. No 60MU 16.10.57. Converted to GA Mk 11. A&AEE 1966. Returned to HAL 24.04.70. Converted to F Mk 58 (J-4121).
WT809 ff 13.04.55	Delivered 28.06.55. No 66 Sqn 'G' 15.03.56. FWS 'G' 12.12.56. Linton-on-Ouse Stn Flt. Converted to GA Mk 11 '737'. No 738 Sqn 1964. No 764 Sqn Lee-on-Solent. To FRADU '867/VL'. Crashed near Ilchester. W/O 14.06.88.
WT810 ff 19.04.55	Delivered 18.05.55 No 54 Sqn. No 247 Sqn 'W'. Converted to GA Mk 11. Collided with WT712. W/O 25.10.65.
WT811 ff 23.04.55	Delivered 24.05.55 No 111 Sqn 'H'. No 222 Sqn 'D' 28.11.56. No 5MU Kemble 01.11.57. Converted to GA Mk 11. Broken up HSA 1964.

Second production batch of eighty-five aircraft built by Hawker Aircraft Ltd, Kingston-upon-Thames. Contract No SP/6/6867/CB.7a dated 09.05.51.

WV253 ff 29.04.55	Delivered 26.05.55 No 4 Sqn. Returned to HAL. Converted to T Mk 7 standard. Delivered 12.06.62 for spinning trials ETPS '24' Farnborough. Fitted with nose instrumentation boom. Crashed off Lyme Regis. W/O 15.07.68.
WV254 ff 27.04.55	Delivered 20.05.55 No 247 Sqn No 229 OCU. Scrapped 06.61.
WV255 ff 30.04.55	Delivered 24.05.55. No 26 Sqn 'X' 11.07.55. No 5MU Kemble 17.05.57.
WV256 ff 05.05.55	Delivered 20.05.55. No 26 Sqn 'D' 13.06.55. No 5MU Kemble 26.04.57. No 229 OCU. Purchased by HAL. Converted to GA Mk 11. Delivered 02.04.63 No 738 Sqn Lossiemouth, then Brawdy '794/BY'. Lossiemouth '690/LM', later '698/LM'. Remained in service until squadron disbanded 1972. ADTU Yeovilton '732/VL' 1972. FRADU '862/VL' 1973. Shawbury 04.92 for storage. G-BZPB 15.01.01. To Delta Jets Kemble 07.02.
WV257 ff 05.05.55	Delivered 24.05.55. No 26 Sqn 'E' 27.06.55. Stalled on landing approach 22.07.55. CAT4. Repaired by HAL and converted to GA Mk 11 19.10.61. No 738 Sqn FAA. Re-purchased by HAL G-9.341 4.70. Converted to Mk 58A (J-4125). Delivered 29.01.73.
WV258 ff 10.05.55	Delivered 01.06.55. No 54 Sqn 1955. No 111 Sqn 'T' 1955. No 222 Sqn 20.11.56. FWS 08.01.57. GIA 7779M. To HAL 27.09.71 G-9.368. Converted to Singapore Mk 74B '539'. Delivered 13.06.73.
WV259 ff 26.04.55	Delivered 23.05.55 No 4 Sqn. Scrapped by Coleys 1963.
WV260 ff 29.04.55	Delivered 24.05.55. No 26 Sqn 'C' 23.06.55. Oldenburg Stn Flt. No 26 Sqn 19.07.56. No 5MU Kemble 08.05.57. Scrapped 06.61.
WV261 ff 23.04.55	Delivered 01.06.55. No 26 Sqn 'Y' 12.07.55. Crash-landed 13.01.56. No 5MU Kemble 15.05.57. GIA 7780M Halton. HAL as G-9.400 03.73. Converted to Mk 58A 'J-4147'. Delivered 06.12.74.
WV262 ff 07.05.55	Delivered 20.05.55 No 247 Sqn 'D'. No 74 Sqn 'K'. No 5MU Kemble. Scrapped 06.61.
WV263 ff 29.04.55	Delivered 01.06.55 No 4 Sqn. No 118 Sqn 15.03.57. No 5MU Kemble 17.09.57. Scrapped 06.61.
WV264 ff 30.04.55	Delivered 16.05.55 No 111 Sqn 'A'. FWS. No 5MU Kemble. Scrapped 06.61.
WV265 ff 06.05.55	Delivered 24.05.55. No 26 Sqn 'V' 07.07.55. No 5MU Kemble 08.05.57. GIA 7684M Melksham. No 9 SoTT Newton.

WV266 ff 02.05.55	Delivered 23.05.55 No 4 Sqn 'T'. No 67 Sqn 'B' 14.03.57. No 130 Sqn 'B' 17.04.57. No 234 Sqn 'B' 02.05.57. No 5MU Kemble 04.07.57. GIA 7781M Halton. HAL 7.72 as G-9.399. Converted to Mk 58A (J-4146). Delivered 24.10.74.
WV267 ff 05.05.55	No 93 Sqn Jever 'R' 18.01.56. No 98 Sqn 22.03.57. No 5MU Kemble 09.07.57. No 247 Sqn. Purchased by HAL. Converted to GA Mk 11. No 738 Sqn FAA Lossiemouth 1963 then Brawdy 1964. FRADU '836' GIA 'A2737' (1) 27.03.69. FRADU as '836/VL'. Bird-strike 29.11.85. GIA 'A2650'. School of Aircraft Handling '836/DD' 06.02.86. North Weald until 02.96, then moved to USA. N267WV. Current status and location unknown.
WV268 ff 16.05.55	Delivered 14.06.55 No 93 Sqn. No 26 Sqn 'T' 29.06.55. No 5MU Kemble 18.05.57. GIA 7701M Manston.
WV269 ff 12.05.55	Delivered 02.06.55 No 54 Sqn. No 74 Sqn 'H'. No 5MU Kemble. Scrapped 1961.
WV270 ff 12.05.55	Delivered 14.06.55 No 54 Sqn. No 26 Sqn 'Z'. No 5MU Kemble. Scrapped 06.61.
WV271 ff 12.05.55	Delivered 21.06.55 No 4 Sqn. Crashed at Oldenburg. Pilot killed. W/O 23.10.55.
WV272 ff 11.05.55	Delivered 06.06.55 No 54 Sqn. No 74 Sqn 'L'. GIA 7782M Halton. To HAL 23.09.71 as G-9.369. Converted to Singapore T Mk 75A '540'. Delivered 18.07.73. Sold to Nick Costin, Bankstown Airport, Sydney, Australia, 1998.
WV273 ff 26.05.55	Delivered 23.06.55. No 67 Sqn 'D' 31.03.56. No 112 Sqn 18.04.57. No 5MU Kemble 21.05.57. Scrapped 06.61.
WV274 ff 12.05.55	Delivered 20.06.55 No 4 Sqn. No 118 Sqn 15.03.57. CAT4, forced-landing 16.07.57. To HAL 18.09.57 but repairs uneconomical. No 47MU 21.11.57. SOC 09.06.58.
WV275 ff 13.05.55	Delivered 15.06.55. No 4 Sqn 'D' Jever. No 111 Sqn No 229 OCU. Scrapped 06.61.
WV276 ff 20.05.55	Delivered 15.06.55. Rolls-Royce. Trials at A&AEE with Avon 121 engine. No 74 Sqn 'P'. GIA 7847M 'D' Halton. RAE Farnborough. OFMC Duxford 12.98. Frontline Aviation Museum, Sandown Airport. Hawker Hunter Aviation, Scampton 5.05. External storage. Shipped by road to Luftfahrt und Technik Museum, Wernigerode, Germany, 22.07.05.
WV277 ff 13.05.55	Delivered 06.06.55 No 14 Sqn 'K'. No 93 Sqn 17.01.56. No 229 OCU 01.04.57. No 111 Sqn No 5MU Kemble. Scrapped 06.61.
WV278 ff 18.05.55	Delivered 30.06.55. No 66 Sqn 'H' 22.03.56. FWS 27.02.57. No 5MU Kemble. Scrapped 06.61.
WV279 ff 20.05.55	Delivered 17.06.55 No 4 Sqn No 229 OCU. No 5MU Kemble. Scrapped 06.61.
WV280 ff 24.05.55	Delivered 16.06.55 RAF Flying College, Manby. No 5MU Kemble. Scrapped 06.61.
WV281 ff 24.05.55	Delivered 14.06.55 No 54 Sqn 'M'. No 74 Sqn 'Q'. No 5MU Kemble. Scrapped 06.61.
WV314 ff 19.05.55	Delivered 20.06.55. No 66 Sqn 10.09.56. Horsham St Faith Stn Flt 29.01.57. No 222 Sqn 'J' 13.05.57. No 43 Sqn 14.10.57. No 92 Sqn 'B'. No 74 Sqn 'JH'.
WV315 ff 24.05.55	Delivered 20.06.55. No 66 Sqn 10.09.56. Metropolitan Sector Commander 21.11.56. No 43 Sqn 'I'. No 5MU Kemble. Scrapped 06.61.
WV316 ff 25.05.55	Delivered 17.06.55 No 4 Sqn. No 98 Sqn 15.03.57. No 5MU Kemble 17.07.57. GIA 7676M. Scrapped 06.64.
WV317 ff 27.05.55	Delivered 23.06.55 No 54 Sqn. No 247 Sqn 'S' Odiham 1955-57. No 74 Sqn 'D'.
WV318 ff 23.05.55	Fitted with wing leading edge extensions. Delivered No 5MU Kemble 16.06.55. Reissued to No 33MU Lyneham 07.55. No 14 Sqn Oldenburg. No 93 Sqn 'G' Jever 27.03.56. HAL repairs 1959. Converted to T Mk 7. Delivered to No 5MU Kemble 01.06.59 for storage. No 111 Sqn 03.60. CFS Type Flt Little Rissington 23.11.60 to 11.03.63. Converted to T Mk 7A 10.64. No 5MU Kemble. No 5 Sqn. No 74 Sqn. No 12 Sqn Honington 02.70 with Buccaneer Force, then to No 15 Sqn and No 16 Sqn Laarbruch 06.71. No 208 Sqn Lossiemouth. Delta Jets Kemble 1996. Flown to Kemble 03.96 as G-FFOX. Post-restoration flight 02.05.96. Fitted with T Mk 8 arrester hook.

WV319 ff 03.06.55	Delivered 28.06.55. CFS Type Flt 07.11.55 to 13.01.59. Crash-landed Kemble 13.01.56. Repaired No 32 MU. Returned to HAL. Converted to T Mk 8. Yeovilton Instrument Flt as 'VL/949'. No 764 Sqn FAA. W/O 26.03.65
WV320 ff 26.05.55	Delivered 28.06.55. No 222 Sqn 'S' 05.12.55. FWS 24.01.57. GIA 7685M Melksham.
WV321 ff 06.06.55	Delivered 11.07.55 No 111 Sqn 'E'. No 222 Sqn 'E' 04.12.56. No 5MU Kemble 01.11.57. No 4 Sqn 'B'. Scrapped 06.61.
WV322 ff 08.06.55	Delivered 28.06.55 No 43 Sqn 'G'. No 92 Sqn. Returned to HAL 1959. Converted to T Mk 8. No 764 Sqn FAA. No 284 (Training) Squadron, Cranwell GIA 9096M. Sold 29.11.00 for £28,000. Ferried to Kemble 14.02.01. Delta Jets, Kemble, G-BZSE. Towerdrive, Exeter, 06.04.
WV323 ff 08.06.55	Delivered 28.06.55. No 66 Sqn 23.03.56. No 92 Sqn 04.04.56. No 43 Sqn 'H'. GIA 7668M Devizes. No 2397 Sqn ATC.
WV324 ff 31.05.55	Delivered 27.06.55 No 43 Sqn 'D'. No 74 Sqn 'C'. GIA 7668M Halton.
WV325 ff 13.06.55	Delivered 15.07.55. CFS Type Flt 'C' 09.11.55 to 10.03.59. Stored No 5MU Kemble. Returned to HAL 19.10.67 as G-9.250. Converted to Jordanian FGA Mk 73A '846'. Delivered 07.10.71.
WV326 ff 07.06.55	Delivered 05.07.55 No 54 Sqn. No 74 Sqn 'R'. GIA 7669M Halton. HAL 01.72 as G-9.396. Converted to Chilean FR Mk 71A (J-735). Delivered Chile 11.01.74. Chilean Air Force Museum, Los Cerrillos.
WV327ff 13.06.55	Delivered 11.07.55. No 111 Sqn 'U' later 'D' 1955. No 222 Sqn 'H' 04.12.56. No 5MU Kemble 01.11.57. GIA 7670M Halton.
WV328 ff 08.06.55	Delivered 30.06.55 No 247 Sqn. Destroyed by fire at Odiham following starter cartridge explosion. W/O 25.01.56.
WV329 ff 02.06.55	Delivered 01.07.55 No 54 Sqn 'D'. No 245 Sqn No 229 OCU. GIA 7671M Halton. To HAL 07.72 as G-9.407. Converted to Mk 58A 'J-4144'. Delivered 10.09.74.
WV330 ff 04.06.55	Delivered 17.07.55 No 54 Sqn. No 247 Sqn 'G'. No 245 Sqn 'D'. No 229 OCU. GIA 7672M Larkhill.
WV331 ff 10.06.55	Delivered 12.07.55. CFS Type Flt Kemble 'B' 07.11.55 to 11.12.58. Stored No 5MU Kemble. GIA 7783M Halton. Flown back to HAL 08.71 as G-9.372. Converted to Singapore Mk 74B '543'. Delivered 16.08.73. 01.00 Hunter Flight, Sunshine Airport, Queensland, 09.98 Archerfield, Queensland Australia. VH-JGP Australian Museum of Flight, Nowra.
WV332 ff 07.06.55	Delivered 27.06.55. No 67 Sqn 'C' and 'X' 12.03.56. No 112 Sqn 18.04.57. No 234 Sqn 'P' 06.05.57. No 5MU Kemble 04.07.57. GIA 7673M Halton. To HAL 7.72 as G-9.406. Converted to T Mk 68 (J-4201). Delivered 02.08.74. No 1254 Sqn ATC Godalming, Surrey. Nose displayed at Tangmere Aviation Museum 09.07.
WV333 ff 10.06.55	Delivered 01.07.55 No 43 Sqn 'A'. Stored by No 47MU 14.03.58.
WV334 ff 14.06.55	Delivered 05.07.55 No 54 Sqn 'C'. No 74 Sqn 'E'. No 5MU Kemble. Scrapped 1961.
WV363 ff 22.06.55	Delivered 08.07.55 No 234 Sqn 'K' Geilenkirchen. CAT4 02.10.56. To HAL for repair. Re-purchased and converted to T Mk 8 for FAA. No 764 Sqn Lossiemouth '701/LM' then '687/LM' from 1965. A&AEE 09.63. No 759 Sqn '801/BY'. Loaned to FRU Hurn 11.12.69. Yeovilton ADTU '747' 26.08.70. FRADU 1972. Placed in store Kemble until 4.81 then to FRADU as '872/VL'. Crashed into Moray Firth off Lossiemouth and not recovered. W/O 15.02.92.
WV364 ff 17.06.55	Delivered 30.06.55. No 26 Sqn 16.09.55. No 98 Sqn 14.11.55. No 118 Sqn 18.01.56. No 93 Sqn 'S' 28.02.56. No 98 Sqn 12.04.57. GIA 7674M Halton. Flown back to HAL 03.71 as G-9.347. Converted to Singapore Mk 74B '530'. Delivered 17.01.73.
WV365 ff 27.06.55	Delivered 23.08.55 No 54 Sqn. No 247 Sqn. Stored at No 5MU Kemble 07.07.58.
WV366 ff 16.06.55	Delivered 05.07.55 No 43 Sqn 'T'. GIA Halton. To HAL 02.71 as G-9.350. Converted to Singapore Mk 74B '537'. Delivered 09.04.73.
WV367 ff 27.06.55	Delivered 08.08.55. No 67 Sqn 'X' 25.01.56. No 130 Sqn 17.04.57. No 234 Sqn 'J' 02.05.57. No 5MU Kemble 16.08.57. GIA 7677M Larkhill.

WV368 ff 24.06.55	Delivered 12.07.55. No 98 Sqn 07.09.55. No 4 Sqn 04.10.55. No 118 Sqn 'F' 20.10.55. No 93 Sqn 18.01.56. No 118 Sqn 'F' 22.05.56. No 5MU Kemble 21.08.57. Scrapped 06.61.
WV369 ff 27.06.55	Delivered 14.07.55. No 26 Sqn 'E' 27.09.55. No 5MU Kemble 17.05.57. GIA 7675M Halton.
WV370 ff 23.06.55	Delivered 05.08.55 No 54 Sqn. No 74 Sqn 'M'. Stored at Kemble 06.58.
WV371 ff 30.06.55	Delivered 11.08.55 No 54 Sqn '0'. No 74 Sqn 'N'. GIA 7678M Larkhill.
WV372 ff 15.07.55	Delivered 02.09.55 No 222 Sqn 'H' Leuchars. Damaged by fire 29.11.56. To HAL. Converted to T Mk 7. Delivered 07.05.59. No 5MU Kemble. Jever Stn Flt 08.01.60. Gutersloh Stn Flt 'T' 12.62. No 2 Sqn 'R' 05.67. Wildenrath 07.70. RAE 30.03.71. 4FTS Valley. No 208 Sqn. No 237 OCU '372' Lossiemouth as Buccaneer crew trainer. To FAA 10.84 as '879/VL'. Returned to FAA as '877/VL'. Placed in storage at Shawbury 1988. School of Aircraft Handling, School of Flight Deck Operations, until 1996. Delta Jets, owned by 'Fox-One consortium', registered as G-BXFI.
WV373 ff 29.06.55	Delivered 14.07.55 No 98 Sqn. No 118 Sqn. Crashed at Ahlhorn. Pilot killed. W/O 08.12.55.
WV374 ff 04.07.55	To RAE 05.08.55. No 67 Sqn 'L' 16.10.56. No 112 Sqn 18.04.57. No 5MU Kemble 16.05.57. Converted to GA Mk 11. Delivered 20.08.62 No 738 Sqn FAA '647'. 'Rough Diamonds' team at Brawdy. To HAL 24.04.70 as G-9.339. Converted to F Mk 58A 'J-4123'. Delivered to Swiss AF 19.01.73.
WV375 ff 30.06.55	Delivered 27.09.55 No 54 Sqn 'A'. GIA 7679M Weeton.
WV376 ff 04.07.55	Delivered 23.08.55. No 222 Sqn 'W' 11.11.55. FWS 'I' 24.01.57. GIA 7680M Weeton.
WV377 ff 04.07.55	Delivered 09.08.55 No 14 Sqn. No 26 Sqn 07.11.55. No 93 Sqn 'V' 16.03.56. No 222 Sqn 'W' 17.06.57. No 5MU Kemble 01.11.57. GIA 7681M Larkhill.
WV378 ff 05.07.55	Delivered 17.08.55. No 222 Sqn 'C' 02.12.55. No 43 Sqn 'H' 19.12.56. GIA 7682M Weeton.
WV379 ff 15.07.55	Delivered 23.08.55. No 111 Sqn 'V' 1955. No 222 Sqn 28.11.56. FWS 08.01.58. GIA 7683M Larkhill.
WV380 ff 07.07.55	Delivered 11.08.55 RAF Manby. To HAL 1960. Converted to GA Mk 11. Delivered 29.05.62 No 738 Sqn FAA Lossiemouth. To HAL 24.04.70. Converted to Mk 58A (J-4119). Delivered 03.11.72.
WV381 ff 03.08.55	Delivered 17.08.55. No 222 Sqn 'Q' 02.12.55. FWS 'F' 24.01.57. Converted to GA Mk 11. Delivered 07.62 Lossiemouth. FRU. FRADU '732'. No 5MU Kemble. W/O 01.11.72. GIA UKAEA Lightning Studies Unit Culham, Oxfordshire, 1998. Nose to RAF Benson 10.06.07, to Kemble 25.06.07.
WV382 ff 11.07.55	Delivered 15.08.55. No 67 Sqn 'V' and 'C' 08.03.56. No 112 Sqn 18.04.57. No 5MU Kemble 20.05.57. Re-purchased by HAL. Converted to GA Mk 11. Delivered 14.08.62 No 738 Sqn FAA Lossiemouth, then Brawdy '784/BY' 01.66. FRU Hurn 04.69 '830'. Yeovilton 20.10.72. Units merged '830/VL'. No 5MU Kemble for six years in storage then Shawbury. Lee-on-Solent 05.02.85. GIA 'A2730'. Sold to group at Long Marston.
WV383 ff 12.07.55	Delivered 23.08.55. RAF Manby 19.12.55. To HAL 1958. Converted to T Mk 7. Delivered 03.07.59. No 79 Sqn 06.01.60. Gutersloh Stn Flt 8.01.60. Jever Stn Flt 14.04.61. No 4 Sqn 16.05.61. No 19MU St Athan 05.06.61. No 28 Sqn 1964. No 20 Sqn 'T' Tengah 05.01.67. No 5MU Kemble 04.12.69. RAE Farnborough 07.77. A&AEE. RAE Heritage Centre 13.04.00.
WV384 ff 10.08.55	Delivered 02.09.55. No 71 Sqn 25.04.56. Landing accident. Pilot killed. W/O 05.04.57.
WV385 ff 05.08.55	Trials with Avon 115. Delivered 17.10.55. No 66 Sqn 'L' 04.04.56. AFDS 01.11.56. No 229 OCU. No 3 CAACU Exeter 05.60. GIA 7775M.
WV386 ff 12.08.55	No 222 Sqn 'G' Leuchars 02.12.55. No 229 OCU 30.08.57. No 19MU St Athan 17.03.59. Stored until 10.12.62. GIA 7775M. No 1 SoTT Halton. Purchased by HAL 02.71. Converted to Singapore T Mk 75A '532'. Delivered 15.12.72. SOC 1994. Registered in US as N81827 with Adrenalin Airsports.

WV387 ff 11.08.55	Delivered 05.09.55 No 43 Sqn 'Q'. No 67 Sqn 10.01.57. No 71 Sqn 08.02.57. No 5MU Kemble 20.05.57. Broken up Kemble 09.58.
WV388 ff 16.08.55	Delivered 16.09.55. No 222 Sqn 05.12.55. Crashed into sea E of Leuchars. Pilot killed. W/O 25.03.5.
WV389 ff 16.08.55	Delivered 07.09.55. No 222 Sqn 'H' 29.11.55. No 66 Sqn 'L' 10.09.56. HQ 2TAF 30.11.56. No 98 Sqn 14.12.56. No 229 OCU 26.07.57. No 247 Sqn. To HAL 1968. Converted to Abu Dhabi FGA Mk 76 '703'. Delivered 06.11.70.
WV390 ff 16.08.55	Delivered 07.09.55 No 20 Sqn 'G'. No 26 Sqn 13.05.57 to 12.06.57. RAF Flying College Manby 20.09.57. Crashed at Strubby. W/O 29.03.58.
WV391 ff 23.08.55	Delivered 13.09.55 No 20 Sqn 'Z'. No 26 Sqn 09.05.57. No 229 OCU 'RS-E' 31.05.57. Stalled on approach to Chivenor. Pilot killed. W/O 05.06.58.
WV392 ff 19.08.55	Delivered 09.09.55 Linton-on-Ouse Stn Flt. No 66 Sqn 13.03.56. Crashed 2 miles E of Filey. W/O 17.07.56.
WV393 ff 19.08.55	Delivered 09.09.55 Linton-on-Ouse Stn Flt. No 66 Sqn 'D' 14.03.56. AFDS 01.11.56. Returned to HAL 1.73. Converted to Swiss Mk 58A 'J-4151'. Delivered 19.02.75.
WV394 ff 23.08.55	Delivered 21.09.55. No 20 Sqn 'E' 16.12.55. No 26 Sqn 29.04.57. No 229 OCU 12.08.57. Broken up HAL 09.64.
WV395 ff 02.09.55	No 20 Sqn 'W' 09.12.55. CAT4 after flying accident 13.03.57. GIA Cosford. Re-purchased by HSA 01.08.75. SOC 01.08.75. Donated by BAe to Aviodome, Netherlands. Combined with parts from E-410.
WV396 ff 24.08.55	No 5MU Kemble 23.09.55. No 20 Sqn 'F' Oldenburg 04.01.56. No 26 Sqn 09.05.57. No 5MU Kemble. No 229 OCU 12.08.57. Returned to HAL 1963. Converted to T Mk 8C. No 759 Sqn FAA Brawdy '809/BY' 01.66. Air Direction Training School and Yeovilton Stn Flt '748'. No 5MU Kemble. Returned as '871'. Withdrawn from use 1985. Returned to service as '879/VL' 1986. Retired 1995. To Shawbury until 1997. Valley gate guard 9249M.
WV397 ff 25.08.55	No 5MU Kemble 27.09.55. No 20 Sqn 'B' 28.11.55. No 26 Sqn 9.05.57. No 5MU Kemble 31.05.57. No 229 OCU. To HAL 1963. Converted to T Mk 8C. No 759 Sqn FAA. Crashed into Irish Sea. Pilot killed. W/O 23.11.65.
WV398 ff 29.08.55	No 5MU Kemble 19.09.55. No 20 Sqn 'A' 24.11.55. No 26 Sqn 09.05.57. No 229 OCU 20.09.57. Gate guard at Spitalgate. Converted to Swiss T Mk 68 'J-4203'. Delivered 04.03.75.
WV399 ff 26.08.55	No 5MU Kemble 23.09.55. No 222 Sqn 'B' 14.11.55. No 229 OCU 30.08.57. Scrapped Lasham 1963.
WV400 ff 30.08.55	No 5MU Kemble 20.9.55. No 66 Sqn 'K' 14.04.56. AFDS 01.11.56. No 229 OCU. HAL 13.06.68. Converted to Abu Dhabi FR Mk 76A '710'. Delivered 19.01.71.
WV401 ff 01.09.55	No 20 Sqn 'T' 28.11.55. No 26 Sqn 01.07.57. No 229 OCU 23.10.57. To HAL 02.11.67. Converted to Jordanian FGA Mk 73B '851'. Delivered 13.11.71.
WV402 ff 06.09.55	Delivered 07.10.55. No 222 Sqn 'T' 14.11.55. CAT4. HAL 15.11.56. No 229 OCU. To HAL 28.08.68. Converted to Abu Dhabi FGA Mk 76 '704'. Delivered 05.12.70.
WV403 ff 02.09.55	Delivered 27.09.55. No 67 Sqn 'J' 19.01.56. No 5MU Kemble 16.04.57. Broken up Kemble 09.58.
WV404 ff 15.09.55	Delivered 27.09.55 No 54 Sqn. No 247 Sqn 'T'. No 229 OCU. GIA 7768M. To HAL 02.05.72. Converted to Swiss Mk 58A 'J-4137'. Delivered 05.06.74.
WV405 ff 09.09.55	Delivered 23.09.55. No 222 Sqn 'A' 23.11.55. No 229 OCU 19.10.57. Transferred to FAA, HMS *Condor*, 01.66. To HAL 1970. Converted to Swiss Mk 58A 'J-4106'. Delivered 23.03.72.
WV406 ff 09.09.55	No 222 Sqn 'F' Leuchars 02.12.55. No 229 OCU 19.10.57. No 3 CAACU Exeter.
WV407 ff 14.09.55	No 5MU Kemble 04.10.55. No 20 Sqn 'O' 09.12.55. CAT4 11.04.56. RAF Flying College Manby. To HAL 14.08.68. Converted to Jordanian FGA Mk 73A '849'. Delivered 21.12.71.

WV408 ff 19.09.55	No 20 Sqn 'U' 24.11.55. No 26 Sqn 09.05.57. No 229 OCU 12.08.57. To HAL 16.11.67. Converted to Jordanian FGA Mk 73A '845'. Delivered 26.08.71.
WV409 ff 20.09.55	Delivered 07.10.55. No 66 Sqn 'N' 06.03.56. No 234 Sqn 'K' 14.12.56. No 229 OCU 'ES-A' 26.04.57. Collided with XG210 on ground. W/O 03.61.
WV410 ff 21.09.55	No 5MU Kemble 18.10.55. No 20 Sqn 'X' 29.12.55. No 26 Sqn 'B' Gutersloh 09.05.57. No 229 OCU 10.09.57. W/O 20.03.59.
WV411 ff 26.09.55	No 5MU Kemble 10.10.55. No 20 Sqn 'D' 29.12.55. No 229 OCU 31.05.57. Transferred to FAA, HMS *Condor*, 01.66. Returned to HAL 1970. Converted to Swiss Mk 58A 'J-4108'. Delivered 27.04.72.
WV412 ff 23.09.55	No 5MU Kemble 10.10.55. No 112 Sqn 'A' 12.04.56. Aircraft crashed Sylt, Germany. W/O 13.09.56.

Third production batch of three aircraft built by Hawker Aircraft Ltd, Kingston-upon-Thames. Contract No SP/6/7144/CB.7a dated 19.07.51. Instruction to Proceed 12.07.51. Remainder of contract to Blackpool.

WW589 ff 06.10.55	No 5MU Kemble 09.11.55. No 20 Sqn 04.01.56. No 26 Sqn 15.05.57. No 229OCU 12.09.57. GIA 7943M Halton. Re-purchased by HAL 02.71. Converted to Swiss Mk 58A 'J-4130'. Delivered 13.04.73.
WW590 ff 06.10.55	No 5MU Kemble 09.11.55. No 20 Sqn 'C' 09.12.55. No 26 Sqn 01.07.57. No 229 OCU 29.09.57. HAL 04.72. Converted to Swiss Mk 58A 'J-4138'. Delivered 19.06.74. SOC 18.05.93.
WW591 ff 04.10.55	Delivered 02.11.55. Transferred to Denmark order as first Danish Mk 51 'E-401'.

First production batch of twenty aircraft built by Hawker Aircraft (Blackpool) Ltd under Contract No SP/6/8435/CB.7a dated 15.08.52.

WW646	Delivered 11.03.55 A&AEE for check on manufacturer standard of production. No 111 Sqn 'P'. No 222 Sqn 'P' 04.12.56. CAT3. No 60MU 05.09.57. Scrapped 02.02.59.
WW647	Delivered 29.03.55 No 5MU Kemble. No 4 Sqn 26.09.55. No 247 Sqn 'A'. No 98 Sqn 'C' 03.10.55. No 5MU Kemble 10.07.57. Returned to HSA 12.05.61. Scrapped 1964.
WW648	Delivered 15.03.55 No 33MU. No 247 Sqn 'C'. No 98 Sqn 'D' 19.04.55. No 229 OCU 22.03.57. Returned to HSA 12.05.61. Scrapped 1964.
WW649	Delivered 15.03.55 No 33MU. No 98 Sqn 'E' 15.04.55. No 5MU Kemble 10.07.57. Returned to HSA 08.09.61. Scrapped Dunsfold 08.64.
WW650	Delivered 16.03.55 No 33MU. No 98 Sqn 'F' 19.04.55. No 222 Sqn 'R' 16.05.57. No 5MU Kemble 05.11.57. Returned to HSA 08.09.61. Scrapped Dunsfold 08.64.
WW651	Delivered 28.03.55 No 5MU Kemble. No 43 Sqn 'C' Leuchars 1954. No 111 Sqn 'F' 1955-56. No 222 Sqn 28.11.56. No 98 Sqn. Overshot runway at Luqa, Malta. W/O 27.12.56. SOC 13.02.57.
WW652	Delivered 29.03.55 No 5MU Kemble. No 247 Sqn. No 98 Sqn 'G' 19.04.55. No 222 Sqn 'H' 29.03.57. No 229 OCU 29.07.57. Stored No 5MU Kemble 12.57. Returned to HSA 1961. Scrapped Dunsfold.
WW653 ff 30.03.55	Delivered 01.04.55 No 33MU. No 98 Sqn 21.04.55. No 118 Sqn 03.10.55. No 98 Sqn 04.10.55. No 5MU Kemble 08.04.57. GIA 7784 Halton 12.11.62. Re-purchased by HAL 1.72. Converted to Chilean FGA Mk 71 'J-737'. Delivered Chile 11.01.74.
WW654	Delivered 28.03.55 No 5MU Kemble. No 4 Sqn 26.09.55. No 98 Sqn 'L' Jever 04.10.55. CAT4 07.03.56. No 229 OCU 14.11.57. Re-purchased by HSA 1961. Converted to GA Mk 11. Delivered FAA Lossiemouth 04.09.62. No 738 Sqn until 1970. To storage. To FRADU '833/VL'. Withdrawn from use late 1986. FRADU '833'. GIA A2664 (2). To Culdrose 11.02.87 using rear fuselage of GA.11 XF368. SAH 'A2754' '834/DD'. Withdrawn 03.93. To Portsmouth by 15.12.93. On display former RNAS Ford site.
WW655	Delivered 30.03.55 No 5MU Kemble. No 98 Sqn 'M' 21.04.55. No 5MU 15.07.57. CAT4 14.01.57. Placed in store. WFU 07.57. Returned to HSA 14.03.61. Scrapped 1964.
WW656	Delivered 30.03.55 No 5MU Kemble. No 98 Sqn 'N' Jever 20.04.55. No 5MU Kemble 08.04.57. Returned to HSA 23.03.61. Scrapped Dunsfold 08.64.

WW657	Delivered 30.03.55 No 5MU Kemble. No 118 Sqn 'A' then 'G' 28.04.55. CAT4. To No 71MU 14.01.57. Stored No 47MU Hawarden 01.58. Scrapped 01.01.59.
WW658	Delivered 06.04.55 No 33MU. No 98 Sqn 'O' 20.04.55. No 229 OCU 'RS-5' 08.04.57. Returned to HSA 07.04.61. Scrapped Dunsfold 06.64.
WW659	Delivered 01.04.55 No 33MU. No 247 Sqn 'V'. Returned to HAL 1960. Converted to GA Mk 11. FAA 28.06.61. No 738 Sqn FAA. Returned to HAL 24.04.70. Converted to Mk 58A 'J-4126'. Delivered 07.02.73. SOC 05.11.91.
WW660	Delivered 06.04.55 No 33MU. No 118 Sqn 'B' 27.04.55. No 229 OCU 01.07.57. W/O 06.11.57. Sold as scrap No 5MU Kemble 01.10.58.
WW661	Delivered 07.04.55 No 33MU. No 98 Sqn 21.04.55. CAT4 03.10.56. SOC 24.02.58. Returned to HAL 24.02.58. Converted to T Mk 8. Transferred to FAA 28.02.59. Delivered RNAS Lossiemouth 12.03.59. No 764 Sqn FAA. Yeovilton Stn Flt '728/VL'. Crashed near Yeovilton. W/O 31.03.69.
WW662	Delivered 27.04.55 No 54 Sqn. Returned to HAL 20.09.55. Converted to Peruvian Mk 52 '632'. ff 09.12.55. Delivered Peru 04.56.
WW663	Delivered 27.04.55 No 5MU. No 14 Sqn 'H' 24.05.55. No 5MU 18.04.57. Returned to HSA 14.03.61. Broken up at HSA Chilbolton 1962.
WW664	Delivered 03.05.55. No 26 Sqn 'C' 08.06.55. No 26 Sqn. Forced-landing at Ahlhorn 28.05.56. CAT 4. Returned to HAL. Converted to T Mk 8B (prototype) 30.03.58. CA aircraft Boscombe Down. No 764 Sqn FAA Lossiemouth. No 700 Sqn '86/LM' FAA Lossiemouth 05.61 to 09.62. Converted to Singapore T Mk 75 '514'. Delivered 1969. Original F Mk 4 nose section restored Newark-on-Trent.
WW665	Delivered 03.05.55 No 5MU Kemble. No 247 Sqn 'E'. No 245 Sqn. Scrapped 02.02.59 No 49MU Colerne.

Second production batch of 100 aircraft built by Hawker Aircraft (Blackpool) Ltd. Contract No SP/6/9817/CB.7a, dated 24.08.53.

XE657	Delivered 03.05.55 No 5MU Kemble. No 14 Sqn 'Y' 13.06.55. Broken up Dunsfold 11.64.
XE658	Delivered 03.05.55 No 5MU Kemble. No 54 Sqn 'R'. No 74 Sqn. Reduced to spares 05.61.
XE659	Delivered 04.05.55 No 33MU. No 54 Sqn. No 66 Sqn. No 92 Sqn 'J'. No 229 OCU 'RS-23'. GIA 7785M Halton. To HAL 03.72. Converted to Mk 58A 'J-4149'. Delivered 10.01.75.
XE660	Delivered 18.05.55 No 5MU Kemble. No 247 Sqn 'R' later 'H' 02.06.55. Crashed near Lasham. Pilot killed. W/O 05.12.56.
XE661	Delivered 06.06.55 No 33MU. No 54 Sqn 'B'. No 74 Sqn 'F'. CAT4 28.05.57. To HAL for repair but sold for scrap 01.01.59.
XE662	Delivered 18.05.55 No 5MU Kemble. No 247 Sqn 'Z'. No 74 Sqn 'S'. Crashed short of runway at Horsham St Faith, W/O 25.05.57.
XE663	Delivered 01.06.55 No 5MU Kemble. No 4 Sqn 'V' 20.07.55. No 43 Sqn 'V' 21.03.57. Scrapped Dunsfold 08.64.
XE664	Delivered 23.05.55 No 5MU Kemble. No 26 Sqn 'F' 08.06.55. CAT3 22.08.56. To HAL for repair. No 43 Sqn Leuchars. To HAL 1958. Converted to T Mk 8. Delivered 12.03.59. No 764 Sqn FAA Lossiemouth '708/LM' later '699/LM'. Crash-landed Milltown 22.05.68. To HAL. Converted to Singapore T Mk 75 '516'. Delivered 09.70. Singapore Air Defence Command with No 229 OCU Chivenor as '500'. Sold to David Phillips, Ardmore, New Zealand. Nose and cockpit of original XE664 in Gloucestershire Aviation Collection.

XE665	Delivered 24.05.55 No 5MU Kemble. To Wing Leader Jever 20.07.55. No 118 Sqn 'A' 17.04.56. CAT4 13.06.56. To HAL 1957. Converted to T Mk 8. Delivered. 20.04.59 No 764 Sqn FAA Lossiemouth. Yeovilton Stn Flt and Flag Officer Flying Training (FOFT) 1965. To No 759 Sqn Brawdy '802/BY' until 1969. Placed in store. Transferred to FRADU '878/VL' 1974. WFU 1978. No 5MU Kemble. No 237 OCU 05.80. FRADU '876/VL'. To Shawbury for storage. Sold and registered as G-BWGM. In open store at Exeter with Hunter Flying Club 11.98.
XE666	Delivered 06.06.55 No 5MU Kemble. No 111 Sqn. No 4 Sqn 'X' 11.07.55. No 245 Sqn 'R' 21.03.57. No 229 OCU 'ES-D'.
XE667	Delivered 06.06.55 No 5MU Kemble. No 4 Sqn 'Z' 12.07.55. No 98 Sqn 'Z' 15.03.57. No 5MU Kemble 09.07.57. Broken up Dunsfold 11.64.
XE668	Delivered 06.06.55 No 5MU Kemble. No 4 Sqn 'X'. No 26 Sqn 'A' & 'G' 23.06.55. No 5MU Kemble 01.05.57. WFU 1961. To HAL and converted to GA Mk 11. No 738 Sqn FAA Lossiemouth 11.64. Brawdy from 1966. FRU Hurn 31.07.69. Yeovilton '832/VL' 1972. SAH Culdrose 'A2733'. Handed over as '832/DD' 08.02.85. Predannack Royal Navy Fire School 1993. Sold to Barry Pearson, Bristol Outdoor Pursuits Centre, Hamburger Hill, Bristol 19.09.07. Gate guard.
XE669	Delivered 16.06.55 RAE. No 98 Sqn 23.07.56. CAT5, engine fire after take-off. Pilot killed. W/O 04.12.56.
XE670	Delivered 21.06.55 No 5MU Kemble. No 26 Sqn 12.09.55. No 93 Sqn 'D' 09.02.56. No 5MU Kemble 29.03.57. GIA 7762M (later 8585M). No 71MU Bicester 11.07.63. Nose section at RAF Museum, Cosford.
XE671	Delivered 06.06.55 No 33MU. No 54 Sqn 'K'. 'Black Knights' aerobatic team 1955. Aircraft crashed E of Odiham. W/O 01.05.56.
XE672	Delivered 06.06.55 No 33MU. No 112 Sqn 'B' 10.04.56. No 5MU Kemble 13.05.57. Reduced to spares 19.05.61.
XE673	Delivered 27.06.55 No 33MU. No 112 Sqn 'C' 12.04.56. No 234 Sqn 'E' 02.05.57. No 5MU Kemble 04.07.57. Converted to GA Mk 11. Delivered Lossiemouth 14.06.62.
XE674	Delivered 21.06.55 No 33MU. No 112 Sqn 'D'. No 234 Sqn 'D' 13.04.57. No 5MU Kemble 10.05.57. Converted to GA Mk 11. Delivered 5.06.62 No 738 Sqn FAA. Purchased by HAL 27.04.70. Converted to Mk 58A 'J-4124'. Delivered 19.01.73.
XE675	Delivered 27.06.55 No 33MU. No 26 Sqn 'H' 19.09.55. No 98 Sqn 25.11.55. No 93 Sqn 'E' 18.01.56. Fitted with wing leading edge extensions. No 229 OCU 01.04.57. No 111 Sqn.
XE676	Delivered 01.07.55 No 5MU Kemble. No 222 Sqn 'U' 17.01.56. CAT3. No 60MU 15.07.57. FWS. Scrapped 11.62. Brawdy dump 09.67.
XE677	Delivered 11.07.55 No 5MU Kemble. Lyneham 29.07.55. No 4 Sqn 'M' 02.09.55. No 118 Sqn 17.10.55. No 93 Sqn 'Q' 17.01.56. Fitted with wing leading edge extensions. No 229 OCU 29.03.57. SOC 07.07.58. Returned to HAL 04.01.61. Refurbished and presented to Loughborough College as G-HHUN 30.10.89. Restored 21.01.94 and operated by Jet Heritage Bournemouth. Crashed Dunsfold. Pilot killed. W/O 05.06.98.
XE678	Delivered 04.07.55 No 5MU Kemble. No 222 Sqn 'X' 28.02.56. FWS 14.01.57. GIA Halton. Returned to HAL 06.72. Converted to Mk 58A 'J-4145'. Delivered 07.10.74.
XE679	Delivered 04.07.55 No 5MU Kemble. No 111 Sqn 'B' 1955-56. No 222 Sqn 'Y' 04.12.56. No 5MU Kemble 05.11.57. GIA 7887M Halton. Returned to HAL 9.71. Converted to Singapore Mk 74B '541'. Delivered 13.06.73.
XE680	Delivered 11.07.55 No 5MU Kemble. No 130 Sqn 'Y' 21.03.56. No 234 Sqn 'L' 02.05.57. No 5MU Kemble 04.07.57. Converted to GA Mk 11. Delivered 29.04.63. 738 Squadron '784/BY' Brawdy 1967-69. 'Rough Diamonds' aerobatic team. Crashed at Rhayader-Llangurig. W/O 03.03.69.
XE681	Delivered 01.07.55 No 33MU. No 66 Sqn 'J' 21.03.56. FWS 15.12.56. Reduced to spares 19.05.61.

XE682	Delivered 05.07.55 No 33MU. No 118 Sqn 'Z' Jever 07.03.56. No 5MU Kemble 22.10.57. Converted to GA Mk 11. Delivered 01.07.63 No 738 Sqn FAA Lossiemouth then Brawdy '786/BY'. 'Rough Diamonds' aerobatic team 1967. FRU Hurn as '835'. Yeovilton 1972. FRADU '835/VL' with 'Blue Herons' 1975-80. WFU Shawbury 1985. Fuselage and wings to Culdrose 22.11.85. Culdrose dump 28.11.86.
XE683	Delivered 20.07.55 No 5MU Kemble. No 54 Sqn 'S'. No 74 Sqn 'G'. Scrapped Coleys 1963.
XE684	Delivered 20.07.55 No 33MU 20.12.55. No 98 Sqn 29.08.55. No 4 Sqn 03.10.55. No 118 Sqn 20.12.55. No 93 Sqn 18.01.56. No 229 OCU 22.03.57. No 111 Sqn. Reduced to spares 05.61.
XE685	Delivered 18.07.55 No 33MU. No 98 Sqn Jever 24.08.55. No 93 Sqn 'B' Jever. CAT4 03.05.56. Re-purchased by HSA 1960. Converted to GA Mk 11. Delivered 06.05.63 No 764 Sqn FAA Lossiemouth. Placed in store No 5MU Kemble during 1966. Yeovilton Stn Flt '708/VL' 1967. No 764 Sqn '694/LM' until 1972. Yeovilton ADTU '731/VL' 1973. FRADU '861/VL'. Retired 1994. Sold as G-GAII. Hunter Flying Club, Exeter, owned by Alan Fowles.
XE686	Delivered 22.07.55 No 5MU Kemble. CFE and AFDS 1955-56. No 247 Sqn 'H'. No 245 Sqn 'O'. No 229 OCU 'ES-C'. Scrapped 11.62.
XE687	Delivered 22.07.55 No 5MU Kemble. No 118 Sqn 'F' 28.11.55. No 93 Sqn 'C' 22.05.56. No 98 Sqn 'G' 22.03.57. No 5MU Kemble 25.07.57. Scrapped 11.62.
XE688	Delivered 24.08.55 No 45MU. No 222 Sqn 'V' 14.11.55. No 74 Sqn 'A' 17.01.57. Reduced to spares 19.05.61.
XE689	Delivered 22.07.55 No 5MU Kemble. No 67 Sqn 'W' 02.03.56. No 130 Sqn 17.04.57. No 234 Sqn 'K' 02.05.57. No 5MU Kemble 04.07.57. Re-purchased by HAL 1961. Converted to GA Mk 11. Delivered 04.07.63 No 764 Sqn FAA Lossiemouth '695/LM' until 1973. Fitted with PR.11 nose. FRADU '864/VL'. Retired 1994. Placed in store Shawbury. Sold 1995 to Barry Pover, Classic Jets. In open storage, Exeter, as G-BWGK.
XE702	Delivered 22.07.55 No 5MU Kemble. A&AEE trials. No 92 Sqn 'E'. No 43 Sqn 'B'. GIA 7794M Sealand 11.07.63. To HAL 09.71. Converted to T Mk 68 'J-7204'. Delivered 15.04.75.
XE703	Delivered 10.08.55 No 5MU Kemble. No 4 Sqn 'J' 02.09.55. No 93 Sqn 'P' 03.02.56. No 118 Sqn 'B' 09.04.57. No 5MU Kemble 19.08.57. SOC 11.62. Scrapped Dunsfold.
XE704	Delivered 07.10.55 No 45MU. No 118 Sqn 'S'. No 112 Sqn 'S' 10.04.56. No 5MU Kemble 21.05.57. GIA 7788M Halton. To HAL 01.72. Converted to Chilean T Mk 72 'J-736'. Delivered 15.02.74.
XE705	Delivered 24.08.55 No 45MU. No 111 Sqn. No 92 Sqn 'F'. No 43 Sqn 'P'. W/O 03.10.56.
XE706	Delivered 11.10.55 No 45MU. No 43 Sqn 'P'. No 66 Sqn 'P' 26.03.56. No 92 Sqn 'L' 05.04.56. Scrapped 11.62.
XE707	Delivered 01.09.55 No 5MU Kemble. No 118 Sqn 16.11.55. CAT3 20.01.56. No 93 Sqn 'T' 19.04.56. No 98 Sqn 22.03.57. No 5MU Kemble 18.07.57 for storage. CFS Type Flt Kemble 'A' 11.11.57 to 14.11.58. To HAL. Converted to GA Mk 11. No 738 Sqn FAA Lossiemouth 08.64. No 764 Sqn 11.64. To FRU '833' 13.10.72. Yeovilton. No 5MU Kemble for storage early 1973. FRADU '865/VL' 1981. SOC 1994. Rear fuselage to XE673. G-BVYH 31.01.95. Sold to George Lazik, Santa Monica, as N707XE. Chris Perkins Hunter Flying Club 01.04. Sold to USA.
XE708	Delivered 29.08.55 No 5MU Kemble. No 14 Sqn 18.10.55. No 229 OCU 15.04.57. GIA Halton.
XE709	Delivered 05.09.55 No 45MU. No 222 Sqn 'D' 02.12.55. No 43 Sqn 'Q' 04.01.57. Scrapped, SOC 11.62.
XE710	Delivered 02.09.55 No 5MU Kemble. No 14 Sqn 'R' 18.10.55. No 5MU 25.04.57. No 229 OCU. Scrapped Dunsfold, SOC 11.62.
XE711	Delivered 11.10.55 No 45MU. RAF Flying College Manby. JHSS 'A' Flt Benson. Scrapped Dunsfold, SOC 11.62.

XE712 Delivered 02.09.55 No 5MU Kemble. No 222 Sqn 'E' 06.02.56. CAT3 22.11.56. No 43 Sqn 'D'. Converted to GA Mk 11. Delivered 06.04.62 No 738 Sqn FAA. Predannack Fire School 08.95.

XE713 Delivered 18.10.55 No 45MU. No 66 Sqn 'E' 14.03.56. No 5MU Kemble 07.01.57. No 67 Sqn. Scrapped Dunsfold 09.64.

XE714 Delivered 19.10.55 No 5MU Kemble. No 67 Sqn 12.01.56. Bruggen Stn Flt 'ELM' 07.02.56. No 67 Sqn 04.03.57. No 112 Sqn 18.04.57. No 234 Sqn 'C' 02.05.57. No 5MU Kemble 04.07.57. Scrapped 11.62.

XE715 Delivered 01.11.55 No 45MU. No 71 Sqn. No 3 Sqn 'L' 22.06.56. No 130 Sqn 'L' 12.12.56. No 112 Sqn 'L' 06.05.57. No 5MU Kemble 21.05.57. GIA 7807M.

XE716 Delivered 28.10.55 No 5MU Kemble. No 67 Sqn 'A' 19.01.56. No 112 Sqn 18.04.57. No 5MU Kemble 14.05.57. Re-purchased by HAL. Converted to GA Mk 11. Delivered 06.07.62 Lossiemouth. FRU Hurn '834' 01.70. Yeovilton 1972. FRADU '834/VL'. Crashed into sea off Portland Bill. W/O 16.05.83.

XE717 Delivered 28.10.55 No 5MU Kemble. No 67 Sqn 'G' 06.03.56. No 112 Sqn 18.04.57. No 234 Sqn 'G' 25.05.57. No 5MU Kemble 04.07.57. Converted to GA Mk 11. Delivered 05.06.63 No 738 Sqn Brawdy '794/BY' 08.66. Yeovilton Stn Flt. Returned to HAL 24.04.70. Converted to Mk 58A 'J-4122'. Delivered 14.02.72.

XE718 Delivered 02.11.55 No 5MU Kemble. No 93 Sqn 'A' 12.03.56. No 98 Sqn 'A' 27.03.57. No 5MU Kemble 09.07.57. Scrapped 11.62.

XF289 Delivered 02.11.55 No 5MU Kemble. No 67 Sqn 'Q' Bruggen 25.01.56. CAT3. Returned to HAL 14.12.56. Converted to T Mk 7 and later T Mk 8. No 764 Sqn Lossiemouth '709/LM'. Air Direction Training School 08.70. Yeovilton Stn Flt '738/VL' 1972. FRADU '872' 1974. '875' 1982-94. G-BVYI 31.01.95. Deregistered 06.06.95. To USA as N289XF 10.07.95. European Warbirds Org, Chino, CA, USA. Northern Lights, Montreal, Canada.

XF290 Delivered 29.11.55 No 5MU Kemble. No 67 Sqn 'F' 24.02.56. Crashed after engine failure at Bruggen. CAT 5. W/O 14.08.56.

XF291 Delivered 29.11.55 No 5MU Kemble. No 67 Sqn 'E' 06.03.56. No 112 Sqn 18.04.57. No 5MU Kemble 21.05.57. Converted to GA Mk 11. Delivered 05.06.63. No 738 Sqn FAA 08.64. To HAL. Converted to F Mk 58A 'J-4118'. Delivered overland and assembled at Emmen.

XF292 Delivered 12.12.55 No 5MU Kemble. No 130 Sqn 'A' 09.04.56. No 112 Sqn 'A' 21.05.57. No 5MU Kemble. Scrapped 09.64.

XF293 Delivered 08.12.55 No 5MU Kemble. No 112 Sqn 'N' 10.04.56. No 234 Sqn 'N' 02.05.57. No 5MU Kemble 04.07.57. Scrapped Feltham 11.62.

XF294 Delivered 08.12.55 No 5MU Kemble. No 130 Sqn 'B' 06.04.56. No 112 Sqn 'B' 06.05.57. No 5MU Kemble 27.05.57. Scrapped Feltham 10.03.61.

XF295 Delivered 29.11.55 No 5MU Kemble. No 130 Sqn 'C' 09.04.56. No 112 Sqn 'E' 06.05.57. No 5MU Kemble 19.06.57. FTU Benson. Scrapped Feltham 11.62.

XF296 Delivered 09.01.56 No 5MU Kemble. No 112 Sqn 06.04.56. No 67 Sqn 'Z' Bruggen 12.04.56. No 130 Sqn 17.04.57. No 234 Sqn 'H' 02.05.57. No 5MU Kemble 04.07.57. Scrapped Dunsfold 11.62.

XF297 Delivered 08.12.55 No 5MU Kemble. No 130 Sqn 'D' 12.04.56. No 234 Sqn 'D' 02.05.57. No 5MU Kemble 04.07.57. Converted to GA Mk 11. Delivered 14.08.62. No 738 Sqn FAA 08.64. 'Rough Diamonds' aerobatic team '781/BY' Brawdy 08.66 to 1967.

XF298 Delivered 29.12.55 No 5MU Kemble. No 130 Sqn 'E' 10.04.56. No 5MU Kemble 02.05.57. Scrapped 11.62.

XF299 Delivered 29.12.55 No 5MU Kemble. No 43 Sqn 'O'. Scrapped Dunsfold 09.64.

XF300	Delivered 09.01.56 No 5MU Kemble. No 71 Sqn 14.05.56. No 234 Sqn 'P' 30.05.56. No 130 Sqn 'W' 27.12.56. No 5MU Kemble 06.05.57. Re-purchased by HAL and converted to GA Mk 11. Delivered 12.02.63 No 738 Sqn FAA Lossiemouth. Brawdy 08.64. No 764 Sqn Lossiemouth '694/LM' later '690/LM'. No 5MU Kemble and placed in store. FRADU '860/VL' 06.03.80 until 05.95, then Shawbury 12.06.95. G-BZPC 15.01.01. Delta Jets, Kemble, 07.02.
XF301	Delivered 30.12.55 No 5MU Kemble. No 43 Sqn 'L'. Re-purchased by HAL. Converted to GA Mk 11. Delivered 19.10.62 No 738 Sqn FAA Lossiemouth. Brawdy '791/BY' 1964. 'Rough Diamonds' aerobatic team. Lossiemouth No 764 Sqn '679/LM' 1970-72. No 5MU Kemble, reserve aircraft for FRADU. FRADU '834/VL' 1983 until 08.93. Shawbury. FAA. Registered as N301XF to Moscos F. Garrett, Villa Park, CA.
XF302	Delivered 30.12.55 No 5MU Kemble. No 43 Sqn 'S'. GIA 7774M Halton. Returned to AWA 16.12.71. Converted to Chilean FGA Mk 71 'J-733'. Delivered 15.02.74. Antofagasta, Cerro Moreno AB, Chile 04.98.
XF303	Delivered 09.01.56 No 5MU Kemble. No 66 Sqn 'A' 05.03.56. CAT4. HAL 23.08.56. Repaired and transferred to FAA. GIA A2565 HMS *Condor*, Arbroath. Returned to HAL 04.02.70. Converted to Mk 58A 'J-4105'. Delivered 10.03.72. Scampton for OFMC/HHA as G-BWOU.
XF304	Delivered 16.01.56 No 5MU Kemble. No 66 Sqn 'B' 06.03.56. Caledonian Sector Commander 21.11.56. Turnhouse Stn Flt. No 43 Sqn.
XF305	Delivered 09.01.56 No 5MU Kemble. No 67 Sqn 21.03.56. CAT5 at Sylt after simultaneous firing of two starter cartridges. W/O 28.11.56.
XF306	Delivered 09.01.56 No 5MU Kemble. No 112 Sqn 'E' 16.04.56. No 5MU Kemble 23.04.57. GIA 7776M. Returned to HAL 06.72. Converted to Mk 58A 'J-4133'. Delivered 22.02.74.
XF307	Delivered 16.01.56 No 5MU Kemble. No 112 Sqn 'F' 20.04.56. No 5MU Kemble 21.05.57. Stored Kemble 05.66.
XF308	Delivered 25.01.56 No 5MU Kemble. No 130 Sqn 'F' 09.04.56. No 5MU Kemble 16.04.57. GIA 7777M. Returned to HAL 08.72. Converted to Mk 58A 'J-4135'. Delivered 27.03.74.
XF309	Delivered 10.01.56 No 5MU Kemble. No 112 Sqn 'C' 12.04.56. No 5MU Kemble 17.04.57. GIA 7771M Finningley. Returned to HAL 5.73. Converted to Kenyan Mk 80 '805'. Delivered. 08.12.74.
XF310	Delivered 20.07.56. Fairey Aviation Limited 06.56. Used for Fairey Fireflash missile trials installation. Sold back to HAL for resale 1958. Converted to T Mk 7. ff 24.04.59. Delivered 29.05.59. No 20 Sqn 'T' Tengah 06.11.61. No 5MU Kemble 06.03.67. No 1TWU Brawdy. No 2TWU '01' Lossiemouth. One of six T Mk 7s transferred to Royal Navy in 1982. Yeovilton, FRADU '876/VL' then '869/VL'. Culdrose 03.93 with Flight Deck school until 1996. Shipped to Australia as VH-HMS 1970. Sold to Hank Ebes, Melbourne, 08.98.
XF311 ff 16.12.55	Delivered 10.01.56 No 5MU Kemble. No 234 Sqn 'B' Geilenkirchen 08.05.56. No 130 Sqn 'V' Bruggen 01.01.57. No 112 Sqn 'V' Bruggen 06.05.57. No 5MU Kemble 21.06.57. Issued to FAA HMS *Condor*, Arbroath, as GIA A2566, transferred 18.06.63. SOC 22.08.67. Sold to HSA 1970. Refurbished. Transferred as GIA to Singapore 31.01.70. To Singapore Air Force Technical College as 'SAF-TECH-11' 31.01.70. Sold to Queensland Museum, Australia, for restoration, arriving at Caloundra 07.06.89.
XF312	Delivered 18.01.56 No 5MU Kemble. No 71 Sqn 14.05.56. No 112 Sqn 'T' 08.01.57. No 26 Sqn 17.05.57. No 5MU Kemble 10.09.57. Purchased by HSA 6.72. Converted to Mk 58A 'J-4150'. Delivered. 30.01.75.
XF313	Delivered 18.01.56 No 5MU Kemble. No 71 Sqn 'G' 08.05.56. No 112 Sqn 29.04.57. CFS Type Flt Kemble 04.05.57 to 04.02.59. Scrapped 09.64. To Staravia, Lasham.
XF314	Delivered 27.01.56 No 5MU Kemble. No 43 Sqn 'N'. No 229 OCU 'RS-24' 09.57. CFS. To Staravia, Lasham. Spares recovery. Scrapped by 09.64.

XF315	Delivered 27.01.56 No 5MU Kemble. No 93 Sqn. No 118 Sqn 'A' 07.03.56. Jever Stn Flt 17.04.56. No 5MU Kemble 23.08.57. No 43 Sqn.
XF316	Delivered 24.01.56 No 33MU. HQ Flt 2TAF. No 71 Sqn 22.06.56. No 112 Sqn 'R' 02.01.57. GIA 7778M Halton. Purchased by HSA 06.72. Converted to Mk 58A 'J-7134'. Delivered 12.03.74.
XF317	Delivered 30.01.56 No 33MU. No 67 Sqn 'U' 08.03.56. No 5MU Kemble 16.04.57. GIA 7773M Halton. Purchased by HSA 12.71. Converted to Chilean FR Mk 71A 'J-734'. Delivered 11.01.74. Chilean Air Force Museum, Los Cerrillos, Chile.
XF318	Delivered 27.02.56 No 33MU. No 118 Sqn 'Z'. No 130 Sqn 'Z' 19.04.56. No 5MU Kemble 16.04.57. GIA A2567, HMS *Condor*, Arbroath. Purchased by HSA 01.04.70. Converted to Mk 58A 'J-4110'. Delivered 06.06.72.
XF319	Delivered 25.01.56 No 5MU Kemble. No 66 Sqn 'F' 15.03.56. No 112 Sqn 'A' 04.12.56. No 5MU Kemble 17.04.57.GIA 'B' Halton. Gray Tuplin Ltd, Southall, as G-BTCY 22.01.91. Deregistered by CAA 10.03.99. To Ed Stead, Manchester NH, USA.
XF320	Delivered 09.02.56 No 33MU. No 247 Sqn 'R'. No 229 OCU 'RS-4'. Scrapped 09.64. To Staravia, Lasham.
XF321	Delivered 27.02.56 No 33MU. No 130 Sqn 'G' 10.04.56. No 5MU Kemble 16.04.57. No 56 Sqn 1956. No 130 Sqn 1957-58. Returned to HSA 1958. Converted to T Mk 7. No 43 Sqn. No 8 Sqn. No 1417 Flt 'TZ' Khormaksar 09.08.65 to 30.09.67. RAE Farnborough 1976. Unintended wheels-up landing at RAE Bedford. W/O 27.07.84. RNEC, Manadon, as GIA A2734. Phoenix Aviation, Bruntingthorpe, 28.05.04. Nose to Cockpit Collection, RAF Coltishall.

T Mk 7 XF321 operated in the Middle East before being assigned to the RAE at Farnborough. Pictured during a visit to Binbrook, the aircraft was painted in an overall Light Grey colour scheme. After being written off in a landing accident at Bedford during 1984, the aircraft was eventually used as a spares source by Classic Aviation at Exeter. The nose section has been preserved. *Terry Senior*

XF322	Delivered 15.02.56 No 45MU. No 112 Sqn 'O' 20.04.56. No 5MU Kemble 17.04.57. Returned to HSA 1957. Converted to T Mk 8 for FAA Brawdy '780/BY' 08.66. No 764 (Training) Sqn '704/LM' Lossiemouth. Crashed into St Georges Channel SW of Brawdy following mid-air collision with T Mk 8 XF938. W/O 15.12.67.
XF323	Delivered 17.02.56 No 45MU. RAF Flying College Manby. Purchased by HSA 30.10.71. Converted to Chilean FGA Mk 71 'J-732'. Delivered 07.09.73.
XF324	Delivered 05.03.56 No 33MU. No 92 Sqn 'D'. Linton-on-Ouse 1956. No 222 Sqn 'K' 06.05.57. CAT3. No 60MU 13.09.57. Scrapped at Lasham 09.63. To Staravia, Lasham.
XF357	Delivered 01.03.56 No 33MU. No 130 Sqn 'T' Bruggen 10.04.56. No 5MU Kemble 16.04.57. Returned to HSA 1959. Converted to T Mk 8. No 764 Sqn FAA Lossiemouth 1959. No 738 Sqn Lossiemouth '634/LM' by 1962. Brawdy as '779'/BY', later 777/BY', 1964. FRADU as '870/VL' early 1970s. Placed in store 1975-76. Back to FRADU 1978 as '877/VL'. In store at Shawbury 1985. Back to FRADU '871/VL' 1986 until retired 1994. Operated by Elvington Events Ltd. Stichting Hawker Hunter Foundation 'N-321'. Flown to Leeuwarden 15.05.07.
XF358	Delivered 06.03.56 No 5MU Kemble. No 112 Sqn 'P' 10.04.56. No 5MU Kemble 17.04.57. Returned to HSA 1959. Converted to T Mk 8C Lossiemouth. Delivered to FAA 21.07.59. No 700 Sqn FAA 1961. Tengah, Singapore, 06.11.61. Royal Naval Air Holding Unit. No 759 Sqn FAA at Brawdy '804/BY' until 1969. ADTU and Stn Flt '728/VL'. Yeovilton, FRADU as '875/VL' later '870/VL' 1973. Shawbury for storage. Cockpit section to Northern Lights in Canada, mated to F Mk 58 fuselage J-4095. ff Quebec 04.06.
XF359	Delivered 27.02.56 No 33MU. No 130 Sqn 'L' 19.04.56. No 5MU Kemble 12.12.56. Scrapped Lasham 09.63.
XF360	Delivered 05.03.56 No 33MU. No 130 Sqn 09.04.56. No 234 Sqn 'B' 02.01.57. No 3 Sqn 01.05.57. No 229 OCU 'ES-K' 06.06.57. No 5MU Kemble 1962/67. Purchased by HSA 19.08.71. Converted to Singapore FR Mk 74B '542'. Delivered 13.06.73.
XF361	Delivered 01.03.56 No 33MU. No 130 Sqn 06.04.56. CAT4 07.08.56. Returned to HAL. No 229 OCU 'RS-22'. Purchased by HSA 20.06.68. Converted to Mk 58A 'J-4117'. Delivered 30.10.71.
XF362	Delivered 27.02.56 No 33MU. No 112 Sqn 'Q' 24.04.56. No 71 Sqn 29.01.57. CAT4 04.02.57. In store at No 5MU Kemble 1961/62. Purchased by HSA 19.9.68. Converted to Abu Dhabi FGA Mk 76 '705'. Delivered 05.12.70.
XF363	Delivered 01.03.56 No 33MU. No 66 Sqn 10.09.56. No 3 Sqn 14.12.56. No 5MU Kemble 06.06.57. No 229 OCU. No 234 Sqn 'ES-65' 1960. Transferred to FAA.
XF364	Delivered 06.03.56 No 33MU. No 130 Sqn 'X' 10.04.56. No 234 Sqn 'J' 11.01.57. No 3 Sqn 07.05.57. No 5MU 21.05.57. No 229 OCU. Purchased by HSA 06.06.68. Converted to Jordanian FGA Mk 73A '843'. Delivered 15.07.71.
XF365	Delivered 19.03.56 No 33MU. No 71 Sqn 'B' 25.04.56. No 5MU Kemble 16.04.57. No 229 OCU. Purchased by HSA 8.69. Converted to Mk 58A 'J-4109'. Delivered 11.05.72.
XF366	Delivered 07.03.56 No 33MU. No 112 Sqn 'R' 10.04.56. No 71 Sqn 28.12.56. No 5MU Kemble 16.04.57. No 229 OCU. Stored at Kemble 1966.
XF367	Delivered 19.03.56 No 33MU. No 4 Sqn RAF Flying College Manby. No 71 Sqn 'B' 26.04.56. Returned to HAL 21.08.68. Converted to Abu Dhabi FGA Mk 76 '706'. Delivered 05.12.70.
XF368	Delivered 07.03.56 No 33MU. No 5MU 06.06.57. No 71 Sqn. No 4 Sqn 'C' 27.04.56. No 3 Sqn 'C' 15.03.57. No 118 Sqn 'N'. No 229 OCU Chivenor. Purchased by HSA in 1960. Converted to GA Mk 11 1962. No 764 Sqn FAA Lossiemouth '695' later '693' until 1972. Yeovilton ADTU '733/VL'. FRADU as '863/VL' 1973 until 1995. Shawbury for storage. Cape Town, based with Thunder City fleet.
XF369	Delivered 08.03.56 No 33MU. No 71 Sqn 26.04.56. No 234 Sqn 'M' 04.05.56. No 5MU Kemble 09.07.57. RAF Flying College Manby. GIA 7941M Halton. Purchased by HSA 02.71. Converted to Singapore FR Mk 74B '538'. Delivered 09.04.73. Sold to Geoff Moesker, Queensland, Australia, 08.98.

XF370	Delivered 15.03.56 No 33MU. No 4 Sqn 'Z' 27.04.56. No 118 Sqn 'C' 14.03.57. APS Sylt Stn Flt 26.08.57. No 19MU St Athan 22.06.59. RAF Staff College Strubby. GIA 7772M. Purchased by HSA 04.70. Converted to Mk 58A 'J-4136'. Delivered. 30.04.74.

Third production batch of fifty-six aircraft built by Hawker Aircraft (Blackpool) Ltd. Contract No SP/6/10344/CB.7a.

XF932	Delivered 23.03.56 No 45MU. No 234 Sqn 'A' 10.05.56. Crashed on approach to Klein-Brogel, Belgium, W/O 21.09.56.
XF933	Delivered 23.03.56 No 33MU. No 71 Sqn 25.04.56. CAT4 17.01.57. Returned to HAL for repair. CFS Type Flt Kemble 31.08.60 to 06.09.61. Purchased by HSA 03.72. Converted to Mk 58A 'J-4132'. Delivered 28.01.74.
XF934	Delivered 03.04.56 No 5MU Kemble. No 234 Sqn 'C' 10.05.56. No 5MU Kemble 06.05.57. CFS Type Flt Kemble 'C' 13.02.59 to 15.06.61. Scrapped Kemble 03.07.63. Staravia, Lasham.
XF935	Delivered 23.03.56 No 45MU. No 234 Sqn 'D' 10.05.56. No 5MU Kemble 14.05.57. CFS Type Flt Kemble 'D' 28.11.58 to 15.03.61. Purchased by HSA 06.09.68. Converted to Abu Dhabi FGA Mk 76 '707'. Delivered 09.01.71.
XF936	Delivered 11.04.56 No 33MU. No 234 Sqn 'E' 08.05.56. No 5MU Kemble 03.06.57. Purchased by HSA 15.11.67. Converted to Jordanian FGA Mk 73A '844'. Delivered 26.08.71. Sultan of Oman AF '844' 1976.
XF937	Delivered 23.03.56 No 5MU Kemble. No 112 Sqn 'T' 24.04.56. No 71 Sqn 08.01.57. No 130 Sqn 16.04.57. No 234 Sqn 'A' 02.05.57. No 5MU Kemble 09.07.57. RAF Flying College Manby. Purchased by HSA 16.10.68. Converted to Swiss Mk 58A 'J-4116'. Delivered 19.09.72.
XF938	Delivered 27.03.56 No 5MU Kemble. No 71 Sqn 'D' 25.04.56. No 5MU Kemble 15.04.57. No 229 OCU. Returned to HSA 1960. Converted to T Mk 8B. Brawdy 06.62. Crashed into St Georges Channel following mid-air collision with T Mk 8 XF322. W/O 15.12.67.
XF939	Delivered 04.04.56 No 5MU Kemble. No 71 Sqn 10.05.56. No 5MU Kemble 12.04.57. No 229 OCU. Returned to HSA 1960. Converted to T Mk 8B. Upgraded to T Mk 8C. No 759 Sqn FAA. W/O 15.02.65.
XF940	Delivered 06.04.56 No 5MU Kemble. To 2ATAF 17.05.56. No 71 Sqn 24.05.56. No 5MU Kemble 15.06.57. Flying accident 24.10.57. Returned to No 71MU. To HAL 14.11.57 for re-wire as CAT4(R) 02.12.57 until 13.10.58. Returned to No 5MU Kemble 16.10.58. Indefinite loan to ETPS '25' Farnborough 23.02.59. Loaned to HAL Dunsfold 23.08.60. Crashed on approach to Farnborough, W/O 13.10.61. Cockpit section used by RAE for ground procedures training, currently being restored by Mark Templeman.
XF941	Delivered 19.04.56 No 45MU. No 71 Sqn 27.07.56. No 5MU Kemble 15.04.57. No 229 OCU. GIA 8006M. No 2030 ATC Sqn Birmingham 18.03.68. Purchased by HSA 03.72. Converted to Swiss Mk 58A (J-4139). Delivered 07.74.
XF942	Delivered 27.03.56 No 5MU Kemble. No 71 Sqn 25.04.56. No 5MU Kemble 15.04.57. No 229 OCU. Returned to HSA 1964. Converted to T Mk 8C. No 759 Sqn FAA '801/BY'. Crashed in Pembrokeshire during aerobatics 06.11.68.
XF943	Delivered 06.04.56 No 5MU Kemble. No 234 Sqn 'F' 08.05.56. No 5MU Kemble 06.05.57. CFS Type Flt Kemble 'B' 30.10.58 to 27.06.62. Hit wall at Kemble while overshooting. W/O 29.06.62.
XF944	Delivered 09.04.56 No 45MU. No 234 Sqn 'G' 10.05.56. No 5MU Kemble 07.05.57. CFS Type Flt Kemble 'A' 28.10.58 to 11.03.63. Purchased by HSA 03.72. Converted to Swiss Mk 58A 'J-4142'. Delivered 21.08.74.
XF945	Delivered 11.04.56 No 5MU Kemble. No 234 Sqn 'H' 17.05.56. Crashed during night operations 15 miles SE of Aachen. W/O 18.11.56.

XF946	Delivered 06.04.56 No 5MU Kemble. No 234 Sqn 'L' 24.05.56. No 3 Sqn 'E' 01.05.57. No 5MU Kemble 12.06.57. GIA 7804M No 71MU mobile static display aircraft. Jordanian War Memorial, Amman, Jordan.
XF947	Delivered 20.04.56 No 5MU Kemble. No 3 Sqn 'S' 06.06.56. No 5MU 06.06.57. No 229 OCU. Transferred to FAA. GIA A256, HMS *Condor*, Arbroath. Purchased by HSA 1970. Converted to Swiss Mk 58A 'J-4104'. Delivered 02.02.72.
XF948	Delivered 09.04.56 No 5MU Kemble. No 3 Sqn 'J' 06.06.56. No 5MU 12.06.57. No 229 OCU. Aircraft crashed near Lundy Island. Pilot killed. W/O 07.11.57.
XF949	Delivered 24.04.56 No 33MU. No 3 Sqn 'C'. Two start cartridges fired simultaneously at Geilenkirchen and aircraft destroyed by fire. W/O (CAT5) 13.01.57.
XF950	Delivered 17.04.56 No 33MU. No 234 Sqn 25.05.56. Geilenkirchen Stn Flt 12.06.56. No 3 Sqn 'M' 14.08.56. No 5MU 02.05.57. Purchased by HSA 20.03.71. Converted to Singapore T Mk 75A '536'. Delivered 13.04.73. Sold to Geoff Moesker, Brisbane, Australia, as VH-XHH.
XF951	Delivered 01.05.56 No 33MU. No 111 Sqn. No 3 Sqn 'P' 06.06.56. No 229 OCU 06.06.57. Purchased by HSA 05.72. Converted to Swiss T Mk 68 'J-4202'. Delivered 12.74.
XF952	Delivered 23.04.56 No 33MU. No 234 Sqn 'N' 29.05.56. No 5MU Kemble 18.06.57. No 229 OCU. Purchased by HSA 18.07.68. Converted to Jordanian FGA Mk 73A '848'. Delivered 07.10.7l.
XF953	Delivered 26.04.56 No 5MU. RAF Flying College Manby. Damaged beyond repair after bird-strike. W/O 10.11.59.
XF967	Delivered 01.05.56 No 5MU Kemble. No 3 Sqn 'Z' 14.06.56. No 5MU 21.06.57. Returned to HSA 1963. Converted to T Mk 8C. RN Changi 01.66. No 208 Sqn 04.94. No 237 OCU 'XC'. Servicing Instructional Flight Scampton 'T' GIA 9186M. Sold to Exeter as G-BZRI 29.11.00. Sold to Mike Beachyhead and based with the Thunder City fleet. Registered as ZU-CTN.
XF968	Delivered 02.05.56 No 5MU Kemble. No 3 Sqn 'R' 06.06.56. No 5MU 07.05.57. No 229 OCU. Purchased by HSA 25.07.68. Converted to Jordanian FGA Mk 73A '847'. Delivered 21.12.71. Sultan of Oman AF '847' 1976. Gate guard RAFO Salalah, Oman.
XF969	Delivered 01.05.56 No 5MU Kemble. No 3 Sqn 'D' 06.06.56. HAL (CAT4) 10.10.56. ETPS '26'. Purchased by HSA 02.71. Converted to Singapore FR Mk 74B '529'. Delivered 15.12.72. Alan Arthur, Tocumwal Airfield, NSW, Australia, 09.98.
XF970	Delivered 02.05.56 No 33MU. A&AEE tropical trials, Idris 18.06.56. No 130 Sqn 16.10.56. No 234 Sqn 'P' 27.12.56. No 26 Sqn 14.05.57. No 5MU Kemble 10.09.57. No 229 OCU. Purchased by HSA 02.71. Converted to Singapore T Mk 75A '528'. Delivered 10.11.72. 2004 Hunter Flight, Maroochydore, Queensland, Australia. VH-RHO.
XF971	Delivered 02.05.56 No 33MU. No 3 Sqn 'K' 06.06.56. HAL (CAT3) 17.12.56. Purchased by HSA 02.10.68. Converted to Abu Dhabi FR Mk 76A '709'. Delivered 19.01.71.
XF972	Delivered 15.05.56 No 33MU. No 3 Sqn 'V' 22.06.56. No 26 Sqn 06.06.57. No 5MU Kemble 12.09.57. No 71 Sqn. No 14 Sqn. No 229 OCU. GIA 7948M. Purchased by HSA 02.73. Converted to Kenyan FGA Mk 80A '804'. Delivered Kenya 08.12.74. Delivered Zimbabwe AF 19.03.81. Thornhill AB, Gweru, Zimbabwe.
XF973	Delivered 28.05.56 No 5MU Kemble. No 66 Sqn 'M' 17.09.56. No 71 Sqn 10.01.57. No 5MU Kemble 15.04.57. CFS Type Flt Kemble 27.08.57 to 11.03.63. Purchased by HSA 03.72. Converted to Swiss Mk 58A 'J-4143'. Delivered 18.09.74.
XF974	Delivered 15.05.56 No 33MU. No 3 Sqn 'G' 27.06.56. No 26 Sqn 16.05.57. No 229 OCU 10.05.57. GIA Halton. Registered to Gray Tuplin Ltd as G-BTCX 22.01.91. Deregistered by CAA 10.03.99. To Ed Stead, Manchester, NH, USA 09.91.

XF975	Delivered 17.05.56 No 5MU Kemble. No 3 Sqn 'W' 22.06.56. No 229 OCU 'RS-27' 21.06.57. GIA 7945M St Athan. Purchased by HSA 05.73. Converted to Kenyan FGA Mk 80 '806'. Delivered 13.01.75.
XF976	Delivered 14.05.56 No 5MU Kemble. No 111 Sqn. No 3 Sqn 'B' 14.06.56. No 229 OCU 'RS-15' 06.06.57. Converted to Swiss Mk 58A 'J-4112'. Delivered 29.06.72.
XF977	Delivered 17.05.56 No 33MU. No 118 Sqn 'A' Jever 23.07.56. APS Sylt Stn Flt 14.10.57. No 19MU St Athan 22.06.59. Returned to HSA 1962. Converted to GA Mk 11. Lossiemouth 02.10.62. No 738 Sqn Lossiemouth. No 764 Sqn Lossiemouth '691/LM'. Fitted with F.95 reconnaissance cameras and re-designated PR.11. Returned to Lossiemouth '696/LM'. ADTU Yeovilton '735/VL'. FRADU '865/VL'. Crashed into English Channel 12 miles SW of Start Point. W/O 17.03.81.
XF978	Delivered 17.05.56 No 33MU. No 20 Sqn 'A' 04.07.56. No 26 Sqn 09.05.57. CAT3 23.08.57. Returned to HSA 1963. Converted to T Mk 8B. Lossiemouth '110/LM' 10.64. No 764 Sqn FAA '687/LM'. Crashed into Moray Firth. W/O 19.07.67.
XF979	Delivered 28.05.56 No 5MU Kemble. RAF Flying College Manby 'A'. Staff College Strubby 'A'. Purchased by HSA 22.10.68. Converted to Jordanian FGA Mk 73B '850'. Delivered 13.11.71.
XF980	Delivered 28.05.56 No 33MU. CFS Type Flt Kemble 25.07.56 to 24.08.56. Mid-air collision with Javelin XA644 over Wotton-under-Edge. W/O 24.08.56.
XF981	Delivered 05.06.56 No 5MU Kemble. No 71 Sqn 06.07.56. No 5MU Kemble 17.04.57. RAF Flying College Manby. Purchased by HSA 18.11.67. Converted to Swiss Mk 58A 'J-4114'. Delivered 10.08.72.
XF982	Delivered 01.06.56 No 5MU Kemble. No 43 Sqn 'D'. No 229 OCU. Purchased by HSA 02.72. Converted to Chilean FR Mk 71A '738'. Delivered 15.02.74.
XF983	Delivered 01.06.56 No 33MU. No 20 Sqn. No 26 Sqn 04.07.56. Oldenburg Stn Flt 09.07.56. No 26 Sqn 15.05.57. No 229 OCU 12.09.57. Returned to HSA 1963. Converted to T Mk 8C. No 759 Sqn FAA. Crashed at Brawdy, W/O 29.09.66.
XF984	Delivered 05.06.56 No 5MU Kemble. No 4 Sqn 09.07.56. No 71 Sqn 25.03.57. No 5MU Kemble 16.04.57. No 229 OCU 'ES-55'. Transferred to FAA. GIA A2570, HMS *Condor*, Arbroath. Purchased by HSA 1970. Converted to Swiss Mk 58A 'J-4113'. Delivered 12.07.72.
XF985	Delivered 06.06.56 No 33MU. No 26 Sqn 14.05.57. No 5MU Kemble 10.09.57. No 71 Sqn 20.07.56. No 5MU Kemble 10.04.57. Placed in storage. Returned to HSA 1963. Converted to T Mk 8B. No 759 Sqn FAA Lossiemouth, then Brawdy '811/BY'. Placed in storage. ADTU Yeovilton '746'. FRADU fleet '877/VL' by 11.74 until 1990. Re-coded '873/VL' 1984. Crashed at Roman Farm near Charminster, Dorset. W/O 10.04.90.
XF986	Delivered 12.06.56 No 33MU. No 112 Sqn 'A' 17.10.56. No 234 Sqn 'H' 07.01.57. No 5MU Kemble 20.05.57. No 229 OCU 'RS-11'. Crashed at Milton Demeral, North Devon. W/O 07.08.59.
XF987	Delivered 06.06.56 No 33MU. No 93 Sqn 'O' 23.07.56. No 118 Sqn 01.04.57. No 229 OCU 20.09.57. No 3 CAACU. Converted to Jordanian FGA Mk 73A '842'. Delivered 15.07.71. Sultan of Oman AF '842' 1976. Returned by RAFO to RJAF 1994. Preserved at King Hussein College, Mafraq AB, Jordan.
XF988	Delivered 14.06.56 No 5MU Kemble. RAF Flying College Manby. Scrapped 07.63.
XF989	Delivered 20.06.56 No 33MU. No 26 Sqn 'A' 10.09.56. No 229 OCU Chivenor 'ES-B' marked as No 145 (Reserve) Sqn 10.09.57. No 5MU Kemble. Scrapped 07.63.
XF990	Delivered 22.06.56 No 33MU. No 111 Sqn. No 3 Sqn 'K' 27.07.56. Wing Leader Geilenkirchen 14.08.56. No 3 Sqn 'K' 25.02.57. No 229 OCU 'RS-16' 6.06.57. Purchased by HSA 08.72. Converted to Swiss Mk 58A 'J-4139'. Delivered 29.07.74.

XF991	Delivered 31.08.56 No 5MU Kemble. No 234 Sqn 'A' Geilenkirchen 23.10.56. No 5MU Kemble 07.05.57. No 229 OCU Chivenor '4' marked as No 145 (Reserve) Sqn. Returned to HSA 1963. Converted to T Mk 8C. No 759 Sqn Brawdy 1964. No 764 Sqn Lossiemouth '688/LM' 1965. To Yeovilton 1973-74, FRADU '874/VL'. Crashed during take-off. W/O 25.05.78.
XF992	Delivered 23.07.56 No 45MU. No 43 Sqn 'C'. Returned to HSA 1963. Converted to T Mk 8B. Re-purchased by HSA 01.68 and cannibalised for spares. Converted to Swiss F Mk 58A 'J-4129'.
XF993	Delivered 17.07.56 No 45MU. No 43 Sqn. No 229 OCU. No 74 Sqn. Engine seized during ground run, W/O 1960. Colerne Fire Dump 1961.
XF994	Delivered 23.07.56 No 33MU. No 66 Sqn 17.08.56. AFDS 01.11.56. No 229 OCU Chivenor. Placed in store. Returned to HSA 1962. Converted to T Mk 8B 1963. No 759 Sqn '805/BY' Brawdy. Yeovilton, with Naval Flying Standards Flt/Stn Flt '718/VL' 1969. Placed in store at Kemble until 10.80. After Yeovilton FRADU '718', later '874', later '873/VL'. Retired 1995. Shawbury for storage. Boscombe Down during 1998 for restoration. GIA use with Apprentices School. Boscombe Down 2001.
XF995	Delivered 20.07.56 No 33MU. No 247 Sqn. No 229 OCU. Returned to HSA 1963. Converted to T Mk 8C. Delivered 11.06.64. Laarbruch Stn Flt 1980-81. No 5MU Kemble. No 12 Sqn Servicing Instructional Flight (SIF) Scampton. GIA 9237M 'K'. Sold by Phillips at auction 29.11.00. Ferried to Delta Jets, Kemble, 16.02.01. G-BZSF. Deregistered by CAA 11.04.07. MoD/HHA contract work 2008.
XF996	Delivered 06.11.56 No 5MU Kemble. No 98 Sqn. No 247 Sqn. No 229 OCU 'ES-72' 1959. Aircraft crashed at Saunton Sands on approach to Chivenor. W/O 06.05.59.
XF997	Delivered 20.07.56 No 5MU Kemble. No 43 Sqn 'F'. Force-landed on mudflats in River Eden estuary due to bird ingestion. CAT5. W/O 28.06.56.
XF998	Delivered 20.07.56 No 33MU. No 54 Sqn 'A'. No 245 Sqn No 229 OCU. GIA 7950M. Purchased by HSA 02.72. Converted to Swiss Mk 58A 'J-4140'. Delivered 17.07.74.
XF999	Delivered 20.07.56 No 45MU. CFS. Crashed 3 miles E of Kemble when undercarriage jammed partly down in landing circuit. Pilot killed. W/O 24.10.56.
XG341	Delivered 20.07.56 No 5MU Kemble. No 43 Sqn 'H'. RAF Flying College Manby. Purchased by HSA 20.06.68. Converted to Abu Dhabi FGA Mk 76 '702'. Delivered 06.11.70.
XG342	Delivered 20.07.56 No 5MU Kemble. No 66 Sqn. No 111 Sqn 'A'. No 247 Sqn. No 222 Sqn 'W' 11.12.56. No 229 OCU. No 234 Sqn 'ES-69' Chivenor 30.08.57. No 5MU Kemble. Scrapped 07.63.

Hawker Hunter F Mk 5

One production batch of 105 aircraft built by Armstrong Whitworth Aircraft Ltd, Coventry. Second part of Contract No SP/6/6315/CB.7a.

WN954 ff 19.10.54	A&AEE Boscombe Down. Trials and used for target towing.
WN955	Flight trials with Sapphire AS Sa.7 engines.
WN956	A&AEE for fuel system trials. No 1 Sqn 'P'. No 41 Sqn 'G'.
WN957	RAE Farnborough miscellaneous flight trials. North Weald. Stafford 7407M. TEE Llanbedr. SOC 20.02.57. Nose section to Mike Rolfe, Stockport.
WN958	A&AEE trials with various external stores. SOC 14.02.60.
WN959	Delivered 26.04.55 No 5MU Kemble. No 1 Sqn 'A'.
WN960	Delivered 26.04.55 No 5MU Kemble. Damaged and placed in storage at No 45MU.
WN961	Delivered 05.05.55 No 5MU Kemble. RAE Farnborough on trials. No 257 Sqn 08.07.55. No 41 Sqn 'G' 11.07.55.
WN962	Delivered 28.04.55 No 5MU Kemble. No 1 Sqn 'N'. No 41 Sqn.
WN963	Delivered 20.05.55 No 5MU Kemble. No 257 Sqn 'T'. No 41 Sqn 'F' 08.07.55. No 34 Sqn, W/O 03.09.57. Aircraft to HSA for fatigue testing 03.62.

WN964	Delivered 11.05.55 No 5MU Kemble. No 41 Sqn 'C'.
WN965	Delivered 16.05.55 No 5MU Kemble. No 41 Sqn 'P'. Crashed at Wilmington near Dartford. Pilot killed. W/O 17.04.56.
WN966	Delivered 28.04.55 No 5MU Kemble. No 41 Sqn 'M'.
WN967	Delivered 20.05.55 No 5MU Kemble. No 34 Sqn. No 41 Sqn 'A'. Fatigue tested by HSA, SOC 3.62.
WN968	Delivered 25.03.55 No 5MU Kemble. No 41 Sqn. No 56 Sqn 'P' Waterbeach. Force-landed near Wattisham after engine flame-out, damage severe, not repaired. W/O 29.09.55.
WN969	Delivered 11.05.55 No 5MU Kemble. No 41 Sqn 'N'. Crashed 2 miles E of Biggin Hill after loss of power, W/O 07.06.57.
WN970	Delivered 16.05.55 No 45MU. No 34 Sqn 'L' Tangmere. No 54 Sqn 'M'. No 56 Sqn Waterbeach. No 49MU Colerne 01.12.58. SOC. Purchased by HSA and scrapped 1963.
WN971	Delivered 30.03.55 No 45MU. No 56 Sqn 'T'. W/O 14.10.55.
WN972	Delivered 28.03.55 No 45MU. No 263 Sqn. No 41 Sqn 'U'.
WN973	Delivered 18.04.55 No 45MU. No 1 Sqn 'B'.
WN974	Delivered 26.05.55 No 5MU Kemble. No 1 Sqn 'H'. SOC and scrapped 01.59.
WN975	Delivered 18.05.55 No 5MU Kemble. No 263 Sqn. No 1 Sqn 'C'. SOC and scrapped 01.59.
WN976	Delivered 28.03.55 No 45MU. No 263 Sqn 'P'. Wattisham Stn Flt. No 56 Sqn 'P'. Severe damage during emergency landing at Duxford after engine failed. W/O 27.08.58.
WN977	Delivered 26.05.55 No 5MU Kemble. No 1 Sqn 'P'. Crashed into sea off Selsey after collision with WP137. Pilot killed. W/O 05.05.58.
WN978	Delivered 29.05.55 No 5MU Kemble. No 34 Sqn 'B' Tangmere. No 1 Sqn 'F'. SOC 1963.
WN979	Delivered 29.03.55 No 5MU Kemble. No 56 Sqn 'E' Waterbeach. No 49MU Colerne 05.11.58. SOC and scrapped 1963.
WN980	Delivered 31.03.55 No 45MU. No 263 Sqn. No 1 Sqn 'F'. SOC and scrapped 05.58. Cockpit restored by Skysport.
WN981	Delivered 16.03.55 No 45MU. No 263 Sqn 'V'.
WN982	Delivered 31.03.55 No 5MU Kemble. No 56 Sqn 'H'. No 1 Sqn.
WN983	Delivered 16.03.55 No 45MU. No 263 Sqn. No 41 Sqn 'T'.
WN984	Delivered 31.05.55 No 5MU Kemble. No 56 Sqn 'O'. No 41 Sqn 'C'.
WN985	Delivered 16.03.55 No 45MU. AFDS tactical evaluation trials. No 34 Sqn 'T'. No 208 Sqn 15.01.58 to 02.04.58. No 19MU St Athan. No 41 Sqn 'P'.
WN986	Delivered 22.04.55 No 45MU. No 1 Sqn 'P'. No 56 Sqn 'H'. No 41 Sqn 'M'.
WN987	Delivered 25.04.55 No 45MU. No 56 Sqn 'C' Waterbeach. No 49MU Colerne 05.11.58. Purchased by HSA and scrapped 1962.
WN988	Delivered 13.04.55 No 45MU. No 1 Sqn 'A'. No 41 Sqn 'H'.
WN989	Delivered 06.04.55 No 45MU. No 263 Sqn. Crashed E of RAF Bawdsey. W/O 03.08.55.
WN990	Delivered 06.04.55 No 45MU. No 263 Sqn 'F'. No 56 Sqn 'U' Waterbeach. No 49MU Colerne 01.12.58. Wattisham Stn Flt. Purchased by HSA and scrapped 1962.
WN991	Delivered 23.03.55 No 45MU. AFDS tactical evaluation trials 04.55. No 34 Sqn 'D'. No 208 Sqn 15.01.58 to 05.02.58. No 5MU Kemble.
WN992	Delivered 25.04.55 No 45MU. No 56 Sqn 'D'.
WP101	Delivered 14.04.55 No 45MU. No 56 Sqn 'N'. Crashed 5 miles NW of RAF Waterbeach following undercarriage fracture during landing. W/O 15.03.58.

WP102	Delivered 28.04.55 No 45MU. No 257 Sqn. No 263 Sqn. No 56 Sqn 'Q' Waterbeach. No 49MU Colerne 01.12.58. Purchased by HSA and scrapped 1962.
WP103	Delivered 25.04.55 No 45MU. No 1 Sqn 'C'. No 41 Sqn. No 263 Sqn. No 56 Sqn 'J' Waterbeach. No 49MU Colerne 05.11.58. Scrapped HSA 1962.
WP104	Delivered 28.04.55 No 5MU Kemble. No 56 Sqn 'A' Waterbeach 1954. No 49MU Colerne 01.12.58. Purchased by HSA and scrapped 1963.
WP105	Delivered 06.04.55 No 45MU. No 263 Sqn. No 1 Sqn 'R'.
WP106	Delivered 26.04.55 No 45MU. No 56 Sqn 'F'.
WP107	Delivered 01.04.55 No 5MU Kemble. No 263 Sqn 'A'. Crashed near Wymeswold after engine failure, W/O 25.01.56.
WP108	Delivered 02.04.55 No 5MU Kemble. No 263 Sqn 'T'. No 41 Sqn 'S'.
WP109	Delivered 06.04.55 No 5MU. No 56 Sqn 'B'.
WP110	Delivered 21.04.55 No 5MU Kemble. No 56 Sqn 'M'.
WP111	Delivered 14.04.55 No 45MU. No 34 Sqn 'M'. No 208 Sqn 15.01.58 to 02.04.58. No 19MU St Athan.
WP112	Delivered 28.04.55 No 45MU. No 1 Sqn 'D'. No 34 Sqn 'E' Tangmere 1956. No 41 Sqn. Scrapped 1961.
WP113	Delivered 19.04.55 ASM. No 34 Sqn 'V' Tangmere. No 1 Sqn 'Q'.
WP114	Delivered 19.04.55 ASM. Test bed for various Sapphire engines.
WP115	Delivered 22.04.55 No 45MU. No 56 Sqn 'G'. Crashed 3 miles from Soham. Pilot killed. W/O 04.02.57.
WP116	Delivered 28.04.55 No 45MU. No 1 Sqn 'D'. No 56 Sqn 'W'.
WP117	Delivered 22.04.55 No 45MU. No 1 Sqn 'R'. No 41 Sqn 'N'.
WP118	Delivered 18.05.55 No 5MU Kemble. No 257 Sqn 'W'. No 1 Sqn 'S'. Crashed near East Grinstead after controls failed in dive. W/O 05.12.55.
WP119	Delivered 25.04.55 No 45MU. No 1 Sqn 'T'. No 41 Sqn 'L'. Aircraft fatigue tested to destruction HSA 03.62.
WP120	Delivered 2.05.55 No 5MU Kemble. No 56 Sqn 'S'. No 1 Sqn 'R'. Aircraft fatigue tested to destruction HSA 03.62.
WP121	Delivered 10.06.55 No 45MU. No 1 Sqn 'W'.
WP122	Delivered 10.06.55 No 45MU. No 41 Sqn 'A'.
WP123	Delivered 10.05.55 No 5MU Kemble. No 41 Sqn 'O'. No 56 Sqn 'B' Waterbeach. No 49MU Colerne 05.11.58. Purchased by HSA and scrapped 1963.
WP124	Delivered 26.05.55 No 45MU. No 34 Sqn Tangmere. Belly-landed on runway at Nicosia, Cyprus, after losing power on approach. W/O 29.10.56.
WP125	Delivered 12.05.55 No 5MU Kemble. No 56 Sqn 'R'. Scrapped 1962.
WPl26	Delivered 16.05.55 No 5MU Kemble. No 1 Sqn. No 34 Sqn 'R'. No 208 Sqn 15.01.58 to 03.03.58. No 5MU Kemble.
WP127	Delivered 10.05.55 No 5MU Kemble. No 1 Sqn 'G. No 34 Sqn 'O'. No 208 Sqn 15.01.58 to 03.03.58. No 5MU Kemble.
WP128	Delivered 13.06.55 No 45MU. No 41 Sqn 'C'. Undershot runway on approach to Biggin Hill. W/O 26.11.55.
WP129	Delivered 10.06.55 No 45MU. No 41 Sqn 'J'.
WP130	Delivered 06.06.55 No 5MU Kemble. No 34 Sqn 'S' Tangmere. No 56 Sqn 'S'. W/O 28.02.57.
WP131	Delivered 12.05.55 No 5MU Kemble. No 1 Sqn 'V'. Swung off runway when undercarriage leg collapsed on landing at Tangmere. W/O 28.02.57.
WP132	Delivered 19.05.55 No 5MU Kemble. No 34 Sqn 'T' Tangmere. Crashed on overshoot at Nicosia, Cyprus. W/O 29.08.56.

WP133	Delivered 02.05.55 No 45MU. No 41 Sqn 'D'. No 34 Sqn 'L'. No 208 Sqn 15.01.58 to 02.04.58. No 19MU St Athan.
WP134	Delivered 23.05.55 No 5MU Kemble. No 263 Sqn. No 41 Sqn. Crashed 2 miles SW of North Weald. W/O 02.09.56.
WP135	Delivered 02.06.55 No 45MU. No 41 Sqn 'E'. Crashed NW of Valley, Anglesey, when force-landed on beach after engine failed on approach. W/O 05.05.56.
WP136	Delivered 30.06.55 No 45MU. No 34 Sqn 'N' Tangmere. No 56 Sqn 'D' Waterbeach. No 49MU Colerne 05.11.58. Purchased by HSA and scrapped 1962.
WP137	Delivered 25.05.55 No 5MU Kemble. No 1 Sqn 'S'. Collided with WN977 and crashed into the sea off Selsey, near Isle of Wight. Pilot killed. W/O 05.05.58.
WP138	Delivered 24.05.55 No 5MU Kemble. No 34 Sqn 'T' Tangmere. No 1 Sqn. Crashed at Chivenor. W/O 21.06.56.
WP139	Delivered 26.05.55 No 45MU. No 34 Sqn '1' Tangmere. No 56 Sqn 'O' Waterbeach. No 49MU Colerne 05.11.58. Purchased by HSA and scrapped 1962.
WP140	Delivered 22.06.55 No 45MU. No 34 Sqn 'R' Tangmere. No 71MU, CAT5 after accident 30.08.57. GIA 7550M No 12 SoTT Melksham 07.12.57, later cancelled.
WP141	Delivered 19.05.55 No 45MU. No 41 Sqn 'B'. No 1 Sqn 'GRC'.
WP142	Delivered 18.05.55 No 5MU Kemble. No 34 Sqn 'L', later 'W', Tangmere. Crashed on take-off from Biggin Hill. CAT5. W/O 02.09.57.
WP143	Delivered 23.06.55 A&AEE. No 1 Sqn. No 257 Sqn 'M'. RAE. Crashed after suffering ricochet damage during RAE gun-firing trials on Pendine Range 20.10.55. W/O 1956. DBR. GIA 7292M Pembrey Ranges 31.01.56.
WP144	Delivered 24.05.55 No 5MU Kemble. No 1 Sqn 'PJS'. No 34 Sqn 'JRG'. Crashed at East Stratton, Hampshire, following collision with Royal Navy Whirlwind helicopter XK908. W/O 10.09.57.
WP145	Delivered 10.06.55 No 45MU. No 34 Sqn 'X' Tangmere. W/O 11.12.56.
WP146	Delivered 11.05.55 No 45MU. No 1 Sqn 'H'.
WP147	Delivered 13.06.55 No 45MU. No 41 Sqn 'L'. No 1 Sqn 'G'.
WP148	Delivered 13.06.55 No 45MU. No 41 Sqn 'K' Biggin Hill 1955. No 56 Sqn 'G' Waterbeach. No 49MU Colerne 05.11.58. SOC 1963.
WP149	Delivered 22.06.55 No 45MU. No 34 Sqn 'H' Tangmere. No 1 Sqn. No 56 Sqn 'H'. SOC with distorted airframe following loss of the canopy in flight. SOC 24.04.58.
WP150	AWA and A&AEE on miscellaneous trials.
WP179	Delivered 15.06.55 No 45MU. No 263 Sqn 'W'.
WP180	Delivered 16.06.55 No 45MU. No 41 Sqn 'H'. No 1 Sqn 'F'. No 34 Sqn. Destroyed on ground when EOKA terrorist explosive device placed in nose wheel bay, Nicosia, Cyprus. W/O 12.11.56.
WP181	Delivered 30.06.55 No 5MU Kemble. No 263 Sqn. No 41 Sqn 'P'. No 1 Sqn Tangmere. Undercarriage collapse while landing at Tangmere. W/O 28.02.57.
WP182	Delivered 23.06.55 No 45MU. No 34 Sqn 'C' Tangmere. No 1 Sqn 'V'.
WP183	Delivered 30.06.55 No 5MU Kemble. No 56 Sqn 'AW', later 'V', 1958.
WP184	Delivered 28.06.55 No 45MU. No 34 Sqn 'A' Tangmere.
WP185	Delivered 04.07.55 No 5MU Kemble. No 1 Sqn 'P' Tangmere. No 34 Sqn 'E'. To Akrotiri 05.56 for Operation 'Musketeer'. Returned to Tangmere 24.12.56. No 208 Sqn 15.01.58. Surplus to requirements 29.08.1958. No 45MU Henlow for storage as 7583M. RAF Museum Hendon 1973 until 11.89. Abingdon as gate guard 1990. Bought by Paul & Andy Wood, Great Dunmow, Essex.
WP186	Delivered 19.07.55 No 5MU Kemble. No 56 Sqn 'AW' Waterbeach 01.58. Waterbeach Stn Flt 01.11.58. No 49MU Colerne 19.11.58.

WP187	Delivered 8.07.55 No 5MU Kemble. No 41 Sqn 'R'. Aircraft caught fire and DBR during start-up at Biggin Hill. W/O 10.12.55.
WP188	Delivered 12.07.55 No 5MU Kemble. No 1 Sqn 'X'.
WP189	Delivered 19.07.55 No 45MU. Metropolitan Sector Cdr aircraft. No 263 Sqn Wattisham.
WP190	Delivered 22.07.55 No 5MU Kemble. No 1 Sqn 'K' Tangmere 13.08.55. Detached to Nicosia, Cyprus, during Suez Crisis 08.56 to 01.57. GIA 7582M Bircham Newton. SOC 11.07.58. GIA 8473M Stanbridge. Hunter Restoration Flight, Quedgeley, 01.96 for restoration to museum static display condition. Donated to Tangmere Aviation Museum 06.02.
WP191	Delivered 03.08.55 No 5MU Kemble. No 1(F) Sqn 'Z'. No 34 Sqn.
WP192	Delivered 03.08.55 No 45MU. No 34 Sqn 'D', later 'O', Tangmere. Crashed on take-off from Biggin Hill 02.09.57. To RAF Fire Rescue School, Sutton-on-Hull, 01.10.57 for crash rescue training as GIA 7471M. To Newton 25.11.58. SOC 07.02.64.
WP193	Delivered 18.08.55 No 45MU. No 34 Sqn Tangmere. Crashed 4 miles W of Blandford Forum, Dorset. W/O 23.02.56.
WP194	Delivered 29.07.55 No 5MU Kemble. No 41 Sqn. No 56 Sqn 'H'.

Hawker Hunter F Mk 6

First production batch of seven aircraft with pre-Mod 228 Mk 4 wings built by Hawker Aircraft Ltd, Kingston-upon-Thames. Instruction to Proceed dated 12.07.51. Second part of Contract No 7144.

WW592 ff 23.05.55	HAL & A&AEE performance trials 1955-56. Re-purchased by HSA 1968. Converted to Abu Dhabi FR Mk 76A '708'. Crashed. W/O 1971.
WW593 ff 19.08.55	Delivered. 23.05.55. HAL and A&AEE for miscellaneous trials. Returned to HSA 1960. Converted to FR Mk 10. Delivered 26.04.61 No 19MU St Athan. No 4 Sqn 25.02.65. No 2 Sqn 13.01.70. No 14 Sqn 'H'. No 2 Sqn. SOC as CAT5 15.04.70. Repainted for display at Gutersloh by 07.72. To fire dump 07.75.
WW594 ff 23.09.55	HAL P.1109A aerodynamic test aircraft for DH Firestreak installation – missiles not fitted. Returned to HSA 25.01.60. Converted to FR Mk 10. Delivered 05.05.61 No 19MU St Athan. No 229 OCU '11'. Purchased by HSA 07.74. Converted to Lebanese FGA Mk 70A 'L.182', later 'L.282'. Delivered 08.12.75.
WW595 ff 29.09.55	HAL Blackpool trials. HSA Dunsfold 13.10.59. Converted to FR Mk 10. Delivered 22.12.60 No 19MU St Athan. No 4 Sqn 'G' Gutersloh 03.02.61. Crashed near Furstenberg. Pilot killed. W/O 23.01.67.
WW596 ff 10.10.55	CA aircraft trials at Blackpool. DFCS. Returned to HSA 14.10.59. Converted to FR Mk 10. Delivered 02.03.61 No 19MU St Athan. No 2 Sqn 'N' Gutersloh 08.08.61. No 229 OCU 14.03.66. No 2 Sqn 'N' 09.05.67. Purchased by HSA 02.03.71. Converted to Indian T Mk 66E 'S.1390'. Delivered 13.06.73.
WW597 ff 10.10.55	No 19MU St Athan. Stored 1955-58. Became Jordanian F Mk 6 '712'. Delivered 12.11.58.
WW598 ff 31.12.55	HAL trials Dunsfold. To RAE Farnborough and Bedford High Speed Flt. Converted to P.1109A and employed for tropical low-altitude gust investigation and TSR2 cockpit trials. RAE Llanbedr. Re-purchased by HSA 14.05.74. Converted to Lebanese FGA Mk 70A 'L-280'. Delivered 08.12.75.

Second production batch of 100 aircraft built by Hawker Aircraft Ltd, Kingston-upon-Thames. Not initially fitted with wing leading edge extensions (later retrofitted).

XE526 ff 11.10.55	Delivered 11.01.56 No 5MU Kemble. No 63 Sqn 'C' Waterbeach. Returned to HAL 1957. Converted to Swiss FGA Mk 58 'J-4008' ff 8.9.58. Delivered 03.10.58.
XE527 ff 29.12.55	Delivered 11.01.56 No 19MU St Athan. Returned to HAL 1957. Converted to Swiss FGA Mk 58 'J-4006'. ff 26.08.58. Delivered 02.10.58. Canadian Warplane Heritage, Hamilton, Ontario, Canada.
XE528 ff 28.12.55	Delivered 17.01.56 No 5MU Kemble. Returned to HAL 1957. Converted to Swiss FGA Mk 58 'J-4009'. ff 09.09.58. Delivered 08.11.58.

XE529 ff 25.12.55 Delivered 17.01.56 No 5MU Kemble. Returned to HAL 1957. Converted to Swiss Mk .58 'J-4005'. ff 25.07.58. Delivered 25.09.58.

XE530 ff 28.12.55 CA aircraft RR. Delivered 03.02.56. Trials with various Avon engines 1956. No 14 Sqn 09.05.58. No 26 Sqn 'A' 09.06.58. Returned to HSA 01.02.61. Converted to FGA Mk 9. No 208 Sqn 'A' 21.05.62 to 15.09.64. No 19MU St Athan. No 43 Sqn. Purchased by HSA 11.67. Converted to Kuwaiti T Mk 67 '220'. Delivered 22.05.69.

XE531 ff 09.01.56 CA aircraft RR. Tropical trials 1956. Returned to HSA 1959. Converted to FGA Mk 9. Converted to Mk 12. Trials with head-up display and vertical nose camera RAE 1965. Crashed on take-off, W/O 17.03.82.

The unique Hunter Mk 12 XE531 was converted from a Mk 6 single-seater, and was to be one of a fleet of aircraft fitted with a head-up display (HUD) for TSR2 training. When the TSR2 programme was later cancelled, the sole Mk 12 was assigned to research duties mostly connected with HUD development. The aircraft crashed on take-off from Farnborough during March 1982.

XE532 ff 10.01.56 CA aircraft RR. Delivered 03.02.56. Trials with various Avon engines 1956. No 92 Sqn 'L' Middleton St George. 'Blue Diamonds' aerobatic tram 09.61. Returned to HSA 05.11.64. Converted to FGA Mk 9. No 208 Sqn 'K', later 'A', 23.11.64 to 17.11.66. No 19MU St Athan. No 8 Sqn 'D'. Crashed 3 Miles WSW of Dubai on Dubai gunnery range. 2TAF. No 208 Sqn. W/O 06.05.68. SOC 11.06.68.

XE533 ff 12.01.56 Delivered 04.02.56 No 5MU Kemble. Returned to HAL 1957. Converted to Swiss Mk 58 'J-4002'. ff 14.06.58. Delivered 25.07.58.

XE534 ff 12.01.56 Delivered 04.02.56 No 5MU Kemble. Converted to Lebanese F Mk 6 'L-172' 22.05.62.

XE535 ff 17.01.56 Delivered 04.02.56 No 5MU Kemble. No 20 Sqn 'D' 27.01.58. No 26 Sqn 'F' 11.06.58. Returned to HSA 01.02.61. Converted to FGA Mk 9. No 28 Sqn 'C'. Crashed into Lion Rock Ridge, Hong Kong. Pilot killed. W/O 28.12.62.

XE536 ff 27.01.56 Delivered 23.02.56 No 5MU Kemble. Returned to HAL 1957. Converted to Swiss Mk 58 'J-4001'. ff 29.03.58. Delivered 03.04.58.

XE537 ff 18.01.56 Delivered 23.02.56 No 5MU Kemble. Returned to HAL 1957. Converted to Indian Mk 56 'BA233'. ff 21.01.58. Delivered 06.02.58.

XE538 ff 20.01.56 Delivered 14.02.56 No 5MU Kemble. Returned to HAL 1957. Converted to Indian Mk 56 'BA234'. ff 17.01.58. Delivered 06.02.58.

XE539 ff 20.01.56 Delivered 14.02.56 No 5MU Kemble. Returned to HAL 1957. Converted to Indian Mk 56 'BA235'. ff 27.01.58. Delivered 06.02.58.

XE540 ff 08.02.56 Delivered 23.02.56 No 5MU Kemble. Returned to HAL 1957. Converted to Indian Mk 56 'BA236'. ff 29.01.58. Delivered 05.03.58.

XE541 ff 09.02.56	Delivered 23.02.56 No 5MU Kemble. Returned to HAL 1957. Converted to Swiss Mk 58 'J-4003'. ff 09.07.58. Delivered 10.58.
XE542 ff 02.02.56	Delivered 14.02.56 No 5MU Kemble. Returned to HAL 1957. Converted to Swiss Mk 58 'J-4004'. ff 17.10.58. Delivered 10.58. Crashed at Schwendibach. W/O 08.10.58.
XE543 ff 13.02.56	No 19MU St Athan. A&AEE Boscombe Down for armament trials. Became Jordanian F Mk 6 '707' 12.11.58.
XE544 ff 27.01.56	Delivered 26.03.56 No 5MU Kemble. No 66 Sqn 'V' 06.12.57. Returned to HSA 11.05.59. Converted to FGA Mk 9. No 208 Sqn 'L' 28.04.60 to 01.11.61. No 19MU St Athan. No 208 Sqn 17.07.62. Abandoned take-off at Nairobi, Kenya. W/O 17.09.62. SOC 18.09.62.
XE545 ff 14.02.56	Delivered No 5MU Kemble. No 4 Sqn. Returned to HAL 1957. Converted to Swiss Mk 58 'J-4007'. ff 27.08.58. Delivered 03.10.58.
XE546 ff 16.02.56	Delivered 08.03.56 No 5MU Kemble. No 93 Sqn 'Z' 29.01.58. No 26 Sqn 'M' 18.06.58. Returned to HSA 01.02.61. Converted to FGA Mk 9. No 208 Sqn 'B' 12.10.67 to 20.12.67. No 19MU St Athan. No 43 Sqn 'B'. 1TWU Brawdy 'C' 10.76.
XE547 ff 09.02.56	Delivered 08.03.56 No 5MU Kemble. Returned to HAL 1957. Converted to Indian Mk 56 'BA237'. ff 03.01.58. Delivered 03.58.
XE548 ff 06.02.56	Delivered 23.02.56 No 5MU Kemble. No 4 Sqn 'H' 12.02.58. No 5MU Kemble 12.01.61. Converted to Rhodesian FGA Mk 9 '120'.
XE549 ff 01.02.56	Delivered No 5MU. Returned to HAL 1957. Converted to Indian Mk 56 'BA238'. ff 09.01.58. Delivered 03.58.
XE550 ff 23.02.56	Delivered 14.02.56 No 5MU Kemble. No 93 Sqn 'R' 12.02.58. HSA 03.01.61. Converted to FGA Mk 9. Delivered 05.10.61 No 43 Sqn 'X'. Kuwaiti AF 28.12.67.
XE551 ff 14.02.56	Delivered No 5MU Kemble. A&AEE for miscellaneous trials 1956-57. Became Jordanian F Mk 6 '700' 07.11.58.
XE552 ff 23.02.56	Delivered 26.03.56 No 5MU Kemble. No 65 Sqn Duxford 'U'. No 263 Sqn. Returned to HSA 1960. Converted to FGA Mk 9. No 208 Sqn 'D' 31.10.61 to 23.02.64. No 19MU St Athan. No 208 Sqn 17.06.71 to 23.06.71. St Athan. No 54 Sqn 'M'. 1TWU, No 234 Sqn 'R' Brawdy. No 2TWU. Crashed into sea 20 miles N of Lossiemouth during Air Combat Training. Pilot killed. W/O 23.02.81.
XE553 ff 28.02.56	Delivered 26.03.56 No 5MU Kemble. Returned to HAL 1957. Converted to Swiss Mk 58 'J-4012'. ff 04.10.58. Delivered 12.12.58.
XE554 ff 29.02.56	Delivered 08.03.56 No 19MU St Athan. Returned to HAL 1957. Converted to Swiss Mk 58 'J-4010'. ff 18.09.58. Delivered 10.11.58.
XE555 ff 15.02.56	Delivered 14.03.56 No 5MU Kemble. Returned to HAL 1957. Converted to Swiss Mk 58 'J-4011'. ff 22.09.58. Delivered 04.12.58.
XE556 ff 24.02.56	Delivered 14.03.56 No 19MU St Athan. No 208 Sqn 'B' 10.02.58 to 18.08.58. No 103MU. Returned to HSA 1961. Converted to FR Mk 10. No 2 Sqn 'W' Gutersloh 09.11.61. Purchased by HSA 02.03.71. Converted to Indian T Mk 66E 'S.1391'. Delivered 13.06.73.
XE557 ff 01.03.56	Delivered 26.03.56 No 5MU Kemble. No 263 Sqn '1'. No 19 Sqn 'O'. Purchased by HSA 16.03.70. Converted to Chilean Mk 71 '722'. Delivered 03.06.71. Antofagasta, Cerro Moreno AB, Chile.
XE558 ff 14.03.56	Delivered No 5MU Kemble. A&AEE for air firing trials 1956-57. Became Jordanian F Mk 6 '701' 01.11.58. W/O 01.07.60.
XE559 ff 02.03.56	Delivered 04.04.56 No 33MU. No 74 Sqn 'D'. Converted to Rhodesian FGA Mk 9 '116'.
XE560 ff 01.03.56	CA aircraft HAL for various trials. No 43 Sqn 'E'. No 65 Sqn. No 66 Sqn 'G' 08.08.60. No 19MU St Athan. Converted to Rhodesian FGA Mk 9 '126'.
XE561 ff 09.03.56	Delivered 04.04.56 No 33MU. No 19 Sqn 'Z'. No 43 Sqn 'D'. No 54 Sqn. No 111 Sqn. No 1 Sqn. No 229 OCU. Purchased by HSA 16.03.70. Converted to Chilean FGA Mk 71 '726'. Delivered 04.05.71. Antofagasta, Cerro Moreno AB, Chile.

XE579 ff 15.03.56	Delivered 04.04.56 No 19MU St Athan. No 208 Sqn 'A' 11.02.58 to 06.08.59. To No 19MU St Athan. Returned to HSA 15.02.60. Converted to FR Mk 10. Delivered 15.03.61 No 19MU St Athan. Delivered from Kuwait to Bahrain. No 8 Sqn 21.07.61. Flew into ground near Zinjibah, Eastern Aden Protectorate. Pilot killed. W/O 03.08.61.
XE580 ff 15.03.56	Delivered 04.04.56 No 19MU St Athan. No 208 Sqn 'F' 05.03.58 to 13.04.59. No 19 Sqn. Returned to HSA 28.09.59. Converted to FR Mk 10. Delivered 01.12.60 No 19MU St Athan. No 4 Sqn 'D' 25.01.61. Purchased by HSA 04.06.70. Converted to Chilean FGA Mk 71 '730'. Delivered 02.09.71.
XE581	No 33MU 1956. No 247 Sqn 'T'. Returned to HSA 25.02.59. Converted to FGA Mk 9. Delivered 30.10.59 No 19MU St Athan. No 8 Sqn 'D' Khormaksar 29.01.60 to 22.11.61. Crashed 28 Miles SW of Doha, Qatar. Pilot killed. W/O 22.11.61.
XE582	No 33MU 1956. No 247 Sqn. No 66 Sqn 'J' 06.12.57. Returned to HSA 16.11.60. Converted to FGA Mk 9. No 20 Sqn 'J' Tengah 10.11.61. No 5MU Kemble 31.01.70. 1TWU Brawdy '48'.
XE583	No 33MU 1956. No 19 Sqn 'D'. Crashed off Skydstrup, Danish coast. Pilot killed. W/O 12.09.61.
XE584 ff 20.03.56	Delivered 10.04.56 No 19MU St Athan. CFE. DFLS 'G'. No 53 Sqn 'W'. No 263 Sqn 'W' Stradishall 07.58. No 1 Sqn 'W' Waterbeach 1962. Converted by RAF to Interim FGA Mk 9, retaining Rolls-Royce Avon 203 engine. No 208 Sqn 18.05.61 to 07.06.61. No 5MU Kemble. No 8 Sqn. No 229OCU 'W'. Re-purchased by HSA 02.76. Awaiting final disposal 1979. Stored at Bitteswell. Macclesfield Historical Aviation Society, Barton Aerodrome, UK. Cockpit section owned by G. Sparkes at North-West Aviation Heritage Museum, Hooton Park, Cheshire.
XE585 ff 06.04.56	Delivered 1.05.56 No 5MU. CFE. DFLS 'Q' West Raynham. Returned to HAL 16.09.59. Converted to FR Mk 10. Delivered 28.11.60 No 19MU St Athan. No 4 Sqn 'A' 05.01.61. No 2 Sqn 'E' 27.05.70. Purchased by HSA 27.05.70. Converted to Indian T Mk 66E 'S.1392'. Delivered 13.12.73.
XE586 ff 19.03.56	Delivered 17.04.56 No 5MU Kemble. No 263 Sqn. Crashed in inverted spin during aerobatic practice, hitting house in Ipswich. W/O 02.08.57.
XE587 ff 23.03.56	Delivered 01.05.56 No 19MU St Athan. Demonstrations in Switzerland and tail parachute development. F Mk 6 with tail parachute never converted to Interim status. ETPS 1978. A. N. McNeil, Augusta, NJ, USA, 1997. Hunter One, Wilkes Barre/Scranton Airport, USA. Registered as N587XE.

Like the rather better-known XE601, F Mk 6 XE587 was partially modified to FGA 9 standard but not re-designated as such. Similarly, it also spent most of its operational career on test and evaluation duties. It remains in airworthy condition under civilian ownership in the USA. *Terry Senior*

XE588 ff 06.04.56	Delivered 15.05.56 No 19MU St Athan. Demonstrations in Switzerland 1957. Spinning trials HAL and A&AEE. Abandoned over Gorley Common during spinning trials. Crashed 3 miles N of Ringwood, Hampshire. W/O 09.11.57.
XE589 ff 03.04.56	Delivered 17.04.56 No 33MU. No 74 Sqn. Returned to HSA 1962. Converted to FR Mk 10. Delivered 24.02.61 No 8 Sqn 'V' Khormaksar 28.04.61 to 26.04.63. No 1417 Flt 'JM' 26.04.63 to 26.03.64. No 5MU Kemble for refurbishment. No 1417 Flt 'RC' 01.07.64 to 30.08.66. CAT4 09.05.66. Sold to HSA 02.68. Converted to Abu Dhabi FGA Mk 76 '701'. Delivered 09.03.70.
XE590 ff 06.04.56	Delivered 07.05.56 No 33MU. No 19 Sqn 'R'. No 263 Sqn Jever Stn Flt 21.05.58. No 93 Sqn 04.06.58. No 4 Sqn 'V' 10.07.58. CAT5 following bird-strike near Jever. W/O 09.11.60.
XE591 ff 28.03.56	Delivered No 33MU. No 74 Sqn 'G'. No 65 Sqn. No 229 OCU. Remained as F Mk 6 and was not upgraded to Interim status. Re-purchased by HSA 04.04.66. Converted to Saudi Arabian F Mk 60 '60/602'. Delivered 02.05.66.
XE592 ff 05.04.56	Delivered 07.05.56 No 33MU. No 43 Sqn. No 111 Sqn 'P'. 'Black Arrows' aerobatic team. Returned to HSA 14.07.59. Converted to FGA Mk 9. No 54 Sqn 'Y'. No 8 Sqn 'F' Khormaksar 26.11.62 to 16.10.64. Crashed into sea 3 miles W of RAF Masirah, Oman. Pilot killed. W/O 16.10.64.
XE593 ff 06.04.56	Delivered 07.05.56 No 19MU St Athan. No 65 Sqn 'P'. DBR in starter explosion, Duxford, W/O 23.01.61.
XE594 ff 09.04.56	Delivered 07.05.56 No 5MU Kemble. No 66 Sqn 'J' 11.10.56. No 63 Sqn 07.11.56. No 56 Sqn 'N' Waterbeach 11.58. No 229 OCU. Crashed after colliding with XF433 during formation aerobatics, Hartland Point. W/O 07.03.63.
XE595 ff 09.04.56	Delivered No 33MU. No 66 Sqn 03.10.56. No 65 Sqn 'P' Duxford 08.11.56. Crashed after pilot abandoned take-off at Jever. CAT4 26.07.57. Changed to CAT5 and SOC 01.09.57.
XE596 ff 09.04.56	Delivered No 33MU. No 66 Sqn 04.10.56. No 63 Sqn 07.11.56. No 56 Sqn 'M' Waterbeach 11.58. Returned to HSA 14.10.59. Converted to FR Mk 10. Delivered 31.05.61 No 19MU St Athan. No 229 OCU Chivenor '13'. Crashed 7 miles S of Paderborn, Germany. Pilot killed. W/O 19.03.70.
XE597 ff 16.04.56	Built at Kingston-upon-Thames. Delivered No 19MU St Athan 31.08.56. No 66 Sqn Linton-on-Ouse 09.10.56. No 63 Sqn 'A' Waterbeach 07.11.56. No 56 Sqn 'A' Waterbeach 11.58. Returned to HSA 06.05.59. Converted to FGA Mk 9. Delivered No 19MU St Athan 01.03.60. No 66 Sqn. No 208 Sqn 'G' Stradishall 01.04.60. No 208 Sqn Eastleigh, Kenya, 08.04.60. Khormaksar until 29.09.61. Ministry of Aviation 1961-63. No 54 Sqn Stradishall 18.03.63. West Raynham 14.08.63. HSA for repairs following CAT4 damage 20.12.63. No 1 Sqn 'A' West Raynham 06.04.65. No 1 Sqn Wittering 18.07.69. Returned to West Raynham Stn Flt 30.10.69. No 229 OCU Chivenor 16.03.70. 1TWU 02.09.74. Brawdy 'F' 11.07.77. No 2TWU 'F' Lossiemouth 09.06.78. No 1TWU Brawdy 14.04.81. No 79 Sqn. Remained at Brawdy until 1985. GIA 8874M No 1 SoTT Halton 05.87. Cockpit section removed, remainder of airframe placed on dump 06.89. Cockpit section purchased by Wolverhampton Aviation Group, transferred to Finstall, Bromsgrove, 19.03.02.
XE598 ff 16.04.56	Delivered 11.05.56 A&AEE. Trials with gun-blast deflectors. Returned to HAL and refurbished as Lebanese F Mk 6 'L-170' 31.10.58.
XE599 ff 19.04.56	Delivered 01.05.56 A&AEE. Intensive gun firing trials. No 208 Sqn 'C' 03.03.58 to 22.04.59. No 74 Sqn 'D'. HAL and A&AEE trials with wing leading edge extensions. Returned to HSA 09.12.59. Converted to FR Mk 10. Delivered 15.03.61 No 19MU St Athan. No 131MU as MEAF Command Reserve 03.07.61 to 08.61. No 8 Sqn 'X' Khormaksar 08.61 to 1962. Returned to MEAF reserve. No 1417 Flt 'JD' Khormaksar 26.04.63 to 01.11.63. No 19MU St Athan for refurbishment 27.04.63 to 02.06.64. No 1417 Flt 'RJ' Khormaksar 03.06.64 to 1965, recoded 'GC' 01.67 and 'DW' until 08.09.67. Purchased by HSA 10.71. Converted to Singapore Mk 74B '535'. Delivered 14.03.73. Alan Arthur, Tocumwal Airfield, NSW, Australia, 1998.

XE600 ff 20.04.56 Delivered 15.05.56 A&AEE. Damaged in accident and returned to HAL 1957. Converted to Indian Mk 56 'BA239'. ff 14.02.58. Delivered 03.58. Refused acceptance by Indian AF and returned to HSA 15.04.59. Reverted to XE600. HSA trials with nose wheel brake. Converted to FGA Mk 9. No 8 Sqn 'G' 20.01.60 to 25.06.62. Crashed following beat-up of Khormaksar range. Pilot killed. W/O 25.06.62.

XE601 ff 02.05.56 CA aircraft HAL 1956. Built as F Mk 6 and progressively modified through F Mk 6A to FGA Mk 9 although never officially designated Mk 9. 'A' Sqn at A&AEE Boscombe Down, various trials including biological warfare spraying tests. Porton Down. Retired after last flight at Boscombe Down 15.07.99 and stored at Boscombe Down Museum. Bought by Northern Lights Group, Canada 04.04. Registered as G-ETPS. Skyblue Aviation, Exeter, 01.05. Owned by Andy Foan.

XE601, illustrating the unique spray tanks only fitted to this aircraft. *Terry Senior*

XE602 ff 04.05.56 Delivered 14.08.56 No 5MU Kemble. No 66 Sqn 06.09.56. No 92 Sqn 12.09.56. No 56 Sqn 'G' Waterbeach 11.58. No 229 OCU '41'. Crashed into barrier on landing at Chivenor. W/O 08.03.61.

XE603 ff 15.05.56 Delivered 21.06.56 No 5MU Kemble. CFE, AFDS and DFLS. No 19 Sqn 'R'. Purchased by HSA 03.66. Converted to Jordanian FGA Mk 73 '832'. Delivered 13.06.68.

XE604 ff 05.06.56 Delivered 28.06.56 No 33MU. No 263 Sqn 'D'. North Weald Stn Flt. No 1 Sqn 'D' Stradishall. Converted by RAF to Interim FGA Mk 9. No 1 Sqn 'Z' Stradishall. Crashed into sea after stall at Cowden range. Pilot killed. W/O 02.03.61.

XE605 ff 14.05.56	Trials HAL and A&AEE. Returned to HSA 04.12.59. Converted to FR Mk 10. No 2 Sqn 'O' Gutersloh 10.02.67. Returned to HSA 05.71. Converted to Singapore FR Mk 74B '523'. Delivered 15.08.72.
XE606 ff 17.05.56	Delivered 21.06.56 No 33MU. AFDS. No 54 Sqn. No 229 OCU. Brawdy 10.76. Laarbruch 09.98. St Athan for preparation for display 2000. Displayed outside No 4 Sqn Hangar RAF Cottesmore. Marked as 'XJ673'.
XE607 ff 25.05.56	Delivered 21.06.56 No 5MU Kemble. No 263 Sqn. Returned to HSA 1963. Converted to FGA Mk 9. No 66 Sqn. No 208 Sqn 'F' 22.03.60 to 29.03.62. No 8 Sqn. Crashed at Khormaksar after failing to recover from supersonic dive during open day as result of runaway tailplane trim fault. Pilot killed. W/O 30.03.62.
XE608 ff 08.06.56	Delivered 03.07.56 No 33MU. DFLS and AFDS. DFCS 1962. No 229 OCU. Brawdy 10.76. Battle Damage Repair Training, GIA 8717M Bruggen.
XE609 ff 10.06.56	No 33MU. No 54 Sqn 'R'. Returned to HSA 1959. Converted to FGA Mk 9. No 8 Sqn 'A'. No 66 Sqn. No 208 Sqn 'E', later 'H', 23.03.60 to 19.06.62. To No 5MU Kemble. No 8/43 Sqn 'A'. Engine fire at Khormaksar. DBR. W/O 07.04.66.
XE610	CA aircraft Bitteswell. No 5MU Kemble. No 74 Sqn 'J' Coltishall. Returned to HSA 07.11.60. Converted to FGA Mk 9. Delivered 17.07.61 No 20 Sqn 'C'. Crashed after engine cut out in South China Sea 40 miles S of Kuantan, Malaysia. W/O 26.06.69.
XE611	CA aircraft Bitteswell. No 5MU Kemble. DFCS. Returned to HSA 1960. Converted to FGA Mk 9. No 208 Sqn 'C' 28.11.66 to 18.07.67. No 8/43 Sqn 'Q'. No 208 Sqn 'C'. Brakes failed on landing at Lyneham. Purchased by HSA 25.03.69. Converted to Swiss Mk 58A 'J-4103'. Delivered 17.01.72. Purchased by Dr Dan Lavigna, Tarzana, California. Based at Van Nuys Airport, Los Angeles, as N4103Y. Acquired by Northern Lights Combat Air Support, Canada.
XE612	CA aircraft Bitteswell. No 5MU Kemble. No 74 Sqn 'M'. Lost power on approach at Horsham St Faith. W/O 17.05.60.
XE613	CA aircraft Bitteswell. No 5MU Kemble. No 74 Sqn 'C'. Converted to Rhodesian FGA Mk 9 '118'. Zimbabwe Air Force Museum, Gweru.
XE614 ff 03.03.56	No 5MU Kemble. No 263 Sqn 'R' Wattisham then Stradishall. Returned to HSA 01.07.60. Converted to FR. Mk 10. Delivered 27.02.61. No 1417 Flt 'PL', later 'GC', Khormaksar 1961-66. No 8 Sqn 'W', later 'Z'. Purchased by HSA 06.71. Converted to Singapore FR Mk 74A, later modified to FR Mk 74B '533'. Delivered 21.02.73 No 141 'Merlin' Sqn. Modified to FR Mk 74BS 09.87. SOC 06.89. Donated to Queensland Air Museum by Greg Ackman 01.06.06.
XE615 ff 22.05.56	Delivered 21.06.56 No 263 Sqn 'A'. Returned to HSA 1959. Converted to FGA Mk 9. No 1 Sqn 'E', later 'A'. No 54 Sqn 'U'. CAT5. Sold to HSA 30.10.69. Converted to Singapore FGA Mk 74 '508'. Delivered 14.01.71.
XE616 ff 05.05.56	Delivered 25.06.56 No 5MU Kemble. No 263 Sqn 'E'. Converted to Interim Mk 9. No 1 Sqn 'E', later 'P', Stradishall. Crashed into sea off Holkham, Norfolk. Pilot killed. W/O 21.05.69.
XE617 ff 11.06.56	Delivered 03.07.56 No 65 Sqn 'D'. No 66 Sqn 03.09.56. No 92 Sqn 'D' 06.09.56. Converted to FGA Mk 9. No 208 Sqn 'K'. Delivered 03.03.66. Crashed at Muharraq, Bahrain, W/O 07.05.66. Wreckage to Muharraq dump.
XE618 ff 12.06.56	Delivered 04.07.56 Church Fenton Stn Flt. No 66 Sqn 'ES' 15.09.58. No 56 Sqn 'H' Waterbeach 11.58. Returned to HSA 01.05.59. Converted to FGA Mk 9. Delivered 04.02.60. No 208 Sqn 'D' 04.02.60 to 28.11.61. To No 19MU St Athan. No 5MU Kemble. No 8 Sqn 'D' Khormaksar 1963. No 208 Sqn 'L' 02.07.65 to 18.07.67. No 8/43 Sqn. Kuwaiti AF 28.12.67.
XE619 ff 07.06.56	Delivered 28.06.56 No 263 Sqn 'S'. No 1 Sqn. Forced-landing at Stradishall, diverted to Honington. Pilot Fg Off. W/O 17.02.59.
XE620 ff 12.06.56	Delivered. 04.07.56 No 111 Sqn. No 263 Sqn. Returned to HSA 1959. Converted to FGA Mk 9. No 208 Sqn 'H' 15.08.66 to 05.08.67. No 8 Sqn 'F'. No 43 Sqn. Purchased by HSA 04.68. Converted to Indian Mk 56A 'A967'. Delivered 31.07.69.

XE621 ff 17.06.56	Delivered 04.07.56 No 66 Sqn 29.08.56. No 92 Sqn 06.09.56. No 111 Sqn. Collided with XF525 at North Weald 07.06.57. Made high-speed landing at Stansted. CAT4. Returned to HAL, Blackpool. Returned to service 02.59. To HAL 06.08.59. Converted to FR Mk 10. Delivered 22.12.60. No 2 Sqn 'H' 08.02.61. CAT5, stalled and dived into ground at Papenburg. Pilot killed. W/O 30.01.62.
XE622 ff 07.06.56	Delivered 02.07.56 No 263 Sqn. No 1 Sqn 'Y' 09.59. No 66 Sqn 'N' 08.04.60. No 1 Sqn 'E' 28.10.60. Returned to HAL. Converted to FGA Mk 9. No 28 Sqn 'A'. DBR after starter explosion Kai Tak. W/O 12.07.66.
XE623 ff 12.06.56	Delivered 03.07.56 No 263 Sqn 'G'. Returned to HSA 07.04.59. Converted to FGA Mk 9. No 66 Sqn. No 208 Sqn 'C' 23.03.60 to 07.04.62. No 19MU St Athan. No 43 Sqn 'Q'. Crashed 1 mile NE of Khormaksar. W/O 11.08.64.
XE624 ff 17.06.56	Delivered 03.07.56 No 263 Sqn. Converted to FGA Mk 9 02.60. No 1 Sqn 'B'. No 1TWU 'G' Brawdy. No 79 Shadow Sqn 10.76. SOC 25.07.84. Last GIA 8875M. Gate guard Brawdy. Sold summer 2004 to Steve Petch, Sileby, Leicestershire. Moved to buyer's location 20.01.05.
XE625 ff 17.06.56	Delivered 11.07.56 No 263 Sqn. Converted to FR Mk 10. No 2 Sqn 'I' 28.02.61. No 4 Sqn 'F' 12.02.70. Purchased by HSA 04.06.70. Converted to Chilean FGA Mk 71 (J-729). Delivered 22.07.71. Antofagasta, Cerro Moreno AB, Chile, 2001.
XE626 ff 17.06.56	Delivered 20.07.56 No 263 Sqn 'P'. Returned to HSA 28.09.59. Converted to FR Mk 10. Delivered 02.12.60 No 4 Sqn 'E' 25.01.61. No 4 Sqn 'U' 06.12.66. No 229 OCU '9' 18.05.70. Purchased by HSA 22.08.72. Converted to Kenyan FGA Mk 80 '803'. Delivered 24.06.74.
XE627 ff 13.06.56	Delivered 20.07.56. No 65 Sqn 'T' Duxford 1957. No 92 Sqn. No 229 OCU. No 54 Sqn 'D'. Horsham St Faith Stn Flt. No 54 Sqn. No 1 Sqn. No 54 Sqn. No 229 OCU '61'. No 1 Sqn. No 229 OCU. No 1TWU Brawdy. Preserved at Imperial War Museum, Duxford, Cambs.
XE628 ff 20.06.56	Delivered 11.07.56 No 111 Sqn. No 1 Sqn 'X' 09.59. No 263 Sqn. No 66 Sqn 'M' 08.04.60. No 1 Sqn 'X' 21.09.60. Returned to HAL. Converted to FGA Mk 9. No 1 Sqn 'G'. Crashed into sea off Tobruk, Libya, while flying out of El Adem. Pilot killed. W/O 24.04.63.
XE643 ff 26.06.56	Delivered 17.07.56 No 45MU. No 66 Sqn 08.10.56. No 63 Sqn 08.01.57. No 56 Sqn 'S' Waterbeach 11.58. Returned to HAL. Converted to FGA Mk 9. No 66 Sqn. No 208 Sqn 'K'. No 8/43 Sqn 03.05.60. Abandoned take-off at Mombasa, Kenya, 09.12.61. GIA 8586M. Scrapped HSA 05.62. Nose recovered by No 71MU at Aldergrove Air Training Corp, RAF Aldergrove, N Ireland.
XE644 ff 21.06.56	Delivered 20.07.56 No 5MU Kemble. No 65 Sqn 'B'. No 66 Sqn 29.08.56. No 92 Sqn 06.09.56. No 229 OCU. Purchased by HSA 23.05.70. Converted to Chilean FGA Mk 71 '728'. Delivered 07.07.71.
XE645 ff 28.06.56	Delivered 20.08.56 No 33MU. No 66 Sqn 29.08.56. No 63 Sqn 05.11.57. No 92 Sqn. No 56 Sqn 'A' Waterbeach 11.58. Returned to HSA 12.03.59. Converted to FGA Mk 9. No 56 Sqn 'H'. No 54 Sqn 'P' 1960. No 208 Sqn 'M' 26.11.62 to 30.06.65. No 19MU St Athan. No 8/43 Sqn 'U', later 'SW', Khormaksar. Became Jordanian FGA Mk 73 '827/M'.
XE646 ff 21.06.56	Delivered. 20.07.56 No 33MU. No 263 Sqn. Returned to HSA 1960. Converted to FGA Mk 9. No 1 Sqn 'V' West Raynham. Crashed into Leconfield village. W/O 30.12.66. SOC 31.12.66.
XE647 ff 28.06.56	Delivered 20.08.56 No 5MU Kemble. No 92 Sqn. No 63 Sqn 'E' Waterbeach 1958. No 56 Sqn 'E' Waterbeach 11.58. Returned to HSA 14.05.59. Converted to FGA Mk 9. No 66 Sqn. No 208 Sqn 'O' 11.05.60 to 12.10.61. No 19MU St Athan. No 208 Sqn 'H' 01.06.62. Crashed 10 miles SE of Dassa Island, Arabian Gulf, after colliding with XK139. W/O 30.06.64.
XE648 ff 29.06.56	Delivered 22.08.56 No 33MU. No 66 Sqn 05.10.56. No 63 Sqn 16.01.57. No 56 Sqn 'H'. Crashed at Nicosia, Cyprus, 09.09.59. W/O 16.10.59.

XE649 ff 30.06.56 Delivered 14.08.56 No 5MU Kemble. No 66 Sqn 03.09.56. No 92 Sqn 27.09.56. No 65 Sqn 'N' Duxford. Returned to HSA 23.02.59. Converted to FGA Mk 9. No 208 Sqn 'S', later 'J' and 'N', 13.07.67 to 30.07.71. No 5MU Kemble. No 1TWU 'P' Brawdy 10.76. Crashed following engine failure 15 miles SE of Aberystwyth, Wales. W/O 12.05.82.

XE650 ff 02.07.56 Delivered 20.07.56 No 33MU. No 263 Sqn. No 1 Sqn 'U' Waterbeach 1962. Interim FGA Mk 9. Later upgraded to FGA Mk 9. No 208 Sqn 'A' 18.10.70 to 17.06.71. No 8 Sqn 'A'. No 208 Sqn 'G'. Returned to HSA 02.76. Stored at Bitteswell awaiting disposal 02.80.

XE651 ff 05.07.56 Delivered 14.08.56 No 33MU. No 63 Sqn 'R'. No 66 Sqn 02.10.56. No 63 Sqn 08.01.57. No 56 Sqn 'R' Waterbeach 11.58. Returned to HSA 10.03.59. Converted to FGA Mk 9. No 8 Sqn 'M' Khormaksar 21.01.60 to 18.10.62. No 1 Sqn 'L'. No 1TWU 'W' Brawdy 10.76. Crashed into sea 40 miles S of Brawdy. W/O 13.05.77.

XE652 ff 16.08.56 Delivered 04.09.56 No 5MU Kemble. No 66 Sqn 09.10.56. DFLS 22.01.57. Returned to HSA 1960. Converted to FGA Mk 9. No 20 Sqn 'A'. Purchased by HSA 06.04.70. Converted to Singapore FGA Mk 74 '519'. Delivered 20.05.71.

XE653 ff 06.07.56 Delivered 24.08.56 No 5MU Kemble. No 43 Sqn. No 111 Sqn 'S'. 'Black Arrows' aerobatic team. No 229 OCU 6.64. 1TWU 1977. Registered G-BVWV 22.12.94. Sold to Mike Beachyhead, registered on 21.03.95. Cape Town, South Africa, for the Yello Summer display team registered as ZU-AUJ.

XE654 ff 08.08.56 Delivered 31.08.56 No 5MU Kemble. No 65 Sqn 'G'. No 92 Sqn. Returned to HSA 12.03.59. Converted to FGA Mk 9. No 19MU St Athan. No 8 Sqn 'E' 25.01.60 to 29.08.62. To HSA for refurbishment. No 8 Sqn 'LJ' Khormaksar 02.02.63 to 30.07.63. No 131MU for repair. No 208 Sqn 'J' Khormaksar 28.05.64 to 14.07.66. No 8 Sqn. Crashed at Al Khaburah, Oman. Pilot killed. W/O 20.11.67.

XE655 ff 08.08.56 Delivered 24.08.56 No 5MU Kemble. No 63 Sqn. No 92 Sqn. No 56 Sqn 'F' Waterbeach 11.58. Returned to HSA 01.04.59. Converted to FGA Mk 9. No 43 Sqn 1964. No 8 Sqn 'H'. No 43 Sqn 'L' Khormaksar 28.12.63 to 12.10.64. To HAL. Became Jordanian FGA Mk 73 '817' 26.01.68.

XE656 ff 09.08.56 Delivered 24.08.56 No 19MU St Athan. No 65 Sqn 'R' Duxford. No 92 Sqn 'B'. DFLS. No 229 OCU. GIA 8678M. Auto & Technik Museum, Speyer, Germany 1998. Displayed in front of museum 2003.

Third production batch of 110 aircraft built by Hawker Aircraft Ltd, Kingston-upon-Thames, partially transferred to Armstrong Whitworth Aircraft Ltd, Coventry (XG150-XG168, indicated by 'Built AW' in following list).

XG127 ff 18.07.56 Delivered 24.08.56 No 19MU St Athan. No 66 Sqn 05.10.56. No 63 Sqn 18.01.57. No 56 Sqn 'P' Waterbeach 11.58. Returned to HSA 1962. Converted to FR Mk 10. Delivered 30.01.61 No 2 Sqn 'Y'. No 19MU. Purchased by HSA 10.06.68. Converted to Swiss Mk 58A 'J-4101'. Delivered to Swiss AF and assembled at Emmen 07.12.71. ff 12.09.72. Fitted with Swedish target towing. SOC Meiringen 16.12.94. Flygfvapnet Museum, Såtenäs AB, Sweden, as SE-DXC. Advancerad Flygträning, Hässlö, Sweden, 2004. Owned by Northern Lights Combat Air Support, Canada, registered as C-GJMP. Sold to Hunter Aviation International, Wilmington, DE. Registered as N323AX 09.03.05.

XG128 ff 14.08.56 Delivered 04.10.56 No 33MU. No 20 Sqn 'Y'. No 65 Sqn 'S'. DFLS. Returned to HSA for conversion to FGA Mk 9. No 8 Sqn 'Q' Khormaksar 07.05.60 to 13.01.61. Flew into rising ground in Wadi Yahar near Aden. Pilot killed. W/O 13.01.61.

XG129 ff 14.08.56 Delivered 03.10.56 No 19MU St Athan. No 43 Sqn. No 111 Sqn 'F'. 'Black Arrows' aerobatic team. Returned to HSA 06.09.67. Converted to Indian Mk 56A 'A936'. Delivered 15.11.68.

XG130 ff 14.08.56 Delivered 03.09.56 No 19MU St Athan. No 66 Sqn 10.10.56. No 63 Sqn 'U', later '1', 18.01.57. No 56 Sqn 'U' Waterbeach 11.58. Returned to HSA 02.04.59. Converted to FGA Mk 9. No 1 Sqn 'E' Waterbeach 1962. No 54 Sqn 'X'. No 208 Sqn 'C' 04.11.69 to 01.07.70. No 5MU Kemble. No 45 Sqn '61' Wittering 1971. Crashed near Melton Mowbray. Some wreckage still buried deep in impact area. W/O 17.06.74.

XG131 ff 16.08.56	Trial installation of wing-tip tanks with HAL. Delivered 10.10.56 No 5MU Kemble. No 14 Sqn 'N' 07.06.57. No 19MU St Athan 07.03.63. No 229 OCU. Flew into high ground in cloud at Hawkridge. Pilot killed. W/O 19.03.71.
XG132 ff 14.08.56	Delivered 12.09.56 No 5MU Kemble. AFDS. Royal Jordanian AF 06.62 to 10.62.
XG133 ff 20.08.56	Delivered 24.09.56 No 5MU Kemble. No 263 Sqn. No 19 Sqn 'A' Leconfield 1957. Crashed on take-off at Duxford. Pilot killed. W/O 07.09.58.
XG134 ff 21.08.56	Delivered 14.09.56 No 19MU St Athan. No 63 Sqn. No 56 Sqn 'O' Waterbeach 11.58. Returned to HSA 1960. Converted to FGA Mk 9. No 66 Sqn 'A'. No 208 Sqn 'A' Eastleigh 22.03.60. No 208 Sqn 'A' Kuwait 11.07.61. Crashed at Mutla Ridge, Kuwait. Pilot killed. W/O 11.07.61.
XG135 ff 24.08.56	Delivered 24.09.56 No 5MU Kemble. No 19 Sqn. No 263 Sqn. Returned to HSA 1960. Converted as prototype FGA Mk 9. No 8 Sqn 'P' Khormaksar 21.01.60 to 04.11.63. No 5MU Kemble 10.66. No 208 Sqn 'A' 19.05.67 to 24.06.69. No 19MU St Athan. No 45 Sqn 'B'. Crashed at Barnack, near Wittering. W/O 09.04.73.
XG136 ff 28.08.56	Delivered 17.09.56 No 5MU Kemble. AFDS. Returned to HSA 1960. Converted to FGA Mk 9. No 8 Sqn 'C' Khormaksar 21.01.60 to 08.05.62. No 43 Sqn 'G' Khormaksar 06.05.63 to 17.04.64. Crashed at Haref, Libya. Pilot killed. W/O 17.04.64.
XG137 ff 27.08.56	Delivered 18.09.56 No 19MU St Athan. DFLS 'D'. No 54 Sqn 'E'. No 92 Sqn 'E'. No 229 OCU. Became Jordanian FGA Mk 73 '813' 02.68.
XG150	Built AW. Delivered No 5MU Kemble. HAL 06.11.57. Converted to Indian Mk 56 'BA247'. ff 28.03.58. Delivered. 19.05.58.
XG151	Built AW. No 19MU St Athan. No 54 Sqn 'H'. F Mk 6/Interim FGA Mk 9. Converted to FGA Mk 9. No 54 Sqn 'H' Waterbeach. No 1TWU 'H' Brawdy. No 2TWU 'H' Lossiemouth. Crashed short of runway at Lossiemouth. Cockpit removed to Lossiemouth fire dump. W/O 03.04.81.
XG152	Built AW. No 19MU St Athan. DFLS. No 19 Sqn 'X'. No 229 OCU. Luftwaffen Museum, Berlin (Gatow), 17.09.05.
XG153	Built AW. No 5MU Kemble. No 66 Sqn 'L' 20.12.56. No 92 Sqn APS. Returned to HSA 14.11.60. Converted to FGA Mk 9. No 20 Sqn 'L' Tengah 11.11.61. No 5MU Kemble 12.01.70. Purchased by HSA 05.71. Converted to Singapore FR Mk 74B '520'. Delivered 08.06.72.
XG154	Built AW. No 19MU St Athan. No 66 Sqn 'T' 04.01.57 to 20.06.60. To Horsham St Faith for conversion to Interim F Mk 6/FGA Mk 9 12.59. No 43 Sqn. Returned to HAL. Converted to FGA Mk 9. No 208 Sqn 'V' Khormaksar 07.05.62 to 17.05.62. No 19MU St Athan for refurbishment. No 8 Sqn 'H' Khormaksar 26.02.63 to 08.63, 'E' until 31.08.64. No 54 Sqn. No 208 Sqn 'J' 09.66 to 1968. No 8 Sqn 'E'. No 43 Sqn 'B'. No 208 Sqn 'L' 1969. No 229 OCU Chivenor. No 1TWU. No 79 Shadow Sqn '54', later 'M', Brawdy 10.76. GIA 8863M. RAF Museum collection, Hendon.
XG155	Built AW. No 19MU St Athan. Initially converted to F Mk 6/Interim/FGA Mk 9. Converted to FGA Mk 9. No 54 Sqn 'T' 1962-63. No 2TWU 'J' Lossiemouth. Zimbabwe AF '1807'. Delivered 10.04.84. Thornhill AB, Gweru, Zimbabwe.
XG156	Built AW. No 5MU Kemble. No 54 Sqn 'K'. Initially Interim Mk 9. No 54 Sqn 'F'. No 229 OCU. No 79 Sqn 'E'. Ran off runway and crashed Gibraltar. Pilot killed.. W/O 09.10.71. Wreckage still present 11.74. Later scrapped.
XG157	Built AW. No 33MU. No 54 Sqn 'H'. No 1 Sqn 'Y'. No 229 OCU. Crashed into hill at Challacombe, Devon. Pilot killed. W/O 16.06.66.
XG158	Built AW. No 19MU St Athan. No 65 Sqn 'O'. Pendine, Dyfed, Wales. Removed during range clear-up prior to 2002 and scrapped.
XG159	Built AW. No 5MU Kemble. No 19 Sqn. No 263 Sqn. No 5MU Kemble. No 56 Sqn 'P' 25.01.60. No 92 Sqn. 'Blue Diamonds' aerobatic team. No 229 OCU Chivenor. No 5MU Kemble 04.67. Became Jordanian FGA Mk 73 '717' 04.67.

XG160	Built AW. No 5MU Kemble. No 43 Sqn. No 111 Sqn 'U'. 'Black Arrows' aerobatic team. No 229 OCU. No 1TWU 10.76. 8831M. Auctioned at Scampton 12.94. To Royal Jordanian Historic Flight, Bournemouth. Bournemouth Museum. Registered as G-BWAF.
XG161	Built AW. No 33MU. AFDS. DFLS 'P'. Stored by No 5MU Kemble. No 229 OCU. Flew into sea off Boscastle, Cornwall. Pilot killed. W/O 14.02.74.
XG162	Built AW. No 5MU Kemble. No 111 Sqn DFLS 'W'. Crashed on approach to West Raynham. Pilot killed. W/O 07.11.57.
XG163	Built AW. No 5MU Kemble. Returned to HAL 1957. Converted to Indian Mk 56 'BA248'. Delivered 09.06.58.
XG164	Built AW. No 33MU. No 111 Sqn. No 74 Sqn 'H'. West Raynham Stn Flt. Brawdy 10.76. No 5MU Kemble. GIA 8681M '31' Halton. Dismantled and placed in store Shawbury. Gate guard, Military Aircraft Spares, Poole. Fitted with starboard wing of XL623.
XG165	Built AW. No 33MU. No 208 Sqn 'GW'. Middleton St George Stn Flt. Crashed at Bowley, Co Durham. W/O 18.04.58.
XG166	Built AW. Delivered 03.12.56 No 5MU Kemble. No 14 Sqn 'K' 29.07.57. No 19MU St Athan 17.12.62. No 229 OCU. Crashed into sea off Hartland Point. Pilot killed. W/O 17.02.64.
XG167	Built AW. Delivered 18.02.57 No 5MU Kemble. No 19 Sqn 'F'. Supplied to Lebanon as 'L-174' 13.10.58.
XG168	Built AW. Delivered 02.04.57 No 5MU Kemble. No 208 Sqn 'GW' 20.03.58 to 20.04.59. No 66 Sqn 21.04.59. Horsham St Faith. Returned to HSA 06.08.59. Converted to FR Mk 10. Delivered 30.01.61 No 2 Sqn 'N' 28.02.61. No 19MU St Athan 01.09.62. No 2 Sqn 'V' 18.11.66. No 229 OCU '10' 22.05.70. To Royal Jordanian AF as '852'.
XG169 ff 21.08.56	Delivered 14.09.56 No 19MU St Athan. No 19 Sqn 'Q'. Returned to HSA 1960. F Mk 6/Interim/FGA Mk 9. Converted to FGA Mk 9. No 8 Sqn 'B' Khormaksar 28.01.60 to 04.04.63. No 208 Sqn 'C' Khormaksar then Muharraq 24.01.64 to 16.02.66. No 19MU St Athan. No 43 Sqn 'K' Khormaksar then Sharjah 28.11.66 to 16.01.68. No 19MU St Athan. No 8/43 Sqn 'B'. No 229 OCU. No 79 Sqn 'N'. Crashed between Buckland Brewer and Parkham, Devon. W/O 05.06.73.
XG170 ff 30.08.56	Delivered 03.10.56 No 19MU St Athan. No 43 Sqn. No 111 Sqn 'G'. 'Black Arrows' aerobatic team. No 234 Sqn. No 229 OCU '52' Chivenor. 1TWU Brawdy. Purchased by HSA 06.09.67. Converted to Indian Mk 56A 'A940'. Delivered 27.01.69.
XGI71 ff 23.08.56	Delivered 24.09.56 No 5MU Kemble. No 43 Sqn. No 111 Sqn 'E'. 'Black Arrows' aerobatic team. To Royal Jordanian AF 06.62.
XG172 ff 27.08.56	Delivered 19.09.56 No 5MU Kemble. No 19 Sqn 'B'. No 263 Sqn. No 229 OCU. No 1TWU Brawdy 10.76. GIA 8832M Scampton. Owned by City of Norwich Aviation Museum, marked as XG168.
XG185 ff 04.09.56	Delivered 03.10.56 No 19MU St Athan. No 3 Sqn. No 19 Sqn 'T'. No 5MU Kemble in storage 1966-67. No 4 FTS 1976. Aircraft crashed at Maltreath Sands, Anglesey, W/O 21.04.76.
XG186 ff 02.09.56	Delivered 17.09.56 No 5MU Kemble. No 19 Sqn. No 66 Sqn 15.10.56. DFLS 18.01.57. No 92 Sqn 'J'. 'Blue Diamonds' aerobatic team. Purchased by HSA 06.10.67. Converted to Indian Mk 56A 'A941'. Delivered 22.03.69.
XG187 ff 27.08.56	Delivered 17.09.56 No 5MU Kemble. No 66 Sqn 10.10.56. No 63 Sqn 18.01.57. No 56 Sqn 'V' Waterbeach 11.58. Became Jordanian F Mk 6 '811' 06.62.
XG188 ff 12.09.56	Delivered 03.10.56 No 19MU St Athan. No 19 Sqn 'C'. Crashed at Thixendale, Yorkshire. W/O 15.05.61.

XG189 ff 04.09.56 Delivered 3.10.56 No 19MU St Athan. No 111 Sqn 'D'. 'Black Arrows' aerobatic team. No 92 Sqn. Purchased by HSA 4.10.67. Converted to Indian Mk 56A 'A942'. Delivered 18.04.69.

XG190 ff 31.08.56 Delivered 10.10.56 No 19MU St Athan. No 43 Sqn. No 111 Sqn 'C'. 'Black Arrows' aerobatic team. No 92 Sqn 'N'. 'Blue Diamonds' aerobatic team. Purchased by HSA 13.09.67. Converted to Indian Mk 56A 'A939'. Delivered 27.01.69.

XG191 ff 02.09.56 Delivered 03.10.56 No 19MU St Athan. No 19 Sqn 'E' Church Fenton. No 263 Sqn. No 229 OCU 05.64. No 1TWU Brawdy. Flew into sea 30 miles SW of Brawdy. Pilot killed. W/O 16.08.76.

XG192 ff 21.09.56 Delivered 17.10.56 No 19MU St Athan. DFLS 'F' West Raynham 1960. FTU Benson. DFLS. Crashed on take-off, Nicosia, Cyprus. W/O 16.01.62.

XG193 ff 06.09.56 Delivered 10.10.56 No 19MU St Athan. No 43 Sqn. No 111 Sqn 'A'. 'Black Arrows' aerobatic team. Collided with Hunter XG200 during aerobatic practice. Pilot killed. W/O 10.06.60.

XG194 ff 06.09.56 Delivered 03.10.56 No 19MU St Athan. No 43 Sqn. No 111 Sqn 'N'. 'Black Arrows' aerobatic team. No 92 Sqn. Converted to FGA Mk 9 by RAF 3.65. No 1 Sqn 'Q'. No 229 OCU. No 1TWU, No 79 Sqn 'A' Brawdy 10.76. GIA 8839M Cosford. North Luffenham, Leics, for EOD training.

XG195 ff 04.09.56 Delivered 03.10.56 No 19MU St Athan. No 19 Sqn 'V'. Initially converted to F Mk 6/Interim/FGA Mk 9 then converted to FGA Mk 9. No 1 Sqn 'C' Waterbeach 1962. No 208 Sqn 'K' 16.03.70 to 20.01.71. No 5MU Kemble for storage. Purchased by HSA 02.76. Stored at Bitteswell 02.80. Hemswell Aircraft Museum with nose of GA.11. Nose sold to Lakeside & Leisure Complex, Aeroventure, Doncaster, S Yorkshire. Rear fuselage to scrap dealer in Scunthorpe, Lincs, 09.05.

XG196 ff 10.09.56 Delivered 03.10.56 No 19MU St Athan. No 19 Sqn 'D'. No 234 Sqn '31' 09.64. No 1TWU. No 5MU Kemble. Joint Services Command & Staff College, Mychett, Surrey. GIA 8702M.

XG197 ff 11.09.56 Delivered 23.10.56 No 5MU Kemble. No 63 Sqn DFLS 'A'. DFCS 'A'. Converted to F Mk 6A. No 1TWU, No 79 Sqn Brawdy 10.76. Crashed at Tintagel, Cornwall. W/O 06.07.79.

XG198 ff 24.09.56 Delivered 30.10.56 No 19MU St Athan. No 74 Sqn 'Q'. No 111 Sqn. No 263 Sqn Eastern Sector Commander's aircraft 19.01.58. No 92 Sqn. No 229 OCU Chivenor 24.03.61. Part of 'Black Dragons' aerobatic team1963. Crashed near Kidwelly. Pilot killed. W/O 04.09.67.

XG199 ff 11.09.56 Delivered 02.10.56 No 33MU. No 19 Sqn 'T'. No 229 OCU 3.64. Purchased by HSA 2.12.69. Converted to Chilean FGA Mk 71 '724'. Delivered 18.02.71. Museum at Los Cerrillos, Santiago, Chile.

XG200 ff 13.09.56 Delivered 02.10.56 No 33MU. No 43 Sqn. No 111 Sqn 'Q'. 'Black Arrows' aerobatic team. Collided with XG193 10.06.60. FTU Benson. No 229 OCU '57' 07.64. Collided with XG235 and crashed into sea 3 miles NW of Tintagel, Cornwall. Pilot killed. W/O 15.05.67.

XG201 ff 22.09.56 Delivered 23.10.56 No 5MU Kemble. No 43 Sqn. No 111 Sqn 'B'. 'Black Arrows' aerobatic team. No 92 Sqn 'B'. Converted to F Mk 6 but not upgraded to Interim status. Purchased by HSA 06.09.67. Converted to Indian Mk 56A 'A937'. Delivered 15.11.68.

XG202 ff 18.09.56 Delivered 11.10.56 No 19MU St Athan. No 66 Sqn 'E' 04.10.57. Crashed near Morpeth. W/O 13.12.57.

XG203 ff 18.09.56 Delivered 17.10.56 No 5MU Kemble. No 43 Sqn. No 111 Sqn 'H'. 'Black Arrows' aerobatic team. Bounced on landing at North Weald and exploded. W/O 30.04.57.

XG204 ff 25.09.56 Delivered 25.10.56 No 19MU St Athan. DFLS 'B'. DFCS West Raynham 1960. No 4 FTS Valley '70' 1968. Crashed Rhosneigr, N Wales. Pilot killed. W/O 15.08.69.

XG205 ff 17.09.56	Delivered 03.10.56 No 19MU St Athan. No 247 Sqn 'Y'. No 43 Sqn 'E'. No 229 OCU. Converted to F Mk 6/Interim/FGA Mk 9. Converted to FGA Mk 9. No 54 Sqn 'E'. No 208 Sqn 'K' 20.01.68 to 01.04.70. Purchased by HSA 05.04.70. Converted to Singapore Mk 74A '506'. Delivered 24.06.71. Alan Arthur, Tocumwal Airfield, NSW, Australia.
XG206 ff 28.09.56	Delivered 22.10.56 No 19MU St Athan. DFLS. DFCS 'E'. CFE. Crashed off St Abbs Head. Pilot killed. W/O 01.06.65.
XG207 ff 04.10.56	Delivered 29.10.56 No 19MU St Athan. No 93 Sqn. Converted to F Mk 6/Interim/FGA Mk 9. Converted to FGA Mk 9. Delivered 01.60 No 1 Sqn 'F' Waterbeach 1962. No 54 Sqn 'C' 10.65. No 2TWU Lossiemouth. No 45 Sqn 'R' Wittering 1972-74. No 1TWU '46'. To Zimbabwe AF '1088'. Delivered 10.04.84. Thornhill AB, Gweru, Zimbabwe. Gate guard, King George VI Barracks, Harare, Zimbabwe.
XG208 ff 02.10.56	Delivered 22.10.56 No 19MU St Athan. No 93 Sqn 'C' 09.04.57. No 26 Sqn 'J' 11.06.58. CAT5, crashed 3 miles W of Gutersloh during aerobatic display. W/O 24.03.59.
XG209 ff 27.09.56	Delivered 25.10.56 No 33MU. No 14 Sqn. No 111 Sqn. DFLS. DFCS 'C'. 8709M Halton. Nose to Glen Newman, Chelmsford, Essex. 08.98.
XG210 ff 05.10.56	Delivered 01.11.56 No 33MU. No 66 Sqn. No 14 Sqn 'A' 09.04.57. No 19 Sqn. HAL 17.02.58. CFE. BAe, Hatfield. DRA Bedford. Private owner, Beck Row, Suffolk, fitted with starboard wing of XL572 and port wing of XL623.
XG211 ff 24.09.56	Delivered 31.10.56 No 5MU Kemble. No 92 Sqn 'A'. No 111 Sqn. Purchased by HAS 12.10.67. Converted to Indian Mk 56A 'A943'. Delivered 18.04.69.
XG225 ff 20.10.56	Delivered 01.11.56 No 5MU Kemble. No 20 Sqn 'P'. No 74 Sqn. No 92 Sqn 'M' Middleton St George. No 229 OCU Chivenor '58'. No 63 Sqn 08.08.73. No 229 OCU 9.64. No 1TWU Brawdy 10.76. No 5MU Kemble. GIA 8713M No 2 SoTT Cosford. Gate guard, Cosford, 16.03.2005.
XG226 ff 28.09.56	Delivered 29.10.56 No 5MU Kemble. No 92 Sqn. No 66 Sqn 'B' 08.59. No 92 Sqn 1960. No 229 OCU. No 1TWU Brawdy 10.76. GIA 8800M Catterick. No 1242 Sqn ATC, Faversham, Kent. Nose to Manston History Museum 18.06.05. Centre section on composite at Long Marston.
XG227 ff 28.09.56	Delivered 15.10.56 No 5MU Kemble. No 111 Sqn. No 92 Sqn 'B'. Crashed into North Sea off Middlesbrough. Pilot killed. W/O 31.12.58.
XG228 ff 02.10.56	Delivered 25.10.56 No 5MU Kemble. No 92 Sqn 'C'. Converted to FGA Mk 9 1965. No 229 OCU '56' Chivenor. No 1TWU, No 79 Sqn 'E' Brawdy 10.76.
XG229 ff 03.10.56	Delivered 22.10.56 No 5MU Kemble. No 74 Sqn. No 92 Sqn 'F' Linton-on-Ouse 1957. No 56 Sqn 'M'. No 229 OCU '42' 1965-66. Converted to F Mk 6 but not upgraded to Interim status. Crashed at Merton, Devon. W/O 27.08.71.
XG230 ff 04.10.56	Delivered 22.10.56 No 5MU Kemble. No 92 Sqn. Crashed near Easingwold, Yorkshire. W/O 15.11.56.
XG231 ff 17.10.56	Delivered 30.10.56 No 5MU Kemble. No 66 Sqn 15.11.56. No 92 Sqn 'A' 18.11.56. No 111 Sqn. No 74 Sqn. Purchased by HSA 19.04.66. Converted to Jordanian F Mk 6 '715'. Delivered 13.06.67.
XG232 ff 13.10.56	Delivered 30.10.56 No 5MU Kemble. No 92 Sqn 'G' Middleton St George 1960. A&AEE 1964. Purchased by HSA 21.06.66. Converted to Chilean FGA Mk 71 'J-714'. Delivered 19.06.68.
XG233 ff 05.10.56	Delivered 30.10.56 No 5MU Kemble. No 66 Sqn 'J' 20.12.56. Crashed 6 miles SE of Famagusta, Cyprus. Pilot killed. W/O 20.08.58.
XG234 ff 11.10.56	Delivered 25.10.56 No 5MU Kemble. No 74 Sqn. No 92 Sqn 'G', later 'E'. No 229 OCU Chivenor. 'Black Dragons' aerobatic team 1963. Purchased by HSA 09.05.68. Converted to Jordanian FGA Mk 73A '830'. Delivered 22.07.69.
XG235 ff 20.10.56	Delivered 05.11.56 No 5MU Kemble. No 66 Sqn 30.11.56. No 92 Sqn 'R' 20.12.56. No 56 Sqn 'O'. No 74 Sqn. No 229 OCU '44' 05.64. Collided with XG200 and crashed into sea 3 miles NW of Tintagel, Cornwall. W/O 15.05.57.

XG236 ff 19.10.56	Delivered 09.11.56 No 5MU Kemble. No 66 Sqn 'N' 29.11.56. Loaned to Iraqi AF by RAF 07.05.57. Returned to No 66 Sqn. Crashed near Kielder Reservoir, Northumberland, near Scottish border. Pilot killed. W/O 14.02.58.
XG237 ff 05.11.56	Delivered 16.11.56 No 5MU Kemble. No 66 Sqn 'C' 04.12.56. Converted to F Mk 6/Interim/FGA Mk 9. No 43 Sqn 17.06.60. Converted to FGA Mk 9. No 8 Sqn 'T' Khormaksar 15.10.62 to 1963, recoded 'M' from 1963 to 12.01.65. No 43 Sqn 'C'. No 208 Sqn 'D' 18.08.66 to 22.02.67. CAT3 22.02.67. Sold to HSA 15.01.68. Became Jordanian FGA Mk 73A '828/I'. Delivered 22.07.69.
XG238 ff 16.10.56	Delivered 09.11.56 No 5MU Kemble. No 74 Sqn. No 66 Sqn 29.11.56. No 92 Sqn 'F' 20.12.56. Crashed into sea off Pomos Point, Cyprus. Pilot killed. W/O 05.05.61.
XG239 ff 20.10.56	Delivered 26.11.56 No 5MU Kemble. No 92 Sqn 'MH'. Crashed on take-off. W/O 11.01.58.
XG251 ff 20.10.56	Delivered 26.11.56 No 5MU Kemble. No 66 Sqn 'G' 28.12.56. Loaned to Iraqi AF 07.05.57. No 103MU, Akrotiri, Cyprus, 01.07.57. No 66 Sqn 'G'. No 14 Sqn 'E' 21.09.59. No 5MU Kemble 17.12.62. Returned to HSA 1960. Converted to FGA Mk 9. No 1 Sqn 'J'. Purchased by HSA 30.10.69. Converted to Singapore FGA Mk 74 '507'. Delivered 10.12.70.
XG252 ff 25.10.56	Delivered 26.10.56 No 5MU Kemble. No 66 Sqn 'D' Linton-on-Ouse 06.12.56. Converted to F Mk 6/Interim/FGA Mk 9. Horsham St Faith 06.09.60. No 54 Sqn 'U' Stradishall 28.10.60. HAL 1960. Converted to FGA Mk 9. No 54 Sqn 'U' Stradishall. Loaned to MoA for trials 05.61 to 06.06.61. Stradishall Stn Flt until 01.01.62. No 54 Sqn 'W' Waterbeach 11.12.62 to 16.05.63. No 5MU Kemble for servicing 25.06.63. CAT3 flying accident. No 60MU Dishforth. Returned to No 54 Sqn West Raynham 03.10.63. CAT3, repaired at El Adem, Libya. Wittering 16.03.70. St Athan 05.06.70. No 8 Sqn Bahrain 04.02.71. No 5MU Kemble 30.09.71. No 45 Sqn West Raynham 30.08.72. No 45 Sqn 'H' Wittering 09.72 until 04.06.76. No 5MU Kemble for servicing. No 1TWU '45' Brawdy 27.07.76. No 2TWU Lossiemouth 24.01.80 to 1981. No 5MU Kemble. No 1TWU 26.10.81 to 1982. GIA 8840M No 2 SoTT Cosford 25.10.84. Gate guard Hereford 02.02.88. To France, owned by John St Clair-Quentin, Gloucestershire.
XG253 ff 01.11.56	Delivered 27.11.56 No 5MU Kemble. No 66 Sqn 'A' 28.12.56. Converted to F Mk 6/Interim/FGA Mk 9. To HAL. Converted to FGA Mk 9. No 54 Sqn 'W'. No 1 Sqn 'X' Waterbeach 1962. No 8 Sqn Khormaksar. Crashed at Khormaksar. W/O 28.10.62.
XG254 ff 01.11.56	Delivered 26.11.56 No 5MU Kemble. No 54 Sqn 'A'. Converted to F Mk 6/Interim/FGA Mk 9. Converted to FGA Mk 9. No 54 Sqn 'A' Waterbeach 1962-63. No 1TWU, No 63 Sqn 'K'. GIA 8881M Coltishall, Norfolk, on dump. Preserved at Norfolk & Suffolk Aviation Museum, Flixton.
XG255 ff 23.10.56	Delivered 14.11.56 No 5MU Kemble. No 66 Sqn 'S' 04.12.56. Converted to F Mk 6/Interim/FGA Mk 9. No 43 Sqn 21.06.60. HAL. Converted to FGA Mk 9. No 8 Sqn 'G' Khormaksar 18.07.62 to 30.05.64. No 208 Sqn 'E' 02.02.65 to 18.05.67. No 8/43 Sqn. No 208 Sqn 'E' 01.08.67 to 12.10.67. No 43 Sqn. No 8 Sqn 'G'. No 208 Sqn. Damaged in start-up fire at Muharraq 1967. No 19MU St Athan. HSA. Became Jordanian FGA Mk 73 '825/K' 12.67.
XG256 ff 27.10.56	Delivered 26.11.56 No 5MU Kemble. No 66 Sqn 'K' 03.01.57. Converted to F Mk 6/Interim/FGA Mk 9. HAL 04.59. Converted to FGA Mk 9. No 208 Sqn 'B' 01.08.66 to 12.10.67. No 43 Sqn 'G' Nicosia 1961 to 20.02.63. No 19MU St Athan. No 8 Sqn 'H' Khormaksar 28.08.63 to 06.12.65. No 8 Sqn 'B' Khormaksar 01.12.67 to 29.06.68. No 208 Sqn 17.06.71 to 21.06.71. No 19MU St Athan. No 229 OCU 'P'. Crashed on Bodmin Moor after hitting wires supporting TV mast. Abandoned near Liskeard. W/O 27.03.73.
XG257 ff 25.10.56	Delivered 27.11.56 No 5MU Kemble. No 66 Sqn 'H' 04.12.56. CAT4, Nicosia, Cyprus 24.05.57. No 93 Sqn 'C' 01.01.60. No 5MU Kemble 04.01.61. Transferred to Royal Jordanian AF '812' 06.62 to 10.62.
XG258 ff 31.10.56	Delivered 03.01.57 No 5MU Kemble. No 93 Sqn 'B' 08.04.57. Jever aerobatic team. Crashed during aerobatics at Spangdahlen. Pilot killed. W/O 17.05.57.

XG259 ff 06.11.56	Delivered 30.11.56 No 5MU Kemble. No 54 Sqn 'G'. Extensively damaged in night taxiing collision with bus at Akrotiri, Cyprus, 14.07.58. Fuselage repaired and shipped to Jordan for ground instruction use.
XG260 ff 26.10.56	No 54 Sqn. Converted to F Mk 6/Interim/FGA Mk 9.Converted to FGA Mk 9 01.60. No 54 Sqn 'B'. No 229 OCU 'T'. Purchased by HSA 13.09.69. Converted to Singapore FGA Mk 74 '501'. Delivered 12.11.70. Paya Lebar AB, Singapore. Gate guard.
XG261 ff 19.11.56	Delivered 03.12.56 No 5MU Kemble. No 43 Sqn 'G'. Converted to FGA Mk 9 11.60. No 54 Sqn 'C' Waterbeach 1962-63. No 208 Sqn 'L' Muharraq 05.06.67 to 21.07.69. No 19MU St Athan. No 45 Sqn '64' Wittering 1971-74. No 1TWU '40'. No 2TWU 'J'. Crashed near Dufftown, Grampian. W/O 28.05.80.
XG262 ff 08.11.56	Delivered 03.01.57 No 5MU Kemble. No 4 Sqn 'Z' 19.03.57. CAT4. Returned to HSA 13.06.58. Converted to Jordanian FR Mk 10 '712'. Delivered 11.60.
XG263 ff 01.11.56	Delivered 03.12.56 No 5MU Kemble. No 4 Sqn 'G' 11.03.57. No 19MU St Athan 20.02.61. Transferred to Royal Jordanian AF '805' 06.62 to 10.62.
XG264 ff 09.11.56	Delivered 30.11.56 No 5MU Kemble. No 111 Sqn. Returned to HSA 1960. Converted to F Mk 6/Interim/FGA Mk 9. Converted to FGA Mk 9. No 54 Sqn 'D' Waterbeach 1962-63. No 1TWU 'T' Brawdy 10.76. GIA 8715M 02.02.82. Target on Otterburn Ranges 08.97. Scrapped.
XG265 ff 31.10.56	Delivered 27.11.56 No 5MU Kemble. No 66 Sqn 'P' 03.01.57. Returned to HSA 04.10.60. Converted to FGA Mk 9. Delivered 17.07.61. No 20 Sqn 'K' 25.09.61. Crashed at Labuan, Borneo. Pilot killed. W/O 01.03.64.
XG266 ff 07.11.56	Delivered 26.11.56 No 5MU Kemble. No 66 Sqn 'R' 31.12.56. Returned to HSA 16.11.60. Converted to FGA Mk 9. Delivered 17.08.61. No 20 Sqn 'N' 26.09.61. Purchased by HSA 6.05.71. Converted to Singapore FR Mk 74B '521'. Delivered 08.06.72. Purchased by Steve Appleton, Adrenalin Airsports, Boise, Idaho.
XG267 ff 14.11.56	Delivered 03.12.56 No 5MU Kemble. No 4 Sqn 'F' 28.02.57. No 5MU Kemble 12.01.61. Transferred to Royal Jordanian AF '801' 06.62 to 10.62.
XG268 ff 09.11.56	Delivered 03.12.56 No 5MU Kemble. No 4 Sqn 'E' 28.02.57. No 5MU Kemble 21.01.61. Transferred to Royal Jordanian AF '806' 06.62 to 10.62.
XG269 ff 14.11.56	Delivered 4.12.56 No 5MU Kemble. No 4 Sqn 'W' 28.02.57. No 5MU Kemble 12.02.61. Transferred to Royal Jordanian Air Force '807' 06.62 to 10.62.
XG270 ff 10.11.56	Delivered 4.12.56 No 5MU Kemble. No 4 Sqn 'D' 28.02.57. CAT5 after forced-landing at Isle of Baltrum. W/O 31.05.57.
XG271 ff 19.11.56	Delivered 03.12.56 No 5MU Kemble. No 54 Sqn 'F'. Converted to F Mk 6/Interim/FGA Mk 9. Converted to FGA Mk 9. No 54 Sqn 'F'. Take-off accident at Sylt. W/O 13.07.61.
XG272 ff 04.12.56	Delivered 09.01.57 No 5MU Kemble. No 20 Sqn. No 93 Sqn 'A' 02.04.57. Returned to HSA 04.01.61. Converted to FGA Mk 9. Delivered 11.01.61. No 20 Sqn 'H' 01.07.62 to 23.09.69. No 5MU Kemble. Purchased by HSA 19.03.70. Converted to Swiss Mk 58A 'J-4111'.
XG273 ff 15.11.56	Delivered 30.11.56 No 5MU Kemble. No 54 Sqn 'L'. No 66 Sqn 'E' 01.04.60. Converted to F Mk 6/Interim/FGA Mk 9. No 1 Sqn 28.10.60. HAL. Converted to FGA Mk 9. No 54 Sqn 'U' Waterbeach 1962-63. Collided with XF446 on approach to El Adem and crashed 4 miles S of Tobruk. W/O 18.04.67.
XG274 ff 15.11.56	Delivered 03.01.57 No 5MU Kemble. No 66 Sqn. No 14 Sqn 'P'. No 229 OCU. No 5MU Kemble. No 4 FTS Valley '71' 1967. GIA 8710M Halton. R. J. Everett, Ipswich, Suffolk.
XG289 ff 26.11.56	Delivered 06.02.57 No 33MU. No 93 Sqn 'Z' 29.03.57. CAT 5 after crashing 1 mile NE of Sylt. W/O 29.11.57.

XG290 ff 20.11.56	Delivered 18.12.56 No 19MU St Athan. Laarbruch Stn Flt 05.80 to 10.80. A&AEE miscellaneous trials. GIA 8711M Halton. Bentley Priory. Swanton Morley. Bournemouth. Air Defence Collection. Nose/cockpit preserved by Tony Dyer at Boscombe Down (DERA) Museum. Rear fuselage at Delta Jets, Kemble, for spares. Kemble dump with T Mk 7 front fuselage and starboard wing from G-PSST 2005.
XG291 ff 21.11.56	Delivered 03.01.57 No 19MU St Athan. No 14 Sqn 'O' 01.05.57. No 5MU Kemble 17.12.62. Converted to FGA Mk 9. No 28 Sqn 'C' Kai Tak, Hong Kong. No 20 Sqn 'Z' 05.01.67 to 08.01.70. No 5MU Kemble. No 45 Sqn 'K' Wittering 1971-74. No 1TWU 'Y' Brawdy 10.76.
XG292 ff 27.11.56	Delivered 02.01.57 No 19MU. No 14 Sqn 'R' 01.05.57. No 26 Sqn 'D' 09.06.58. Returned to HSA 30.01.61. Converted to FGA Mk 9. No 208 Sqn 'M' 29.06.67 to 21.07.69. No 19MU St Athan. No 8 Sqn 'A'. Purchased by HSA 05.04.70. Converted to Singapore FR Mk 74A '512'. Delivered 04.08.71. Steve Appleton, Boise, Idaho, USA.
XG293 ff 03.12.56	Delivered 09.01.57 No 33MU. No 4 Sqn 'V' 28.02.57. No 20 Sqn 'D'. No 26 Sqn 'G' 11.06.58. HSA 02.02.61. Converted to FGA Mk 9. Delivered 23.01.62. No 20 Sqn 'F' 01.07.62. Crashed at Tengah. W/O 21.04.64.
XG294 ff 29.11.56	Delivered 24.01.57 No 5MU Kemble. No 93 Sqn 'W' 22.03.57. No 2 Sqn 'W' 01.01.61. No 5MU Kemble 10.02.61. Converted to Rhodesian FGA Mk 9 '122'.
XG295 ff 27.11.56	Delivered 18.12.56 No 19MU St Athan. No 14 Sqn 'S' 08.04.57. Collided with XJ643 and crashed 11.05.57. Repaired. No 5MU Kemble 05.01.61. Converted to Rhodesian FGA Mk 9 '121'.
XG296 ff 30.11.56	Delivered 29.01.57 No 33MU. No 93 Sqn 'X' 19.03.57. Returned to HSA 03.01.61. Converted to FGA Mk 9. Delivered 20.10.61 No 43 Sqn 'A' Nicosia and Khormaksar from 16.12.61 to 11.03.65. No 229 OCU 'B'. Purchased by HSA 22.10.69. Converted to Singapore FGA Mk 74 '510'. Delivered 04.03.71.
XG297 ff 03.12.56	Delivered 03.01.57 No 33MU. No 4 Sqn 'Y' 15.03.57. To HSA 03.01.61. Converted to FGA Mk 9. Delivered 19.09.61 No 20 Sqn 'B'. Tengah 10.11.61. No 28 Sqn 'B' Kai Tak 25.08.62. No 20 Sqn 'Y' 05.01.67. No 5MU Kemble 19.02.70. Purchased by HSA 8.03.76. Stored at Bitteswell awaiting final disposal 07.79. Nose section at Aeroventure, Doncaster, South Yorkshire.
XG298 ff 29.11.56	Delivered 03.01.57 No 5MU. No 4 Sqn 'X' 13.03.57. HSA 02.01.61. Converted to FGA Mk 9. Delivered 25.10.61 No 43 Sqn 'E' Nicosia and Khormaksar from 05.12.61 to 28.01.64. No 208 Sqn 'E' Khormaksar and Muharraq from 28.01.64 to 15.09.64. No 19MU St Athan. Transferred to Royal Jordanian AF as FGA Mk 9 '826/L' 07.09.67.

Fourth production batch of forty-five aircraft built by Hawker Aircraft Ltd, Kingston-upon-Thames. Fitted with wing leading edge extensions from outset; gun blast deflectors retrofitted.

XJ632 ff 03.01.57	Delivered 24.01.57 No 5MU Kemble. No 93 Sqn 'H' 22.03.57. No 26 Sqn 'H' 03.07.58. Returned to HSA 30.01.61. Converted to FGA Mk 9. Delivered 29.11.61 No 208 Sqn Khormaksar 1957, 'K' 26.03.62 to 26.03.64. To No 5MU Kemble for refurbishment. No 208 Sqn 'B' 21.11.67 to 22.09.69. No 19MU St Athan. Purchased by HSA 31.10.69. Converted to Singapore FGA Mk 74 '505'. Delivered 10.12.70.
XJ633 ff 04.12.56	Delivered 15.02.57 No 5MU Kemble. No 93 Sqn 'T' 26.03.57. CAT 4. Returned to HSA for repair 01.05.57. No 65 Sqn 'J'. No 66 Sqn 'X' 04.05.59. Returned to HSA 01.10.59. Converted to FR Mk 10. No 4 Sqn 'H', later 'K', 29.01.60 to 03.02.61. No 2 Sqn 'S' 29.01.70. Purchased by HSA 02.03.71. Converted to Singapore FR Mk 74B '534'. Delivered. 21.02.73.
XJ634 ff 03.01.57	Delivered 24.01.57 No 33MU. No 93 Sqn 'V'. No 229 OCU. No 1TWU Brawdy 10.76. Sold to R. J. Everett, Ipswich, Suffolk. To Royal Saudi Air Force Museum as 60.602.
XJ635 ff 03.01.57	Delivered 01.03.57 No 5MU. No 93 Sqn 'F' 29.03.57. No 5MU Kemble 04.01.61. Returned to HSA for conversion to FGA Mk 9. Delivered 27.07.65. No 208 Sqn 22.05.70 to 14.09.71. No 19MU St Athan. Rebuilt by HSA 1976. No 1TWU Brawdy 'B', later 'I', 1976. Crashed near Aberystwyth. Pilot killed. W/O 04.05.76.

XJ636 ff 18.12.56	Delivered 22.01.57 No 33MU. No 4 Sqn 'K' 11.03.57. No 26 Sqn 'K' 24.07.58. No 14 Sqn 'S' 02.01.61. No 19MU 17.02.62. No 1 Sqn 'D' 1965. No 208 Sqn 'F' 18.09.67 to 20.10.69. No 5MU Kemble. No 208 Sqn 'K' 15.01.71 to 14.09.71. No 5MU Kemble. No 1 TWU Brawdy. Crashed on take-off. Pilot killed. W/O 25.10.76. Wreckage to Brawdy fire dump.
XJ637 ff 03.01.57	Delivered 28.01.57 No 5MU Kemble. No 4 Sqn 'U' 15.03.57. No 2 Sqn 04.01.61. No 19MU St Athan 22.02.61. No 229 OCU 1964. 1TWU Brawdy 10.76. Crashed at Talfarm, Pembrokeshire. W/O 14.03.79.
XJ638 ff 10.01.57	Delivered 27.05.57 No 5MU Kemble. No 4 Sqn 'D' 02.07.57. No 5MU Kemble 12.01.61. Converted to Rhodesian FGA Mk 9 '123'.
XJ639 ff 18.12.56	Delivered 11.03.57 No 33MU. No 4 Sqn 'C' 22.03.57. No 2 Sqn 'C' 04.01.61. No 19MU St Athan 21.03.61. No 229 OCU 12.64. No 1TWU. GIA 8687M Cranwell. Stored dismantled at Ipswich 6.95. Sold to private buyer. Sunk by owner in lake near Doncaster for scuba dive training.
XJ640 ff 03.01.57	Delivered 21.01.57 No 33MU. No 4 Sqn 'T' 12.03.57. No 5MU Kemble 04.01.61. Returned to HSA 1963. Converted to FGA Mk 9. Delivered 02.07.65 No 1 Sqn 'F'. Harrier OCU 'F'. Purchased by HSA 15.01.75. Converted to Lebanese FGA Mk 70A 'L-285'. Delivered 01.06.77.
XJ641 ff 03.01.57	Delivered 5.02.57 No 33MU. No 93 Sqn 'G' 02.04.57. CAT5 when lost at night off Frisian Islands. Pilot presumed drowned, aircraft never located. W/O 11.11.59.
XJ642 ff 15.01.57	Delivered 29.01.57 No 5MU Kemble. No 14 Sqn 'A' 08.04.57. No 19MU St Athan 17.12.62. No 54 Sqn 'L'. Converted to FGA Mk 9. No 4 Sqn 'L'. Purchased by HSA 09.02.70. Converted to Singapore FGA Mk 74 '518'. Delivered 20.05.71.
XJ643 ff 15.01.57	Delivered 04.03.57 No 5MU Kemble. No 14 Sqn 'B' 08.04.57. CAT4 after collision with XG295 11.05.57. Returned to HSA 22.08.57. Converted to FGA Mk 9. No 66 Sqn. No 208 Sqn 'M' Eastleigh and Khormaksar 26.04.60 to 22.11.61. No 19MU St Athan. No 28 Sqn. No 20 Sqn 'W' 05.01.67 to 20.10.69. No 5MU Kemble. Purchased by HSA 06.02.70. Converted to Singapore FGA Mk 74 '515'. Delivered 24.03.71.
XJ644 ff 09.01.57	Delivered 11.02.57 No 33MU. No 14 Sqn 'C' 09.04.57. No 19MU St Athan 17.12.62. Converted to FGA Mk 9. No 54 Sqn 'K'. No 4 Sqn 'K'. Purchased by HSA 20.01.75. Converted to Lebanese FGA Mk 70A 'L-284'. Delivered 17.05.77.
XJ645 ff 10.01.57	Delivered 05.02.57 No 33MU. No 93 Sqn 'E' 29.03.57. No 2 Sqn 'E' 01.01.61. No 5MU Kemble 10.02.61. Returned to HAS. Converted to FGA Mk 9. No 208 Sqn 'A' 02.06.66 to 05.08.67. No 43 Sqn 'N'. Purchased by HAS 11.04.68. Converted to Jordanian FGA Mk 73A '831/Q'. Delivered 19.05.69. Gift to Sultan of Oman AF '831' 1975. With No 6 Sqn Thumrait. Gate guard, RAFO Seeb, Oman.
XJ646 ff 09.01.57	Delivered 26.01.57. No 19MU St Athan 05.02.57. 2TAF 04.04.57. No 14 Sqn 'D' 15.04.57. CAT3 after flying accident 03.09.57. No 14 Sqn 23.09.57. CAT3R 13.10.58. No 14 Sqn 20.11.58. No 14 Sqn 02.05.60. No 14 Sqn 13.07.61 No 19MU St Athan 17.12.62. Converted to FGA Mk 9 27.08.65. No 5MU Kemble 06.09.65. No 8/43 Sqn 'P' Khormaksar 13.10.65. No 8 Sqn 05.09.67. No 19MU St Athan 29.11.67. Purchased by HSA 25.04.68. Converted to Indian Mk 56A 'A968'. Delivered 31.07.69. Preserved at Kalaikunda, India. Ta Qali, Malta, for restoration.
XJ673 ff 07.01.57	Delivered 11.02.57 No 33MU. No 14 Sqn 'E' 15.04.57. CAT 4. Returned to HSA 18.08.59. No 66 Sqn. Returned to HSA 08.60. Converted to FGA Mk 9. No 20 Sqn 'XX' 07.09.61. Crashed into sea off Manila. W/O 02.04.69.
XJ674 ff 16.01.57	Delivered 11.02.57 No 5MU Kemble. No 4 Sqn 'A' 19.03.57. No 26 Sqn 'L' 11.06.58. No 5MU Kemble 11.01.61. To HSA. Converted to FGA Mk 9. Delivered 11.06.65. No 20 Sqn 'O' 11.01.66. Crashed near Tengah. W/O 22.07.68.
XJ675 ff 15.01.57	Delivered 15.02.57 No 5MU Kemble. No 93 Sqn 'T' 22.03.57. CAT 5 after crash on take-off near Aurich. W/O 08.01.60.

XJ676 ff 06.02.57	Delivered 01.03.57 No 5MU Kemble. No 93 Sqn 'S'. No 2 Sqn 04.01.61. No 229 OCU 10.98. Imperial War Museum, Duxford. Leavesden Studios for filming of James Bond movie 10.99. Scrapped 2002.
XJ677 ff 25.01.57	Delivered 04.03.57 No 5MU Kemble. Presented to Iraq by UK Government '394' 15.04.57.
XJ678 ff 10.01.57	Delivered 05.02.57 No 33MU. Presented to Iraq by UK Government '395' 15.04.57.
XJ679 ff 17.01.57	Delivered 20.03.57 No 5MU Kemble. Presented to Iraq by UK Government '396' 15.04.57.
XJ680 ff 18.02.57	Hugh Merewether Delivered 26.03.57 No 5MU Kemble. No 20 Sqn 'A' 03.05.57. Returned to HSA 03.01.61. Converted to FGA Mk 9. Delivered 11.10.61. No 43 Sqn 'F' Nicosia then Khormaksar 05.12.61 to 04.64. No 8 Sqn 'C'. Purchased by HSA 16.12.69. Converted to Singapore FGA Mk 74 '511'. Delivered 04.03.71.
XJ681 ff 24.01.57	Delivered 21.02.57 No 5MU Kemble. Presented to Iraq by UK Government '397' 15.04.57.
XJ682 ff 16.01.57	Delivered 26.03.57 No 5MU Kemble. Presented to Iraq by UK Government '398' 15.04.57.
XJ683 ff 07.02.57	Delivered 29.05.57 No 5MU Kemble. No 93 Sqn 'D'. No 5MU Kemble. Returned to HSA for conversion to FGA Mk 9. No 43 Sqn 'L' Nicosia and Khormaksar 15.01.62 to 11.07.63. Returned to HAL for rebuild following fire damage 08.63. No 20 Sqn 'F' Tengah 15.05.64. No 1TWU 'X'. CAT3 at Gibraltar 24.05.77. To UK in C-130K to No 5MU Kemble for rebuild. No 1TWU Brawdy. Rhodesian/Zimbabwe AF. Thornhill AB, Gweru. Manyame AB, Zimbabwe.
XJ684 ff 11.02.57	Delivered 26.03.57 No 5MU Kemble. No 20 Sqn 'B' 03.05.57. Returned to HSA 03.01.61. Converted to FGA Mk 9. Delivered 26.10.61. No 43 Sqn 'D' Nicosia and Khormaksar 16.12.61 to 01.11.64. Wheels-up landing at Bahrain. No 19MU St Athan for refurbishment 1964-65. No 8 Sqn 'H' Khormaksar. Purchased by HSA 16.12.69. Converted to Singapore FGA Mk 74 '513'. Delivered 12.11.70.
XJ685 ff 08.02.57	Delivered 01.04.57 No 5MU Kemble. No 20 Sqn 'C' 06.05.57. Returned to HSA 31.01.61. Converted to FGA Mk 9. Delivered 09.01.62. No 20 Sqn 'R' 01.07.62 to 28.07.69. Badly damaged in Far East. Fuselage purchased by HSA 04.11.69. Converted to Singapore FGA Mk 74 '502'. Delivered 12.11.70. Geoff Moesker, Archerfield Airfield, Brisbane, Queensland, Australia.
XJ686 ff 22.01.57	Delivered 26.03.57 No 5MU Kemble. No 20 Sqn 'D' 03.05.57. Returned to HSA 31.01.61. Converted to FGA Mk 9. Delivered 11.12.61. No 20 Sqn 'B' 01.07.62. No 5MU Kemble 19.02.70. No 58 Sqn '41' Wittering 1973. No 1TWU '41' Brawdy 10.76.
XJ687 ff 26.02.57	Delivered 12.08.57 No 5MU. No 66 Sqn 'H' Acklington 12.09.57. Returned to HSA 17.04.59. Converted to FGA Mk 9. No 208 Sqn 'B', later 'G', Eastleigh then Khormaksar 23.03.60 to 26.05.62. No 8 Sqn 'E' Khormaksar 26.05.62 to 18.10.62. No 208 Sqn 'F' 25.11.63 to 12.05.66. No 19MU St Athan. No 208 Sqn 'E' 05.08.67 to 17.11.68. No 19MU St Athan. No 208 Sqn 'J' 30.06.69 to 07.06.71. No 1TWU 'O' Brawdy 10.76.
XJ688 ff 08.02.57	Delivered 26.03.57 No 5MU Kemble. No 26 Sqn 'E' 09.06.58. HSA 31.01.61. Converted to FGA Mk 9. Delivered 09.01.62. No 208 Sqn 'G' Khormaksar and Muharraq 22.03.62 to 25.03.65. No 19MU St Athan. No 208 Sqn 27.09.67 to 06.11.67. No 8 Sqn. No 208 Sqn 'E' 06.12.68 to 18.05.71. No 8 Sqn. No 1TWU, No 79 Sqn 'B' Brawdy.
XJ689 ff 06.02.57	Delivered 20.03.57 No 5MU Kemble. No 14 Sqn 'F' Gutersloh 01.05.57. No 19MU St Athan 17.12.62. No 66 Sqn. Returned to HSA 30.01.64. Converted to FGA Mk 9. Delivered 14.12.64 No 66 Sqn. No 8 Sqn 'B' Khormaksar 1966-67. Purchased by HSA 05.04.70. Converted to Singapore FR Mk 74A '517', one of four for Singapore under contract HSAL/68/SG/078. Delivered 04.08.71. Retired from service 1992. Jet Imports Ltd, David Phillips, Ardmore, New Zealand, as ZK-JIL. Wheels-up landing at Ardmore.

XJ690 ff 12.02.57	Delivered 26.03.57 No 5MU Kemble. No 14 Sqn 'H' 08.05.57. No 5MU Kemble 17.12.62. Converted to FGA Mk 9. Delivered 27.10.64. No 20 Sqn 'G' 31.03.65. No 5MU Kemble 19.02.70. Purchased by HSA 02.76. Stored at Bitteswell awaiting final disposal 1979. Nose fitted to G-ERIC on gate guard at Bournemouth.
XJ691 ff 08.02.57	Delivered 26.03.57 No 5MU. No 14 Sqn 'M' 01.05.57. No 19MU 17.12.62. No 66 Sqn. Converted to FGA Mk 9. No 208 Sqn 'G' 02.02.65. Crashed into sea 18 miles SE of Muharraq, Bahrain. Pilot killed. W/O 27.04.67.
XJ692 ff 18.02.57	Delivered 08.05.57 No 5MU Kemble. No 20 Sqn 'X' 05.06.57. Returned to HSA 31.01.61. Converted to FGA Mk 9. Delivered 07.12.61. No 43 Sqn Nicosia and Khormaksar 01.01.62 to 14.08.64. No 208 Sqn 'B'. Purchased by HSA 05.04.68. Converted to Indian Mk 56A 'A969'. Delivered 23.09.69.
XJ693 ff 11.02.57	Delivered 08.05.57 No 5MU Kemble. No 20 Sqn 'W' 13.06.57. CAT5 after wheels-up landing at Gutersloh. W/O 03.10.60.
XJ694 ff 20.02.57	CA aircraft Dunsfold 1957. No 208 Sqn 'D' 05.03.58 to 10.04.59. Horsham St Faith. Returned to HSA 27.11.59. Converted to FR Mk 10. Delivered 14.04.61. No 2 Sqn 'Z' 09.09.63. No 4 Sqn 'Z' 12.08.63. Pooled between No 2 and No 4 Sqns until 16.02.71. Purchased by HSA 17.02.71. Converted to Indian T Mk 66E 'S.1389'. Delivered 13.06.73.
XJ695 ff 11.02.57	Delivered 17.04.57 No 5MU Kemble. No 20 Sqn 'U' 31.05.57. No 14 Sqn 'L' 07.02.61. To HSA. Converted to FGA Mk 9. Delivered 15.01.65. No 20 Sqn 'K' 27.05.65. No 5MU Kemble 19.02.70. No 45 Sqn Wittering. No 58 Sqn Wittering. No 1TWU 'Z' Brawdy 10.76. Flew into flock of birds and force-landed at Brize Norton 16.11.78. DBR. GIA 8677M Kemble. Forward and centre fuselage 8738M at CTE for fire training at Manston, Kent, 05.02.82. Rear fuselage from XF519. On Manston fire dump 04.99. Burned and scrapped by 03.02.
XJ712 ff 22.02.57	Delivered 05.57 No 5MU Kemble. No 20 Sqn 'J' and 'F' 31.05.57. No 14 Sqn 'L' and 'B' 02.01.61. No 1 Sqn 14.02.63. Purchased by HSA 01.04.66 as G-9.210. Converted to Saudi Arabian F Mk 60 '60/601'. Delivered 1966.
XJ713 ff 22.02.57	Delivered 05.57 No 5MU Kemble. No 20 Sqn 'G' 20.05.57. No 14 Sqn 'B' 02.01.61. No 1 Sqn. No 229 OCU '55'. Purchased by HSA 10.09.69. Converted to Chilean FGA Mk 71 '722'. Delivered 21.12.70. Chilean Air Force Museum, Los Cerrillos, Chile.
XJ714 ff 13.02.57	Delivered 26.03.57 No 5MU Kemble. Tropical trials Idris 06.57. No 208 Sqn 'B' 13.10.58 to 13.04.59. Horsham St Faith. Returned to HSA 21.08.59. Converted to FR Mk 10. Delivered 28.11.60. No 4 Sqn 'B' 1961. No 8 Sqn 'V'. Purchased by HSA 29.06.71. Converted to Singapore FR Mk 74B '531'. Delivered 17.01.73. Alan Arthur, Tocumwal Airfield, NSW, Australia 26.04.98.
XJ715 ff 22.02.57	Delivered 30.05.57 No 5MU Kemble. No 111 Sqn 'H'. 'Black Arrows' aerobatic team. No 229 OCU. Returned to HSA 06.04.66. Converted to Saudi Arabian F Mk 60 1966.
XJ716 ff 20.02.57	Delivered 05.57 No 5MU Kemble. No 20 Sqn 31.05.57. No 19MU St Athan 04.01.61. To HSA. Converted to Rhodesian FGA Mk 9 '1284'. Rebuilt from two air-damaged aircraft XJ716 and XE559.
XJ717 ff 25.02.57	Delivered 27.05.57 No 5MU Kemble. No 20 Sqn 'Z' 31.05.57. No 26 Sqn 'Z' 09.06.58. No 14 Sqn 'R' 12.01.61. No 54 Sqn 14.02.62. Purchased by HSA 05.07.67. Converted to Chilean FR Mk 71A 'J-717'. Delivered 24.09.68.
XJ718 ff 20.03.57	Delivered 08.05.57 No 5MU Kemble. No 118 Sqn. No 93 Sqn 'B' 29.05.57. No 5MU Kemble 05.01.61. Returned to HAS. Converted to Rhodesian FGA Mk 9 '117', later '1817'. Second RRAF FGA Mk 9. Delivered 15.12.62. Thornhill AB, Gweru, Zimbabwe.

Fifth production batch of fifty-three aircraft built by Hawker Aircraft Ltd, Kingston-upon-Thames. Original contract was for 153 aircraft but was subsequently reduced by 100.

XK136 ff 21.03.57	Delivered 19.08.57 No 5MU Kemble. No 74 Sqn. Returned to HSA 1960. Converted to FGA Mk 9. No 20 Sqn 'O' 08.09.61. Aircraft crashed near Batu Pahat in Johore. W/O 19.10.64.

XK137 ff 11.03.57	Delivered 08.05.57 No 5MU. No 20 Sqn 'V' Furstenfeldbeck 17.06.57. Returned to HSA 03.01.61. Converted to FGA Mk 9. No 43 Sqn 'P' Nicosia then Khormaksar 16.11.61 to 02.09.63. No 54 Sqn 'E' 1965. No 208 Sqn 'F' 04.11.69 to 19.01.71. No 45 Sqn '42', later 'D', Wittering 1971-74. No 1TWU '42' Brawdy.
XK138 ff 21.03.57	Delivered 27.05.57 No 5MU Kemble. No 14 Sqn 'Y'. No 20 Sqn 24.06.57. Returned to HSA 03.01.61. No 20 Sqn to Tengah 10.11.61. No 5MU Kemble 10.02.70. No 45 Sqn Wittering 1973. No 1TWU, No 79 Sqn 'I' Brawdy. Transferred to Chile as 'J-746' 22.05.82. Museo Aeronautico, Santiago.
XK139 ff 21.03.57	Delivered 12.08.57 No 5MU Kemble. No 66 Sqn 'G' 09.09.57. Returned to HAL 27.04.59. HSA 02.01.61. Converted to FGA Mk 9. Delivered 19.09.61 No 1 Sqn 'X'. No 208 Sqn 'J' Khormaksar and Muharraq 27.11.61 to 30.06.64. Crashed 10 miles SE of Dassa Island after mid-air collision with XE647. Pilot killed. W/O 30.06.64.
XK140 ff 29.03.57	Delivered 08.57 No 5MU Kemble. No 74 Sqn. Returned to HSA 15.04.59. Converted to FGA Mk 9. Delivered 04.02.60 No 54 Sqn 'E'. No 8 Sqn 'D'. No 208 Sqn 'E' 02.11.61 to 05.02.64. No 19MU St Athan for refurbishment. No 208 Sqn 'H' 20.08.64 to 04.01.66. No 19MU St Athan. No 208 Sqn 'D' 26.04.67 to 30.05.67. No 19MU St Athan. No 5MU Kemble for storage 08.12.71. No 45 Sqn 'P' 1971-74. No 2TWU 'N'. Crashed into the sea near Lochinver. W/O 03.07.79.
XK141 ff 15.04.57	Delivered 12.08.57 No 5MU Kemble. No 74 Sqn 'F'. No 54 Sqn '33'. No 229 OCU '55' 1961. No 1TWU Brawdy 10.76. No 2TWU Lossiemouth 1979. Sold to Chile as attrition replacement FGA Mk 9 'J-754'. Delivered 06.01.83.
XK142 ff 08.04.57	Delivered 11.9.57 No 5MU Kemble. No 74 Sqn 'L' Coltishall. Returned to HSA 28.11.60. Converted to FGA Mk 9. Delivered 7.9.61 No 20 Sqn 'P' Tengah, flown out 23.10.61, delivered 06.11.61. No 5MU Kemble 19.02.70. Purchased by HSA 5.05.71. Converted to Singapore FR Mk 74B '522'. Delivered 18.08.72.
XK143 ff 16.05.57	Delivered 08.57 No 5MU Kemble. Presented to Iraq by UK Government '400' 19.12.57.
XK144 ff 25.04.57	Delivered 08.57 No 5MU Kemble. Presented to Iraq by UK Government '401' 19.12.57.
XK145 ff 15.04.57	Delivered 21.08.57 No 5MU Kemble. Presented to Iraq by UK Government '402' 19.12.57.
XK146 ff 26.04.57	Delivered 08.57 No 19MU St Athan. Presented to Iraq by UK Government '403' 19.12.57.
XK147 ff 24.04.57	Delivered 08.57 No 19MU St Athan. Presented to Iraq by UK Government '404' 19.12.57.
XK148 ff 09.05.57	Delivered A&AEE Boscombe Down. No 5MU Kemble for storage. Purchased by HSA 13.07.67. Converted to Chilean FR Mk 71A 'J-715'. Delivered 15.07.68.
XK149 ff 20.05.57	Delivered 26.09.57 No 5MU Kemble. AFDS tactical evaluation 1958. No 14 Sqn 'T' Gutersloh 1962. No 1 Sqn 03.63. No 54 Sqn 07.63. No 229 OCU Chivenor '44'. No 79 Sqn. No 1TWU, 79 Sqn '44' Brawdy 10.76. GIA 8714M '34' Cranwell 10.85.
XK150 ff 30.05.57	Delivered 09.57 No 5MU Kemble. AFDS Venom Replacement Evaluation Trials (VRET) Aden 1958. Returned to HSA. Converted to FGA Mk 9. No 8 Sqn 'A' Khormaksar 19.01.60 to 02.02.63. To UK for refurbishment. No 208 Sqn 'B' Khormaksar then Muharraq 13.02.64 to 14.07.66. No 19MU St Athan. Transferred to Royal Jordanian AF 26.01.68.
XK151 ff 16.06.57	Delivered. 09.57 No 5MU Kemble. AFDS Venom Replacement Evaluation Trials (VRET) Aden 1958. Last Hunter fighter issued to RAF. Returned to HSA. Converted to FGA Mk 9. No 8 Sqn 'X' Khormaksar 19.01.60 to 03.62, recoded 'J' from 03.62 to 02.02.63. No 19MU St Athan for refurbishment. No 208 Sqn 'D' Khormaksar and Muharraq 20.02.64 to 04.08.66. No 19MU St Athan for refurbishment. No 208 Sqn 'D' 11.09.69 to 17.06.71. No 1TWU '49' Brawdy 10.76. No 2TWU Lossiemouth. Crashed on Isle of Skye. Pilot killed. W/O 12.02.80. Wreckage to Lossiemouth dump and scrapped by 1987.

XK152 ff 03.06.57 Delivered 17.09.57 No 19MU St Athan. Presented to Iraq by UK Government '405' 19.12.57.

XK153 ff 07.06.57 Delivered 17.09.57 No 19MU St Athan. Presented to Iraq by UK Government '406' 19.12.57.

XK154 ff 20.06.57 Delivered 23.08.57 No 19MU St Athan. Presented to Iraq by UK Government '407' 19.12.57.

XK155 ff 05.07.57 Delivered 23.09.57 No 19MU St Athan. Presented to Iraq by UK Government '408' 19.12.57.

XK156 ff 09.07.57 Delivered 01.10.57 No 19MU St Athan. Presented to Iraq by UK Government '409' 19.12.57.

Indian Government order for 160 F Mk 6 Hunters from HAL in 1957 with designation F Mk 56. Contract No 6/Acft/12132/CB.7(a) for fifty aircraft from HAL (Blackpool) cancelled completely. Contract No 6/Acft/12133/CB.7(a) for 153 aircraft built at Kingston-upon-Thames cut by 100 aircraft and thirty-two of original contract diverted to initial part of Indian contract. Some aircraft were in advance stages of construction when diverted and retained RAF serials alongside Indian serials until delivery.

XK157 ff 09.07.57 Completed as Mk 56 (BA201). ff 11.10.57. Dunsfold for training. Delivered to Benson 17.12.57. No 7 Sqn IAF Ambala.

XK158 ff 08.07.57 Completed as Mk 56 (BA202. ff 11.10.57. Dunsfold for training. Delivered to Benson 17.12.57. No 7 Sqn IAF Ambala.

XK159 ff 11.07.57 Completed as Mk 56 (BA203). ff 14.10.57. Dunsfold for training. Delivered to Benson 17.12.57. No 17 Sqn IAF Poona.

XK160 ff 15.08.57 Completed as Mk 56 (BA204). ff 15.10.57. Dunsfold for training. Delivered to Benson 17.12.57. No 7 Sqn IAF Ambala.

XK161 ff 17.09.57 Completed as Mk 56 (BA205). ff 21.10.57. Delivered to Istres 25.10.57. No 7 Sqn IAF Ambala.

XK162 ff 17.09.57 Completed as Mk 56 (BA206). ff 21.10.57. Delivered to Istres 25.10.57. No 7 Sqn IAF Ambala.

XK163 ff 20.09.57 Completed as Mk 56 (BA207/XK163). ff 13.11.57. Delivered to Benson as BA207 26.11.57. No 7 Sqn IAF Ambala. No 17 Sqn IAF Poona.

XK164 ff 26.11.57 Completed as Mk 56 (BA208/XK164). ff 08.11.57. Delivered to Benson as BA208 26.11.57. No 17 Sqn IAF Poona.

XK172 ff 14.10.57 Completed as Mk 56 (BA216/XK172). ff 22.11.57. Delivered to Benson as BA216 06.12.57. Crashed after mid-air collision with BA212 during formation aerobatics. W/O 1958.

XK173 ff 14.10.57 Completed as Mk 56 (BA217/XK173). ff 04.11.57. Delivered to Benson as BA217 06.12.57. No 17 Sqn IAF Poona.

XK174 ff 01.11.57 Completed as Mk 56 (BA218/XK174). ff 21.10.57. Delivered to Benson as BA218 06.12.57. No 17 Sqn IAF Poona.

XK175 ff 18.10.57 Completed as Mk 56 (BA219/XK175). ff 01.11.57. Delivered to Benson as BA219 31.12.57. No 7 Sqn IAF Ambala.

XK176 ff 24.10.57 Completed as Mk 56 (BA220/XK176). ff 01.11.57. Delivered to Benson as BA220 06.12.57. No 17 Sqn IAF Poona.

XK213 ff 01.11.57 Completed as Mk 56 (BA221/XK213). Delivered to Benson as BA221 20.12.57. No 17 Sqn IAF Poona.

XK214 ff 06.12.57 Completed as Mk 56 (BA222/XK214). Delivered to Benson as BA222 06.12.57. No 7 Sqn IAF Ambala.

XK215 ff 24.10.57 Completed as Mk 56 (BA223/XK215). Delivered to Benson as BA223 06.12.57. No 7 Sqn IAF.

XK216 ff 01.11.57 Completed as Mk 56 (BA224/XK216). Delivered to Benson as BA224 06.12.57. No 7 Sqn IAF Ambala.

XK217 ff 23.11.57 Completed as Mk 56 (BA225/XK217). Delivered to Benson as BA225 10.12.57. No 7 Sqn IAF Ambala.

XK218 ff 07.11.57 Completed as Mk 56 (BA226/XK218). Delivered to Benson as BA226 10.12.57. No 17 Sqn IAF Poona.

XK219 ff 28.11.57 Completed as Mk 56 (BA227/XK219). Delivered to Benson as BA227 23.12.57. No 7 Sqn IAF Ambala.

XK220 ff 28.11.57 Completed as Mk 56 (BA228/XK220). Delivered to Benson as BA228 23.12.57. No 7 Sqn IAF Ambala.

XK221 ff 22.11.57 Completed as Mk 56 (BA229/XK221). Delivered to Benson as BA229 20.12.57. No 17 Sqn IAF. Crashed 04.06.58.

XK222 ff 11.12.57 Completed as Mk 56 (BA230/XK222). Delivered to Benson as BA230 23.12.57. No 7 Sqn IAF Ambala.

XK223 ff 19.12.57 Completed as Mk 56 (BA231/XK223). Delivered to Benson as BA231 31.12.57. No 7 Sqn IAF Ambala.

XK224 ff 19.12.57 Completed as Mk 56 (BA232/XK224). Delivered to Benson as BA232 31.12.57. No 7 Sqn IAF Ambala.

Single production batch of 100 aircraft built by Armstrong Whitworth Aircraft Ltd, Coventry. Contract No SP/6/9818/CB.7a. Aircraft retrospectively fitted with wing leading edge extensions and gun blast deflectors. Production completed 31.12.56.

XF373 RR engine trials Wymeswold 1956. Supplied to Jordan '703' under US Offshore Procurement 1958.

XF374 CA aircraft Dunsfold 1956-59. Fuel system trials with early 230-gal drop tanks. Converted to Rhodesian FGA Mk 9 '127'.

XF375 CA aircraft MoS and AWA Wymeswold and Bitteswell 1956-57. To English Electric Warton 01.59. ETPS '6' 23.04.63. RAE Farnborough 1976. Cranwell 8736M. Registered as G-BUEZ with OFMC Duxford 03.04.92. Moved to Spanhoe 2004. Last noted 01.09.

XF376 No 5MU Kemble. A&AEE trials 1956-57. Returned to HSA 08.05.59. Converted to FGA Mk 9. No 66 Sqn 26.02.60. No 208 Sqn 'J' Eastleigh and Khormaksar 07.04.60 to 31.01.61. No 8 Sqn 'Q' Khormaksar 31.01.61 to 26.07.61. No 131MU for repairs. No 208 Sqn 'M', later 'P', 19.09.61 to 05.04.63. No 5MU Kemble for refurbishment. No 8 Sqn 'K' Khormaksar 07.10.63 to 25.03.66. No 208 Sqn 'H' 05.08.67 to 29.01.69. No 19MU St Athan. No 1TWU Brawdy 'D' 10.76.

XF377 CA aircraft AWA Bitteswell 1956-57. Supplied to Lebanon 'L-173' under US Offshore Procurement arrangement 1958. Crashed on Bekka Plain while flying from Rayak, Lebanon. Pilot killed. W/O 26.03.62.

XF378 CA aircraft HAL Dunsfold. P.1109B aircraft with complete De Havilland Firestreak AAM missile system installed. Trials at Hatfield and Valley 1957. Cannibalised after fuselage fire. W/O 1959.

XF379 CA aircraft HAL Dunsfold 1956-57. Trials of lateral fuselage air brakes from WB188. Supplied to Jordan '709' under US Offshore Procurement arrangement 1958.

XF380 No 19MU St Athan. CA aircraft Dunsfold 1957. Supplied to Jordan '710' under US Offshore Procurement arrangement 1958.

XF381 A&AEE Boscombe Down for miscellaneous trials 1957. Supplied to Jordan '711' under US Offshore Procurement arrangement 1958.

XF382 No 65 Sqn 'Q'. No 63 Sqn. No 65 Sqn. No 92 Sqn Linton-on-Ouse 09.56. DFCS 'R'. No 229 OCU. No 1TWU, No 79 Sqn Brawdy 10.76. Fitted with Starboard Oblique nose camera for anti-drug-smuggling and interdiction in Wales. SOC 08.07.84. Midland Air Museum, Coventry Airport, Baginton.

XF383	No 263 Sqn 'F'. No 111 Sqn. No 65 Sqn 'V'. No 4 FTS. No 237 OCU. No 216 Sqn. No 12 Sqn. No 229 OCU/No 145 Reserve Sqn '14' Chivenor 1961. No 5MU Kemble. No 12 Sqn Honington 1984. GIA 8706M Wittering. Nose preserved at Kidlington. Sold to Julian Mitchell, Oxford. Gloster Aviation Club.
XF384	No 66 Sqn 13.09.56. No 92 Sqn 18.09.56. No 111 Sqn. DFLS. DFCS 'U' West Raynham. No 229 OCU. No 5MU Kemble in storage. No 4 FTS Valley from 03.67. Collided with XF387 W of Valley on approach. Aircraft to dump. W/O 10.08.72.
XF385	No 66 Sqn 28.08.56. No 92 Sqn 07.09.56. No 65 Sqn 'M'. No 229 OCU. Overshot at Chivenor. W/O 20.02.63.
XF386	No 66 Sqn 09.56. No 92 Sqn 'C' 10.56. No 65 Sqn 'C' 1960. No 229 OCU 4.61. Returned to HAS. Converted to GA.11. Sold to Thunder City, Cape Town, South Africa.
XF387	No 5MU Kemble. No 66 Sqn 11.10.56. No 63 Sqn 'D' 07.11.56. No 56 Sqn 'D' Waterbeach 11.58. No 234 Sqn. No 229 OCU. No 4 FTS. Collided with XF384 at Valley on approach. Pilot killed. W/O 10.08.72.
XF388	RR engine trials. CA aircraft Bitteswell 1957. No 65 Sqn. Returned to HSA 13.05.59. Converted to FGA Mk 9. Delivered 5.04.60 No 54 Sqn 'Z'. No 208 Sqn 'B' Khormaksar 24.10.61 to 23.02.64. No 19MU St Athan for refurbishment. No 8 Sqn 'C'. Crashed 50 miles SE of Dubai. Wreckage to Muharraq. W/O 26.06.68.
XF389	No 92 Sqn. No 56 Sqn 'K' Waterbeach 11.58. DFLS. DFCS 'S' West Raynham 1960. No 229 OCU. F Mk 6 but not upgraded to Interim status. Purchased by HSA 23.05.68. Converted to Jordanian FGA Mk 73A '829'. Delivered 19.05.69.
XF414	No 19MU St Athan. No 56 Sqn 'P'. No 63 Sqn 'P'. Returned to HSA 1961. Converted to FGA Mk 9. No 20 Sqn 'E'. Crashed at Layang near Tengah. W/O 20.02.67.
XF415	No 19MU St Athan. Stored 1957-58. No 26 Sqn 'J' 09.04.59. No 19MU St Athan 04.01.61. Transferred to Royal Jordanian AF 06.62 to 10.62.
XF416	No 33MU. No 43 Sqn. No 111 Sqn 'T'. 'Black Arrows' aerobatic team. Converted to FGA Mk 9. No 20 Sqn 'U' 05.01.67. No 5MU Kemble 13.02.70. 1TWU 'U' Brawdy 10.76. No 2TWU. Zimbabwe AF '1098' 10.04.84. Thornhill AB, Gweru, Zimbabwe.
XF417	No 14 Sqn 09.05.58. No 26 Sqn 'B' 09.06.58. No 19MU St Athan 04.01.61. Stored. Transferred to Royal Jordanian AF 06.62 to 10.62. Hermeskeil, Germany.
XF418	No 33MU. AFDS. DFLS 'T'. DFCS. No 229OCU No 234 Sqn and TWU Brawdy 10.76. To Laarbruch for BDRT 1984. No 92 Sqn Wildenrath for display. On closure of Wildenrath sold to Flugausstellung Leo Junior Museum at Hermeskeil.
XF419	No 19MU St Athan. No 74 Sqn 'C' Coltishall. Converted to FGA Mk 9. No 1 Sqn 'Q', later 'X'. No 58 Sqn 'L' Wittering 1973. No 1TWU '47' Brawdy. Zimbabwe AF '8112'. Thornhill AB, Gweru, Zimbabwe.
XF420	No 5MU Kemble. DFLS 'R'. DFCS 'P' West Raynham 1960. No 5MU Kemble for storage 1961-62. No 54 Sqn 11.62. No 229 OCU. Crashed following engine failure over Chawleigh, Devon. W/O 27.07.73.
XF421	No 5MU Kemble. No 247 Sqn 12.03.57. No 54 Sqn 'P' 26.03.57. Returned to HSA 27.04.59. Converted to FGA Mk 9. No 66 Sqn. No 208 Sqn 'H' Eastleigh and Khormaksar 07.04.60 to 04.11.61. No 5MU Kemble. No 8 Sqn 'C' Khormaksar 16.06.62 to 17.01.75. Overshot into sea at Khormaksar. W/O 22.02.67.
XF422	No 5MU Kemble. Stored 1957-58. No 208 Sqn 'X' 17.03.58 to 09.04.59. No 19 Sqn. Returned to HSA 04.12.59. Converted to FR Mk 10. Delivered 11.04.61. No 2 Sqn 'M' 30.06.67. No 229 OCU 20.12.67. No 2 Sqn 'H'. Purchased by HSA 02.03.71. Converted to Singapore FR. Mk 74B '524'. Delivered 22.08.72.
XF423	No 5MU Kemble. No 33MU for storage 1957-58. No 93 Sqn 08.03.60. No 19MU St Athan 17.01.61. Transferred to Royal Jordanian AF 06.62 to 10.62.

XF424	No 5MU Kemble. No 43 Sqn 'O' Leuchars. No 111 Sqn. No 247 Sqn 'V'. Returned to HSA 1959. Converted to FGA Mk 9. ff 25.09.59. Delivered 03.11.59. No 8 Sqn 'S' Khormaksar 18.01.60. DBR Sharjah 29.03.60. No 131MU 07.04.60. SOC CAT5 10.12.60.
XF425	No 5MU Kemble. No 33MU for storage 1958. No 74 Sqn 'H' Coltishall. Collided with XF502 over Cantley near Norwich. W/O 26.08.59.
XF426	No 5MU Kemble. Returned to HSA 1960. No 208 Sqn 'Y' 17.03.58 to 14.04.59. No 103MU. Converted to FR Mk 10. Delivered 24.01.61. No 229 OCU '12' Chivenor 1965. Presented to Jordan '853' 1976, later passed to Oman. Presented by Sultan of Oman to RAF Museum, Hendon.
XF427	No 5MU Kemble. No 54 Sqn. Crashed into ground 2 miles E of Yarmouth. Pilot killed. W/O 13.03.57.
XF428	No 5MU Kemble. No 208 Sqn 'S' 05.03.58 to 10.04.59. Horsham St Faith. Returned to HSA 1960. Converted to FR Mk 10. ff 02.11.60. Delivered 21.04.61. No 4 Sqn 'C' 25.01.61. No 2 Sqn 20.02.65. Purchased by HSA 07.07.71. Converted to Singapore FR Mk 74B '525'. Delivered 11.10.72.
XF429	No 33MU. Returned to HAL 1957. Converted to prototype FR Mk 1. ff 07.11.59. Delivered 07.03.66. No 1417 Flt as 'PN', later 'KS', from 14.03.67 to 30.09.67. No 8 Sqn 'X' Khormaksar then Muharraq from 30.09.67 to 1968. Purchased by HSA 04.71. Converted to Swiss Mk 58A 'J-4131'. Delivered 10.01.74.
XF430	No 5MU Kemble. No 43 Sqn. No 111 Sqn 'V'. 'Black Arrows' aerobatic team. No 5MU Kemble for storage 1960-64. Returned to HSA 1964. Converted to FGA Mk 9. Delivered 09.07.65. Harrier OCU 'N' Wittering. Purchased by HSA 16.01.75. Converted to Lebanese FGA Mk 70A 'L-283'. Delivered 17.05.77.
XF431	No 5MU Kemble. No 54 Sqn 'O'. No 66 Sqn 07.04.60. No 43 Sqn 17.06.60. Converted to F Mk 6/Interim/FGA Mk 9. Converted to FGA Mk 9. No 208 Sqn 'L' Khormaksar 10.62 to 07.67. No 8/43 Sqn. No 208 Sqn 'M' 1970-71. No 45 Sqn Wittering 1971-73. No 1TWU '43' Brawdy. Transferred to Zimbabwe AF as '1180'. Delivered 30.09.87. Thornhill AB, Gweru. Stored at Manyame (New Sarum). WFU 02.98.
XF432	No 5MU Kemble. No 208 Sqn 'E' 05.03.58 to 06.04.59. Horsham St Faith. Returned to HSA 14.10.59. Converted to FR Mk 10. ff 11.60. Delivered 21.12.60. No 2 Sqn 'S' 09.02.61. No 4 Sqn 'K' 29.01.70. Purchased by HSA 17.05.71. Converted to Singapore FR Mk 74B. Registered as G-BABM and used by HSA as demonstrator aircraft without FR nose. To Singapore as FR Mk 74B '526'. Tocumwal Airfield, NSW, Australia.
XF433	No 19MU St Athan. No 65 Sqn 'G'. No 263 Sqn. No 229 OCU. Collided with XE594 over Hartland Point, Devon. W/O 07.03.63.
XF434	No 5MU Kemble. No 247 Sqn. No 43 Sqn. Crashed 15 miles SSE of Nicosia, Cyprus, while flying out of Akrotiri. W/O 09.04.60.
XF435	No 5MU Kemble. No 247 Sqn. No 43 Sqn 'P'. Converted to F Mk 6/Interim/FGA Mk 9. Converted to FGA Mk 9. No 8 Sqn 'E' Khormaksar 15.10.62 to 02.63, recoded 'J' 02.63 to 31.08.64. No 208 Sqn 'G' 1969-71. No 229OCU. No 1TWU, No 79 Sqn 'L', later '52', Brawdy 10.76. GIA 8880M PEE Pendine, Wales.
XF436	No 5MU Kemble. No 208 Sqn 'H' 05.03.58 to 19.03.59. No 103MU. Returned to HSA 14.11.59. Converted to FR Mk 10. Delivered 24.02.61. No 8 Sqn 'U' 28.04.61 to 26.04.63. Transferred to No 1417 Flt 'SB' Khormaksar 26.04.63 to 15.07.63. No 19MU St Athan. No 1417 Flt 'JM' Khormaksar 17.04.64 to 04.07.65, recoded 'PN' 04.65 to 17.02.67. No 19MU St Athan for refurbishment. No 8 Sqn 'W' Khormaksar and Muharraq 30.09.67 to 1968. CAT4 when landing at Jeddah 29.11.69. Purchased by HSA 27.03.70. Converted to Swiss Mk 58A 'J-4115'. Delivered 01.09.72.
XF437	Delivered 10.01.57 No 5MU Kemble. No 247 Sqn 'A'. No 43 Sqn. Converted to FGA Mk 9. No 20 Sqn 'V' Tengah 05.08.66. No 5MU Kemble 22.12.69. Purchased by HSA 06.04.70. Converted to Singapore FR Mk 74A '503'. Delivered 24.06.71. Sold to Geoff Moesker, Archerfield Airfield, Queensland, Australia, 09.98. Registered as VH-FRA.

XF438	No 5MU Kemble. No 208 Sqn 'R' 05.03.58 to 20.04.59. Horsham St Faith. Returned to HSA 14.10.59. Converted to FR Mk 10. ff 11.60 Duncan Simpson. Delivered 11.01.61. No 4 Sqn 'E' 21.02.61. No 229 OCU Chivenor. 'Black Dragons' aerobatic team 1963. Purchased by HSA 3.06.70. Converted to Swiss Mk 58A 'J-4102'. Delivered 24.12.71.
XF439	No 5MU Kemble. No 247 Sqn. No 43 Sqn 'B'. No 19 Sqn. No 54 Sqn. No 1 Sqn. No 63 Sqn '3'. No 229 OCU Chivenor.
XF440	No 5MU Kemble. No 247 Sqn 'D'. No 43 Sqn 'D'. Converted to FGA Mk 9. No 8 Sqn 'S' Khormaksar 15.10.62 to 1963, recoded 'L' 1963 to 25.08.65. No 43 Sqn. Crashed at Al Ittihad, Qatar. W/O 20.02.67.
XF441	No 5MU Kemble. No 19MU St Athan. No 208 Sqn 'P' 05.03.58 to 20.04.59. Horsham St Faith. Returned to HSA 02.11.59. Converted to F Mk 6/Interim/FGA Mk 9. Converted to FR Mk 10. Delivered 07.02.61. No 2 Sqn 'O' 27.03.61. No 1417 Flt 'JD' 20.12.63 to 01.65, recoded 'FG' 01.65 to 25.08.65. No 19MU St Athan for refurbishment. No 8 Sqn 'Y' 13.03.66. No 5MU Kemble. Purchased by HSA 01.10.71. Converted to Singapore FR Mk 74B '545'. Delivered 16.08.73. Alan Nick Costin, Bankstown Airport, Sydney, Australia.
XF442	No 19MU St Athan. No 247 Sqn 'Z'. No 43 Sqn. Returned to HSA 01.06.59. Converted to FGA Mk 9. ff 10.03.60. Used for air firing trials with 68mm SNEB Matra rocket projectiles 1962. No 1 Sqn 'D'. Flew under upper span of Tower Bridge, London, 05.04.68. No 8 Sqn 1971. No 1TWU, No 234 Sqn 'V' Brawdy. No 2TWU 1980.
XF443	No 5MU Kemble. No 65 Sqn 'E' Duxford. No 66 Sqn 03.09.56. No 92 Sqn 06.09.56. No 229 OCU '51'. Crashed at Framington 1 mile SE of Chivenor. W/O 03.08.67.
XF444	No 5MU Kemble. Supplied to Jordan '705' under US Offshore Procurement arrangement 14.11.58.
XF445	No 5MU Kemble. No 263 Sqn. No 1 Sqn 'Z'. Converted to F Mk 6/Interim/FGA Mk 9. Converted to FGA Mk 9. No 1 Sqn 'Z'. No 208 Sqn 'O' 29.08.62 to 28.02.63. No 8 Sqn 'O' Khormaksar 28.02.63 to 31.07.63. No 208 Sqn 18.05.67 to 17.11.68. No 19MU St Athan. No 208 Sqn 'A' 13.06.69 to 18.05.71. No 5MU Kemble. No 1TWU 'Q' Brawdy. Returned to HSA for refurbishment and passed to Chile as 'J-747'.
XF446	No 19MU St Athan. No 111 Sqn 'R'. 'Black Arrows' aerobatic team. No 5MU Kemble for storage 1960-64. Returned to HSA 12.64. Converted to FGA Mk 9. No 54 Sqn 'B'. Damaged in collision with XG273 and swung off runway on landing at El Adem, Libya. CAT4 18.04.67. CAT5 18.10.67. Remained at El Adem until purchased by HSA 06.68. Converted to Indian Mk 56A 'A1010'. Delivered 23.09.69.
XF447	No 33MU. No 65 Sqn 'H'. No 66 Sqn 11.09.56. No 92 Sqn 18.09.56. No 229 OCU Chivenor. 'Black Dragons' aerobatic team 1963. Purchased by HSA 17.09.69. Converted to Chilean FGA Mk 71 'J-723'. Delivered 31.12.70.
XF448	No 5MU Kemble. No 19MU St Athan for storage 1958. No 74 Sqn 'F'. Crashed into sea off Winterton, Norfolk. Pilot killed, wreckage never recovered. W/O 21.08.58.
XF449	No 5MU Kemble. No 263 Sqn. No 19 Sqn 'S' Church Fenton 1957. No 92 Sqn 'S'. 'Blue Diamonds' aerobatic team. CFE. Fire while taxiing at Binbrook. DBR. W/O 06.06.63.
XF450	No 5MU Kemble. DFCS. No 74 Sqn. No 229 OCU. Returned to HSA 01.04.66. Converted to Saudi Arabian F Mk 60 '60/603' 1966. To RJAF as '723'. Given to Oman 1975. Gate guard at Thumrait AB.
XF451	No 5MU Kemble. No 247 Sqn. No 229 OCU. Borrowed by No 92 Sqn. 'Blue Diamonds' aerobatic team 1961. Suffered loss of power on overshoot at Chivenor. W/O 20.07.62.
XF452	No 19MU St Athan. CA aircraft AWA Bitteswell 1957. A&AEE miscellaneous trials 1958. Supplied to Jordan '708' under US Offshore Procurement arrangement 12.11.58.
XF453	No 5MU Kemble. No 54 Sqn 'S'. No 247 Sqn. DFCS 'V' West Raynham. Purchased by HSA 14.07.67. Converted to Chilean FR Mk 71A 'J-716'. Delivered 01.08.68.

XF454 No 19MU St Athan. No 247 Sqn. No 43 Sqn. Returned to HSA 17.03.59. Converted to FGA Mk 9. ff 24.11.59. Delivered 1960. No 1 Sqn 'E'. No 208 Sqn 'F' Khormaksar 29.11.61 to 23.11.63. No 5MU Kemble. No 8/43 Sqn 'A' 1965. Transferred to Royal Jordanian AF '816' 12.01.68.

XF455 No 5MU Kemble. No 247 Sqn 'B'. No 43 Sqn. Converted to FGA Mk 9. ff 07.11.59. Delivered 03.12.59. No 8 Sqn 'T' Khormaksar 05.02.60 to 18.10.62. No 28 Sqn 'E'. No 20 Sqn 10.03.64. Crashed into sea S of Changi, Singapore. Pilot killed. W/O 19.09.64.

XF456 No 5MU Kemble. No 247 Sqn. No 43 Sqn 'A'. Converted to F Mk 6/Interim/FGA Mk 9. Converted to FGA Mk 9. No 43 Sqn 'P' Khormaksar and Sharjar 01.64 to 04.03.65. No 43 Sqn 'S' Khormaksar 1967-68. No 5MU Kemble. No 229 OCU. No 79 Sqn 'S'. Purchased by HSA 09.10.69. Converted to Singapore FGA Mk 74 '509'. Delivered 14.01.71.

XF457 No 66 Sqn 'Y' 24.04.59. No 19MU St Athan 06.08.59. No 5MU Kemble for storage. No 208 Sqn 'V' 14.03.58 to 14.04.59. Horsham St Faith. Returned to HAL and converted to FR Mk 10. Delivered 30.01.61. No 2 Sqn 'A', later 'T', 15.03.61. No 2 Sqn 26.02.70. No 4 Sqn 'U'. No 5MU Kemble. Purchased by HSA 23.07.74. Converted to Lebanese FGA Mk 70A 'L.281'. Delivered 08.12.75.

XF458 No 5MU Kemble. Stored 1958. No 208 Sqn 'Z' 17.03.58 to 21.04.59. No 130MU. Converted to FR Mk 10. No 2 Sqn 'W'. No 4 Sqn 'J' 01.09.65. No 2 Sqn 'I' 11.05.66. Purchased by HSA 17.05.71. Converted to Singapore FR Mk 74B '527'. Delivered 11.10.72. RSingAF Museum, Paya Lebar, Singapore.

XF459 No 5MU Kemble. No 208 Sqn 'T' 12.03.58 to 20.04.59. No 19 Sqn. Returned to HSA 1960. Converted to FR Mk 10. ff 19.09.60. Delivered 01.12.60. No 4 Sqn 'F' 03.02.61. No 2 Sqn 'T' 22.02.70. Purchased by HSA 02.03.71. Converted to Indian T Mk 66E 'S.1393'. Delivered 13.12.73.

XF460 Delivered 06.02.57 No 5MU Kemble. No 208 Sqn 'T'. Converted to FR Mk 10. Delivered 28.03.61. No 131MU Khormaksar, Command Reserve, 16.08.62 to 09.03.63. No 8 Sqn 'X' Khormaksar 09.03.63 to 26.04.63. No 1417 Flt 'AR' Khormaksar 26.04.63 to 1964, recoded 'GT' until 28.03.66. No 19MU St Athan for refurbishment. No 1417 Flt 'DB' Khormaksar 26.10.66 to 30.09.67. No 8 Sqn 'Y' Khormaksar then Muharraq 30.09.67 to 1968. No 5MU Kemble. Re-purchased by HSA 10.71. Converted to Singapore FR Mk 74B '546'. Delivered 11.10.73.

XF461 No 5MU Kemble. Supplied to Lebanon 'L-171' under US Offshore Procurement arrangement 12.11.58.

XF462 No 5MU Kemble. No 66 Sqn 'N' 12.02.58, later '66', Acklington 20.09.58. Returned to HSA 17.04.59. Converted to FGA Mk 9. ff 30.12.59. Delivered 04.02.60. No 1 Sqn 'Y'. No 208 Sqn 'C' Khormaksar 20.10.61 to 05.02.64. No 5MU Kemble for refurbishment. No 208 Sqn 'C' 20.08.64 to 17.11.66. No 19MU St Athan. No 8 Sqn 'F'. Purchased by HSA 24.03.70. Converted to Swiss Mk 58A 'J-4107'. Delivered 12.04.72.

XF463 No 5MU Kemble. Returned to HAL 1957. Converted to Indian Mk 56 'BA241'. ff 21.02.58. Delivered 13.05.58. No 27 Sqn IAF Ambala.

XF495 No 5MU Kemble. Supplied to Lebanon 'L-175' under US Offshore Procurement arrangement 13.10.58.

XF496 No 5MU Kemble. Supplied to Jordan '706' under US Offshore Procurement arrangement 12.11.58.

XF497 No 5MU Kemble. Returned to HAL 1957. Converted to Indian Mk 56 'BA242'. ff 27.02.58. Delivered 12.05.58. No 27 Sqn IAF Ambala.

XF498 No 5MU Kemble. Supplied to Jordan '704' under US Offshore Procurement arrangement 12.11.58.

XF499 No 5MU Kemble. Returned to HAL 1957. Converted to Indian Mk 56 'BA243'. ff 10.03.58. Delivered 15.05.58. Crashed. W/O 1959.

XF500 No 5MU Kemble. Returned to HAL 1957. Converted to Indian Mk 56 'BA240'. ff 14.02.58. Delivered 12.05.58. No 27 Sqn IAF Ambala.

XF501	No 5MU Kemble. Returned to HAL 1957. Converted to Indian Mk 56 'BA244'. ff 06.03.58. Delivered 15.05.58. No 27 Sqn IAF Ambala.
XF502	No 5MU. Stored 1958. No 74 Sqn 'K'. Collided with XF502 from Coltishall over Cantley near Norwich. Crashed into wood near Langley, Norfolk. Pilot killed. W/O 26.08.59.
XF503	No 5MU Kemble. Returned to HAL 1957. Converted to Indian Mk 56 'BA245'. ff 27.03.58. Delivered 09.06.58. No 27 Sqn IAF Ambala.
XF504	No 5MU Kemble. Stored 1958. No 74 Sqn 'B'. Converted to Rhodesian FGA Mk 9 '125'. Delivered 19.04.63. Thornhill AB, Gweru, Zimbabwe.
XF505	No 5MU Kemble. Returned to HAL 1957. Converted to Indian Mk 56 'BA246'. ff 21.03.58. Delivered 19.05.58. No 27 Sqn IAF Ambala.
XF506	No 5MU Kemble. No 111 Sqn 'X'. 'Black Arrows' aerobatic team. No 263 Sqn 'J'. No 1 Sqn 'J' Stradishall. Converted to Rhodesian FGA Mk 9 '119'. Delivered 12.03.63. W/O 25.07.82, sabotage at Thornhill. West-Sale, Australia, 09.00.
XF507	No 5MU Kemble. No 66 Sqn 10.10.56. No 65 Sqn 'A' Duxford 28.11.56. Crashed near Thrapston, Northants. W/O 30.05.60.
XF508	No 19MU St Athan. No 54 Sqn. No 111 Sqn. AFDS. Returned to HSA 15.11.60. Converted to FGA Mk 9. Delivered 19.07.61 No 5MU Kemble. No 20 Sqn 'D' 06.10.61. Crashed into sea near Pontain, Johore. W/O 03.02.69.
XF509	No 5MU Kemble. No 54 Sqn 'M'. Bristol, chase plane for Fairey FD-2 17.04.63. No 1 Sqn. No 65 Sqn. No 263 Sqn. No 229 OCU. No 4 FTS 1968. AFDS. No 54 Sqn. CAT3 Larnaca Range, Cyprus. GIA 8708M Chivenor. Purchased by Humbrol Ltd, Kingston-upon-Hull. Derelict 07.
XF510	No 19MU St Athan. North Weald Stn Flt. Crashed on approach to North Weald. Pilot killed. W/O 30.07.57.
XF511	No 19MU St Athan. No 74 Sqn 'P'. Horsham St Faith 1960. No 111 Sqn 'P' 21.06.60. Converted to FGA Mk 9. No 208 Sqn 'F' Muharraq 1966 to 08.67. No 8/43 Sqn. No 208 Sqn 'F' 1969 to 09.71. No 5MU Kemble. No 1TWU, No 79 Sqn 'S' Brawdy 10.76. No 2TWU 'A' Lossiemouth 1981.
XF512	No 33MU. DFLS. Damaged 19.03.57. Repaired. Purchased by HSA 02.12.69. Converted to Chilean FGA Mk 71 '725'. Delivered 24.03.71.
XF513	No 5MU Kemble. No 54 Sqn 'J'. Overshot landing at Nicosia 27.08.58. Returned to Odiham but DBR. W/O Odiham 01.09.58.
XF514	No 5MU Kemble. No 247 Sqn 'P'. No 43 Sqn. DFCS 'F'. Transferred to Royal Jordanian AF 01.68.
XF515	Delivered 04.12.56 No 5MU Kemble. No 247 Sqn 'W'. No 43 Sqn 'R' Leuchars. Upgraded to F Mk 6A. No 229 OCU. Leuchars 1954-59. Brawdy 10.76. Scampton for storage. 8830M. Auctioned to Global Aviation Ltd, UK, 12.94. Registered G-KAXF. Owned by Tim Manna, Kennet Aviation, Cranfield, Beds. Hunter Flying, Exeter, 09.07.
XF516 ff 05.06.56	Delivered 30.10.56 No 19MU St Athan. No 66 Sqn 'E' 27.11.56 to 09.01.57. No 92 Sqn 'N'. No 56 Sqn 'E'. Damaged in accident at Linton 08.02.57. No 229 OCU 17.03.61. Highest flying hours (5,096) of any Hunter in RAF service 02.80. GIA Cranwell. Sold to P. Helier, Maidenhead. Exeter for restoration 10.94. Crashed off Aberdovey as G-BVVC. W/O 01.06.03.
XF517	No 19MU St Athan. No 66 Sqn 'G' 29.11.56. No 92 Sqn 'L' 04.01.57. No 111 Sqn. No 66 Sqn 'G' 07.04.60. Converted to F Mk 6/Interim/FGA Mk 9 at Horsham St Faith. No 54 Sqn 'V' 28.10.60. Returned to HSA 1961. Converted to FGA Mk 9. No 1 Sqn West Raynham. Crashed off Blakeney, Norfolk. Pilot killed. W/O 15.01.69.
XF518	No 33MU. No 66 Sqn 04.12.56. No 92 Sqn 09.01.57. Transferred to Royal Jordanian AF 06.62 to 10.62.

XF519 No 19MU St Athan. No 66 Sqn 'F' 08.58. CAT2 Acklington 10.09.58. Repaired. Converted to F Mk 6/Interim/FGA Mk 9 Horsham St Faith. No 1 Sqn 'Q' 10.60. Converted to FGA Mk 9. No 208 Sqn 'C' 1969 to 09.71. No 8/43 Sqn. No 1TWU '44' Brawdy. No 45 Sqn 'N' 1971-74. To Zimbabwe AF. Delivered. 10.04.84. Thornhill AB, Gweru, Zimbabwe. Crashed near Beatrice. Pilot killed. W/O 19.02.95.

XF520 Armstrong Whitworth at Bitteswell for ground sound suppression trials with engine muffler pens 1956. No 5MU Kemble. No 92 Sqn 'K' Middleton St George. Purchased by HSA 06.66. Converted to Jordanian FGA Mk 73 '814'. Delivered 13.06.68.

XF521 No 5MU Kemble. No 66 Sqn 'E' 27.11.56. No 103MU, Nicosia, 04.07.57. No 92 Sqn 'RD', later 'X', from 10.58 to 10.60. 'Falcons', reserve team for Fighter Command. Purchased by HSA 06.09.67. Converted to Indian Mk 56A 'A938'. Delivered 22.03.69.

XF522 No 5MU Kemble. No 1 Sqn. No 92 Sqn 'D' Middleton St George. No 66 Sqn 27.11.56. No 92 Sqn 20.12.56. 'Blue Diamonds' aerobatic team 09.61. Dismantled for spares. Fuselage at Benson 09.64. GIA Halton. Aylesbury Fire Service. Buckinghamshire Ambulance Service, Aylesbury. Nose preserved, No 1365 Sqn ATC TAVR Centre, Aylesbury, Bucks. No 2532 (Milton Keynes) Sqn ATC.

XF523 No 5MU Kemble. No 54 Sqn 'N' Waterbeach 1962-63. Converted to F Mk 6/Interim/FGA Mk 9. Converted to FGA Mk 9. No 54 Sqn. Dived into ground, Benina airport, Libya. Pilot killed. W/O 24.06.63.

XF524 No 5MU Kemble. No 111 Sqn. No 54 Sqn 'N', later 'S'. Crashed into hill 15 miles SE of Nicosia, Cyprus. Pilot killed. W/O 05.11.57.

XF525 No 5MU Kemble. No 19 Sqn 'O'. No 263 Sqn. No 19 Sqn 'O'. On loan to No 111 Sqn. Collided with XE621 and crashed at North Weald. Pilot killed. W/O 08.06.57.

XF526 Delivered 07.09.56 No 5MU Kemble. No 56 Sqn 'C' Waterbeach 11.58. No 63 Sqn 'C' 1959. On loan to No 66 Sqn 1959. No 63 Sqn 'C' 1959. No 56 Sqn 'C'. No 43 Sqn. No 56 Sqn. No 229 OCU '9'. No 4 FTS '78' and 'E' Valley. Laarbruch Stn Flt. GIA 8679M Halton. Scrapped St Athan. Sold to Graham Revill, Birlingham, Worcs.

XF527 Delivered 17.10.56 No 19MU St Athan. No 19 Sqn '1'. No 66 Sqn 'S'. No 111 Sqn 'F'. AFDS 'T'. Laarbruch Stn Flt. 4FTS. CFE. Church Fenton Stn Flt. Linton-on-Ouse Stn Flt. GIA 8680M No 1 SoTT. Gate guard Halton 1998.

Fifty F Mk 6 aircraft to have been built by Hawker Aircraft (Blackpool) Ltd under Contract No SP/6/13132/CB.7a dated 22.04.55. Contract cancelled 02.57. Aircraft serials allocated as XJ945-XJ959 (fourteen aircraft), XJ971-XJ997 (twenty-seven aircraft) and XK103-XK111 (nine aircraft).

Hawker Hunter T Mk 7

Production batch of forty-five new-build aircraft built by Hawker Aircraft Ltd, Kingston-upon-Thames, under Contract No 6112626/CB.9c. Ten airframes diverted to Admiralty as T Mk 8s.

XL563 ff 11.10.57 CA aircraft HAL. Delivered 19.12.57 A&AEE. Bristol, as chase plane for Bristol Type 188 trials 13.01.61. RAF Institute of Aviation Medicine. 9218M.

XL564 ff 08.01.58 CA aircraft HSA. Delivered 10.03.60 A&AEE. DHF trials A&AEE 05.10.61. Empire Test Pilots School. Crashed at Boscombe Down 10 miles out. W/O 06.08.98. Nose preserved in burnt-out state at Boscombe Down.

XL565 ff 03.02.58 CA aircraft RR. No 8 Sqn West Raynham Stn Flt. No 208 Sqn 'Y' Khormaksar 12.11.60 to 26.04.63. No 19MU St Athan for refurbishment. No 1417 Flt 'Y' Khormaksar 26.04.63 to 01.08.64. No 19MU St Athan for refurbishment. No 208 Sqn 'S' Muharraq. Nos 6/43 Sqns 'TX' Khormaksar 26.04.66 to 06.11.67. No 19MU St Athan. No 4FTS. No 237 OCU. FAA. FRADU, Yeovilton and later Lossiemouth. Shawbury, later Kemble, for storage. Rear fuselage of Swedish 34006, forward fuselage of WT745 and components of ET-271. Ed Stead, Manchester, NH, USA. Crashed during post-restoration test flight. W/O.

XL566 ff 20.02.58	CA aircraft HSA. Delivered 4.58 A&AEE. Tropical trials, Bahrain 20.06.58. No 43 Sqn Leuchars, Nicosia and Khormaksar 'T' 15.12.59 to 02.06.63. To UK for refurbishment. Nos 8/43 Sqns 'TW' Khormaksar 17.04.64 to 29.04.67. No 208 Sqn 'Z', later 'Y', Muharraq 01.68. No 43 Sqn Laarbruch Stn Flt 1980-81. Used during temporary grounding of RAF Buccaneer fleet. 8891M. Scrapped at Bruggen 1995.
XL567 ff 27.02.58	Built by HAL at Kingston-upon-Thames. Delivered 30.05.58 No 229 OCU 'Z'. Instrument Rating Sqn 'Z' West Raynham 1960. No 4FTS '84'. Laarbruch Stn Flt 1980-81. Used during temporary grounding of RAF Buccaneer fleet. 8723M. Classic Jets, Exeter.
XL568 ff 06.03.58	Delivered. 20.10.58 No 74 Sqn 'X' Coltishall. Flew fastest time in Daily Mail London-Paris air race 07.59. Converted to T Mk 7A 1963. No 237 OCU 1971. No 12 Sqn Laarbruch Stn Flt. 1980.81. Used during temporary grounding of RAF Buccaneer fleet. No 12 Sqn 'ZF'. No 208 Sqn 04.94. Lossiemouth. GIA 9224M Cranwell. Held in store for RAF Museum until 28.02.02 then moved to Cosford. Cold War Hall, Cosford.
XL569 ff 03.04.58	Delivered 01.07.58 No 229 OCU 'ES-85'. No 1TWU. No 2TWU. Ministry of Technology. No 12 Sqn. No 237 OCU. No 15 Sqn Laarbruch Stn Flt. No 237 OCU. No 216 Sqn. No 12 Sqn. No 1TWU. GIA 8833M Cosford, Scampton, Abingdon. Purchased by Aeropark Volunteers Association (AVA). East Midlands Airport 03.02.
XL570 ff 15.03.58	Delivered 30.06.58 No 229 OCU. Crashed into sea. Pilots killed. W/O 25.09.58.
XL571 ff 29.03.58	Delivered 22.07.58. IRS 'X'. No 92 Sqn 'V'. 'Blue Diamonds' aerobatic team. No 229 OCU Chivenor 'Q' 1967. No 229 OCU. Brawdy 11.76. Crashed off Strumble Head, Wales. W/O 08.09.77.
XL572 ff 02.04.58	Delivered 01.07.58 No 229 OCU 'ES-83'. Inverted spin, one pilot killed after ejection, second pilot recovered aircraft to base 27.08.59. Fighter Command Instrument Rating Sqn 'W' West Raynham 1960. Nos 1 and 2TWUs. GIA 8834M Cosford. Hunter Wing Ltd, Jersey. Registered as G-HNTR 07.07.89. Deregistered from CAA 11.10.91.
XL573 ff 17.04.58	Delivered 17.07.58 DFLS. AFDS. DFCS 'L'. No 4 FTS. No 237 OCU. No 229 OCU 'WC'. Laarbruch Stn Flt 1980-81. Used during temporary grounding of RAF Buccaneer fleet. Shawbury for storage. North Weald as B-BVGH.
XL574 ff 18.04.58	Delivered 13.11.58 A&AEE. CA aircraft. HSA windscreen rain dispersal and de-icing trials and fatigue tested to destruction 11.61.
XL575 ff 29.04.58	Delivered 07.58 No 229 OCU 'ES-88'. Instrument Rating Sqn 'Y'. No 229 OCU '80' Chivenor 1970. No 4 FTS Valley. Crashed at Devils Bridge, Dyfed, Wales. Pilots killed. W/O 08.11.71.
XL576 ff 30.04.58	Delivered 25.07.58 No 229 OCU 'ES-81'. Brawdy 11.76. Registered with FAA as N576NL, Wilmington, DE, USA 21.07.99. W/O 17.11.99.
XL577 ff 30.04.58	Delivered 15.07.58 No 229 OCU Chivenor 'ES-82'. No 234 Sqn, No 1TWU. No 237 OCU. No 2TWU Lossiemouth. GIA 8676M Cranwell 03.81. Sold to Ronan Harvey, Delta Jets, 1996. Restored as G-BXKF 20.09.2004.
XL578 ff 27.05.58	Delivered 15.07.58 No 229 OCU Chivenor 'ES-89'. No 1TWU, No 145 Sqn Brawdy 11.76. St Athan. Cranfield. Bruntingthorpe. Norwich. Stored 2002.
XL579 ff 13.05.58	Delivered 29.07.58 No 229 OCU Chivenor 'ES-92' 10.58 to 12.65. No 145 Sqn ETPS Farnborough. Fitted with nose probe 27.09.68. Abandoned over Winterbourne Gunner on approach to Boscombe Down. W/O 22.01.76.
XL583 ff 28.05.58	Delivered 01.07.58 No 229 OCU 'ES-84'. Brawdy 11.76. Crashed on approach to Brawdy. W/O 01.12.81.
XL586 ff 27.06.58	Delivered 6.08.58 No 229 OCU 'ES-90'. No 1TWU. No 2TWU. Warton IRS. Shawbury, later Kemble, for storage.
XL587 ff 17.07.58	Delivered 6.08.58 No 145 Sqn/No 229 OCU 'ES-91'. Brawdy '86' 11.76. Laarbruch Stn Flt 1980-81. Used during temporary grounding of RAF Buccaneer fleet. 8807M. Old Flying Machine Co, Duxford, until 13.12.00. G-HPUX. Hawker Hunter Aviation, Scampton.

XL591 ff 06.08.58	Delivered 19.09.58 No 5MU Kemble. DFCS 'M' 22.10.58. No 229 OCU. No 5MU Kemble 23.11.65. No 4 FTS '82' 08.03.67. CAT3 flying accident 07.01.70. No 60MU 26.01.70. No 4FTS 16.04.70. RAE Farnborough. No 4 FTS 04.02.77. No 5MU Kemble 31.05.78. No 4 FTS 4.10.78. No 5MU Kemble 19.06.79. No 237 OCU 27.07.79. No 237 OCU 02.04.80. No 208 Sqn Honington 01.10.80. No 208 Sqn Lossiemouth 14.07.83. Abingdon for major servicing 17.04.84. No 237 OCU 29.10.84. No 208 Sqn Lossiemouth 11.02.87. No 237 OCU 31.03.87. No 208 Sqn 29.04.87. No 237 OCU 02.08.88. No 208 Sqn Shawbury 14.10.88. SOC. Gatwick Aviation Museum.
XL592 ff 30.08.58	Delivered 29.09.58 No 229 OCU Chivenor 'ES-93'. No 1TWU '84' Brawdy 11.76. 8836M Scampton. Hunter Flying Club/Classic Jet Aircraft Company, Exeter. To Delta Jets, Kemble, 07.02.
XL593 ff 30.08.58	Delivered 29.09.58 DFCS. AFDS 'N'. No 229 OCU '96'. No 1TWU Brawdy. Lost power 4 miles NW of Carmarthen, Dyfed. W/O 05.08.82.
XL594 ff 10.09.58	Delivered 06.10.58 IRS. No 19 Sqn Leconfield 13.07.59. No 111 Sqn 20.01.61. Crashed NW of disused airfield at Carnaby, Yorkshire. Pilot killed. W/O 16.04.64.
XL595 ff 04.09.58	Delivered 09.10.58 AFDS. DFCS 'O' Brawdy 11.76. Civil registered G-BTYL by Cubitt Aviation 29.11.91. Crashed at Broomhead Moor, Yorkshire. Pilot killed. W/O 11.06.93.
XL596 ff 17.09.58	Delivered 06.08.58 No 54 Sqn 'LIV'. No 4 FTS. Harrier Conversion Unit '1' 04.06.70. No 4 Sqn 25.08.70. Crashed at Shawbury. Pilots killed. W/O 02.11.73.
XL597 ff 22.09.58	Delivered 31.10.58. No 66 Sqn 'Z' 20.01.59. CAT4. Returned to HAL 25.03.59. No 208 Sqn 'N' Eastleigh and Khormaksar 12.11.60 to 26.04.63. No 1417 Flt 'W' Khormaksar 26.04.63 to 1964. To UK for refurbishment. No 208 Sqn 'Y' Khormaksar 02.02.65 to 10.04.67. No 19MU St Athan. No 208 Sqn 08.02.68 to 06.07.68. No 4 Sqn 03.11.70. No 20 Sqn 28.08.70. Crashed S of Little Saxham, Bury St Edmunds. W/O 29.05.80.
XL600 ff 07.10.58	Delivered 14.11.58 No 65 Sqn Wattisham Stn Flt. No 4 FTS '83' 1967. Laarbruch Stn Flt 1980-81. Used during temporary grounding of RAF Buccaneer fleet. Registered G-BVWN 23.12.94. Deregistered and transferred to G-VETA 02.07.96. Sold to Skyblue Aviation Ltd. Registered 17.02.06.
XL601 ff 15.10.58	Delivered 03.11.58 No 1 Sqn Stradishall. No 19 Sqn Waterbeach and West Raynham. Pilot lowered undercarriage at more than 300 knots, airframe twisted 07.1959. No 233 OCU Wittering. No 4 FTS Valley '88'. No 237 OCU Honington and Lossiemouth 1980. Royal Navy 1985. FRADU '879', later '874'. Culdrose as 'A2617' 03.93 until 1996. Stored at Weelde AB, Belgium. Registered as G-BZSR. Deregistered from CAA register 15.04.2004.
XL605 ff 14.10.58	Built at HAL, Kingston-upon-Thames. Delivered No 5MU 01.12.58. No 92 Sqn 'T' Middleton St George. No 66 Sqn Linton-on-Ouse. No 229 OCU Chivenor 14.10.65. Purchased by HSA 6.04.66. Converted to Saudi Arabian T Mk 70 '70/617'. ff 20.05.66. Delivered to RSAF 07.06.66. Delivered to Royal Jordanian AF '836' 06.68. Returned to RAF in exchange for FR Mk 10 07.74. Delivered to No 229 OCU Chivenor as XX467 '92'. No 1TWU Brawdy 09.74. No 2TWU Lossiemouth 1980. SOC 14.10.83. Hunter Flying Club, rebuilt with components from XF358 G-TVII. Stored at Exeter.
XL609 ff 01.12.58	Delivered 22.12.58. No 56 Sqn 1962. returned to HSA for conversion to T Mk 7A 01.04.65. Contract cancelled and aircraft returned to RAF. No 5MU Kemble. No 4 FTS '80' from 04.67. No 237 OCU. No 216 Sqn. No 12 Sqn. GIA 8866M Lossiemouth. Broken up at Lossiemouth. Cockpit section to South Yorkshire Air Museum. Sold to Roger Fallows, Yarmouth, Isle of Wight 06.06.
XL610 ff 12.11.58	Delivered 23.01.59 No 111 Sqn. 'Black Arrows' aerobatic team. Wattisham Stn Flt. No 56 Sqn. Crashed at North Kelsey, near Sleaford. Pilots killed. W/O 07.06.62.
XL611 ff 11.12.58	Delivered 20.01.59 No 5MU Kemble. No 43 Sqn. Converted to T Mk 7A 1963. No 43 Sqn A&AEE. Crashed on approach to Boscombe Down. Pilots killed. W/O 14.05.68.

XL600 spent some time in the markings of No 15 Squadron, acting as a continuation trainer while the unit's Buccaneers were temporarily grounded. More recently, the aircraft has reappeared in civilian hands as G-VETA. *Terry Senior*

Pictured as the aircraft is about to take-off on an air test, XL609 is seen at Kemble having just emerged from deep servicing. The aircraft has been repainted in a glossy Light Aircraft Grey but has yet to receive 'stick-on' dayglow orange trim. *Ray Deacon*

XL612 ff 12.12.58	Delivered 23.01.59 No 5MU Kemble. No 402 WTS Sylt 19.03.59. No 19MU St Athan 09.06.61. No 1417 Flt 'X' Khormaksar 26.08.63 to 01.05.66. To UK for refurbishment. No 1417 Flt 'TY' Khormaksar 18.05.67 to 27.09.67. No 8 Sqn 'T' Khormaksar and Muharraq 27.09.67 to 1968. ETPS Boscombe Down. Landing accident at Boscombe Down 28.02.84. CAT2. Last flight 10.08.01. Boscombe Down.
XL613 ff 19.12.58	Delivered 23.01.59 No 5MU Kemble. No 8 Sqn 'Z' Khormaksar 22.09.59 to 01.06.60. Loaned to No 208 Sqn Eastleigh. No 8 Sqn 'Z' Khormaksar 29.11.60 to 26.04.63. No 1417 Flt 26.04.63 to 25.06.63. No 8 Sqn 25.06.63 to 29.10.63. No 1417 Flt 'Z' 29.10.63 to 28.10.65. No 43 Sqn. No 208 Sqn 'Z' Muharraq 1968. Laarbruch Stn Flt 1980-81. Used during temporary grounding of RAF Buccaneer fleet. Ferried to Shawbury 1991. Sold to Mike Beachyhead G-BZRH. Departed Exeter 17.06.01. Based with Thunder City fleet. Registered in South Africa as ZU-LEE 23.05.03.
XL614 ff 01.01.59	Delivered 03.02.59 No 5MU Kemble. Converted to T Mk 7A. No 402 WTS Sylt 01.05.59. No 19MU St Athan 08.09.61. No 237 OCU 'O'. To USA as N614XL 05.96. W/O 22.07.03.
XL615 ff 07.01.59	Delivered 27.01.59 No 19MU St Athan. Khormaksar Stn Flt. No 8 Sqn 'Y' Khormaksar 22.09.59 to 01.06.60. Crashed near Little Aden 13 miles W of Khormaksar. Pilots killed. W/O 01.06.60.
XL616 ff 03.01.59	Delivered 03.02.59 No 19MU St Athan. Converted to T Mk 7A. No 402 WTS Sylt 01.05.59. No 19MU St Athan 06.09.61. No 19 Sqn 03.02.66. No 5MU Kemble 20.04.66. Laarbruch Stn Flt 1980-81. Used during temporary grounding of RAF Buccaneer fleet. No 5MU Kemble. No 208 Sqn/No 1TWU Brawdy 1985. 9223M. Registered G-BWIE. To Sweden registered as SE-DXH with KBAFT Avancerad Flygträning, Hässlö, owned by Lansen Ltd.

Still painted in the colours of the RAE to which the aircraft had previously been assigned, XL616 became a continuation trainer at Laarbruch, 'pooled' between the resident Buccaneer squadrons. The Laarbruch VAS motif is applied to the nose. *Terry Senior*

XL617 ff 03.01.59 Delivered 27.01.59 No 5MU Kemble. No 4 Sqn. Jever Stn Flt 'A'. Gutersloh Stn Flt. No 229 OCU Chivenor '95' 07.64. GIA 8837M Cosford. Terry Elder, Grace Aire Inc, Corpus Christi, TX, USA, 05.94. Registered as N617NL. Sold 2005. Santa Fe, NM, USA.

XL618 ff 10.01.59 Delivered 04.02.59 No 5MU Kemble. Gutersloh Stn Flt. Jever Stn Flt. No 14 Sqn. No 229 OCU 04.64. No 1TWU '87' Brawdy 11.76. No 5MU Kemble. Shawbury for storage. GIA 8892M '05' Cottesmore. Caernarfon, Wales.

XL619 ff 07.01.59 Delivered 03.02.59 No 5MU Kemble. Gutersloh Stn Flt. No 2TAF 12.05.59. No 20 Sqn 'S' Tengah 22.05.62. No 5MU Kemble 02.07.70. No 14 Sqn 01.01.61. No 19MU St Athan 29.03.61. No 45 Sqn 'S' Wittering 07.09.72. Hunter Wing 06.02.75. TWU Brawdy 28.07.76. No 2TWU Lossiemouth 1979. No 1TWU Brawdy. Crashed S of Brawdy. W/O 21.10.81.

XL620 ff 13.01.59 Delivered 03.02.59 No 5MU Kemble. No 66 Sqn 'Z' Acklington 02.06.59. Fighter Command Pool 14.10.60. No 74 Sqn 'Z' Coltishall. SOC 31.03.66. Purchased by HSA 31.03.66. Converted to Saudi Arabian T Mk 70 '70.616'. Delivered 02.05.66. Royal Jordanian AF as '835'. Returned to HSA and RAF 07.74 as XX466. No 229 OCU. No 1TWU Brawdy. FAA 2.84. FRADU, Yeovilton, '879/VL' until 10.84. Shawbury for storage until 1986. Culdrose, School of Aircraft Handling 'A2738' 14.05.86. Recoded '830/DD' SAH 07.86. Retired 1993. Predannack 08.93 for fire crash rescue training until 06.96. Destroyed by 1996.

XL621 ff 09.01.59 Delivered 04.02.59 No 5MU Kemble. Gutersloh Stn Flt 'C' 08.59. Jever Stn Flt 04.61. No 4 Sqn 'L'. No 4 FTS '81' 1973. No 237 OCU. RAE Bedford. G-BNCX. Deregistered permanently 01.03.93. Dunsfold Aerodrome permanent exhibit.

XL622 ff 15.01.59 Delivered 04.02.59 No 5MU Kemble. No 93 Sqn 'J' 18.06.59. Jever Stn Flt 'J' 01.01.61. Stn Flt APS Sylt 07.04.61. No 5MU Kemble 18.08.61. No 4 FTS 1967. Crashed at Blaenau Ffestiniog. Pilots killed. W/O 17.05.71.

XL623 ff 17.01.59 Delivered 04.02.59 No 5MU Kemble. No 65 Sqn. No 208 Sqn. No 43 Sqn. No 1 Sqn. No 74 Sqn. No 1TWU '88' Brawdy 11.76. No 92 Sqn. No 19 Sqn. GIA 8770M Newton. Woking, Surrey, mounted on pole outside 'The Big Apple'.

Conversion order for six F Mk 4s to be brought up to T Mk 7 standard placed with Hawker Aircraft Ltd and Armstrong Whitworth Aircraft Ltd during 1957-58.

WV253 ff 29.04.55 Delivered 26.05.55. Converted to T Mk 7. Delivered 12.06.62 ETPS Farnborough. Fitted with nose instrumentation boom. Crashed in Lyme Bay. W/O 15.07.68.

WV318 ff 23.05.55 One of three F Mk 4s with wing leading edge extensions. Delivered 16.06.55 No 5MU Kemble. No 33MU Lyneham 07.55. No 14 Sqn Oldenburg. No 93 Sqn Jever 27.03.56. HAL for repairs 1959. Converted to T Mk 7 and delivered to No 5MU Kemble for storage 01.06.59. On loan to No 111 Sqn 03.60, later to CFS Little Rissington. Converted to T Mk 7A 10.64. No 74 Sqn. No 12 Sqn Honington 02.70. Nos 15 and 16 Sqns Laarbruch 06.71. No 208 Sqn Lossiemouth. Delta Jets, Kemble, as G-FFOX 03.96. Post-restoration flight 02.05.96.

WV372 ff 15.07.55 Delivered 02.09.55. No 222 Sqn Leuchars 02.09.55. Converted to T Mk 7. Delivered 07.05.59 No 5MU Kemble. Jever Stn Flt 08.01.60. Gutersloh Stn Flt 12.62. No 2 Sqn 05.67. Wildenrath 07.70. RAE 30.03.71. 4 FTS Valley. No 208 Sqn. No 237 OCU. FAA 10.84. Overhauled by Airwork at Bournemouth early 1986. Returned to FAA. Shawbury for storage 1988. School of Aircraft Handling, later School of Flight Deck Operations, until 1996. Delta Jets 'Fox-One Consortium', Kemble. Registered G-BXFI.

WV383 ff 12.07.55 Delivered 23.08.55. Manby 19.12.55. Converted to T Mk 7. Delivered 03.07.59. No 79 Sqn 06.01.60. Gutersloh Stn Flt 8.01.60. Jever Stn Flt 14.04.61. No 4 Sqn 16.05.61. No 19MU St Athan 05.06.61. No 28 Sqn Kai Tak 1964. No 20 Sqn Tengah 05.01.67. No 5MU Kemble 04.12.69. Farnborough 07.77. A&AEE/DERA Boscombe Down. RAE Farnborough Air Sciences Trust, RAE Heritage Centre 13.04.00.

XF310	Built as F Mk 4. Delivered 20.07.56. Fairey Aviation Ltd 06.56. Fairey Fireflash missile trials installation. Converted to T Mk 7. ff 24.04.59. Delivered 29.05.59. No 20 Sqn Tengah 06.11.61. No 5MU Kemble 06.03.67. TWU Brawdy. No 2TWU Lossiemouth. FRADU, Yeovilton. WFU 1993. GIA Culdrose 1993-96. Sold to Hank Ebes, Cheltenham, Melbourne, Australia, 08.98.
XF321	Delivered 27.02.56 No 33MU. No 130 Sqn 10.04.56. No 5MU Kemble 16.04.57. No 56 Sqn 1956. No 130 Sqn 1957-58. Returned to HSA 1958. Converted to T Mk 7. No 43 Sqn/No 8 Sqn. No 1417 Flt Khormaksar 1966-67. To RAE Farnborough 1976. W/O 27.07.84. GIA RNEC Manadon. Phoenix Aviation, Bruntingthorpe, 28.05.04. Nose to Cockpit Collection, RAF Coltishall.

Hawker Hunter T Mk 8

Ten aircraft built by HAL at Kingston-upon-Thames. Transferred from T Mk 7 Contract No 12626 to FAA.

XL580 ff 30.05.58	Delivered 30.07.58. No 764 Sqn FAA. Flag Officer Flying Training Yeovilton 06.62. Converted to T Mk 8M by BAe at Brough 06.80. Fitted with Blue Fox radar equipment to be used as trainer for Sea Harrier. Preserved at FAA Museum, Yeovilton.
XL581 ff 30.05.58	Delivered 01.05.58 Lossiemouth. No 736 Sqn FAA '619/LM'. Crashed at Lossiemouth. Pilot killed. W/O 06.08.58.
XL582 ff 17.06.58	Delivered 01.08.58 Lossiemouth. Yeovilton. No 738 Sqn FAA. HMS *Heron* Stn Flt. Crashed on take-off from Yeovilton. W/O 26.01.68.
XL584 ff 08.09.58	Delivered 06.10.58 No 764 Sqn FAA. Flag Officer Flying Training Yeovilton. FRADU '877', later '871/VL'. Crashed into English Channel 5 miles off Portsmouth. Pilot killed. W/O 31.10.84.
1XL585 ff 14.10.58	Delivered 29.10.58 RAE Bedford. No 700 Sqn FAA. No 738 Sqn FAA. No 764 Sqn FAA '689/LM' Lossiemouth. Crashed at Portsoy, Scotland. W/O 30.01.69.
XL598 ff 15.10.58	Delivered 01.12.58. No 764 Sqn FAA. Converted to T Mk 8C. No 738 Sqn FAA. SOC and sold. Registered as G-BVWG 08.12.94. Sold to Beachy Head Aviation and reregistered 21.03.95. Deregistered 12.06.95. Thunder City, South Africa, as ZU-ATH.
XL599 ff 04.11.58	Delivered 09.12.58 No 764 Sqn FAA. Collided with WT701. W/O 23.08.61.
XL602 ff18.111.58	Delivered 30.12.58 No 764 Sqn FAA Lossiemouth. No 759 Sqn FAA Brawdy 08.64. No 764 Sqn FAA '687/LM', later '699/LM', Lossiemouth 1969-72. Lee-on-Solent for storage 1974. Converted to T Mk 8M by BAe Brough with Blue Fox equipment as trainer for Sea Harrier 03.80. Prototype T Mk 8M. ff 1979. Retained by BAe for trials operating from Holme-upon-Spalding Moor, Yorkshire, then BAe Dunsfold for trials work in conjunction with Sea Harrier FRS.1 on avionics integration tests. Target and chase plane during FRS.2 programme. Yeovilton 30.03.94. Sold to Barry Pearson 1995. Registered G-BWFT. Rear fuselage of Omani '801' stored at Hawarden Airport to be mated with XL602 to make a 'big-bore' Hunter.
XL603 ff 06.12.58	Delivered 13.01.59 FAA Lossiemouth. No 800 Sqn Yeovilton Instrument Flt 'VL/949'. No 759 Sqn FAA 08.64. Converted on production line to T Mk 8 for Royal Navy. Delivered to Lossiemouth 01.59. No 764 Sqn FAA. Yeovilton Stn Flt '950/VL' 1961. No 764 Sqn FAA '704/LM'. Navy Flying Standards Flt '738/VL'. No 764 Sqn FAA '677' 1970. Stored 1972. Converted to T Mk 8M by BAe at Brough with Blue Fox equipment as trainer for Sea Harrier 04.80. Radar Research Squadron RAE Bedford. No 899 Sqn FAA '718', later '720/VL'. Recoded '724' 1988 until retirement in 1994. Sold to Global Aviation Inc and shipped to USA. Current status and location unknown.
XL604 ff 10.12.58	Delivered 19.01.59 No 764 Sqn FAA. Converted to T Mk 8C. Delivered 22.11.63 No 759 Sqn FAA. Purchased by HSA 16.02.73. Converted to Kenyan T Mk 81 '802'. Delivered 24.06.74. Transferred to Zimbabwe AF as '1084'.

Conversion order for eighteen F Mk 4s to be brought up to T Mk 8 standard by HAL and AWA Ltd for FAA.

WT701 WT702 WT722 WT745 WT755 WT772 WT799 WV319 WV322 WV363
WW661 WW664 XE664 XE665 XF289 XF322 XF357 XF358

Conversion order for F Mk 4s and T Mk 8s to be brought up to T Mk 8B standard with TACAN equipment.

WW664 XF967 XF978 XF995

Conversion order for F Mk 4s and T Mk 8s to be brought up to T Mk 8C standard with TACAN equipment.

WV396 WV397 XF938 XF939 XF942 XF983 XF985 XF991 XF992 XF994
XL604

Hawker Hunter FGA Mk 9

Conversion contract placed with HAS Ltd to modify forty F Mk 6s to full FGA Mk 9 standard.

XE544 XE552 XE581 XE592 XE597 XE600 XE607 XE609 XE617 XE618
XE620 XE623 XE643 XE645 XE647 XE649 XE651 XE654 XE655 XF376
XF388 XF419 XF421 XF424 XF442 XF454 XF455 XF462 XG128 XG130
XG134 XG135 XG136 XG169 XJ643 XJ687 XK139 XK140 XK150 XK151

Conversion contract undertaken by RAF and CWP at Horsham St Faith in 1959 to modify thirty-six F Mk 6s to Interim FGA Mk 9 standard. Aircraft initially retained Avon 203 engines, but later progressively re-engined with Avon 207 engines to full FGA Mk 9 standard.

XE584 XE604 XE611 XE615 XE616 XE622 XE624 XE628 XE646 XE650
XF431 XF435 XF440 XF455 XF456 XF517 XF519 XF523 XG151 XG154
XG155 XG156 XG195 XG205 XG207 XG237 XG252 XG253 XG254 XG255
XG256 XG260 XG261 XG264 XG271 XG273

Conversion contract placed with HSA Ltd in 1960 to modify thirty-one F Mk 6s to full FGA Mk 9 standard.

XE530 XE535 XE546 XE550 XE582 XE610 XE652 XF414 XF508 XG153
XG265 XG266 XG272 XG292 XG293 XG296 XG297 XG298 XJ632 XJ673
XJ680 XJ683 XJ684 XJ685 XJ686 XJ688 XJ692 XK136 XK137 XK138
XK142

Conversion contract placed with Hawker Siddeley Aviation Ltd in 1964 to modify six Hunter F Mk 6s to full FGA Mk 9 standard.

XE532 XG291 XJ689 XJ690 XJ691 XJ695

Conversion contract placed with Hawker Siddeley Aviation Ltd in 1965 to modify six Hunter F Mk 6s to full FGA Mk 9 standard.

XF430 XF446 XG251 XJ635 XJ640 XJ674

Conversion contract undertaken by RAF at No 5MU Kemble in 1965 to modify nine Hunter F Mk 6s to full FGA Mk 9 standard.

XF416 XF437 XF511 XG194 XG228 XJ642 XJ644 XJ645 XJ646.

Hawker Hunter FR Mk 10

Conversion contract placed with HSA in 1959 to modify thirty-three Hunter F Mk 6s to FR Mk 10 standard.

WW593	WW594	WW595	WW596	XE556	XE579	XE580	XE585	XE589	XE596
XE599	XE605	XE614	XE621	XE625	XE626	XF422	XF426	XF428	XF429
XF432	XF436	XF438	XF441	XF457	XF458	XF459	XF460	XG127	XG168
XJ633	XJ694	XJ714							

Hawker Hunter GA Mk 11

Conversion contract placed with HSA in 1961 to modify forty Hunter F Mk 4s to GA Mk 11 standard.

WT711	WT712	WT713	WT718	WT721	WT723	WT741	WT744	WT804	WT805
WT806	WT808	WT809	WT810	WV256	WV257	WV267	WV374	WV380	WV381
WV382	WW654	WW659	XE668	XE673	XE674	XE680	XE682	XE685	XE689
XE707	XE712	XE716	XE717	XF291	XF297	XF300	XF301	XF368	XF977

Hawker Hunter T Mk 12

Ordered by Ministry of Supply on behalf of RAE Farnborough, converted from F Mk 6 airframe. Powered by Avon 208 engine.

XE531 F Mk 6 converted by HAL to two-seater standard. Initially fitted with head-up display (HUD) for TSR2 trials but later used for research by RAE Farnborough and RRE Bedford in trials for Harrier Avionics, including trials with HUD and vertical nose camera in 1965. Crashed into trees on airfield on take-off from RAE Farnborough after explosion and engine failure caused by disintegration of 11th-stage engine compressor disc. W/O, 17.03.82.

Appendix 3

Description of the airframe by Hawker Aircraft

Since Neville Duke, then chief test pilot, made the first flight of the prototype P.1067 on 20 July 1951, there have been no fewer than 127 variations on the basic theme. Some of them were minor, others were purely for research, but most were aimed at increasing the operational capabilities of the squadron aircraft. It has been shown that the Hunter can carry guided air-to-air weapons, a very heavy external load of fuel for ferrying flights, a wide variety of air-to-ground rockets, bombs of various sizes and honeycomb launchers for unguided air-to-air rockets – all in addition to the hard-hitting four-30mm Aden gun pack. In addition, cameras can be installed in a special nose section for fighter/reconnaissance work, and a two-seat (side-by-side) cockpit was successfully developed for either training or special tactical roles. A trainer version based on the lower-powered F.4 was ordered for the RAF (Mk 7) and Fleet Air Arm (Mk 8) as well as the Danish air force (Mk 53). The more powerful trainer based on the Mk 6, ordered by India as the Mk 66, could climb in less than seven minutes from rest to more than 47,000 feet, and that the throttle could be slammed open and shut at stalling speed at those heights without causing flame-out or surging.

Manoeuvrability at transonic speeds, with the follow-up tailplane, was precise and effective, and a speed of more than 600 knots was reached in level flight at a height of about 100 feet. Touch-down speed was about 110 knots. Swiss trials amply demonstrated the ability of the Hunter to operate from small airfields tucked away high up in narrow valleys, and to deliver 30mm shells, rockets and bombs accurately against mountain targets. The standard rocket armament consisted of the British Mk 12 rail carrying 3in rocket-motors fitted with either 60lb or 25lb heads of various kinds, although the Hunter was also fitted with Oerlikon, Hispano and Bofors rails and rockets. The American 5in HVAR has been carried, as have various bombs of different weights and nationalities. For RAF training purposes the Saro carrier for two 25lb practice bombs was also used. External tankage included jettisonable tanks with a capacity of either 100 or 230 imperial gallons, carried in any symmetrical combination. Up to four 100-gallon napalm tanks could be carried, and these were demonstrated in Switzerland and Iraq.

The permanent armament feature is the pack of four 30mm Aden guns, firing high-velocity ammunition. For re-arming, the pack is lowered from the aircraft, with the aid of three Type C bomb hoists, onto a cradle running on casters, which is in turn picked up by a rubber-tyred trolley fitted with its own hydraulic fork-lift. Several of these can be towed to the armoury by Land-Rover for re-arming; the guns are normally harmonised in the pack away from the airframe. Before the pack is taken out, the barrels are detached from the breech mechanisms and the whole system so arranged that any barrel will fit any gun and any pack will fit any aircraft. Complete interchangability is thus provided, and results showed that a high degree of accuracy is nevertheless maintained. The gunsight and battle camera are harmonised separately with the airframe.

Effective performance of the gun pack has been conclusively demonstrated in the most rigorous tests. Unrestricted firing is allowed in any attitude, at any speed and at any height, considerable positive and negative g being tolerated. A truly marathon firing trial was carried out during development: some 20,000 rounds were fired through one gun pack under the most difficult flight conditions (500 feet and 600 knots). Each time the trigger was held to discharge the full ammunition load in one burst and the aircraft was brought straight back to Dunsfold for re-arming. After the 20,000 rounds the whole installation was inspected, some very slight modifications were incorporated and a second cycle of 20,000 rounds was then fired off without any trouble. This performance, Hawker claimed, was unequalled by that of any other gun-firing aircraft in the world – 30mm or otherwise.

The Aden gun does not require to be mounted right way up, and such an attachment is, in fact, seldom made. In the Hunter gun pack the inner Adens lie on their sides, lower faces inwards, and the outer guns are upside down. Each barrel has a forward steadying bracket, but may be quickly detached

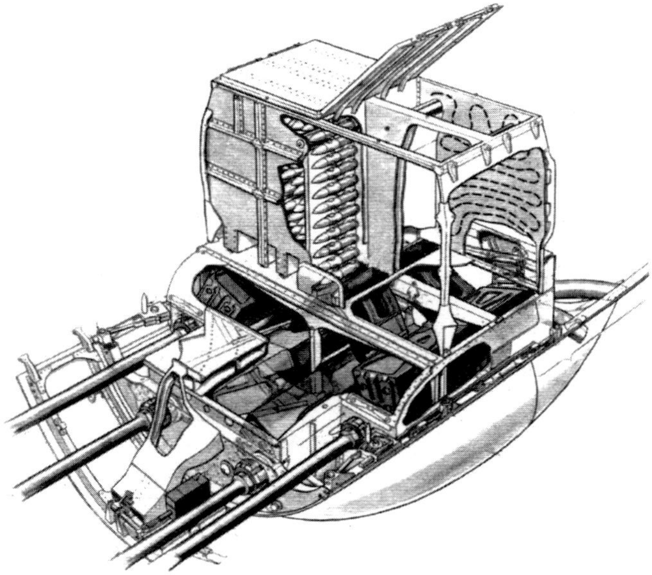

Aden Cannon

from the breech mechanism and removed forwards. The breech mechanism itself is supported at the front in a spherical bearing taking fore-and-aft loads, while the rear end is held by an adjustable link giving torsional support and location. The link is held to the gun pack frame by a captive bolt, which can be adjusted to move the gun body sideways. Vertical adjustment is achieved by rotating an eccentric bush round the captive bolt by means of a worm gear.

Alignment of guns is achieved with the aid of a sighting prism inserted at the muzzle and trained on a special marker board. The inner guns are mounted slightly forward of the outers, pairs of forward mountings being held in two transverse frames, with a central keel structure helping to bear reaction loads. The two pairs of rear mountings are held in two further frames towards the rear of the pack. A decking over these frames carries lugs to which the ammunition tanks are fastened. Ventilating air, intended to prevent the accumulation of a dangerous concentration of gases during firing, is brought in through a duct at the front of the pack, entry being controlled by an electric actuator. Heating air is also led into the gun-chamber from a bleed on the Avon. Six spigots on the airframe pick up fittings on each side of the gun pack and are retained by bolts operated by two rack-and-pinion latch mechanisms on the pack. Two further spigots locate the pack immediately before the main spigots make contact. Before the pack is lowered, small panels forward and aft of it are removed to expose connections, to allow barrel removal and to locate the central bomb hoist near the forward face of the pack. During firing the empty cases are expelled rearwards from the breech mechanisms and dropped overboard through tubes that emerge from the fuselage just aft of the pack.

Some modifications have recently been effected to these outlets to make quite sure that the cases do not strike other parts of the airframe. The links cannot be similarly jettisoned, because they do tend to cause damage; they are therefore collected in two streamlined boxes attached to the gun pack (generally known in the RAF as 'Sabrinas'). They are secured by quick-release fastenings, all designed to avoid the possibility of improper replacement. In addition, blast deflectors have recently been designed for the gun muzzles and are to be fitted to all Hunters. They consist of welded steel boxes mounted in the mouth of each firing tube and standing one or two inches proud of the fuselage contour. Cascades of four vanes in each deflector box turn blast away at right angles to the fuselage, providing about 160lb of upward thrust per gun. This eliminates the nose-down pitching moment that would otherwise be significant when the guns are used at heights above 35,000 feet.

The airframe is built up in standard jigging as a number of self-contained sub-assemblies that can be transported to the makers' airfield at Dunsfold for final assembly. Each wing, for instance, is completed with all equipment, including the undercarriage leg, all tested and painted ready for attachment to any fuselage. The latter, too, is made up in three equipped units, the central one carrying the stub wings. The cockpit section can be detached for transport or repair, and the rear section can be readily detached for engine installation or removal. Extreme nose and tail cones are virtually fairings attached by toggle fasteners. The fin and the single-piece tailplane are mounted on a fin-root structure that is integral with the rear fuselage.

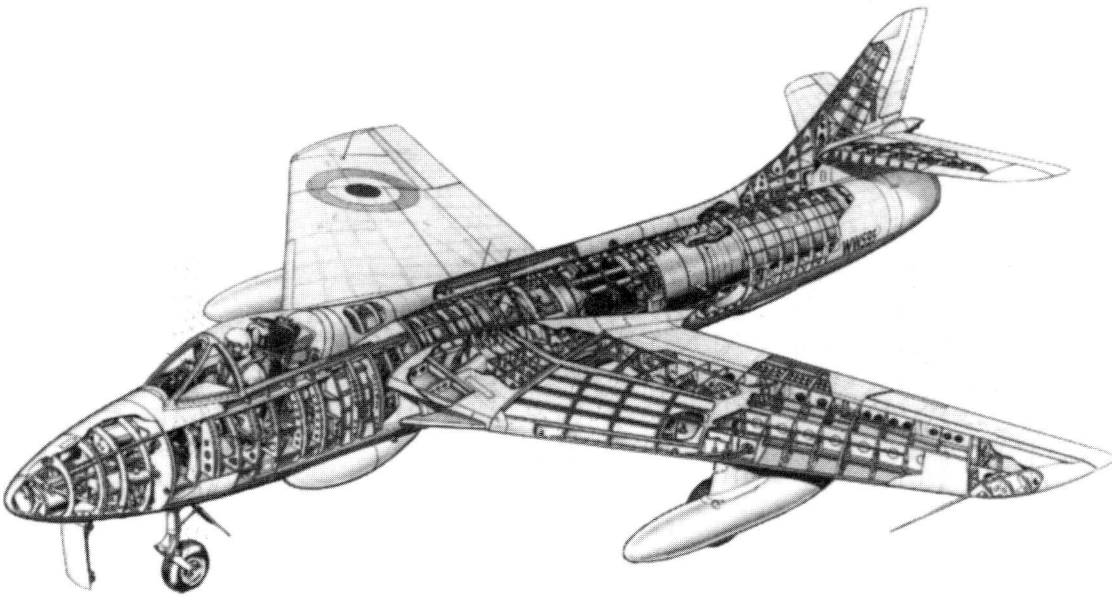

As the cut-away drawing suggests, the structure is extremely strong, but quite conventional in conception. Though many of the skin panels are rolled to a curvature, there is no integral stiffening, and stringers, ribs and frames are built up from formed plate and light extrusions. There are few large forgings, and very little steel is used. Such relative simplicity is a very real advantage in ensuring ease of repair, since there is still no effective way of repairing the massive, integrally stiffened structures now being increasingly used in military aircraft.

Fuselage frames are numbered from 1 to 63, the forward section extending between 1 and 18A and the centre between 19 and 40A. Material is generally DTD 610 (L.72). Although centre and rear sections are built up from closely spaced frames and many stringers, the forward section is based on four heavy longerons, the lower pair being of built-up top-hat section and those above having a pyramid section reinforced by extruded angles. In addition, a keel structure between frames 8 to 11 takes nose wheel drag loads, the retraction jack and surrounding items.

Hunter T Mk 7 instrument panel. *David Griffin*

Hunter F Mk 4 instrument panel.

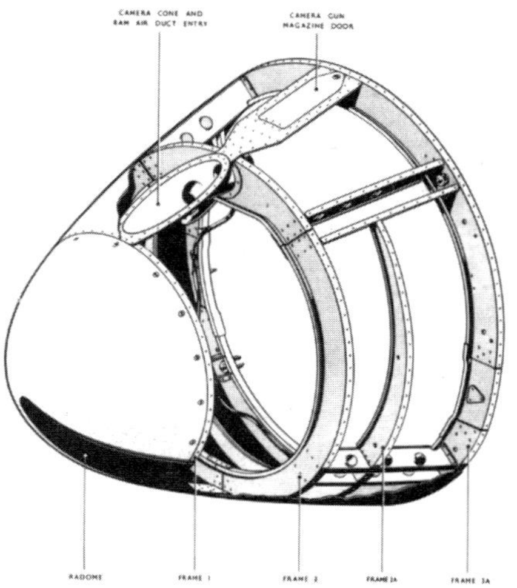

The pressure cabin is contained by bulkheads at frames 6 and 14 and by the cockpit floor, a trunk passing diagonally up through the rear portion to house the control runs. Support for the cockpit coaming and windscreen is provided by the top longerons, they in their turn being assisted in bearing canopy-locking loads by frames 10 and 13, which are doubled for this purpose. The canopy is a normal blown plastic unit running on rails and moved via a chain gear by an electric motor at the rear and to port. It is jettisoned by a cartridge-operated piston jack at each side, initiated either by the pulling of the ejection-seat blind or by a separate manual control. The windscreen is a double-layered unit made up from laminated bullet-proof glass mounted in a rubber channel and clamped down by fairing strips screwed to the supporting pillars. An air-space between thick outer and thin inner glass layers is kept dry by silica-gel capsules. Quarter lights are of 1in-thick Perspex.

Hunter T Mk 8 nose detail, illustrating the Harley light fit and the attached external access ladders.

Hunter FR Mk 10 nose detail. *Ray Deacon*

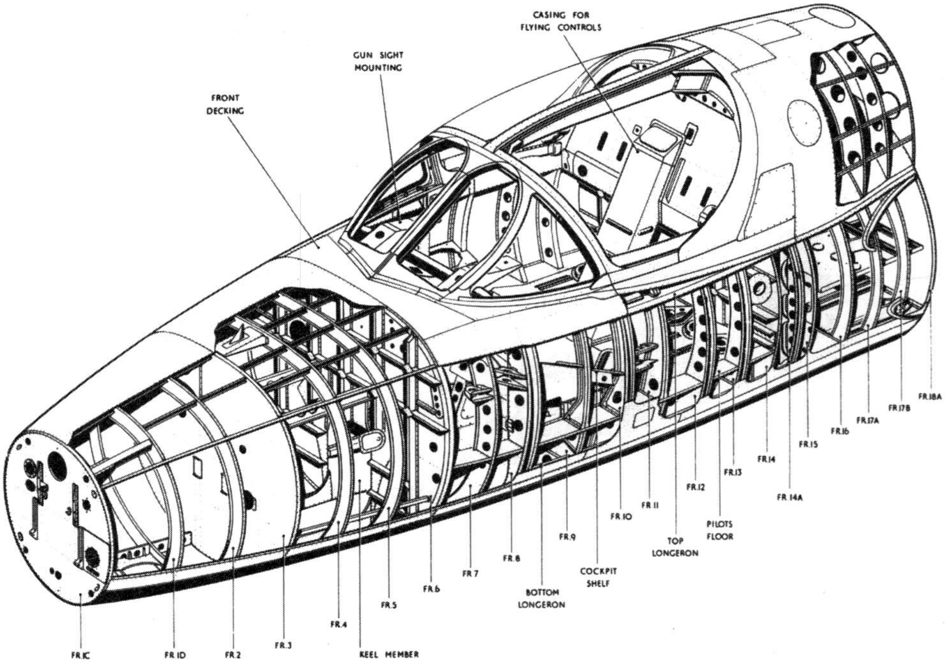
CASING FOR FLYING CONTROLS
GUN SIGHT MOUNTING
FRONT DECKING
FR.18A
FR.17B
FR.17A
FR.16
FR.15
FR.14
FR.14A
FR.13
FR.12
PILOTS FLOOR
FR.11
FR.10
TOP LONGERON
FR.9
COCKPIT SHELF
FR.8
BOTTOM LONGERON
FR.7
FR.6
FR.5
FR.4
KEEL MEMBER
FR.3
FR.2
FR.1D
FR.1C

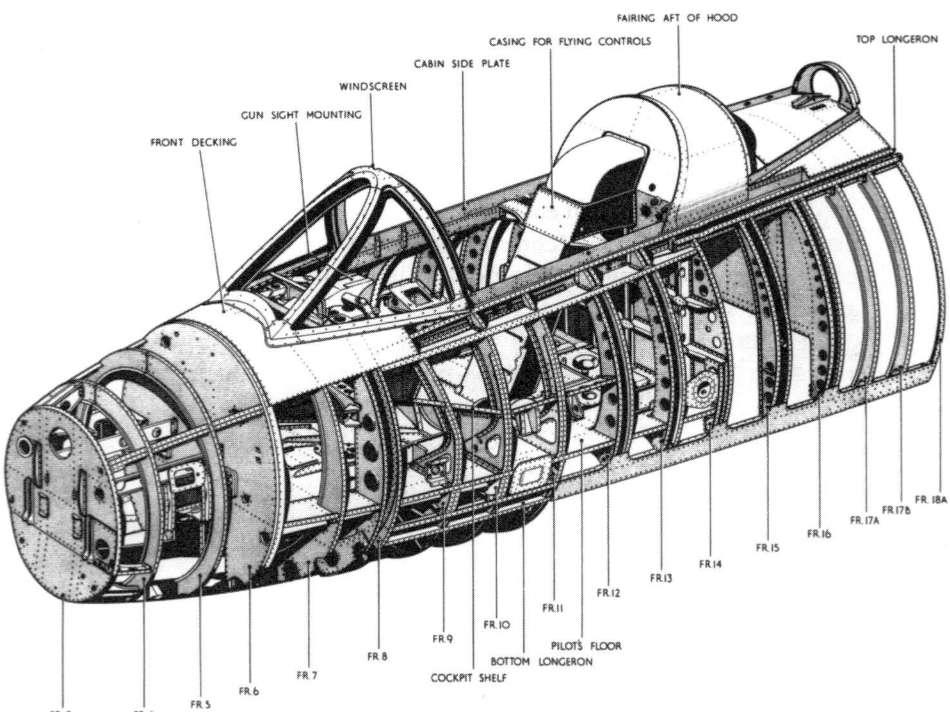
FAIRING AFT OF HOOD
CASING FOR FLYING CONTROLS
TOP LONGERON
CABIN SIDE PLATE
WINDSCREEN
GUN SIGHT MOUNTING
FRONT DECKING
FR. 18A
FR.17B
FR.17A
FR.16
FR.15
FR.14
FR.13
FR.12
FR.11
PILOTS FLOOR
FR.10
BOTTOM LONGERON
FR.9
COCKPIT SHELF
FR.8
FR.7
FR.6
FR.5
FR.4
FR. 3

The fuselage forward of frame 6 houses the retracted nosewheel, the radar-ranging gear and one of the pressurisation valves. Aft of frame 14 lie the ammunition cans, taking up the full fuselage width, and these form a part of the gun pack containing the four Aden in the underside of the fuselage. The next space aft houses radio and electrical-control equipment mounted on racks between frames 16 and 18A. Here the skins are mainly of 16swg, with 14swg where gun-blast has to be resisted and 10swg around the base of the windscreen, where deflection armour qualities are required. Control runs and certain air conditioning components are housed in the dorsal spine, which is an unstressed and detachable fairing in this area.

The centre fuselage carries the main body of the engine with its mountings and the stub wings and, as already indicated, it emerges from the factory as a complete unit ready to receive front and rear fuselages, engine and wings. The general structure comprises closely spaced frames and stringers formed of plate and extrusions, skinned with 14 and 16swg sheet. Frame 19 itself carries the intake frame terminating in spigots bearing the torsional loads from the forward section of the wing. The lips of the intakes are built as part of the centre fuselage and contain the sprung doors giving extra intake area for slow-speed operation as well as the boundary-layer bleed ducts.

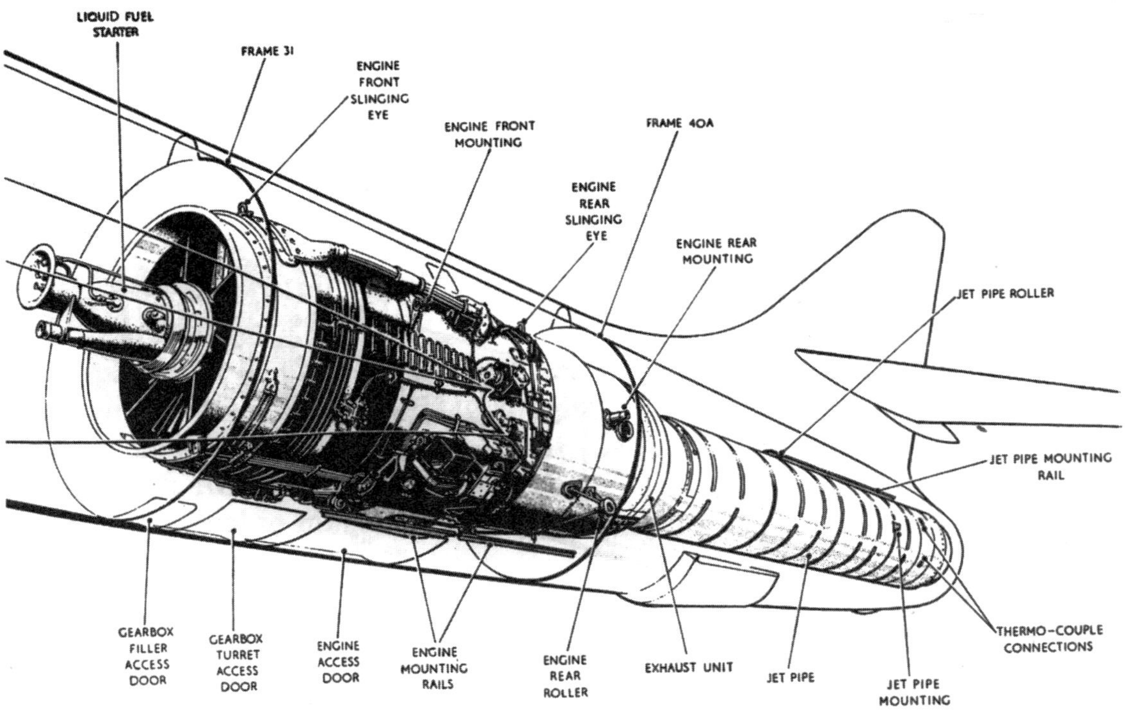

Main wing loads are taken through the double-diaphragm frame 25, which has extruded booms and plate webs surrounding two machined, H-section spar members that actually pick up the spar lugs. The lower of these members is in DTD 364. Wing drag loads are passed into the fuselage at ring frame 32 through steel channel running forward to frame 29 from each pin. Fuselage shear loads are borne by two machined top-hat light-alloy longerons and by the skin, which is continuous round the cylindrical fuselage, inside the stub wing.

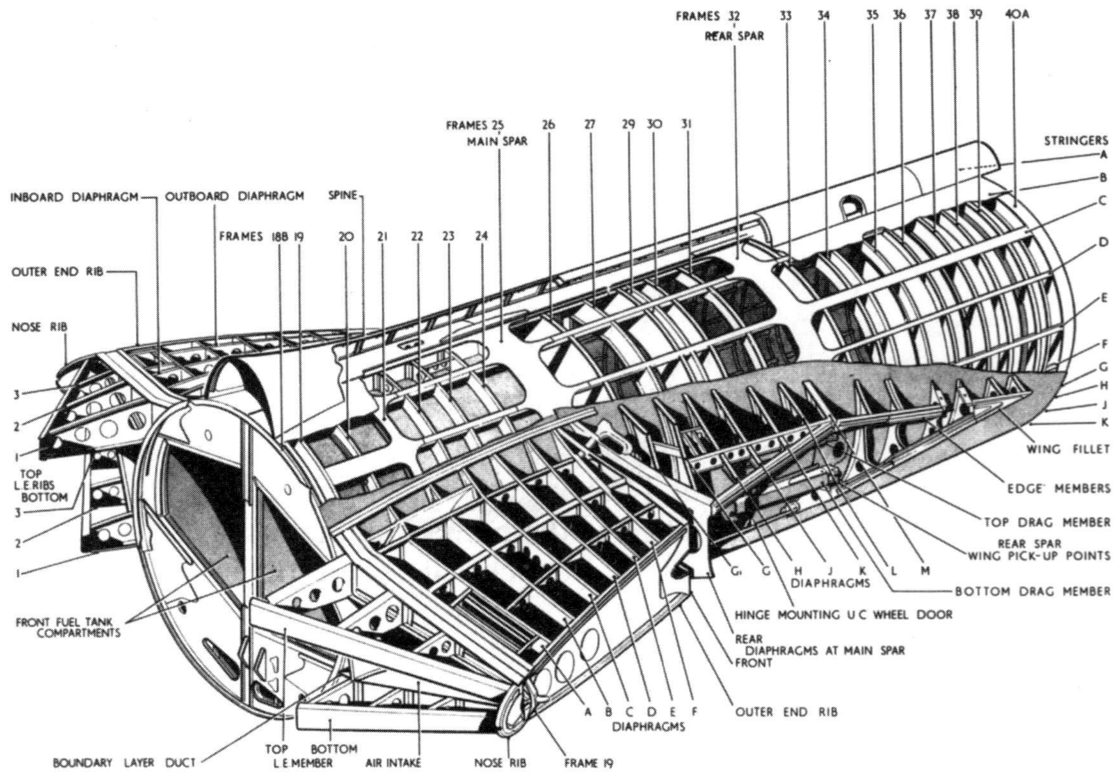

Frame 34 is again reinforced to carry the forward engine mounting. Intermediate frames are mostly divided on the underside to accommodate access panels for the engine and its accessories, although frames 37, 38 and 39 are complete and reinforced in order to transmit some of the shears from the fuselage-break into the continuous upper sections. A thrust member between frames 38 and 40A transmits these loads from the main engine mounting at frame 40A. Fuselage joints both fore and aft of the centre section are of the spigot-and-nut type, though the forward one is not intended for dismantling except for major maintenance and transport.

No structural loads are carried by the main intake trunk, and the intake ring of the engine is pressed against a rubber seal at the end of the duct, level with frame 31.

As might be expected, the rear fuselage is a simpler version of the centre section, and is made up of closely spaced frames and stringers, the main joint being located at frame 40B. Towards this point the light extruded stringers are reinforced by combined bulb and L-section stringers, which terminate at the joint fittings. A few inches aft of this lies frame 41, which forms the forward bulkhead of the after tank bay, reaching from frame 41 to 45. Stainless steel skins protect the alloy-lined bays from jet-pipe heat. Air brake jack loads are taken in an 18swg built-up case between frames 47 and 50, while the hinges are held in 16swg channels between frames 45 and 46. The surface is skinned in 10swg sheet. Aft of frame 46 the dorsal spine begins to take the fin contour on ribs attached to the fuselage decking. Both fin and tailplane are mounted on two posts built integrally with the reinforced frames 52 and 55. The tailplane jack, mounted ahead of frame 52, operates the tailplane beak, which projects through frame 52, the main pivot (and the rear spar of the fin) being held in a steel forging at frame 55. A box structure between the two carries the elevator jack. Between frames 55 and 56 lies the tailpipe support bracket, aft of which the tail cone is attached by four locating spigots and four toggle fasteners. The sloping frame in this area is a relic of the earlier shorter tail-cone. A stainless steel ring surrounds the actual nozzle.

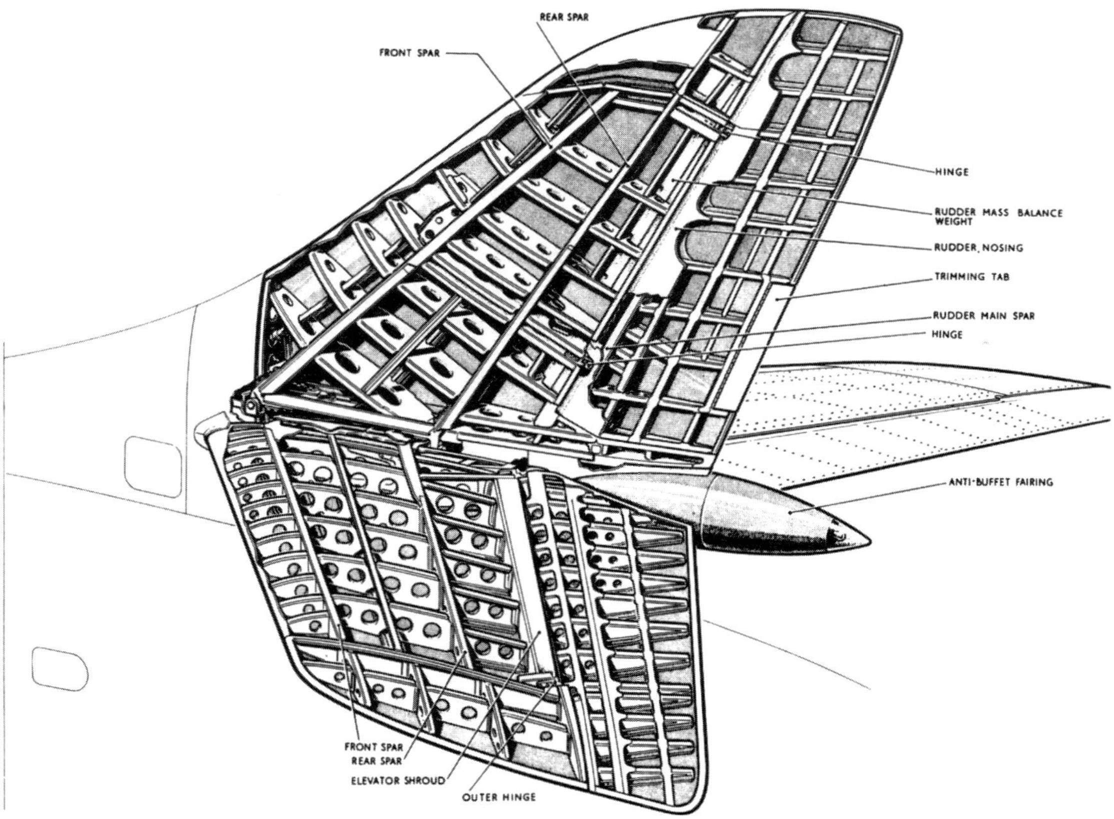

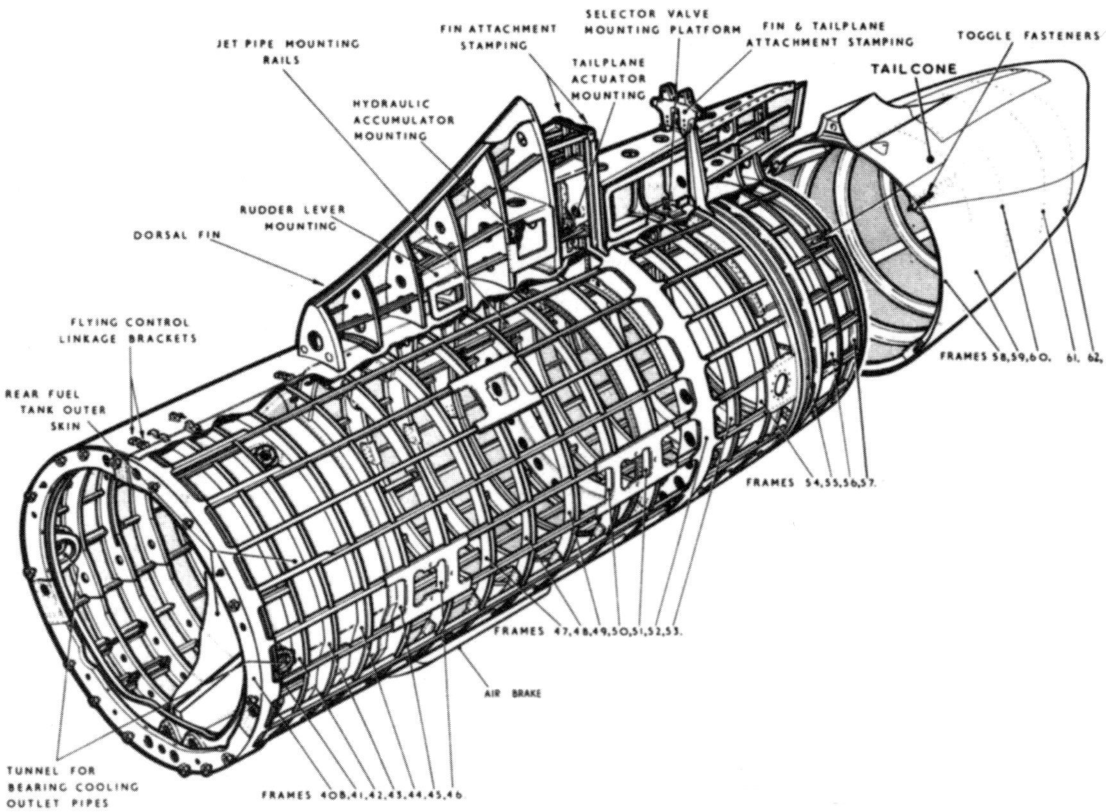

Single-piece tailplane and fin are both built up around two formed plate spars supported by ribs and spanwise stringers. Rudder control push-rods pass up the leading edge and through the fin between closely spaced ribs, while the three central ribs of the tailplane form a box carrying the hinge brackets aft and the operating beak forward. The elevator surfaces are operated by pivot shafts terminating at their inboard ends in a common universal joint holding a single operating lever connected to the jack. Both rudder and elevators have lead mass-balances within shrouds ahead of the hinge line and are conventionally built up from plate spars and ribs.

The Rolls-Royce Avon 200 series engine is installed in the fuselage as a powerplant complete with its firewall, Plessey isopropyl nitrate starter, oil system and fuel-cooled oil cooler. The accessory shaft on the port side is coupled to a shaft system driving a Rotol gearbox mounted as part of the airframe low down to port in the engine bay. From this the two generators and the hydraulic pump are directly driven. They remain in the airframe after engine removal. The ipn starter system is electrically controlled and fed from a tank holding sufficient liquid for three starts.

The firewall, which is mounted on the engine, corresponds with frame 38 and divides the engine bay into two fire zones. Graviner flame-switches are distributed round these two areas and round the tail-pipe and arranged to operate a warning light in the cockpit. The light is set in a button that can be immediately pushed to release foam from two separate bottles into one spray-ring forward of the firewall and another aft of it. An inertia switch will also initiate the extinguisher system after a crash landing.

For removal of the engine, the tail fairing is first detached and the tail-pipe run out on its gantry rail. All leads and fuel lines are separated, hydraulic lines being automatically shut off by Lockheed self-sealing couplings. The rear fuselage spigots are then undone and the whole rear fuselage wheeled away on a special trolley, leaving the rear of the engine exposed. Another trolley is then connected to the centre fuselage so that the engine mountings may be unfastened and the engine rolled along the rails built into the fuselage and straight onto equivalent rails on the trolley. The ipn starter comes away complete with its piping. The reverse of this sequence serves to replace an engine and the aircraft can be ready for ground running in a matter of a few hours.

In addition to the power taken from the engine at the accessory drive, air is bled from the compressor delivery to supply cockpit and fuel-system air, while a further bleed feeds heating air to the gun-pack. NACA-type flush air intakes round the engine bay ensure ventilation, while fuel and other drains have various outlets around the after fuselage. Wing structure follows fairly closely the principles employed in the fuselage, plate and extrusions being combined into both transverse and longitudinal stiffening, covered with large skin panels of relatively heavy gauge rolled to a curvature. Rib stations are lettered and numbered, those ahead of the main spar reading (from the root rib A) 1, 2, 3, G, H, J, K, L, M, N, P, Q1, Q2 and R. The main frame of the wing is formed by the front and rear spars, the undercarriage girder (lying diagonally between the two) and the strong ribs A, G, M and R, which have plate webs and extruded booms. Rib A is the root rib lying against the stub wing; rib G is cranked aft of the main spar to fie normal to the undercarriage girder; rib M carries a butt strap for chordwise wing-skin joints; and rib R closes the main spar, carries a butt strap, and mounts the aileron Hydrobooster and the spar extensions holding the end rib and detachable wingtip and, to port, the pitot boom. Intermediate ribs are simpler flanged plates with lightening holes.

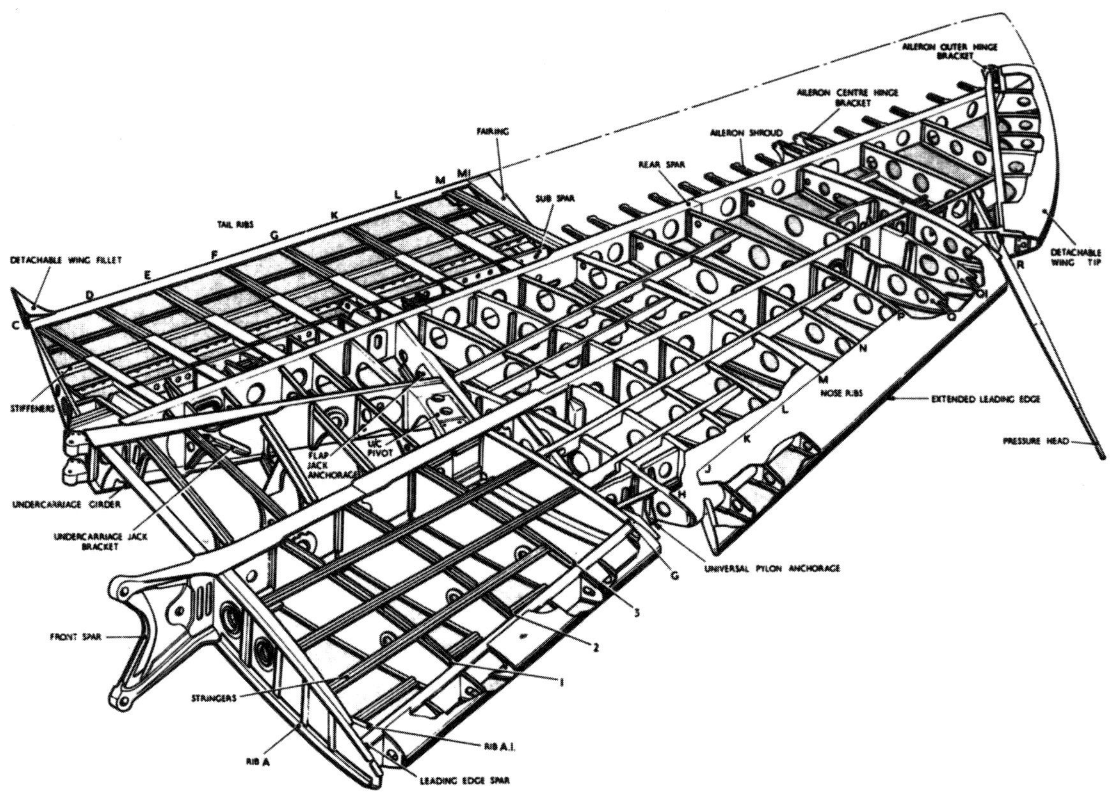

Both main and rear spars are simple plate channel-sections, though extruded booms replace flanges in the main spar inboard of rib J. This structure is progressively strengthened, having steel booms inboard of rib E and terminates where it picks up the steel fittings on fuselage frame 25. The securing pins are tapered steel, with split bushes, fastened by a nut and cup washer. Drag loads from the wing are taken by the top skin and resisted at the rear wing pins on fuselage frame 32. The rear spar joins the undercarriage girder near the pin joint. An undercarriage pivot bracket is built into the space between main spar, undercarriage girder and rib G, the leg and part of the wheel retracting into the space between spar and girder. The decking over the bay is supported by top-hat-section stiffeners interspersed with steel and alloy extrusions. A goodly quantity of hydraulic, electric and fuel system leads are concentrated in the bay within easy reach from the ground. Main wing loads are transferred from the rib-and-spar complex outboard of rib G to the structure ahead of the spar, heavy steel gusset plates serving to take bending loads across the crank in rib G itself.

Aft of the rear spar, the decking over the flaps is held by tapered top-hat ribs mounted on forged brackets and covered by plate reinforcing near their roots. This light decking is skinned both above and below. Brackets for flap hinges and operating and synchronising jacks are of plate and forgings. Similarly, the aileron shroud is supported on forgings, with hinge and actuating brackets of composite construction. The ailerons themselves are made up of plate ribs and stringers based on a channel-section spar, with lead mass-balances in the nose sections. The port aileron also carries the trim tab on a piano hinge, with its electric actuator in the aileron itself.

The flap structure is based on a tubular spar to which are through-bolted forgings forming either roots for the tapering top-hat section ribs, or hinge and operating brackets, or both. Trailing edges are stabilised by doubler strips and a perforated sheet is riveted over the ribs. The flap surface itself is a single skin. Some slot effect is provided for in the geometry of the hinge brackets. To form the main wing-skin, very large panels of 10swg alloy are flush-riveted to ribs and stringers, with a covering over each wing surface out to rib R. They are rolled to contour and butt-jointed along the leading edge, the inner surface of the joint being formed by a butt strap. Outboard of rib R, 14 gauge is used, and the decking over the flaps is skinned with 18 gauge sheet. The pylons, both inboard and outboard, are based on a cast soleplate contacting tank or bomb and supported from extruded flanges, bolted to the wing skin and internal structure, by extruded channel-section stringers. The pylon is covered with a sheet skin.

Suspension of bomb or tank is by a Vickers bomb-slip, which can be lowered to pick up the store by a built-in screw-jack operated from above the wing. A jettison mechanism and the fuel by-pass valve are located within the pylon. Rocket rails of the standard Service Mk 12 type are mounted on posts quickly secured at reinforced cut-outs in both upper and lower wing skins. Two posts hold each rail located between ribs M and R. Bolts retaining the pylons are screwed into blind nuts held in brackets specially positioned next to the relevant ribs. In the case of the outer pylon, rib Q is duplicated ahead of the spar as Q1 and Q2 in order to accommodate these brackets.

The saw-tooth leading edge extension, which is now becoming increasingly familiar on several marks of Hunter, is a simple retrospective modification. It is in fact a single sheet wrapped round short plate ribs and riveted directly onto the wing skin between ribs R and J. The trailing edge of this skin is slightly tapered, but kept thick enough to accommodate rivet heads.

Powered Controls

The powered control system of the Hunter has gone through a number of variations and modifications, each providing some distinct improvement in handling, and resulting finally in a control system that leaves little to be desired. The main changes have brought the horizontal tail control from a boosted elevator and electrically trimmed tailplane, through fully powered elevator, to fully powered elevator combined with a follow-up tailplane. The rudder is purely 'manual', with an electric trim-tab to counteract any unusual asymmetric conditions. This tab may also be connected to an autostabiliser made by Newmark. The surface is hardly used at high speeds. Full power and plain spring feel are applied to the ailerons, with manual reversion and automatic doubling of mechanical advantage as a stand-by. For this eventuality, also, an electric trim tab is provided on the port aileron to relieve out-of-trim forces. With power engaged, lateral trim can be adjusted by offsetting the centre position of the spring feel unit, which is located on the control column itself. Aileron feel has a maximum force of 8lb at full deflection, and the springs are preloaded to provide positive centring and a slight break-out force. The trim tabs are powered by Rotax actuators.

The hydraulic control-surface jacks are Fairey Hydroboosters (one each for ailerons and elevator), and each is anchored to the airframe by a pawl engaging in a slot in the piston rod. The pilot's input rods actuate the valve and the body of the jack moves the surface. The pawl is held in engagement by the main hydraulic supply pressure, and when this drops below 400lb/sq in it automatically disengages under spring force. The pilot then moves the surface by the direct mechanical link through rods, jack valve and body.

When power is selected for ailerons or elevator it is possible for the pawls to fail to engage the slots and simply to clamp the piston rod. Warning doll's-eyes in the cockpit are arranged to show proper engagement by changing from white to black only when the pawls have gone fully home. If the latter simply clamp, the pilot has the positive indication of a white doll's-eye and in this case the controls will revert to manual when the selector switch is released. The misalignment can be remedied by applying a hefty push to the stick to slide the piston past the pawl until the slot is engaged. When such a situation arises during re-engagement in flight, the necessary stick-rocking can be accomplished without producing untoward aircraft responses.

The doubling of mechanical advantage when the ailerons are manually actuated is achieved by varying the length of a telescopic lever in the control-rod circuit. Spring pressure keeps the lever short, thereby providing greater mechanical effectiveness for manual flying, while hydraulic system pressure is applied to lengthen it when power is in use. When loss of hydraulic power automatically allows the lever to shorten again, only half aileron travel is available to the pilot; but this does not additionally hamper him, for aerodynamic loads are in any case too high for him to achieve full deflection.

The horizontal tail-control is rather more complex, because it involves a combination of hydraulic fully powered elevator with electrical follow-up and trimming of the tailplane. Follow-up is achieved by a pantograph linkage from the elevator making contact with micro-switches that control the Rotax motor changing tailplane incidence. Consequent movement of the tailplane adjusts the position of the micro-switch datum to relieve the switches. These are also separated, so that the elevator will move one or two degrees before the linkage contacts a switch. The result is that fine control-adjustments, eg during formation flying, are made with elevator alone, while tailplane follow-up occurs with the larger movements needed during manoeuvring at transonic or approach speeds. The 'neutral' zone also avoids tailplane hunting.

Feel is provided by a standard telescopic spring unit connecting the pilot's control rods and the leading edge of the tailplane. The neutral or trimmed position is therefore reached when the elevator is substantially in line with the tailplane, and stick force remains directly related to stick movement regardless of speed. Overstressing is avoided by jack stalling. A centring effect is provided by a steepening of the rate curve near the neutral position. Trimming is effected quite simply by an electric actuator, which, moving simultaneously with the tailplane actuator, alters the datum of the micro-switches controlling tailplane follow-up. The pilot's control for this purpose is a sliding switch on the top of the control column.

As the Hunter control system is a mixture of hydraulic and electric components, and has many electro-hydraulic elements, the consequences of failure or malfunction of either electric or hydraulic supplies or components can be complex. Careful study of the very clear Hawker technical publication on the subject makes clear the various combinations with which a pilot might have to deal. For the purpose of this description some of the remoter intricacies may be omitted.

The basic stand-by is, of course, manual reversion, which occurs automatically, and hydraulic accumulators in the tail and aileron circuits (respectively at 1,575 and 900lb/sq in) can provide several full control cycles to allow establishment of trimmed conditions for manual flying. Control forces are then very heavy, particularly on the ailerons, but a landing in this condition is perfectly feasible and is a standard training exercise. The tailplane follow-up will in any case continue to operate, though it may sometimes be better to switch it off. For practice in manual flying, power is switched off by closing the electrically actuated cocks governing oil supply to the servos.

In the event of electrical failure and a need to revert to manual control, a stand-by battery can be used to close these cocks. An engine windmilling after a flame-out will supply enough hydraulic pressure to provide adequate powered control, but in most such cases the pilot is recommended to revert to manual before landing. Prior to reversion it is also highly desirable to adjust the tailplane to an angle suitable for trimmed flight at a moderate airspeed and Mach number in order to avoid sudden and heavy out-of-trim stick loads at the moment of disengagement. In the event of tailplane actuator main-motor failure, a second motor integrally mounted with the jack can be brought into operation. Opening of the safety cover that protects the switch provided for this purpose automatically cuts the main motor out of circuit.

The horizontal tail control is now such that accurate gunsight tracking is possible at all speeds. At trans- and supersonic Mach numbers the tailplane is the most effective element and its rate of follow-up to the elevator will allow application of g at a rate of about 0.5g/sec. Stick forces per g are satisfactory, although for high IAS they may feel a little light. Stick forces that vary primarily with stick movement rather than with speed are a noticeable feature of the control system, but in this case the effect is not disconcerting.

Hydraulics

In the Hunter, hydraulic power is used to operate undercarriage, wheel brakes, flaps, air brake and the aileron and elevator Hydroboosters. The main supply is drawn from a reservoir, located in the centre fuselage, through a Micronic filter and into either the main engine-driven pump or a hand-pump located in the engine bay. The latter pump can operate any of the services except the air brake but is not available to the pilot in flight.

The engine-driven pump is a Dowty Vardel two-stage unit producing a working pressure of 3,000lb/sq in, measured by a gauge in the cockpit. If pressure falls below 600lb/sq in a warning light glows and an audio tone is fed into the pilot's earphones. A push-switch is provided to reduce this note if required. Emergency actuation of undercarriage and flaps is by compressed air at 2,000lb/sq in from two spherical bottles mounted immediately behind the ejection seat. Related pressure-gauges are located on the port cockpit console. Emergency air enters the primary hydraulic system, but does not affect the powered control circuits if the normal selectors are in the correct position for the movement required. Oil displaced by the air is dumped overboard.

Emergency wheel-brake operation is by oil at 75lb/sq in stored in two Dunlop accumulators mounted in the nose wheel bay. Most of the valves in the system are electrically operated. The majority of the hydraulic components are supplied by Dowty. A single jack moves each main wheel leg and the three cover doors that are mechanically linked to it. Mechanical sequencing valves then control the operation of the jack, which closes the inboard door and engages a mechanical up-lock for the whole unit. An equivalent arrangement obtains for the nose wheel. A large spigot in each main wheel well locates the leg in the retracted position by engaging a hole in the wheel axle. Undercarriage-mounted micro-switches cut out the retraction circuit when the weight of the aircraft is on the wheels, but this precaution can be overridden by the pilot and retraction-effected in emergencies where a ground run must be converted into a slide. An electric switch with 'up', 'down' and six intermediate positions controls hydraulic flap extension. Synchronisation of the two split flaps is achieved by interconnected idling jacks working at 500lb/sq in, one at each flap. Pressure balance is automatically reset after each retraction. The flaps are intended for use as air brakes, but stressed for normal descents only, while the air brake proper can be used without limitation and is controlled by a sliding switch recessed in the head of the engine-control lever. In order to prevent the air brake scraping on landing, a micro-switch on the undercarriage breaks the circuit on extension and causes the brake to retract automatically. A doll's-eye in the cockpit turns white when the air brake is extended.

The Dunlop wheel-brakes are operated by a hand-grip on the control column, connected with a relay valve on the rudder pedals to allow differential application for steering. A triple-reading gauge shows main supply and individual break pressures. Skidding of the wheels is prevented by Dunlop Maxaret units at each disc brake. Gauges on the port console indicate the pressure of the two emergency-braking accumulators.

Electrical Installations

Electrics intrude into virtually every system of the Hunter, for almost all remote control and indication is by electrical means. Though the ramifications of this system are therefore very considerable, the methods and techniques employed do not differ from the normal.

The main system is of the 28v, single-pole, earth-return, voltage-regulated type, the supply coming from two wide-speed range generators mounted on the accessory gearbox in the engine bay. They are connected in parallel and stabilised by two 24v, 25A-hr Varley-type batteries, which are parallel-connected to give 50A-hr. Voltage regulators for the generators are mounted in the radio bay, where the batteries are also located. For ease of testing and servicing, the earth leads have been concentrated at a number of points conveniently located throughout the airframe. Flight instruments, engine-temperature control, radio and radar-ranging all require AC at 115v, 400c/s, three phase, and this is supplied by two inverters mounted on the cabin floor aft of the ejection seat. No 1 inverter is the main source for flight instruments and engine temperature control, while No 2 acts as a stand-by and as the primary supplier for radar-ranging. If No 1 fails, No 2 automatically takes over and sheds the radar-ranging load. As with most such arrangements, a test switch is provided so that the change-over may be checked by the pilot. A separate low-voltage inverter serves the gun-firing system, and a 1,600c/s single-phase auxiliary inverter is also provided for the radar ranging.

Fuel

Internal fuel tankage in the Hunter F.6 is distributed in each wing, in the forward fuselage and round the exhaust section of the engine, immediately aft of the fuselage break. The inboard universal pylons under each wing can accommodate either Bristol 100-gal plastic drop-tanks or Hawker 230-gal metal drop-tanks, while each outer pylon can take either a bomb or a further 100-gal tank. A very respectable fuel load can be accommodated when long ferry flights are required.

The whole system is divided into port and starboard sections, entirely separate from each other, their delivery to the engine being balanced by a Rotol vane-type fuel-flow proportioner. The front fuselage tanks are the master receivers, collecting fuel from all the other tanks, and they contain recuperators for inverted flight and SPE electric booster pumps. Vapour-release valves ensure that air pressure from fuel boiling or rapid climbs will not block transfer from other tanks.

Fuel is passed from tank to tank by engine-bled air, which is regulated to a constant pressure by a Hymatic reducing valve. The rear tanks first transfer their contents and, after low-level float switches have signalled them to be empty, the outer drop tanks take over. The rear tank supply pipe is then automatically closed and these tanks isolated. Transfer continues through outer to inner pylons, through the groups of wing tanks and into the master receivers. As soon as a drop-tank is jettisoned, its air and fuel pipes are automatically united so that the continuity of the transfer sequence is preserved.

Magnetic indicators in the cockpit show the position of each fuel cock and, although the whole system is normally entirely automatic in operation, overriding switches are provided. In addition, a warning light will show failure of fuel-transfer pressure, and all but the master tank gauges will thereafter immediately read empty. The pilot then has a quantity indication only for the fuel which still remains usable to him, namely that in the master tanks. Failure of one pump, indicated by warning light in the cockpit, causes the other to be switched to high-speed working, thus maintaining supply of fuel to the engine. Indication of both pumps being inoperative – and therefore of low pressure in the delivery line –is also given by warning light. The light will go out when throttle setting is sufficiently reduced to allow the engine to run on the remaining pressure.

The throttle lever in the cockpit operates the high-pressure cock during the first part of its movement forward from the rear stop, a 5-deg segment being reserved for the idling setting. A micro-switch is included to light the undercarriage warning light if the throttle is closed to idling with the wheels up. The low-pressure cock is operated by a separate lever further aft on the port console.

Smiths-Waymouth capacitance gauges are fitted in each internal tank, the total contents per side being registered on separate gauges in the cockpit. The whole fuel system is portrayed on the starboard console in diagrammatic form, gauges giving tank contents, white lines representing fuel pipes and magnetic indicators showing positions of cocks. Control switches are beside the magnetic indicators, and warning lights are suitably placed. Contents of drop-tanks (which drain after the rear tanks and before the wing tanks) are not indicated.

Icing, which may occur in the main engine-fuel filter because of small amounts of water suspended in the fuel, is detected by a Teddington Controls differential-pressure switch, which senses any excessive pressure-drop across the filter. When this occurs, a Plessey priming pump is automatically started by a relay to pump alcohol from a stainless steel one-gallon tank through a 1/16in diameter nozzle into the main fuel delivery pipe.

Normal refuelling is under pressure through an Avery-Hardoll 1.5in nozzle located in the port wheel bay. Switches and controls are set beforehand to ensure correct filling and to prevent backflow and over-pressure. Fuel is fed simultaneously to fuselage and wing tanks, and it passes through the latter to the drop-tanks, if such are fitted. Pressure-relief valves are incorporated in most tanks to help control the fuelling operation. The tanks can also be emptied on the ground through the refuelling connection by setting a Saunders valve reached through the engine-access panel. An external air supply of at least 10lb/sq in is connected to the tank pressurisation system and fuel is thus transferred normally to the master tanks. Suction from a bowser or the master tank booster pumps can bring the fuel overboard to empty the tanks completely.

The fuselage tanks, made by both Marston Excelsior and Fireproof Tanks, are flexible rubber bags with madapolam vulcanised on the outside and covered with three layers of glass cloth finished with fireproof lacquer. Those in the wings are again flexible rubber bags, but have nylon net vulcanised on the outside and covered with fireproof lacquer. Wing and master tanks are introduced into their bays by long cords attached at their suspension points. The cords are pulled through holes in the walls of the bays and are then knotted and cut off short to hold the bags in. The recesses in which the knots lie are plugged with screw-caps. For the master tanks these caps are in the top of the fuselage; for the wing tanks they are behind the front spar and are accessible from the main wheel bay.

Air Conditioning

Pressurisation and air-conditioning of the Hunter cockpit follows what is now established practice. A bleed is taken from the final compressor stage of the Avon and passed through a series of cooling stages and controlling valves into the pressure cabin, the required pressure-differential being achieved by throttling two outlet valves from the cabin. The system is controlled electrically and, after being switched on, automatically builds up pressurisation from 10,000 feet upwards. Under certain flight conditions, and when pressure is switched off, a ram-air intake in the gun-camera sight tube admits fresh air directly into the cabin, through-flow being established by the simultaneous automatic opening of a dumping valve. If cabin altitude exceeds 38,000 feet, or when manually selected by the pilot, an increased engine bleed supply is led directly into the cabin, providing a high flow of virtually uncooled air. This feature is particularly necessary to prevent canopy misting when descending rapidly from very cold upper regions.

A conventional rubber seal closes the gap between sliding canopy and coaming, a differential of 8lb/sq in between cabin and seal pressures being automatically maintained by air drawn from a point well upstream in the conditioning system. Inflation and deflation are controlled by sequence switches and solenoids connected with canopy actuation. A delay is included in the opening sequence to allow partial deflation before the canopy begins to move, and initiation of the jettisoning sequence also causes deflation.

Air initially passes from the compressor bleed through a precooling stage consisting of a Marston Excelsior heat exchanger cooled by ram air drawn from the starboard engine intake trunk and dumped through an exit near the dorsal spine. This pre-cooler is automatically by-passed when either fully hot or flood flow is selected. Further cooling elements consist of a refrigerator made up of a Godfrey boot-strap compressor-turbine unit and heat exchanger, the latter taking cooling air from the port intake trunk. Having switched the system on, the pilot is able to select on a dial automatic control of a temperature between 5 deg C and 30 deg C. He may also select an increase in heat or cold without automatic control and, finally, flood flow. The control unit is a Teddington magnetic amplifier and sensing elements based on a Wheatstone bridge, monitoring an electrically actuated regulating valve. This in turn governs the proportions of hot air from the pre-cooler and cool air from the refrigerator, which are fed into the mixing chamber and through a water extractor, into the cabin via a non-return valve. In the fully hot setting, pre-cooler and refrigerator are cut out, but in the flood setting the flow-control valve and normal ducting as well are largely short-circuited, though the water extractor remains in circuit.

Conditioning air is distributed in the cabin by a series of spray pipes along the lower edges of both canopy and windscreen, and some air is discharged from the cabin near the feet. External connections are fitted for testing the pressure cabin, hood seal inflation and the cabin-pressure controller. Air dumped from the Normalair main cabin-pressure control valve on the forward bulkhead is collected in a shroud and used to cool the radar containers.

Hunter F Mk 4 cockpit interior.

Oxygen

The oxygen system consists of a standard Normalair demand regulator, which passes to the pilot's mask the correct mixture of air and oxygen for all heights according to his breathing rate and strength. Gaseous oxygen is stored at 1,800lb/sq in in two 750-litre cylinders mounted on the starboard side between frames 4 and 6 and retained by wing-nut-tensioned straps and a cable. In addition to the functioning indications of the regulator, the pilot has a total pressure gauge, both these units being mounted on the starboard console. An emergency supply of 55 litres, contained in a bottle stowed in the dinghy pack, can be selected by pulling a toggle, and it is automatically initiated when the seat leaves the aircraft. The main cylinders may be recharged in situ from a connection in the nose wheel bay.

Ejection Seat

The Hunter is fitted with either the Martin-Baker Mk 2H or 3H ejection seat, both being fully automatic and suitable for use at any height above 50 feet. Both seats have the Irvin back-type parachute and duplex drogue-deceleration system, the main difference between the two lying in the fact that the Mk 3H has leg-restraining harness instead of thigh guards. In both cases the pulling of the firing blind first initiates the jettisoning of the canopy and then ejects the seat after a one-second delay. The canopy may nevertheless be jettisoned manually without causing ejection.

Anti-g System

Provision is made in the Hunter for an anti-g suit and pilots are ordered to wear this clothing for all normal operational flying. Air pressure for the suit is stored in two Dunlop spherical air bottles of 110cu in capacity, each at a pressure of between 1,800 and 2,000lb/sq in, mounted behind the ejection seat immediately above the emergency-air bottles for flap and undercarriage operation. A common ground-connection serves for the charging of all four.

Air is automatically fed into the g-suit in proportion to the amount of g being exerted after the pilot has turned the system on with the Hymatic selector valve. It passes through a Dunlop air filter and a Palmer pressure-reducing valve before reaching the Hymatic anti-g valve. On this there is a test button with which the functioning of the system may be verified or the suit inflated intermittently in order to relieve pilot fatigue. A pressure gauge above the controls, which are on the starboard console, indicates the remaining pressure in the bottles. The flexible lead from the system is clipped to the ejection seat and a further lead is connected to the suit itself. Quick-releases effect immediate disconnection on baling out.

Appendix 4

Comparative data summary of nine sample RAF training aircraft

Specification data

	Piston Trainers				Jet Trainers				
	Chipmunk T Mk 10	**Provost T.1**	**Harvard 2.B**	**Balliol T.2**	**Jet Provost T Mk 3 (Mk 4)**	**Vampire T.11**	**Meteor T.7**	**Hunter T.7**	**Gnat T.1**
Engine	Gipsy Major 8	Leonides 25	Wasp	Merlin 35	Viper 102 (105)	Goblin 35	2 Derwent 8	Avon 122	Orpheus 101
Power hp/thrust lb	145	550	550	1,100 of 1,245	1,750 (2,500)	3,500	7,200	7,595	4230
First flown/RAF	1946/50	1950/53	1938 & 44	1947/49	1954/55; 3 in 59 (61)	1950/52	1948	1955/58	1959/62
Numbers to RAF	735	387	2,485 of 5,000*	162	201(185)of 496	535	500+	51	105
Length/height (ft)	25.5/7	29/12	29/12	35/12.5	32.5/10	34.5/6	43.5/13+14	49	38 (31 + pitot)
Span (ft)/aspect ratio	34.5/6.7	35/5.8	42/7.0	39.5/6.1	37/6.4	38/5.5	37/3.9	34/3.3	24/3.3
Wing area (sq ft)	172	214	253	250	214	262	350	349	175
Op wing loading	12	21	21	34	31(33)	42	40	48	46
Take-off weight (lb)	2,000	4,400	5,200	8,400	6,700 (7,100)	11,100	14,100	16,800	8,300
Thrust/weight %**	8.5	14.3	12	14.5	30 (41)	38	58	49/46	59/54
Initial climb(ft/min)	800	1,800-2,200	1,300	1,800-2,500	2,400 (3,300)	4,500	7,600-7,900	8,000	10,500
Initial climb speed (knots/M)	65-70	100	100	120	165	250/.63	300/.63	430/.85	350/.8
Max level/dive (mph)	138	200	180	288	311 (409)	549	585	660	628
Max dive speed (kts)	173	235	225	320	350/.7M	455	510/.76/78	620	1.15M+
Range (miles)	280	648	750	660	525 (690)	853	820/1,000	1,000	950
Service ceiling (ft)	16,000	22,000	22,000	32,000	30,000-35,000	40,000	35,000-39,000***	50,000	48,000

With variations in authoritative sources, followed by inevitable changes in operational procedures, conditions and limitations over time, data, at times approximate, have been simplified selectively – the above figures are comparable for illustrative purposes only.

* Total Harvards over 5,000 British Commonwealth Air Forces (Mk 1 300; Mk II 1,173; Mk IIA 726; Mk IIB 2,485; Mk III 351)

** Thrust/weight ratios for solo flying in a half-fuelled aircraft; piston aircraft a synthetic rating – bhp as percentage of weight in lb

*** Service ceiling was pilot-limited because of lack of pressurisation – pressurised Meteor 8's service ceiling was 44,000ft

Performance Data

	Piston Trainers				Jet Trainers				
	Chipmunk T Mk 10	**Provost T.1**	**Harvard 2.B**	**Balliol T.2**	**Jet Provost T Mk 3 (Mk 4)**	**Vampire T.11**	**Meteor T.7**	**Hunter T.7**	**Gnat T.1**
Engine	Gipsy Major 8	Leonides 25	Wasp	Merlin 35	Viper 102 (105)	Goblin 35	2 Derwent 8	Avon 122	Orpheus 101
Power hp/thrust lb	145	350	550	1,245	1,750 (2,500)	3,500	7,200	7,595	4,230
Stalling speed, clean (kts)	45	65	60	65	73	95	100	140	120-125
With no power, u/c and flaps (kts)	38	60	55	60	65	85	90	130	120-125
Aerobatic speeds (kts)									
Slow or barrel roll	120	140	140	160/180	180	260	270	350	250
Stall turn		120	160	170	220	180	330	290	
Loop	130	165	180	220/250	220	350	375	425	350
Half roll off the top	140	175	200	250/280	250	370	390	450	375
Vertical roll		200		280	300/320	380	400	500	450
Spinning and to 'n' turns	8	OK	OK	4	OK	4 2	4 (no tanks)	n.r	
Controlled descent (A/B+/-flap)				170	250	.7M/250	280	.7M/280	
Glide path			90	95	100	125	130/135	155	160
Base leg	70	90	110	105	100	140-125	140/125	160	160
Reduced airspeed	70	90	75/110	110	90/110	140	160	180/200	170
(flap and power)									
Round out/threshold speed (flap and power) as to fuel wt	55	75	70/75	85/90	85	100	110	130	135/125
Short landing	50	65	70						

Many of the above speeds are fairly agreed, depending on weight and altitude, by most QFIs, although some, as the Gnat stall, are a topic of debate.

Appendix 5

Hunter weaponry

Described by former Hunter Instructor Alan Pollock

The Aden was a superb 30mm-calibre, automatic, gas-operated revolver-type gun developed at the Small Arms Factory, Enfield, from an earlier German concept. Once some early teething problems were cured, this gun, mostly developed down at Pendine near Pembrey, was very reliable. The Aden was deliberately so well designed that it could never be incorrectly assembled after servicing. The Hunter's Aden was regulated to fire at 1,200 rounds per minute. Pneumatically cocked, the electrically fired ammunition was belt-fed from containers within the four-gun-pack system, with disintegrating links collected in 'Sabrina' blister panels – on early firing trials, the links on ejection had damaged or even been ingested into the engine's air intakes. Each gun's revolving drum had five cartridge chambers and, when fired, the gas pressure, acting through a gas port in the barrel via a piston and slide mechanism, revolved the breech cylinder continuously to feed a new round into position for powerful return springs and the slide to move forward to index the cylinder and feed this round into a chamber. Because of their immense power and airframe recoil, only rarely would we fire all four guns at once. In early trials, spent cartridges damaged the fuselage and air brake, so the chutes were redirected much more vertically downwards to eject the empty cartridges straight down. This also had an added benefit, moving the unburned remnant cordite gases much further away from the intakes, preventing flame-outs at high altitude.

The early Hunter's role was as a fighter, its formidable punch from its four 30mm guns evolving once teething problems, mainly at altitude, were cured. In both UK and overseas Commands, when VIPs or overseas visitors arrived on a Hunter station during the mid-to-late-1950s, a formal demonstration would regularly take place on the squadron pans, sometimes as a race between two to four squadron teams. This would display well the aircraft's unique and supremely fast operational turn-round, ably assisted by its quickly replaceable gun packs – a system much faster than its predecessors and contemporaries. Once the 100-gallon, and later the FGA Mk 9's 230-gallon, inboard tanks were in regular use, these dramatically extended the average sortie's flight time on the squadrons, and also the length of refuelling times necessary, so these early demonstration operational turn-round races would become rare, then disappear from normal RAF station and squadron practice.

The speed and usefulness of the 'just possible' ten-minute refuel and re-arm 'Operational Turn-round' from touch-down to taxi out, take-off and airborne wheels up for the next sortie was one of the Hunter's finest attributes. Marshalled into dispersal and facing an Armament Practice Camp sand revetment or safe area over the airfield, the pilot shuts down his Hunter, which is rapidly double-chocked amid a blur of eager armourers and technicians. The procedural maze of orders, grunts and shouts of the turn-round team coalesce into the race of refuellers, riggers, engine mechanics, instrument bashers and electricians, all in one converging swarm of activity around, below, level and above one's gaze. The ladder is up and ejection pins are back in as that other last symphonic engine whine and gradual wind-down reaches one's ears in the open cockpit, with the combined throttle and High Pressure Cock already off. That haunting and distinctively desyncopated Avon music clatters into its final swirl of ever looser compressor blades and the cooling hot metal jet pipe – noisier lower revolutions drop back to a quietened nought.

One's arms are already up, with cape-leather-gloved hands outside on the canopy coaming, clearly visible to the armourers, as one shouts (guns) 'All switches safe' – the wheel well arming circuit safety break plug and flag hanging down are made safe as the front panel drops and fuses are held up. Later Hunters such as the Mk 9 would have a Master Armament Safety Break fitted, which was located in a panel about three feet out from the air intake in the port wing root towards the wheel well, with a Bowden cable and an 18-inch-long red flag hanging down for an even more visible warning. The tail prop must quickly be fixed into and under the rear fuselage to allow the unbalancing gun pack's weight to be dropped out without the rear end sinking down. The empty forked hydraulic lift trolley comes in, with the Sabrina panels already down, grounding with that soft aluminium scraping noise along the pan and

A publicity photo of the FGA Mk 9 weapons fit options. *Hawker*

the rattling of those black gun metal ammunition links being spilled and emptied. The fuselage attachment drop point spindles, port and starboard, are fixed into their holes to be able to hold the lowering and lifting winches, now being hooked on. All four Aden gun barrels have already been unlocked and drawn forward. The removable gun pack, winched up on both sides, is crack-released and wire-wound down, initially with that slightly drunken sway, one way or the other, after the first few inches of ratchet drop takes its weight, eventually nestling onto the quick-kick-adjusted low semicircular cradle, for the pack to settle barely a second before it is drag-wheeled out to the side.

A new replacement gun pack fully loaded – and, in fact, an integral part of the fuselage – comes back into and under that gaping incomplete void, with the trolley repositioned underneath in seconds. The deadly pack of four Aden guns and an appropriate total of 400 or 480 or so practice ball ammo, High Explosive or Armour Piercing 30mm, tray-positioned, interlinked rounds, is within seconds heard once more to be moving those few feet vertically, as the distinctive ratchet whirr and winching noise signals the heavy pack being lifted slowly upwards under control from its cradle into position. The pack comes flush with the rest of the fuselage surfaces and its four locating spigots as the winch notes change with a brief two-or-three-second hesitation before the final crisp whip-cracks that announce positively that the torque limiters have just been exceeded, so the pack is held safely up. The armourers then insert the locking pins into the spigots, before removing the winches and cradle. The barrels are already being twist-locked into place and the front panels secured. More scurrying and watching for the bowser's refuelling hose to be removed, oxygen replenished and trolley gone, starter cartridges (or later, Avpin) topped up, rear fuselage prop gone. The engine checks and turn-round are completed and one's own pre-starting checks are run through as a last look around. The armourer is now holding two or four fuses above his head, ascertaining that all tradesmen have completed their relevant tasks, then shouts loudly 'Stand clear – arming aircraft,' at which stage no one must be in front of the guns. He inserts the fuses and closes securely the rear gun pack panels.

A Form 700 to sign is thrust into the cockpit with a smile and a biro beneath one's nose, as you click in your new spare ciné magazine for the next sortie. With one's helmet back on and the pigtail lead connected, the ejection seat pins are removed, shown and stowed, before the airman descends and removes the red steps from their fuselage trap top. Ready to start, and the thumbs-up interrogation signal is acknowledged, and all in less than eight minutes or so – can't be bad. Press the tit and the cartridge fires (or that acrid, pungent hiss in the case of the F.6's and Mk 9's Avpin), and we're cooking again with Avtur, as the starter vent panel's dzus fasteners are locked once more. The port wheel bay red warning flag is removed from the armament isolation switch and held up. Ready to be on our way. The after-start checks are done in seconds, the radio taxi call, chocks away, brakes off and a check curtsey, and off we go. We'll make 10 minutes from touch-down to take-off wheels-up this time – fancy actually being paid to do this as one's routine day job!

In Aden in 1965-66 we usually kept the outboard pylons on but only had the outboard tanks fitted for any long-range redeployments. This meant that we would only have three 3-inch rocket projectile rails available, one beyond the outboard pylon and two rails spaced inboard between the inboard and outboard pylons. Although one could fit three rockets on each rail, the lowest would not be angled correctly for firing so our normal practice was to have only two 3-inch rockets fitted on each rocket rail for our operations, six on each wing, giving a total of twelve. For fixed gun use we had ball and HE ammunition, almost invariably firstly firing off only our paired inner guns, then the paired outers rather than all four guns at once. For rocketry in Aden and the UK we had the 60lb concrete practice heads, the HE Squashhead R/P or HE SAP (Semi Armour Piercing), and I never actually saw the Anti-Shipping Armour Piercing heads except maybe at SBAC Farnborough, but we did practise ciné attacks on naval vessels for simulating anti-shipping strikes. The high-velocity 68mm SNEB rockets in their Matra pods, which we began to use on No 1(F) Squadron in early 1967, proved a great step forward in the anti-tank role for use against heavy armour. On 24 April I had the privilege of leading what was understood to be the first 'publicly' displayed SNEB attacks on Salisbury Plain, in front of Her Majesty the Queen. With some glee tinged with much relief, on this, my leading attack, time-synchronised just behind a four-ship formation gun-firing fire suppression demonstration, also in front of 16 Para Brigade and sundry VIPs and overseas staffs, I was actually able to see the first SNEB ripple hit and explode exactly into the target tank as I pulled away. One could wish for but not always guarantee this type of 'on the day' weapons performance and to this day I remember well the care with which I had pre-turned to cross the Initial Point with my leading four Hunters and others behind for the run in for such a Royal occasion and audience – this certainly did beat most low-level air display run-ins, gunnery competitions and operational sorties for psyching one up for such an unforgiving arena – however, full marks to the SNEB and one only wished that Eurasian armour would be just as obliging! Further use of operational SNEB warhead attacks was made by us on No 1(F) Squadron on 7 September for the inter-service Exercise 'Unison'.

One pilot's weapons record analysis

Jun 1957	No 26(AC) Sqn RAF Sylt APC, F.4 on first air-to-air, as missed Hunter firing at Pembrey with heavy cold, surprisingly 1/21 on squadron with average only of 16.9% at 10,000ft but paltry 5.9% on the few 20,000ft sorties, when using Sylt & PAI system (firing 100 rounds). Later, using fairer system, scoring 3 out of 4: must fire in all conditions, so allows some harmonisation/sight problems to discourage pick and choose.
March 1958	No IV(AC) Sqn RAF Sylt, F.6, Low Level (10,000ft) 27.2%, Medium (20,000ft: 18-23,000) 25.8%, High Level (25,000ft) 26% (1 shoot).
Jan-Feb 1959	No 26(AC) Sqn RAF Sylt, F.6, all 20,000ft 38.7% (50 rounds, one gun). Fewer shoots, now radar ranging, bad weather, snow/ice and lost flying. Dec 58 and Apr 59, few air-to-ground academic and op strike single pass attacks at Strohen (best 4 of 6, firing 120 rounds), 56.9% (best academic 94.4%). Evasive Ciné at 40,000ft straight off, all to score 82.3%.

Aug 1959	2nd TAF gunnery RAF Sylt (two single-burst attacks only allowed) 39.9% (best two 69% & 55%); told posting brought forward so could not now compete, only fly the C-in-C down to Cazaux, so three poor competition shoots of 17.7% after shaken radar ranging gunsights.
	For three 26(AC) Sqn academic 20,000ft shoots at same time 47.1%.
	After ground and flying instructional tours through 109 DF/GA Course Nov 64/Feb 65 229 OCU Chivenor, all to score Ciné R&T 86.5%, Evasive Medium Level 78.5%, High 72%, Low Level 74%; weapons, best 3 out of 4, in Air to Air Gunnery 58.3%, Air to Ground 90.6%, new R/P 14.5yds, a Hunter F Mk 6, lighter than an FGA.9, fired closer and more accurately.
March 1965-November 1966	No 43(F) Sqn RAF Khormaksar Aden, FGA.9. No air-to-air and mix of academic and operational 2-pass attack firing: Air to Ground 54%, R/P 9.4 to 10.5yds (best 16 R/P 6.6 yds) with mix of preferred first fixed ring and then the depressed pipper system (and fatigue meters on A/G!).
May 1967	No 1(F) Sqn APC RAF El Adem, Libya, best 12 of 16 Air to Ground 56.7%, 3 in R/P 10.1yds (6 sorties). Bombing 93% (7) new SNEB 52% (4). In Nov 67 new supposedly automatic improved scoring system was faulty. Operational SNEB night firing trial, for 4 senior pilots only, with 'co-ordinated' Canberra Lepus flares. Excellent variety of both academic and two and single pass attacks at a dozen or more ranges in England, Scotland, Wales, Germany, Cyprus and Libya in pairs, 4s or even once 14, with regular overseas and UK exercises and deployments.

Appendix 6

Plan drawings

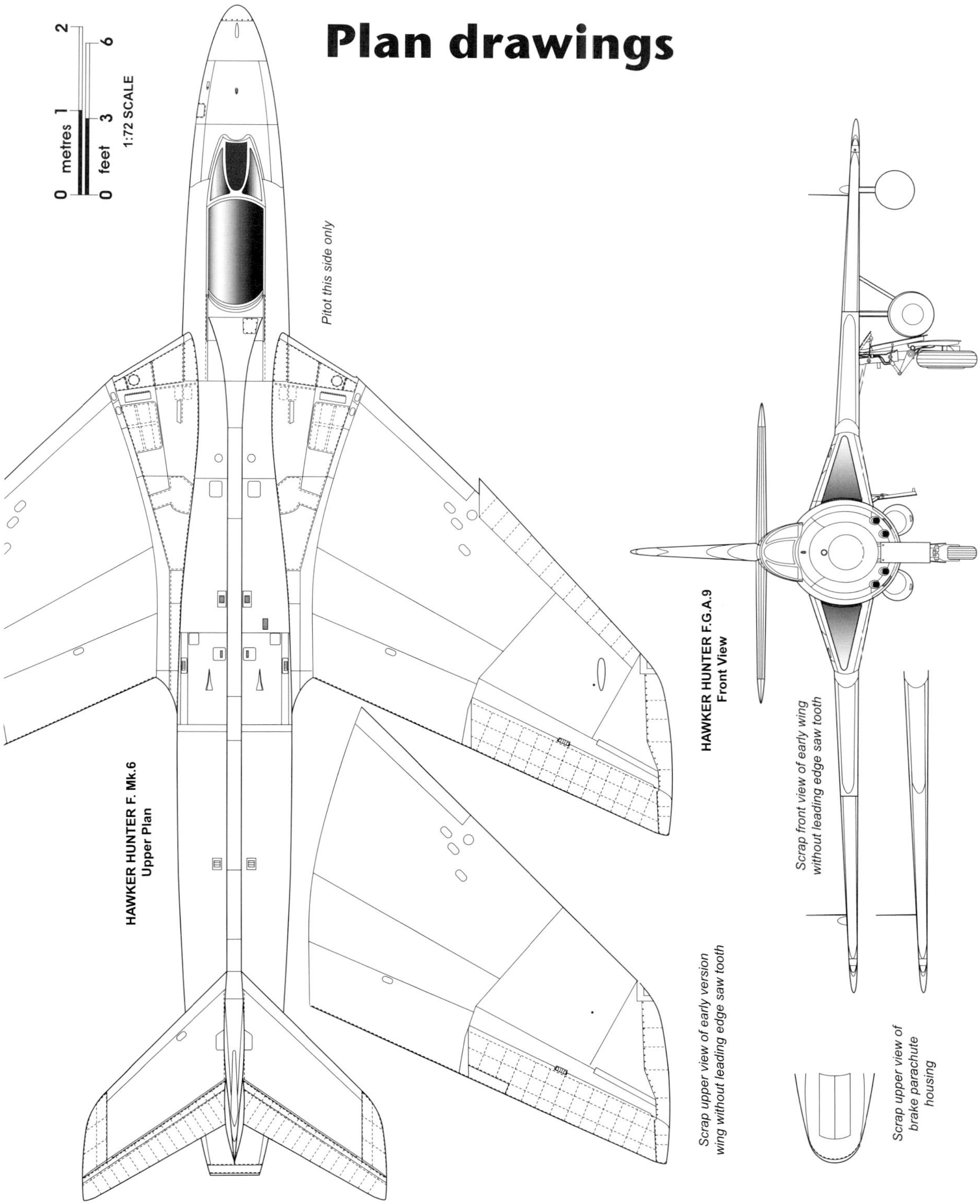

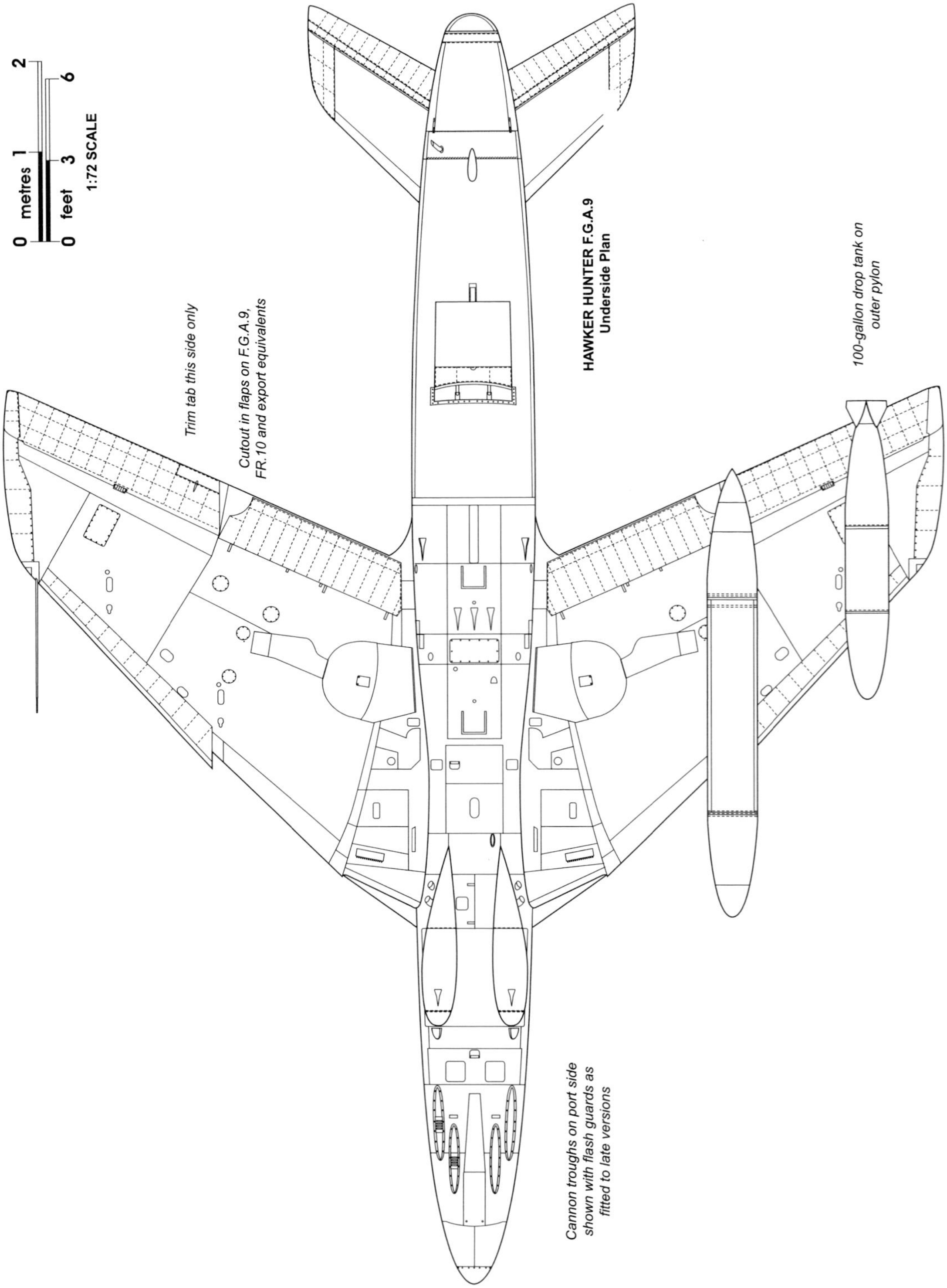
0 metres 1 2
0 feet 3 6
1:72 SCALE
Trim tab this side only
Cutout in flaps on F.G.A.9, FR.10 and export equivalents
HAWKER HUNTER F.G.A.9
Underside Plan
100-gallon drop tank on outer pylon
Cannon troughs on port side shown with flash guards as fitted to late versions

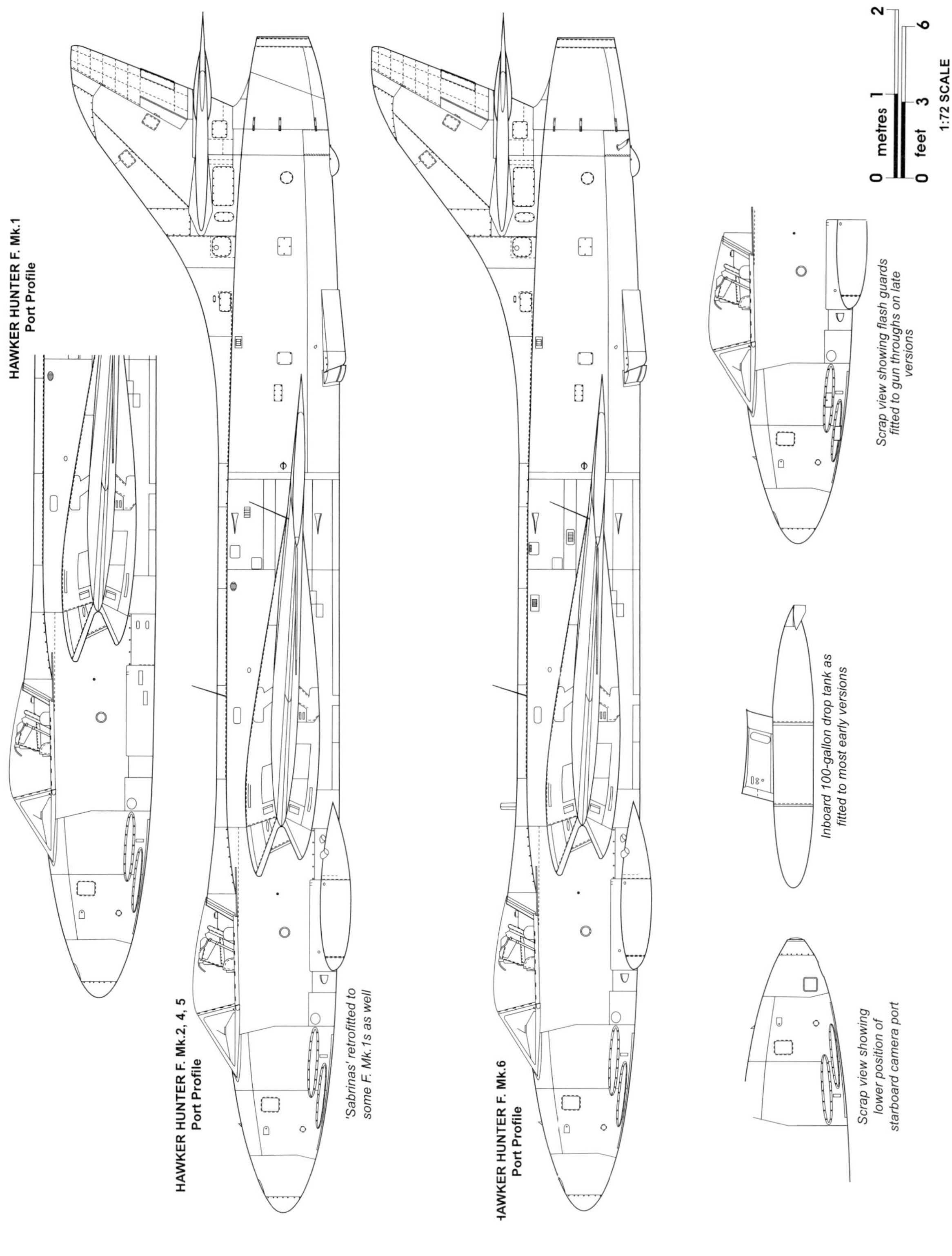
HAWKER HUNTER F. Mk.1
Port Profile
HAWKER HUNTER F. Mk.2, 4, 5
Port Profile
'Sabrinas' retrofitted to some F. Mk.1s as well
HAWKER HUNTER F. Mk.6
Port Profile
0 metres 1 2
0 feet 3 6
1:72 SCALE
Scrap view showing flash guards fitted to gun throughs on late versions
Inboard 100-gallon drop tank as fitted to most early versions
Scrap view showing lower position of starboard camera port

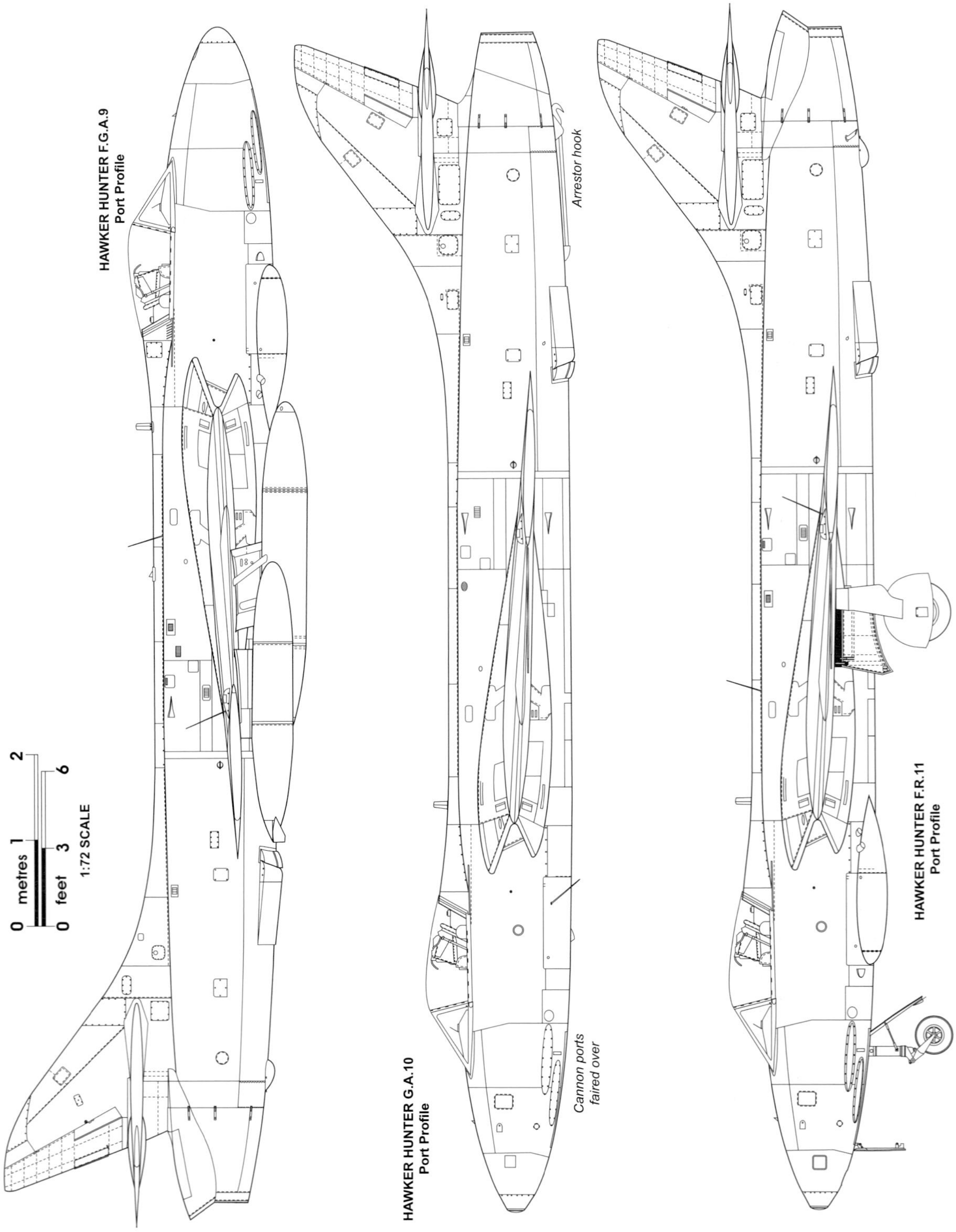
HAWKER HUNTER F.G.A.9
Port Profile
0 metres 1 2
0 feet 3 6
1:72 SCALE
Arrestor hook
HAWKER HUNTER G.A.10
Port Profile
Cannon ports
faired over
HAWKER HUNTER F.R.11
Port Profile

Appendix 7

Detail photographs

Fuselage flush intake and exhaust louvers.

Port wing root and intake fairing.

Underside of starboard wing with aileron hinge detail.

Starboard lower nose illustrating link collector fairing.

Fuselage airbrake looking rearwards.

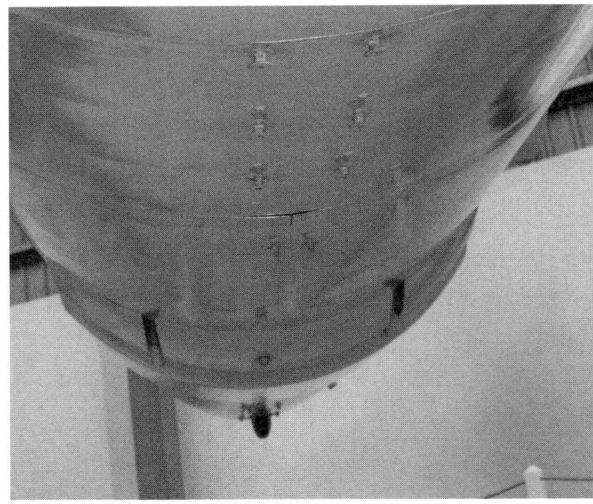

Lower fuselage looking rearwards.

Nose, looking forward (gun link collectors).

Starboard windscreen.

Port main undercarriage bay.

Nose gear door.

Port main landing gear.

Nose landing gear.

Starboard fuselage intake and exhaust louvers.

Port cartridge ejector.

Starboard wing root auxiliary intake.

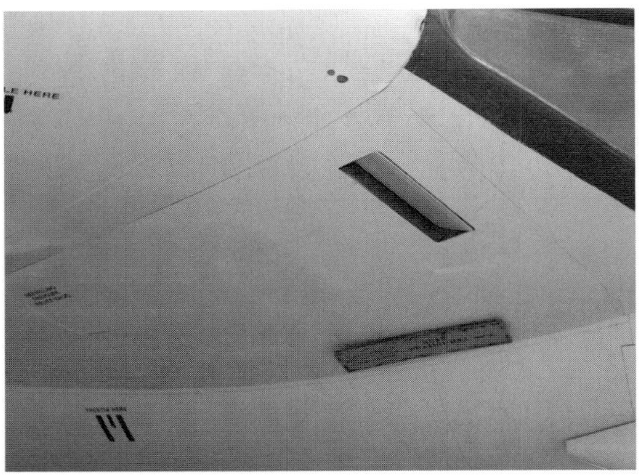

Starboard lower wing root.

Starboard link collector fairing.

Wing pitot tube.

Port rear fuselage.

Tail bullet fairing.

Port wing aileron.

Fuselage airbrake.

Tail fairing and navigation light.

Starboard forward fuselage.

Avon engine cutaway (forward section).

Starboard gun muzzles.

Port main landing gear.

Port main landing gear well.

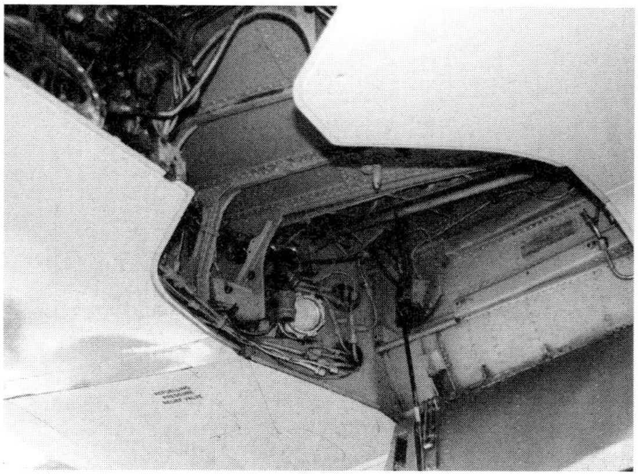

Starboard main gear well.

Nose gear bay looking forward.

Main landing gear link.

Port main landing gear.

F Mk 6 cockpit.

Radar detail, nose cone removed.

Appendix 8

Pilot's Notes

PART I

DESCRIPTIVE

Note: Throughout this publication the following conventions apply:

(a) Words in capital letters indicate the actual markings on the controls concerned.

(b) Unless otherwise stated, all airspeeds and Mach numbers quoted are 'Indicated'.

(c) Fuel poundage figures are all calculated at 7.7lb/gallon (normal AVTAG).

1. Introduction

(a) The Hunter F.6 is a single-seat fighter aircraft powered by a single Avon 203 axial-flow gas turbine engine. (See para 10.)

(b) The aileron and elevator, which are fully powered, both have manual reversion; a variable-incidence tailplane is fitted. Provision is made for fitting an electric flying tail. The cockpit is pressurised and fitted with a Mk 2H or Mk 3H pilot ejection seat which is interconnected with the hood jettison mechanism.

2. Tanks

(a) Fuel is carried in four flexible bag-type tanks in the centre fuselage and in four inter-connected tanks in the forward edge of each mainplane. Provision is made for carrying 2 or 4 x 100-gallon pylon drop tanks, which may be jettisoned electrically. (See para 54.)

(b) Each front tank contains a two-speed electrically-driven immersed booster-pump and a 3½-gallon-capacity fuel recuperator, the contents of which are used when booster-pump pressure falls, eg during inverted flight. The total contents of the recuperators are sufficient to provide for 15 seconds flight in such circumstances at full power at sea level. Two amber warning lights, labelled FUEL LEVEL 650 LB, on the left side of the GGS, are operated by float switches in the front tanks; the lights should come on when the fuel level drops to approximately 650lb (each side). Variations of up to 50lb either side of this figure can be expected.

(c) The tank capacities are:

Weight (at 7.7lb/gal)	Gallons	
Front tanks, including recuperators	1,540	200
Rear tanks	400	52
Wing tanks	1,078	140
Total, internal	3,018	392
2 pylon drop tanks (or 4)	1,540	200
Total all tanks	4,558	592

Pt. 1, Ch. 1—Fuel System

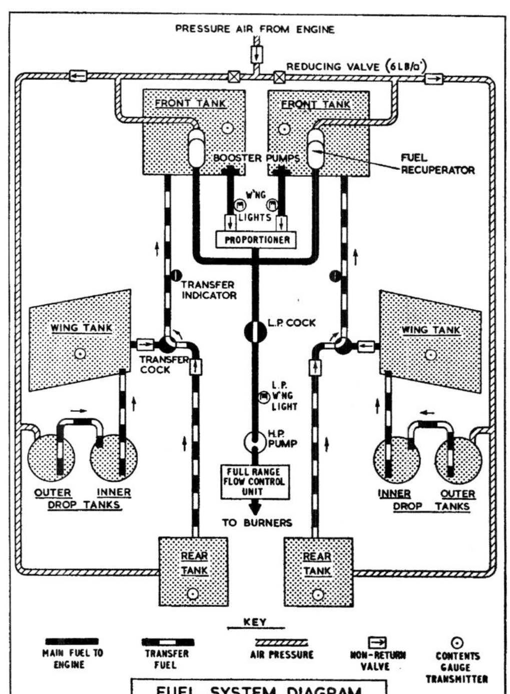

FUEL SYSTEM DIAGRAM

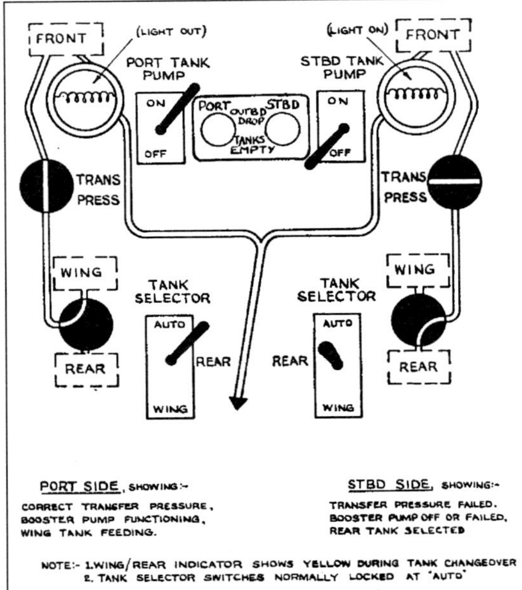

FUEL SYSTEM SWITCHES AND INDICATORS ON STARBOARD SHELF

3. Contents gauges

(a) Two electrical fuel contents gauges indicate the total fuel in the PORT tanks (front, wing and rear) and STBD tanks (front, wing and rear). There are no gauges for the drop tanks.

(b) The fuel gauges read the weight of fuel in pounds. When all gauged tanks are full, each gauge should read 1,500lb approx.

(c) Provision is made for the fitting of a flowmeter on the starboard shelf.

4. Transfer system

(a) Air pressure for the transfer system is taken from a tapping on the engine compressor and is separately fed to the fuselage rear tanks, to the recuperators in the front tanks, and to the drop tanks or the wing tanks, if the drop tanks are not fitted. When drop tanks are fitted, fuel transfer is from outer tank to inner tank to wing tank. When Mod 574 is embodied two magnetic indicators, one for each outboard wing tank, are situated aft of the fuel gauges. Each shows white when all fuel has transferred from its associated outboard drop tank.

(b) Control of the system is by two TANK SELECTOR switches, one for each side of the system. Each switch has three positions, AUTO (forward), REAR (central) and WING (aft). When the switches are set to AUTO the air pressure forces fuel from the rear tanks to the front tanks. When each rear tank is empty a float switch in the tank operates to alter the setting of the transfer cock, shutting off the rear tank and allowing the drop tanks to feed to the wing tanks and the wing tanks to feed to the front tanks. At the same time, the WING/REAR tank indicator operates to show that this is happening. Setting either control switch to WING or REAR causes transfer to take place from the respective tank. When changeover from either wing or rear tanks is taking place the indicator shows yellow and the action of the selector cock motoring over is audible over the R/T.

Although all internal tanks are gauged, and the contents reading should fall when fuel from those tanks is being used, when transfer from the drop tanks (which are un- gauged) is taking place the fuel contents gauge should show a constant reading. Should the air pressure fail, very little fuel will transfer from the drop, wing or rear tanks and TRANS. PRESS indicators will indicate failure by showing cross-line. At the same time the contents gauge transmitters in the rear and wing tanks will become inoperative and the gauges will only indicate the contents of the front tanks, ie the amount of fuel available to the engine. If only one side of the air pressure system fails, the appropriate indicator will show cross-line and the associated gauge will indicate the available fuel contents.

5. Booster-pumps

(a) The booster-pumps, one in each front tank, are controlled by two ON/OFF switches on the starboard shelf, and protected by two circuit breakers. The pumps are designed for two-speed operation. Normally they run continuously at low speed, but each at high speed is capable of supplying the maximum fuel demand to the engine. A circuit breaker for each pump is on the starboard shelf.

(b) High or low speed operation is automatically controlled by pressure switches. If pressure in the pipe-lines falls below 8lb/sq in the pumps are switched to high speed. When pressure rises above 12lb/sq in the pumps are switched back to low speed.

(c) If one pump fails, an adjacent amber warning light comes on and the other pump is switched to high speed. In such circumstances it is preferable to switch off both pumps and accept the feed provided by gravity and air pressure, unless sufficient fuel remains on the side containing the serviceable pump to complete the flight. At high engine power with both pumps off, the low pressure warning light on the starboard quarter panel may come on and the fuel recuperators discharge their contents. In these circumstances no negative-G manoeuvres should be carried out.

(d) A selector switch and an ammeter test socket on the starboard shelf are provided for servicing purposes.

6. Flow proportioner

(a) From the booster-pumps fuel passes to the flow proportioner, which ensures a balanced flow from both sides of the fuel system under normal conditions.

(b) The proportioner consists of a matched pair of vane-type pumps, mounted on a common shaft. Two inlet ports, one for each pump, accept fuel from the associated booster-pump. A single exit port passes the metered fuel through the LP cock to the engine.

(c) Should the proportioner rotor jam, fuel by-passes the proportioner vane pumps via by-passes which contain spring-loaded non-return valves.

(d) Unequal feeding is only likely to occur if inlet pressures to the proportioner differ by more than 2lb/sq in, the greater the pressure difference, the greater the amount of unequal feeding.

7. Low-pressure fuel cock

(a) The LP fuel cock control is on the port shelf and moves forward from OFF to ON. It controls the fuel flow to the engine.

8. Filter de-icing

(a) Since fuel contains a small amount of moisture an automatic de-icing system is fitted to ensure that the fuel filter is not blocked by ice crystals at low OATs. When icing occurs an electric pump is switched on and pumps alcohol to the filter. It is brought into use by the increasing pressure drop through the filter. When the pressure drop returns to normal, the system is switched off.

(b) The de-icing fluid tank, of one gallon capacity, is mounted in the engine bay.

8A. Rear tank explosion suppression system (Mod 582)

(a) General

(i) The explosion suppression system provides an automatic means of suppressing an incipient explosion within a fuel tank. An explosion is not instantaneous but requires a measurable time from the moment of ignition to the development of maximum pressure of a destructive nature. In the initial stages the rate of pressure increase is comparatively slow and hence can be suppressed in this period.

(ii) The basic components of the explosion suppression system are:

A detector unit

A suppressor unit

A power pack

(b) Detection unit

In each rear tank is a highly sensitive photo-electric cell, a detonator firing unit and a suppressor unit. The presence of any light within a fuel tank, such as the initial ignition leading to explosion, causes the photo-electric cell to pass current to the detonator firing unit.

(c) Suppressor unit

(i) When the detonator firing unit operates it causes a detonator in the centre of the suppressor unit to fire. This bursts the outer casing of the suppressor unit to release the suppressant agent. As the suppressor unit casing is suitably scored the metal remains attached to the unit after bursting and no large particles of metal are discharged into the tank.

(ii) The suppressant agent is primarily a fuel with a small quantity of extinguishant. When this mixture is released it has the effect of enriching the fuel/air vapour to a non-explosive ratio, thus terminating the explosion and halting the attendant pressure increase.

(iii) Permanent indication that a unit has fired is given by a match-head fuse within a viewing window at each tank. When a unit fires, the fuse operates to ignite the match-head and to disperse red oxide over the viewing window.

(d) Power pack

A power pack, operated by 115 volt, 400 C/S AC, is fitted forward of the instrument panel. This transforms and rectifies the AC supply to give the required DC voltages to operate the system.

(e) Control of the system

Located on the starboard wall adjacent to the cockpit lights switches is a NORMAL-RESET switch, spring loaded to the NORMAL position and a neon indicator. With the system inoperative the neon indicator glows. When the switch is held down to RESET the power pack becomes active and the indicator is extinguished. The power pack remains active as long as input voltage is maintained. This supply is obtained from the flight instruments (No 1) inverter, or from the No 2 inverter, which acts as a standby (see para 18(f)). If the indicator glows in flight the RESET switch must be operated to bring the system back to use.

9. Pressure refuelling and defuelling

(a) Refuelling is via a connection in the port wheel bay. As each set of tanks is filled, refuelling valves automatically cut off the fuel being supplied to them. During refuelling the LP cock must be OFF, the transfer selector cock switch at AUTO and the defuelling cock OFF. A time switch adjacent to the coupling must be ON in order to energise the refuelling circuit.

(b) Defuelling is via the same coupling. During defuelling the LP cock must be OFF, the transfer selector cock switches at AUTO and the defuelling cock ON. An air pressure of 10lb/sq in is necessary to transfer fuel to the front tanks whence it is either sucked out by the bowser pump or pumped out by the booster-pumps. The air pressure connection is on top of the centre fuselage.

ENGINE CONTROLS

10. Avon 203 engine

(a) General

The engine is a 15-stage axial-flow gas turbine developing 10,000lb static thrust at sea level. The main engine systems include:

Variable swirl vanes. (See para 12.)

A liquid fuel starting system. (See para 13.)

Relighting facilities. (See para 14.)

An anti-icing system. (See para 16.)

A high-pressure fuel system controlled by a flow control unit.

Self-contained oil system.

(b) High pressure fuel pumps

The twin engine-driven HP pumps share a common housing. A servo-control system controls the total pump output and a governor limits overspeeding of the engine.

(c) Flow control unit

The servo control system is operated by the flow control unit. The control unit, which contains the throttle valve, HP cock, barometric pressure control (BPC), acceleration control unit (ACU), and rpm datum control varies fuel flow to the engine according to throttle position, altitude, forward speed and rpm.

(d) Oil system

Oil is carried in the engine sump, the capacity of which is 9½ pints. One pressure and four scavenge pumps maintain a continuous circulation through a cooler and filter to the engine bearings and gears.

11. Throttle/HP fuel cock

The high-pressure fuel cock is controlled by the throttle lever. When the lever is fully back in the HP COCK gate the fuel supply to the engine is cut off. When the lever is moved forward of the gate, sufficient fuel is available to maintain idling rpm. A spring-loaded trigger on the throttle lever prevents the throttle being returned through the HP cock gate, once it is forward of that position. A throttle damper is aft of the control.

12. Automatic variable swirl vanes

(a) The first row of stator blades in the engine compressor consists of variable swirl vanes which assist in imparting swirl to the incoming air. When normal flight rpm are reached, the swirl vanes move progressively to the minimum swirl position, which occurs at about 7,000rpm.

(b) The swirl vanes have no noticeable effect on engine operation. However, until the swirl vanes reach the minimum swirl position at about 7,000rpm the compressor is not operating at maximum efficiency.

13. Starting system

(a) Starting is by means of a liquid fuel starting system. This type of system is similar in operation to a cartridge starter system in that the gases from the burning monofuel drive the starter turbine and thus rotate the engine. Instead of starter cartridges, liquid isopropyl nitrate, stored in a tank, is fed to and ignited within the starter motor.

(b) The main control switches are:

(i) The battery master switch.

(ii) The ignition switch, which controls the current to the igniter plugs. It should normally be locked in the ON position.

(iii) The starter master switch and circuit breaker. These complete the circuit through the starter button, ignition switch and relight switch. When the relight switch is at RELIGHT the starter button is isolated. The starter master switch also controls the electrical supply to the engine instruments, and completes the circuits to No 2 inverter and the booster-pumps. The switch should be ON and the circuit breaker in at all times when the engine is running.

(iv) The starter pushbutton, which initiates the starting sequence.

(c) When the pushbutton is pressed the starting cycle is initiated. For the first two to three seconds the starter motor combustion chamber is purged by air under pressure. Then fuel is fed to the chamber where it is ignited and causes the motor to turn the engine compressor. At the same time the engine high-energy igniter units operate to ignite the fuel spray in the engine combustion chambers. After approximately 7-10 seconds the engine becomes self sustaining and as the starter motor reaches its top speed, a switch operates to shut down the liquid starting system. The engine should then accelerate to the normal idling rpm of 2,500.

(d) Should the starter fuel fail to light up, a pressure-sensitive switch in the starter fuel line operates to cut off the starter system.

(e) It is important that personnel keep clear of the starter exhaust during starting. The exhaust is aft of the engine starter access door beneath the fuselage.

14. Relighting system

A RELIGHT switch is situated on the port shelf, and when held to RELIGHT for two seconds completes the circuit, through a time delay switch, to the high-energy ignition units. The ignition then operates automatically for 30 seconds. The switch is spring-loaded to return to the off position.

15. Top temperature control

Note: On some early aircraft the system is inoperative and the switch is wire-locked OFF.

(a) The top temperature control is brought into use when the switch on the port shelf is set to ON and the nosewheel is unlocked. Should the jet pipe temperature then rise above the maximum for full throttle (685°C), an automatic control reduces the fuel flow to the engine, thereby reducing the jpt. When the switch is OFF the controller is not in operation and jpt must then be controlled by the movement of the throttle lever.

(b) The switch should normally be ON at all times. If it is suspected that the system has failed, the switch should be set to OFF; the jpt must then be controlled by throttle movement.

16. Engine instruments and anti-icing

(a) The engine instruments comprise a jet pipe temperature gauge, an oil pressure gauge and an rpm indicator.

(b) An engine anti-icing OPEN/SHUT switch and indicator are above the starboard shelf. (See para 59.) Hot air is tapped from the engine compressor and with the switch OPEN is ducted to the inlet guide vanes. The system is not intended for de-icing, but to prevent ice forming.

17. Engine fire extinguisher

(a) Two fire-extinguisher bottles, stowed between the air intakes just forward of the engine, are together connected to the engine extinguisher inlet connection. Operation of the system is either by:

(1) A manually-operated pushbutton in the cockpit, on the starboard coaming, or

(2) Two automatically operated inertia switches, one in the gun-bay and one in the radio bay connected in parallel, which operate if a crash landing occurs. Operation of these switches also causes the batteries to be isolated irrespective of the position of the battery master switch.

(b) Twelve resetting flame detector switches are situated round the engine and forward part of the jet pipe. Operation of any of the switches causes the ENGINE FIRE warning light incorporated in the pushbutton to come on, provided that electrical power is available. When the button is pressed the extinguishers discharge their contents through two spray rings, one round the engine compressor and the other round the turbine nozzle box. If the fire is extinguished, the light goes out as the flame switches cool.

(c) The warning light may be tested by operating the switch on the starboard shelf.

(d) When the battery master switch is OFF the fire-extinguishers can be operated only by the inertia switches. The battery master switch must be ON to test the warning light or to operate the system by pushbutton.

MAIN SERVICES

18. Electrical system (24 volt)

(a) Two 6,000-watt engine-driven generators supply the whole of the electrical system and charge two 12-volt aircraft batteries connected in series. Later aircraft will be fitted with two 24-volt batteries connected in parallel, thus materially increasing the available reserve should generator failure occur. The generator failure warning lights situated below the centre of the instrument panel come on only when their associated generator is not supplying power. Both generators should normally cut in at 1,800rpm and cut out at 1,600rpm but this will depend on battery condition.

(b) Control of the batteries is effected by a BATTERY MASTER switch below the centre of the instrument panel. When set to OFF the switch isolates all electrical services from the batteries except the engine fire-extinguisher inertia switches. When the fire-extinguisher inertia switches operate, the batteries are automatically isolated irrespective of the position of the battery master switch.

(c) An external ground supply socket is accessible via the radio bay door and is for use when a ground battery is necessary for servicing purposes.

(d) The accompanying table lists the various electrical warning lights and indicators in the cockpit. Failure in any circuit (except radar ranging) incorporating a 'doll's-eye' magnetic indicator is shown by the indicator showing white.

Electrical warning lights

Service	Indication	Function
Fire warning	1 red light	Gives warning when temperature in engine bay exceeds 300±30°C
Fuel pressure warning	1 red light	Indicates fuel pressure low at engine inlet.
Fuel transfer warning	2 magnetic indicators	Indicate failure of transfer system.
Fuel, front tanks level	2 amber lights	Indicate contents of front tanks below approx 650lb (each)
Wing/rear tank transfer	2 magnetic indicators	Indicate transfer from wing/rear tanks
Generator failure warning	2 red lights	Indicate generator failure due to: (1) Cut-out not closed (2) Fault in circuit
Hydraulic failure warning	1 red lights	Indicates hydraulic pressure below 600lb/sq in
Undercarriage position	3 red or green lights	Indicate position of each U/C unit separately.
No light: unit locked up.		
Red light: unit between locks.		
Green light: unit locked down.		
Undercarriage warning	1 red light	Indicates when throttle closed and U/C not locked up. AL1.
Powered controls	2 white magnetic indicators	Indicate separately disengagement of aileron or elevator hydro-boosters or fault in electrical circuit.
Airbrake position	1 white magnetic indicator	Indicates airbrake not fully in.
Cockpit pressure warning	1 red light	Indicates drop of 0.5lb/sq in in cabin pressure differential.
Telebriefing	1 amber light	Indicates telebriefing in use.
Radar ranging warning	1 white magnetic indicator	Indicates radar ranging in use.
Radar ranging operation	1 blue light	Indicates radar ranging locked on target.

(e) Emergency battery

An emergency battery is provided to supply the emergency lamps, the Manual emergency selector circuit (Mod 502) and, when Mod 488 is embodied, the turn and slip indicator. If fully charged the battery will provide between 7 and 20 hours' continuous use, depending on the load imposed on it.

(f) Instrument supply

(i) The turn and slip indicator is operated by DC whenever electrical supply is available. If electrical supply is lacking the word OFF appears in a window in the face of the instrument. When Mod 488 is embodied the turn and slip indicator may be operated by the emergency battery after setting on the switch below the instrument.

(ii) AC for the Mk 4F compass, artificial horizon, oil pressure gauge and GGS radar ranging is supplied by two inverters. No 1 inverter normally supplies the flight instruments and No 2 inverter normally supplies the radar ranging. The engine starter master switch controls the circuit to No 2 inverter. No 1 inverter is brought into circuit when the undercarriage is not locked down. During taxiing and take-off the AC supply is from No 2 inverter. The RADAR TEST switch may be used to test the inverters before flight (see para 61).

(iii) If No 1 inverter fails in flight, automatic changeover occurs, and No 2 inverter then supplies the flight instruments, the load to the radar ranging equipment, if in use, being shed.

(iv) Located on top of the AC junction box at the aft end of the starboard shelf are two circuit breakers (one for each inverter supply), a magnetic indicator (which gives indication of inverter changeover) and a STANDBY TESTING/NORMAL FLIGHT switch (for servicing purposes only). This latter switch should always be set to NORMAL FLIGHT (inboard). In flight none of these is visible or accessible.

19. Hydraulic System

(a) An engine-driven hydraulic pump maintains a live-line pressure of 2,850±150lb/sq in for the normal operation of the:

Undercarriage and doors
Flaps
Wheel brakes
Aileron and elevator hydroboosters
Airbrake

The main pressure is indicated by the central needle of the triple pressure gauge at the forward end of the port shelf.

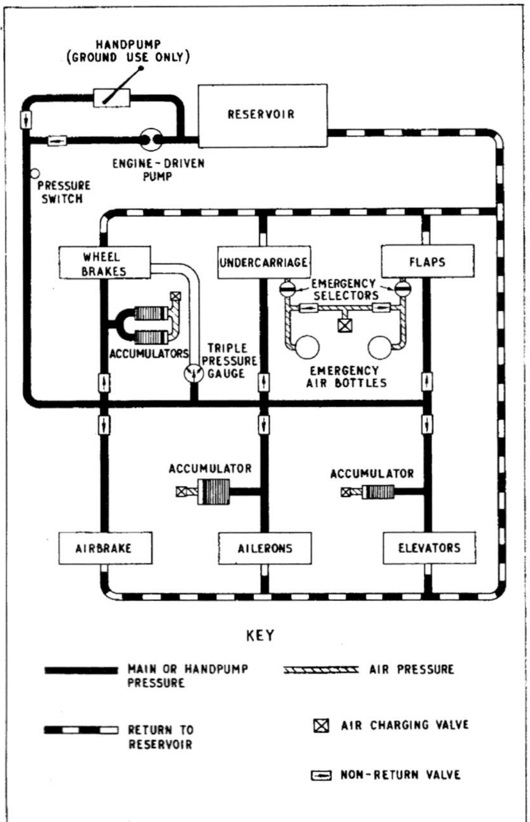

SIMPLIFIED HYDRAULIC SYSTEM DIAGRAM

(b) Five hydraulic accumulators are fitted in the circuit to provide a reserve of power in an emergency. The hydraulic side of the accumulators is connected to the wheel brakes (two), aileron hydroboosters (two), and elevator hydrobooster (one) respectively.

(c) The pressure available in the wheel brakes accumulators is shown on a gauge on the port shelf. The accumulators are charged with air to a pressure of 750lb/sq in via a connection in the nosewheel bay. The aileron accumulators are charged to 900lb/sq in via a connection in the starboard wheel bay and the elevator accumulator is charged to 1,575lb/sq in via a panel in the port side of the fin. The pressure in the elevator accumulator is shown on a gauge visible through a transparent window on the fin; the pressure in the two aileron accumulators is shown on a gauge adjacent to the charging point.

(d) Immediate indication of failure of the live-line system is given audibly by a high-pitched noise over the pilot's headset, and visually by a red light on the port side of the instrument panel, which comes on when pressure in the system has fallen to 600lb/sq in and also by the reading of the triple pressure gauge main needle. The audio warning cut-off switch on the starboard side of the instrument panel enables the noise to be silenced once the warning has been given.

(e) Two high-pressure air bottles, which are charged to 1,800-2,000lb/sq in, are provided for the emergency lowering of the undercarriage and flaps respectively. Air pressure gauges for the bottles are mounted on the port shelf.

(f) A hydraulic handpump, for ground test purposes, is fitted in the engine bay.

POWERED FLYING CONTROLS AND TRIMMERS

20. Flying controls operation

(a) In power

The ailerons and elevators are power-operated, the power being supplied by hydraulic oil under pressure from the aircraft hydraulic system. A hydrobooster consisting of a valve and jack body and jack piston is fitted close to each control surface. The control column is connected direct to the valve control rod so that, depending on the direction of movement of the control column, the valve opens to admit pressure oil to one side of the jack piston and at the same time the other side is opened to return. The piston rod is anchored to the aircraft structure by means of spring-loaded hydraulically operated pawl. The jack body is directly connected to one end of the control surface. When the pawl is in position and hydraulic pressure is fed from the valve to one side of the jack piston, the jack body moves relative to the piston and deflects the control surface. When control column movement ceases, the valve closes causing a hydraulic lock that prohibits further movement of the jack body and control surface.

(b) In manual

(i) Manual operation of the controls may be selected deliberately by operating a switch in the cockpit (see para 21) provided that electrical power is available, or will happen automatically if hydraulic pressure drops below 200lb/sq in.

(ii) When operating in manual, control surface movements are achieved through a mechanical linkage. The spring-loaded hydraulically operated pawls automatically disengage and release the anchored jack piston rods, allowing them and the jack bodies to move freely with the control surfaces. The controls are heavy in Manual except the ailerons.

(c) Hydraulic power reserve

Accumulators are fitted in the power control circuits to provide a reserve of power if the main hydraulic supply fails. This reserve may be sufficient for a maximum of $3^1/_2$ complete cycles of aileron and elevator operation before the controls revert automatically to Manual, but even if no control movement is made, accumulator pressure will not be maintained for a long period due to seepage through the hydraulic components. With some types of hydraulic failure immediate reversion to Manual will result.

21. Controls and Indicators

(a) Pre-Mod 452 aircraft

Two ON/OFF switches are on the instrument panel and control electrically the hydraulic cocks, one for the aileron circuit and the other for the elevator circuit. When the engine is running and a switch is selected ON hydraulic pressure is fed to that circuit and to the pawl, which engages in the jack piston rod.

(b) Post-Mod 452 aircraft

When Mod 452 is embodied a revised wiring and switching arrangement is applied to the aileron and elevator pawl release units so that Manual is automatically selected if any pawl disengages even though momentarily. The aircraft can therefore be flown only with Power correctly engaged or in Manual. The two ON/OFF switches of pre-Mod 452 aircraft are replaced by two ganged MANUAL-Off-POWER selector switches, which are spring-loaded to the central (off) position. Deflection upwards to POWER isolates the fail-safe circuit to allow initial engagement or re-engagement. Deflection downwards immediately selects Manual.

(c) All aircraft

(i) Two magnetic indicators, one for the aileron circuit and the other for the elevator, are mounted beside the switches and show black when the piston rod locking pawls are correctly engaged and white when the appropriate pawl is disengaged or incorrectly engaged, or alternatively when electric power to the indicator is not available.

(ii) An audio warning over the pilot's headset warns the pilot if the pump is losing pressure. Further warning is given by the red warning light and the reading of the hydraulic pressure gauge.

(d) Manual emergency selector buttons (Mod 502)

When Mod 502 is embodied, two yellow and black striped pushbuttons are fitted at the bottom left-hand corner of the instrument panel to permit the emergency selection of Manual of either ailerons or elevators or both, should the aircraft electrical system have failed. When either button is pressed an electrical supply from an independent dry battery is connected to the appropriate selector cock and the associated controls then revert immediately to Manual.

Note: These buttons must not be pressed if normal electrical power is available as in this event it is likely that the control circuit fuse will be blown.

22. Elevator feel and trim

(a) The air loads on the elevator are resisted entirely by hydraulic jack effort, no load being felt on the control column. To provide control feel, a spring is fitted in the control circuit between the control column and the variable incidence tailplane. The spring gives an artificial feel to the elevator control circuit by imposing a force proportional to stick deflection but not airspeed (ie air loads).

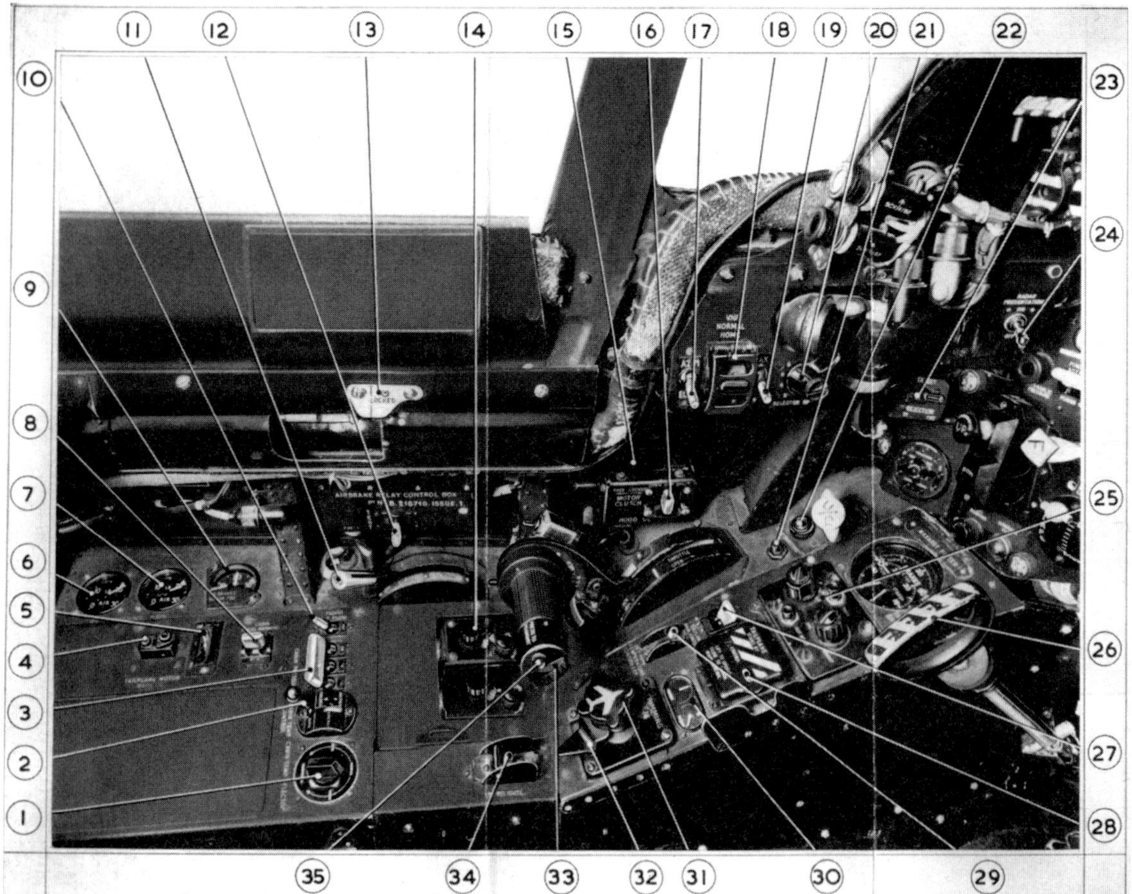

COCKPIT - PORT SIDE

KEY TO FIGURE 1

1. Cabin temperature control
2. Cabin temperature selector
3. Cabin pressure control switch
4. Hood motor circuit braker
5. Inverter selector switch
6. Undercarriage emergency air pressure gauge
7. Flaps emergency air pressure gauge
8. Cabin pressure warning test switch
9. Brakes accumulator pressure gauge
10. Flood flow control switch
11. L.P. fuel cock control
12. Airbrake test switch
13. Hood rails locked indicator
14. V.H.F. control units
15. Hood clutch selector
16. Hood control switch
17. Green Salad AZM/ELEV switch
18. Green Salad VHF NORMAL/HOME switch
19. Green Salad MAX/MIN switch
20. R.P. selector switch
21. Telebriefing push switch
22. Telebriefing warning light
23. Radar ranging target reject switch
24. Radar presentation switch
25. GGS selector dimmer
26. Hood jettison handle
27. Tailplane interconnection switch
28. Tailplane standby motor control
29. J.p.t. control switch
30. Rudder/aileron trim position indicator
31. Rudder/aileron trim control
32. Aileron trim control lock
33. Airbrake control switch
34. Relight switch
35. VHF press-to-transmit switch

KEY TO FIGURE 2

36. Airbrake positon indicator
37. Undercarriage emergency lowering control
38. Undercarriage warning light
39. Undercarriage position indicator
40. Undercarriage selector buttons
41. Flaps emergency lowering control
42. Hydraulic pressure warning light
43. Aileron power selector switch and indicator
44. Elevator power selector switch and indicator
45. Fuel level 650 lb. warning lights
46. Outboard stores jettison pushbutton
47. Inboard stores jettison pushbutton
48. Clear aircraft switch
49. RP Ripple/Normal switch
50. Bomb fusing switch
51. Bombs or inner tanks/RP switch

COCKPIT - FORWARD VIEW

Key to Figure 2—continued

52. Spin, aileron neutral white datum
53. Inverter changeover indicator
54. Mk. 8 GGS
55. Radar ranging lights
56. Fire extinguisher pushbutton and warning light
57. Accelerometer
58. Autostabiliser switch (inoperative)
59. GGS master switch
60. D.M.E. indicator
61. Green Salad indicator
62. Cabin pressure altimeter
63. Cabin pressure warning light
64. Oxygen contents gauge
65. Camera aperture switch
66. Audio warning cut-off switch
67. Fuel low pressure warning switch
68. Turn and slip indicator normal/emergency su switch
69. Generator failure warning lights
70. Ignition switch
71. Engine starter pushbutton
72. Engine starter master switch
73. Pressure head heater switch
74. Camera master switch
75. Rudder pedals adjuster
76. Battery master switch
77. Oil pressure gauge
78. Flaps position indicator
79. Flaps control
80. Radar ranging on/off switch
81. Radar ranging indicator
82. Triple pressure gauge

KEY TO FIGURE 3

83. Port outboard drop tank empty indicator (starbc similar)
84. Port fuel gauge (starboard similar)
85. Mk. 17E oxygen regulator
86. Explosion suppression normal/reset switch
87. Explosion suppression indicator light
88. Emergency lamp switch
89. Compass lamp normal/standby switch
90. Engine anti-icing indicator
91. Engine anti-icing switch
92. U/V lamps on/off dimmer switch
93. Red lamps on/off dimmer switches
94. Navigation lights switch
95. Anti-G system on/off cock
96. Anti-G system test button
97. Anti-G system pressure gauge
98. Port booster pump circuit breaker
99. Starboard booster pump circuit breaker

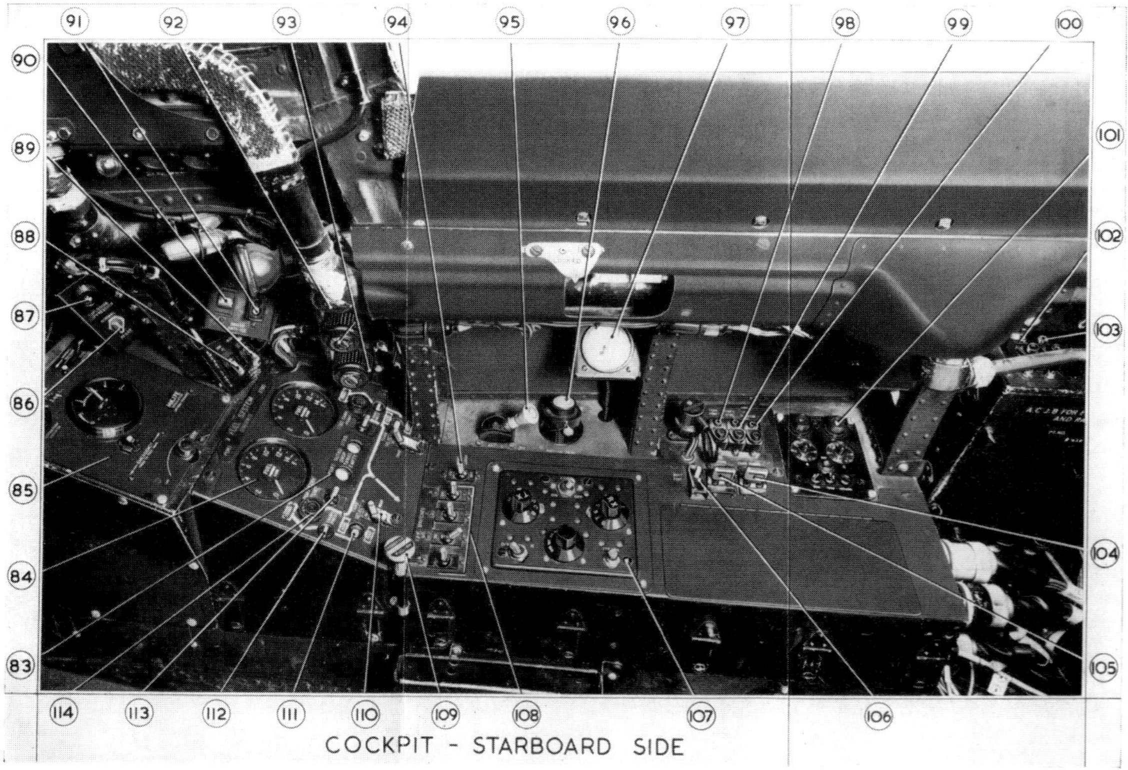

Key to Figure 3—continued

100. **Engine starter circuit breaker**
101. **Mk. 4F compass test panel**
102. **No. 1 inverter circuit breaker**
103. **No. 2 inverter circuit breaker**
104. **Booster pumps test switch**
105. **Fire warning circuit test switch**
106. **Butt test switch**
107. **Rebecca control panel**
108. **I.F.F. switches**
109. **Emergency oxygen manual control**
110. **Port fuel tank selector switch (starboard similar)**
111. **Port fuel tanks wing/rear indicator (starboard similar)**
112. **Port fuel tanks transfer failure indicator (starboard similar)**
113. **Port booster pump on/off switch (starboard similar)**
114. **Port booster pump failure warning light (starboard similar)**

(b) (i) An electrically operated variable-incidence tailplane is provided. It is hinged at the rear end, while a projection on the leading edge is connected to an electrically operated actuator mounted in the dorsal fin below the tailplane. After selecting by feel the required tailplane angle to give zero stick force for a given flight condition, the elevator is trailing with no air load imposed on it. The actuator is operated by one of two electric motors, the main and the standby. The main motor is controlled by a thumb switch on top of the stick and the tailplane operates in the same way as a conventional trimming tab. The tailplane main motor circuit is protected by a circuit breaker on the port shelf.

(ii) The standby motor, which is fitted to meet the case of failure of the main motor, or its electrical supply, is controlled by a switch under a cover on the port shelf. When the cover is raised to enable operation of the standby switch, the main motor circuit is isolated. The standby motor operates at about one-third of the rate of the main motor.

(iii)The setting of the tailplane is shown on an indicator on the port side of the instrument panel.

22A. Electric follow-up tail

(a) General

An electrical interconnection enables the variable-incidence tailplane to follow-up elevator movement. A more powerful elevator booster and a faster-rate tailplane actuator are fitted. Used together they give greater manoeuvrability at high Mach numbers and the booster increases available elevator angle (and hence tailplane angle before jack stalling occurs).

(b) Selectors and indicators

(i) A TAILPLANE ON/OFF switch, by means of which the interconnection may be selected, is fitted on the port shelf. The tailplane indicator is so arranged to indicate when the tailplane is functioning as a flying control, ie the indicator moves when the control column is deflected as well as when the trim switch is actuated.

(ii) The circuit is so arranged that when the cover of the standby trim switch is raised, both the main actuators and the interconnection are isolated.

(c) Function

(i) Whenever the interconnection is switched ON, and electric power is available, it functions irrespective of whether the elevator is in Power or in Manual. It is arranged in the form of a follow-up linkage, so that for a given stick deflection and tailplane trim position a predetermined relationship exists between tailplane and elevator angles. A linkage attached to the elevator carries a switch arm, one end of which moves between two micro-switches. Movement of the elevator relative to the tailplane energises one of these switches, which operates the tailplane actuator to move the tailplane. When the tailplane reaches the required position the micro-switch is de-energised and tailplane movement stops, leaving the elevator deflected relative to the new position of the tailplane.

(ii) To prevent continual hunting of the tailplane actuator, the elevator has a small range of movement over which the micro-switches are not selected.

(iii) A spring-loaded telescopic strut is incorporated in the linkage so that full and unrestricted stick movement is always obtainable The strut telescopes when either the elevator is moved faster than the rate at which the tailplane actuator can follow up or when the tailplane has reached the end of its travel.

(d) Trimming

Trimming is carried out by means of the tab switch on the control column. In addition to altering the tailplane incidence, operation of this switch also resets the datum position of the two micro-switches. During trimming the elevator is kept in line with the tailplane by the centring action of the spring feel strut, the switch arm thus following the resetting of the micro-switches datum position. The datum setting cannot be altered by means of the standby switch.

23. Aileron feel and trim

(a) The air loads are resisted entirely by the hydraulic jack effort and no load is fed back to the control column. To provide control feel, a spring is fitted in the control circuit and gives an artificial stick force, which is proportional only to aileron deflection but not to airspeed. A handwheel on the left-hand side of the control column provides spring feel adjustment and may be used to trim the aircraft when in Power. A white line is inscribed on the handwheel. When the trimmer is neutral the line points forward and the trimmer engages in a spring-loaded detent. Movement from this position is through 300° either to the left or to the right thereby adjusting the zero force position of the stick. This trimmer should not be used in Manual.

(b) A trim tab is fitted on the port aileron. It is only intended for use in Manual and a guard is placed round the combined rudder/aileron trim control to prevent the inadvertent use of the aileron trim when in Power. A trim indicator is forward of the control.

24. Aileron gear change

(a) On reversion to Manual control, the control column/aileron gearing is automatically changed so that for the same stick movement aileron travel in Manual is approximately two-thirds of that obtained in Power.

(b) This automatic changeover is provided for by the insertion of a hydraulic jack in the aileron control linkage. When hydraulic pressure is fed to the jack (via the aileron Power selector cock) the jack extends to alter the linkage effective arm. When hydraulic pressure is lacking the jack retracts under spring pressure.

25. Engaging power controls on the ground

(a) Pre-Mod 452

When hydraulic power and electrical power are available the power selector switches should be switched ON and the control column moved to engage the power controls locking pawls. If, as is likely, the locking pawls are not opposite their slots, they clamp on the side of the piston rods; considerable force is then required to slide the piston rods and engage the pawls in the slots. When all pawls are correctly engaged the control column can be moved freely over its full travel in all directions, and is felt to come up against positive stops at the extreme end of its movement. The indicators will then show black.

(b) Post-Mod 452

To engage Power hold the appropriate switch to POWER with the left hand and move the stick with the right hand, laterally when engaging aileron and fore-and-aft when engaging elevator, until the corresponding magnetic indicator goes black. The switch can then be released. To engage aileron power more easily it is advisable to either:

(i) Select POWER with the stick held hard over to starboard and then move it to port, or

(ii) Attempt engagement immediately after light up before hydraulic pressure builds up

If Power has not been engaged the controls will revert to Manual when the switch is released.

26. Re-engaging power controls in the air

Note: The information on false anchorages applied mainly to pre-Mod 452 aircraft. In post-Mod 452 aircraft false anchorages can still occur when re-engaging Power in the air only so long as the selector switch is held at POWER.

When a re-selection to Power is made in the air, it is possible that the locking pawls will not engage in the slots on the piston rods, but merely clamp on the side of the rods giving a false anchorage: the magnetic indicator remains white. False anchorages can be caused by the pistons 'creeping' to full travel during a lengthy period of Manual flying or when the ailerons are deflected appreciably either by the control column or by the presence of aileron up or down float when selecting Power ON or OFF. False anchorages can give two types of restriction.

(a) One-way restriction (see para 88)

This usually occurs as a result of re-selecting Power with the ailerons deflected, eg when initiating a turn. The locking pawls clamp on the side of the piston rods in opposite senses relative to the slots, eg one rod extended, the other retracted. There will be apparent power-operated movement in one direction due to the clamping of the pawls to the rods. Movement of the control column in the other direction is restricted since power assistance is not available and not only has the friction clamp of both pawls to be overcome but the ailerons have to be deflected manually. This type of restriction can also occur as a result of having one pawl correctly engaged and one pawl out of engagement.

(b) Both-ways restriction (see para 88)

This usually occurs as a result of re-selecting Power ON with the ailerons floating up or down, eg when easing out of a dive. The pawls will grip the piston rods in the same sense relative to their slots, eg both rods extended or both retracted, giving jamming of the control column in the neutral position. Movement of the control column in either direction is restricted by the friction clamp of one pawl and the ailerons having to be deflected manually.

OTHER AIRCRAFT CONTROLS

27. Rudder bar adjustment and rudder trim

(a) The rudder bar is adjustable for leg reach by means of a control at the bottom centre of the instrument panel. When the control is pulled out a plunger is disengaged from a hole in the adjusting shaft, thus allowing the rudder bar arms to be pushed forward against the pressure of a spring or to swing aft by spring pressure. When adjustment is complete, releasing the control to allow the plunger to engage in the nearest hole will fix the rudder bar in the desired position.

(b) The rudder trim tab is actuated electrically by the combined aileron/rudder trim control. The tab position indicator is forward of the control.

28. Flying controls locking gear

(a) Internal locking

The internal locking device consists of four cables which are joined to a clip, designed to attach to the control column. The other end of the cables incorporate hooks which should be attached to the safety harness lugs on the ejector seat and to the outboard edges of the rudder pedals. When fitting the cables the rudder pedals should be in the central position and the cables tensioned by movement of the seat backrest pan.

(b) External locking

Locking clamps are provided for all flying control surfaces.

29. Undercarriage control and position indicator

(a) The tricycle undercarriage is operated hydraulically after electrical selection by either the UP or DOWN pushbutton, on the port side of the instrument panel.

(b) A standard position indicator is fitted to the left of the pushbuttons. A warning light below the pushbuttons comes on if, with the undercarriage NOT locked up, the throttle is closed.

30. Undercarriage emergency operation

(a) Should electrical or hydraulic failure occur, irrespective of the position of the normal selector buttons (but see para 95), the undercarriage may be lowered by pulling the U/C emergency release on the port shelf after first pushing in the central knob. This admits air from the emergency bottle to the wheel unit jacks, forcing them to lower and lock. The available air pressure is shown on gauge at the aft end of the port shelf.

(b) If it is required to retract the undercarriage on the ground only the UP selector push button should be twisted clockwise and then pressed. This is inoperative if the undercarriage emergency system has been used.

31. Flaps control and position indicator

(a) The flaps are selected electrically and operated hydraulically. Selection is by means of a switch on the port side of the instrument panel which provides UP, DOWN (80°) and six intermediate positions (15°, 23°, 30°, 38°, 45°, 60°).

(b) The flaps may be selected to any of the above positions but the extent to which they will lower depends on the air loads. If speed is increased with the flaps extended the angle will be adjusted to suit the air loads.

(c) A flap position indicator is fitted adjacent to the selector switch.

32. Flaps emergency operation

Should electrical or hydraulic failure occur, the flaps may be lowered fully down only by the emergency air system. When the FLAPS EMERGENCY LOWERING control on the instrument panel is pulled, after first pressing in the central knob, the air is directed to the lowering jacks. The available pressure is shown on a gauge at the aft end of the port shelf.

33. Airbrake control

(a) An under-fuselage airbrake is provided, which is electrically selected and hydraulically operated. It will extend fully at any speed.

(b) Control of the airbrake is by means of a three-position switch, spring-loaded to the central off position, on the throttle lever. No position other than fully in or fully out can be selected. A magnetic indicator on the instrument panel shows black only when the airbrake is fully in.

(c) The airbrake is automatically inoperative when the undercarriage is lowered. If the undercarriage is lowered when the airbrake is out, the airbrake will automatically retract. Airbrake IN must not normally be obtained by selecting undercarriage down.

(d) A spring-loaded switch on the port wall enables the airbrake to be tested on the ground. When the switch is held to TEST the airbrake extends through 10° only and then retracts with a slight bump. The indicator shows white momentarily when the airbrake is out.

34. Wheel brakes control

(a) The wheel brakes are operated hydraulically by means of a lever attached to the forward face of the control column and a differential relay controlled by the rudder bar.

(b) The live-line pressure (3,000 ± 150lb/sq in) is shown on a triple pressure gauge together with the pressure at each main wheel (1,500-1,650lb/sq in).

(c) Should the hydraulic system fail, the pressure in the wheel brakes accumulator, which if fully charged should provide sufficient pressure for about 40 operations of the units, will be sufficient for landing but will leave little in hand for subsequent taxiing. The available accumulator pressure is shown on a gauge at the aft end of the port shelf. The brakes will remain effective until accumulator pressure drops to 750lb/sq in.

(d) Maxaret brake units are fitted, which permit the use of full braking when necessary without the danger of wheel locking and tyre damage. The units can come into operation only if the wheels are rotating and in no circumstances should the brakes be applied before touch-down.

COCKPIT EQUIPMENT

35. Entry to aircraft

(a) Access to the cockpit is normally gained via a special ladder, supplied as ground equipment.

(b) A single emergency footstep is located in the port side of the fuselage, below the cockpit.

36. Hood operation

(a) The hood is opened or closed electrically after selection by a three-position OPEN-OFF-SHUT switch on the port wall The hood may be stopped at any position by selecting OFF but in flight the switch must be SHUT, otherwise partial depressurisation may occur. There is a delay of about 5 seconds between selection and operation when the switch is set to OPEN. Above the switch is a clutch lever, operation of which to FREE declutches the actuator, locks the switch at OFF and enables the hood to be moved by hand. A circuit-breaker is fitted at the aft end of the port shelf.

(b) The hood seal is automatically inflated when the hood is fully closed and deflated when either OPEN is selected or the clutch release is set to FREE.

(c) Four pointers are provided, one at each end of the hood rails. Alignment of the pointers with the associated spigots indicates that the hood rails are locked and are safe for flight.

36. Hood jettisoning

(a) The hood is jettisoned by gas pressure from a jettison gun acting on two pistons, which release the hood rails and then push the hood upwards. This is initiated by pulling the handle on the port shelf. This action also operates a micro-switch, which, if electrical power is available, automatically lowers the gunsight.

(b) An external emergency release ring is inside a break panel on the port side of the fuselage below the cockpit. When the release ring is pulled the jettison gun is not fired but the hood rails are released. The hood may then be lifted clear manually.

(c) The hood is also jettisoned automatically whenever either the ejection seat blind handle or the seat-pan alternative firing handle is pulled (see para 40(a)(iv)). In this case the GGS is not automatically lowered.

35. Cockpit pressurisation, heating and demisting

Note: The system is inoperative whenever the hood is open.

(a) Hot air under pressure is bled from the engine compressor to supply the heating, pressurisation and demisting system. The air supply to the cockpit terminates in spray pipes, which provide for windscreen, side panels and hood demisting.

(b) The master ON/OFF switch controls the flow to the cockpit. The temperature control lever permits the selection of AUTO or manual HOTTER-COLDER. The TEMP SELECTOR is for use in AUTO to preselect the desired temperature, which is then maintained by a thermostat. In the manual range, temperature is controlled by first setting the lever to HOTTER or COLDER and then, when the desired temperature is reached, engaging the lever in the central gate.

(c) The cockpit altimeter should indicate in accordance with the following table. Should the cockpit pressure differential drop by more than 0.5lb/sq in a red light on the right of the instrument panel comes on.

Actual altitude (ft)	Cockpit altitude (ft)	Cockpit altitude at which warning light comes on (ft)
20,000	13,000	13,750
30,000	16,500	18,000
40,000	22,500	24,000

(d) (i) A 'flood flow' valve is incorporated in the system, which may be controlled in AUTO or MANUAL according to the setting of the FLOOD AIR FLOW control switch.

(ii) When the switch is set to MANUAL, 'flood flow' is brought into operation and is used primarily for clearing any transparency misting.

(iii) When the switch is at AUTO, should loss of cockpit pressure cause the cockpit altitude to exceed 38,000 feet, an altitude switch operates to supply 'flood' air to the cockpit.

(e) A switch for ground testing the visual warning and flood air supply is aft of the normal controls. It must not be used in flight.

39. Anti-G suit system

(a) The purpose of the system is to provide air at low pressure for the pilot's anti-G suit, the connection for which is on the port side of the ejection seat.

(b) Air under pressure is stored in two air bottles, the contents of which are indicated by a pressure gauge on the starboard wall. When the cock below the gauge is ON and G in excess of approximately 1.5 is applied, a spring-loaded valve operates and allows air to pass to and inflate the anti-G suit. The amount of inflation depends on the amount of G applied.

(c) The system may be tested, with the cock ON, by pressing the ANTIG-TEST button, adjacent to the on/off cock, as gently as possible to avoid severe discomfort due to too rapid inflation.

40. Ejection seat Mk 2H or Mk 3H

Warning:

1. The pilot must ensure that each safety pin is removed and stowed before flight and must lock each handle against the possibility of accidental withdrawal on the ground before leaving the cockpit. All personnel must ensure that each firing handle is locked (ie fabric safety strap passed through handle and secured by its safety pin) before entering the cockpit.

2. If it is necessary to leave the aircraft in an emergency on the ground special care must be taken not to foul the seat pan alternative handle if its safety pin is not in position.

(a) Mk 2H seat

(i) A Mk 2H pilot ejection seat is fitted incorporating a type ZF harness, headrest, footrests, parachute container and a seat well for the dinghy and emergency oxygen supply. A Mk 8 or 13 parachute must be used.

(ii) The height may be adjusted by a lever on the starboard side of the seat; the harness release is also on the starboard side.

(iii) A 60 ft/second or 80 ft/second (Martin Baker Mod 271) ejection gun is fitted. It is important that the pilot is aware of the type of gun fitted, since this affects the minimum ejection height (see para 102(f)).

(iv) The ejection seat gun is fired by pulling the handle above the headrest and, when Martin Baker Mod 295 is embodied, by the alternative firing handle attached to the forward edge of the seat pan between the pilot's legs. When either firing handle is operated, the hood is jettisoned immediately by gas pressure from the hood jettison cartridge; at the same time a delay unit at the back of the headrest is started. This unit withdraws the seat from the seat pan one second after the handle is pulled. The seat is then ejected. If G conditions necessitate using the seat pan alternative firing handle, the pilot must first press his head firmly against the headrest to prevent dislocation of the neck on ejection.

(v) All leads incorporate quick releases which are automatically broken on ejection.

(vi) After ejection, at heights of 10,000 feet and below, a barostat causes an automatic cycle to commence. After three seconds the safety harness is released, as are the face screen, firing handle and headrest pad. An apron attached to the seat drogue then pitches the pilot head first out of the seat, at the same time opening his parachute.

(vii) A manual override D-ring is fitted over the rip-cord D-ring and should be operated to isolate the automatic device if the system has failed.

(b) Leg restraining cords

(i) When Martin Baker Mod 272 is embodied on the Mk 2H seat, leg restraining cords are fitted in lieu of footrests. The thigh guards are retained.

(ii) The leg restraining cords ensure that the occupant's legs are drawn back automatically and restrained close to the seat pan during ejection, thus providing leg clearance and preventing the legs being blown apart after ejection. The cords should be fitted as follows:

1. Secure the leg straps below the knees with the D-rings to the rear.

2. Pass the left nylon cord through the right leg D-ring under the safety harness lap straps and insert the right shoulder harness eye-piece through the loop on the cord. Secure the shoulder harness in the quick release box.

3. Repeat for the other cord, passing the right cord through the left-leg D-ring and attach the loop to the left shoulder harness. Secure the harness.

The restraining cords are fastened to the floor in brackets, the rivets of which shear at 400lb force. The cords pass through snubbing units at the front of the seat pan. These units allow the cords to pass freely down through the unit, but prevent them passing upwards. Thus on ejection the cords are pulled downwards through the unit pulling the legs close to the seat pan. The legs are held there until the safety harness is released. The cords are then pulled through the calf-strap D-rings and free the legs. A release button is provided under each snubbing unit to allow the occupant to adjust the cords to give comfortable leg movement in the aircraft.

(c) Mk 3H seat

The Mk 3H seat incorporates leg restraint cords and the 80ft/secong ejection gun. No thigh guards are fitted. A Mk 9 parachute must be used. Mod 295 introduces the seat pan alternative firing handle. The operation of the seat is as given in (a) (ii)-(vii) and (b) above.

41. Oxygen system

(a) Oxygen is carried in two Mk 5D cylinders. A Mk 17 demand regulator controls the supply to the pilot. Later aircraft may be fitted with a Mk 18 regulator.

(b) A contents gauge is mounted on the right of the instrument panel. The regulator consists of an on/off valve which controls the flow of oxygen, an air inlet NORMAL-100% OXYGEN switch, an emergency three-position switch and a combined flow and blinker unit.

(c) When the on/off valve is on and the inlet switch is at NORMAL an air/oxygen mixture is fed to the pilot's mask, up to a height at which 100% oxygen is automatically delivered. When the inlet switch is at 100% OXYGEN, no air is added, irrespective of the height. This position should be selected if any symptoms of anoxia are present. The emergency switch when moved to either right or left admits oxygen under greater pressure. Normally it should be central but should be offset if cabin pressure failure occurs.

(d) The mask may be tested before flight by firmly pressing in the emergency switch, when in the central position. Oxygen is then supplied under pressure, the firmer the switch is pressed the greater the pressure (up to 5 times that obtained with the switch in either side position). The mask can then be adjusted until no leaks are present.

(e) Emergency bottle

A manual control inboard of the starboard shelf is pulled up to turn on the emergency bottle, providing the safety pin on the emergency bottle is withdrawn.

42. Internal lighting

Three ON/OFF dimmer switches are provided on the starboard wall for the control of the internal lighting. An emergency lamp switch is on the starboard wall above the oxygen regulator. See para 18(e).

43. External lighting

The navigation lights are controlled by an on/off switch on the starboard shelf.

44. Accelerometer

The accelerometer beneath the port coaming indicates all normal accelerations imposed on the aircraft by means of three concentrically mounted pointers. One pointer indicates instantaneous G, the other two register the maximum positive and negative G readings respectively until reset.

45. Pressure head heater

The heater element in the pressure head is controlled by a switch below the centre of the instrument panel.

46. Emergency equipment

(a) Crowbar

This is clipped to the cockpit port wall.

(b) E.2 compass

This is mounted below and to the right of the GGS. Deviations of up to 12° must be expected when the GGS is in the combat position. With the Mk 8 gunsight fitted and GGS master switch on, errors of up to 40° may be experienced on the compass.

(c) Aircraft destructor

Access to this is gained through a quick release panel on the fuselage port side, level with the cockpit.

OPERATIONAL CONTROLS

47. Gyro gunsight Mk 5

Note: Later aircraft will be fitted with a Mk 8 fixed gyro gunsight.

(a) The GGS is housed in a retractable mounting above the instrument panel. Retraction is controlled by an electric motor in circuit with the GGS on/off switch on the right of the sight. An emergency lowering manual control is on the left of the sight.

(b) The sight automatically provides for both manual and radar ranging depending on the position of the throttle twist-grip. Altitude and ballistics units automatically compensate for the time of flight and gravity drop of the particular ammunition in use. The altitude unit is fully automatic but the ballistics unit must be pre-set for the particular ammunition before flight.

(c) The selector-dimmer control on the port shelf contains a dimmer switch and a five-position selector switch. The five positions are:

G (gyro)	Moving graticule only is visible.
F and G (fixed and gyro)	Both graticules are visible.
F (fixed)	Fixed graticule only is visible. The sight may only be used as a reflector sight. Radar ranging not possible.
MRP (Medium RP)	The sight is set for use with SRP (Steep RP) RPs.

48. Radar ranging (ARI 5820)

Note: The aircraft should not be landed with the system switched ON.

(a) The radar ranging ON/OFF switch on the port shelf controls the system; an adjacent magnetic indicator shows white with the system in use and black when its use is discontinued.

(b) Radar ranging automatically provides the GGS with target range information within the limits of 200-800 yards. The radar searches from maximum range down to minimum range until a target is found. If no target is found within the search area the search is automatically repeated When a target is found the radar locks on to it, and a blue light (54) on the right of the GGS lights up to indicate this.

(c) A spring-loaded TARGET REJECT IN-OUT switch is on the port wall. Its function is to unlock the radar from an unwanted target or to check that it is locked on to the correct target. When IN is selected, the radar unlocks from the target and searches inward to minimum range. If no fresh target is found the normal search cycle commences whether the switch is retained at IN or not. When OUT is selected the radar unlocks from the target and searches outwards to maximum range. It will remain at maximum range if the switch is retained at OUT but, when released, the normal search cycle is commenced.

(d) To check that the radar is locked on the correct target set the span knob of the GGS to that of the target type. If the correct target is being ranged the GGS graticule should just encompass the target span. If the target span extends beyond the graticule (over-ranging) or does not extend to the graticule (under-ranging), the wrong target is being ranged and REJECT IN or REJECT OUT respectively should be selected.

Note: (1) When the manual twist-grip is set to a shorter range than that of the target the radar cannot control the graticule until the target comes within the range set by the twist- grip, when the radar takes over the ranging.

(2) When the twist-grip is set to minimum range, the radar is overridden. This is to enable the pilot to partially cage the GGS gyro during violent manoeuvres.

(3) When the desired target is near the ground (below 5,000 feet) and the radar has locked on to the ground it is better to select REJECT IN than REJECT OUT to lock on to the target.

(e) The radar ranging ON/OFF switch on the port shelf controls the system; an adjacent magnetic indicator shows white with the system in use and black when its use is discontinued.

49. VHF equipment

(a) ARI 18064

The two 10-channel VHF controllers, with an adjacent changeover switch, are on the port shelf. A press-to-transmit switch is in the end of the throttle lever handle. The sets are stowed in the starboard equipment bay.

(b) ARI 18012

The telebriefing land-line plug is at the underside of the rear fuselage, forward of the tail bumper. When the plug is connected, the VHF circuit is de-energised and an amber warning light on the port shelf indicates that telebriefing is in use. The pilot's press-to-speak switch is adjacent.

(c) Green salad

The control switches are on the cockpit port wall. The indicator is situated on the starboard side of the instrument panel.

50. IFF Mk 3

(a) The G, G/D F and D switches are grouped on the starboard shelf.

(b) A suppressed aerial is mounted in the leading edge of the fin.

51. DME-Rebecca Mk 7 (ARI 5849) and Mk 8 (ARI 23013)

(a) The control unit is on the starboard shelf. A range and heading meter, which indicates range and left/right heading from the homing beacon, is on the instrument panel.

(b) Two suppressed aerials are fitted in the engine starter and engine access doors.

52. G45 and recorder cameras

(a) The cine and recorder cameras are operated automatically whenever the gun or camera firing switch on the control column is energised, with the camera master switch ON.

(b) The camera master switch is on the switch panel below the instrument panel. The aperture switch is on the starboard side of the instrument panel.

53. Gun firing

(a) The gun firing switch is on the forward face of the control column handgrip. The circuit is automatically isolated when the undercarriage is locked down. A butt test switch is mounted on the starboard shelf and provides an override of the automatic isolation of the gun firing circuit.

(b) When the guns are fired, an electrically operated selector is energised to open the gun bay scavenging flap. This causes air to clear the gases from the gun bay through the link and empty case chutes. When the trigger is released the scoop closes and the air flow is cut off.

54. Bomb/RP release and drop tank jettisoning

(a) Bomb release and jettisoning

(i) The armament control panel is on the port shelf. When the FUSING switch is set to NOSE or TAIL (as required), the BOMBS/RP switch set to BOMBS and the NORMAL/PRACTICE switch set as required the bombs are released by pressing the bomb/RP release pushbutton on top of the control column.

(ii) The bombs may be jettisoned in a safe condition by pressing the CLEAR A/C switch (see (e) below).

(b) RP firing

When the BOMBS/RP switch is set to RP and the RIPPLE/NORMAL and SELECTOR switches set as required, RPs are fired by pressing the bomb/RP release pushbutton. No jettison facilities are available.

(c) Inboard drop tank jettisoning

When the bomb fusing switch is set OFF, both inboard drop tanks are jettisoned by pressing the bomb/RP release pushbutton.

(d) Outboard drop tank jettisoning

When the OUTBOARD PYLON STORES pushbutton is pressed, both outboard tanks are jettisoned.

(e) 'Clear aircraft' switch

When the bomb fusing switch is OFF, all stores, drop tanks or bombs carried on the four pylons are simultaneously jettisoned by pressing the CLEAR A/C switch, on the left of the GGS.

55. Engine limitations-Avon 203

(a)

Power rating	Time limit	Rpm	Jpt °C
*Take-off and Operational Necessity	10 mins (combined)	8,000±50	685
Intermediate	30 mins	7,800	660
Max continuous	Unrestricted	7,600	630
Approach idling	–	4,500 (min)	–
Ground idling	Unrestricted	AL.1 2,500±200	–

* During climbs on the governor the governed speed may be permitted to rise to 8,100rpm. Jpt may be exceeded during hot re-slam accelerations and may be allowed to rise to 740°C for a period not exceeding 10 seconds.

(b) Airspeed limitations

The engine is subject to the following airspeed limitations at 8,000rpm:

Temperature °C	Max S/L Mach No	Min alt for 1.0M	Min alt for 1.2M
-20	1.13	Sea level	3,000 feet
+15	1.0	Sea level	8,500 feet
+35	0.82	7,000 feet	15,500 feet

(c) Oil pressures

Normal at 7,600rpm 35lb/sq in
Minimum at 7,600rpm 25lb/sq in

56. Flying limitations

(a) (i) Intentional spinning is prohibited. Stalling practice is not to be carried out below 25,000 feet nor continued beyond the buffet stage.

(ii) RPs may be carried and fired, with or without drop tanks fitted on the inboard pylons. The maximum diving angle during RP firing must not exceed 60°.

(iii) When carrying external stores, other than empty inboard drop tanks, practice selections of Manual must not be made since the presence of stores increases the difficulties of reselecting Power. Gun ammunition must be carried to maintain the CG within reasonable limits during take-off.

(b) Maximum speeds

In Power	
Clean	620 knots. No Mach limit, but see para 75
With drop tanks:	
on outboard pylons	590 knots or 0.88M
on inboard pylons	590 knots. No Mach limit, but see paras 70(h) and 75.
With RPs	590 knots. No Mach limit, but see para 70(j)
In Manual	0.75M below 15,000 feet
Clean or with stores	0.85M above 15,000 feet
Drop tank jettison	
Drop tanks without fins	250-300 knots
Drop tanks with fins	
inboard	200-450 knots
outboard	200-400 knots
Operation of:	
Undercarriage	250 knots
Flaps over full travel	250 knots (see para 70(d))
Flaps to or from 38°	300 knots or 0.9M (see para 70(d))

Note: The speed for the operation of a service also applies for flight with the service in the extended position.

(c) G limitations

(i) The following accelerometer readings must not be exceeded:

At any height			+7G
If Mod 533 is not fitted	Between 10,000 feet and 30,000 feet	Below 0.9M	The reading at which buffet commences
		Above 0.9M	+4G

(ii) An accelerometer reading of -3.75G must not be exceeded at any height.

(d) Maximum all-up weights

With RPs and inboard tanks	21,735lb
With four drop tanks	21,735lb
With two drop tanks	19,750lb
Clean	17,500lb
Landing	15,600lb

(e) Gun firing

(i) The guns may be fired on the ground.

(ii) In the air, gun firing is permitted at any height and irrespective of the length of the burst subject to the following limitations:

No of guns	Type of ammo	Maximum speed (knots)	Remarks
2	LV or HV	550	Without Mod 646
2	LV or HV	620	With Mod 646
4	LV	400	Without Mods 592 & 598
4	HV	250 below 25,000 feet	ditto
4	HV	350 above 25,000 feet	ditto
4	LV or HV	550	With Mods 592 & 598
4	LV or HV	650	With Mods 592, 598 and 646

Note: 1. If the guns are fired with airbrake OUT at speeds above 550 knots a limited number of ejected case strikes on the airbrake will be experienced; such firing should therefore be confined to cases of operational necessity.

2. When carrying external stores the minimum firing speed is:

350 knots without Mod 645 embodied

150 knots with Mod 645 embodied

(f) Bomb release and jettison

2 x 1,000lb bombs may be carried, on inboard pylons only, released or jettisoned at all speeds and dive angles up to 590 knots and 60° respectively. There is no Mach number limitation, but pilots are warned against committing their aircraft to dives which will result in exceeding 0.9M without sufficient height for pullout, in view of the deterioration in longitudinal control above this speed

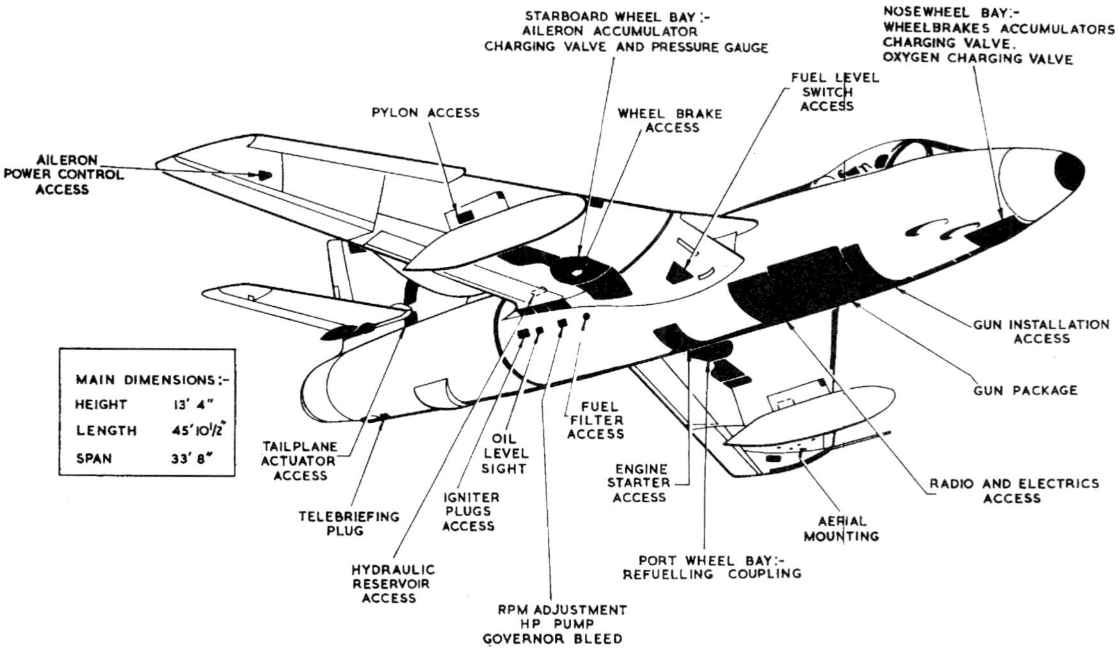

ACCESS PANELS

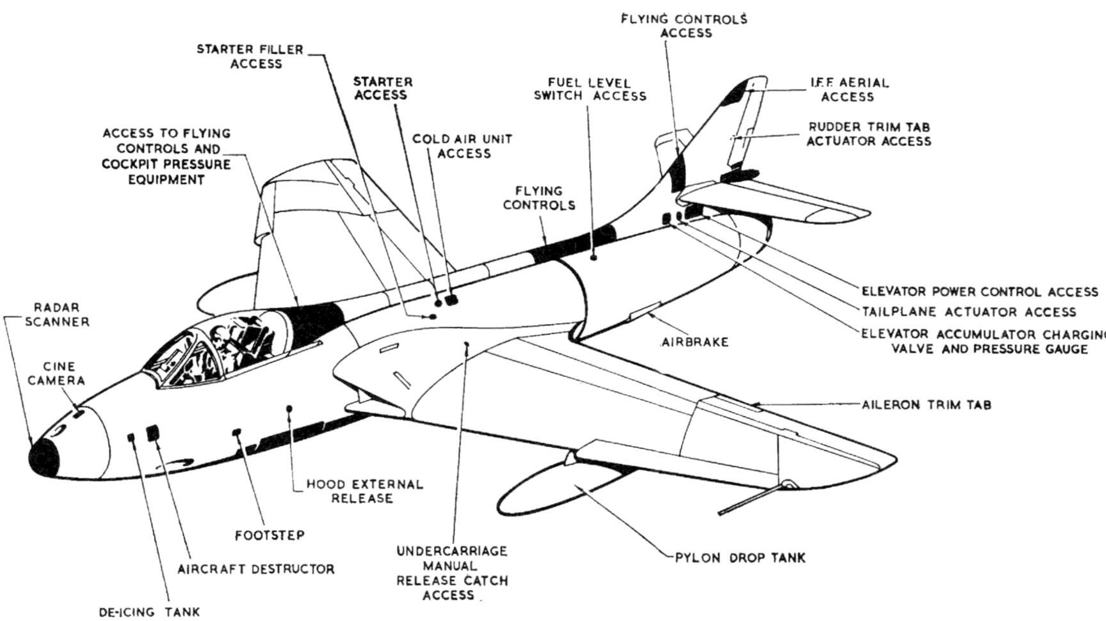

ACCESS PANELS

PART III

HANDLING

MANAGEMENT OF SYSTEMS

57. Management of the fuel system

(a) The HP and LP cocks must be ON and the booster-pumps should normally be ON at all times when the engine is running. The booster-pumps may be left off for starting.

(b) Fuel is supplied to the engine by the two booster-pumps, one of which is situated in the base of each front tank. With the tank selector switches at AUTO, fuel in the rear, wing and drop tanks (if carried) is automatically transferred to the front tanks by air pressure (see para 4 (b)). The order of emptying is:

Rear tanks
Drop tanks (if carried)
Wing tanks
Front tanks

(c) Correct functioning of the system is shown by the fuel contents gauges indicating equal fuel levels, the LP warning lights out and the 'flow plan' indicators reading as follows:

(i) Booster-pump warning lights: Out.

(ii) Transfer indicators: In-line throughout the flight.

(iii) Tank changeover indicator: Reading correctly, ie the indicator should change from REAR to WING when fuel contents reduce to approx 1,300lb and remain at WING for the remainder of the flight.

(iv) Front tank level lights wink for a few seconds and then remain on when the main gauges read 650lb (approx).

(v) When Mod 574 is embodied, the two magnetic indicators on the starboard shelf will show white when the outboard drop tanks are empty

Note: When the rear tanks are empty auto-selection to REAR will occur if the aircraft is inverted for approximately 5 seconds; on resuming normal flight a re-selection to WING will occur automatically. When the aircraft is inverted the level lights will come on irrespective of the fuel state. When normal flight is resumed the lights will go out if the fuel state is above 650lb.

(d) The flow proportioner will maintain accurate fuel balancing with both booster-pumps on or off, provided that the pressures at which the fuel enters the proportioner from the booster-pumps do not differ by more than 2lb/sq in. If unequal emptying occurs with all indicators normal, either the flow proportioner has failed or the inlet differential pressures are greater than 2lb/sq in, in which case fuel balancing must be controlled manually by switching off the pump on the 'low' side until the levels become equal and then switching it on again.

(e) Booster pump failure

If a booster pump fails, indicated by a warning light coming on and the corresponding contents gauge reading high, no fuel will be used from that side if the serviceable pump is left on; in this event reduce rpm to 7,200, switch off both pumps and accept the fuel feed provided by the tank pressurisation and gravity. Then reduce altitude to the limits below according to the aircraft configuration.

With drop tanks	20,000 feet max
Without drop tanks	25,000 feet max

Note: Rpm must thereafter be maintained at 7,200 or below; the altitude limit may be increased in emergency by 10,000 feet if absolute range is essential, but in this event the engine must subsequently be examined for possible damage to the fuel system.

With both booster-pumps off, the recuperators will also discharge, therefore no negative G manoeuvres must be carried out. It is important to land while both sides still contain fuel, because the flow proportioner will maintain out-of-balance levels and the engine will not run with one side empty unless the booster-pump in the side containing fuel is serviceable. The serviceable pump may be switched on for landing if a sufficient head of fuel remains on that side, but it should be remembered that with the pump on, fuel will only be used from that side.

(f) Transfer failure

(i) If one or both transfer indicators show 'cross-line', transfer pressure failure has occurred and steep dives should be avoided due to the possibility of collapsing the tanks. Should an indicator show 'cross-line' before fuel transfer is complete, any fuel remaining in the rear and wing tanks will be unusable and the associated contents gauge will only indicate the usable fuel in the front tanks (770lb max per tank). In these circumstances if the gauge registers more than the total contents of a front tank a faulty gauge should be suspected and only the front tank fuel should be relied on as being available to the engine. If air transfer fails on one side, the booster-pump on the side with the transfer failure should be switched OFF until the contents gauge of the other side indicates an equal amount, then reset the pump ON.

(ii) An indicator may show cross-line when the wing tanks are empty or nearly empty due to an air pressure reducing valve freezing. Since this may only occur when the rear and wing tanks are empty it is not serious from a fuel point of view, but rapid descents must be avoided otherwise the tanks may collapse.

(g) Unusable fuel

(i) If a 'tank empty' float in a rear tank becomes saturated its switch will operate before the tank has emptied. It is important to check that the changeover to wing takes place at approx 1,300lb. If the changeover occurs at a higher reading set the tank selector switch from AUTO to REAR until the contents reduce to below 1,300lb then set the switch to WING. A check that the associated level light comes on at approx 650lb will verify that all fuel has been correctly transferred.

(ii) If changeover to the wing tanks does not function automatically, shown by the magnetic indicator still pointing to REAR at a fuel state less than approx 1,300lb, set the tank selector switch from AUTO to WING. If the magnetic indicator still points to REAR a fault in the changeover system has occurred. The amount shown on the contents gauge will include unusable fuel in the wing tank, ie 539lb, but only the front tank fuel (770lb) should be relied upon as being available to the engine. This condition can be verified by noting at what fuel state the amber low-level lights come on.

(iii) Provided slow forward throttle movements and small changes in attitude are made, the engine will continue to run satisfactorily down to zero gauge readings, but at low level states steep climbs and/or sudden application of power should be avoided, particularly when overshooting, as this will cause fuel in the tanks to move away from the booster-pumps, resulting in a possible fuel starvation and flame extinction. Therefore at low fuel states minimum power consistent with safety should be used, steep attitudes avoided and a glide approach planned. If a flame-out occurs, a relight may be obtained by adopting a nose-down attitude and pressing the relight button. The amount of unusable fuel increases with sudden forward acceleration and/or nose-up attitude: rig tests reveal the following:

Angle between fuel surface and fuselage	Unusable fuel datumlb per side
10°	123
20°	200
30°	230

In the normal unstick attitude, the angle between fuel surface and aircraft datum is approx 23°, at which angle approx 208lb per side are unusable.

(h) Fuel gauge errors

(i) The fuel contents gauges have been found to give erroneous indications due to temperature effects on the electrical gauging system. The magnitude of the error depends on both temperature and time. Low temperatures at high altitude give gauge under- reading; high temperatures at high speeds at low altitude give gauge over-reading.

(ii) To assess the gauge error the readings of the contents gauges should be checked immediately the front tanks low level warning lights come on. The difference between the total gauge readings and 1,300lb indicates the gauge inaccuracy. This figure should be applied to all gauge readings for the remainder of the flight at the same operating conditions. During a descent from altitude, if the assessed inaccuracy was a gauge under- reading, the gauges will progressively become more accurate and may eventually tend to over-read.

58. Engine performance and handling

(a) Take-off

(i) The rpm are governed at 8,000, at which maximum thrust is obtained, but at full throttle this figure will vary with OAT whilst still maintaining maximum thrust. Below are the relevant figures:

+15°C and above	8,000rpm
+10°C	7,985rpm
0°C	7,950rpm
-10°C	7,910rpm

At full power check that the oil pressure is 25lb/sq in minimum.

(ii) Malfunctioning of the swirl vane system will allow the engine to reach maximum rpm without producing maximum thrust. Therefore to verify that maximum thrust is being obtained the following check should be carried out:

1. Max rpm is not obtained at part throttle.
2. The aircraft begins to creep forward against the brakes at 7,000-7,200rpm.
3. The rate of acceleration is normal when the brakes are released.
4. The jpt is more than 560°C.

If, at 17,500lb, the take-off is abandoned at a speed of 100 knots the aircraft will take approximately 750 yards to stop from the point at which the brakes are applied. If it is decided to abandon the take-off and the length of runway remaining is short, the throttle should be set to the HP cock OFF position.

(b) On the climb

(i) High forward speeds at low altitudes may reduce maximum rpm by as much as 100. This, coupled with the effect at (a) (i), may cause the rpm at the commencement of the climb to read as much as 150 less than the static ground figure. This condition will disappear as the climb progresses.

(ii) Maximum rpm may also be reduced if the top temperature controller comes into operation. On some engines the temperature controller reaches the limit of its control at altitudes above 40,000 feet when climbing at full throttle. It therefore becomes necessary to close the throttle slightly to maintain jpt below 685°C.

Note: Until Mod 372 and 438 are embodied, care must be taken to ensure, by throttling back if necessary, that at all altitudes the maximum permissible jpt is not exceeded.

(iii) During a climb at full throttle rpm will increase gradually but must not be allowed to exceed 8,100rpm. When climbing at intermediate power it is necessary to close the throttle gradually as altitude is gained to maintain 7,800rpm.

(c) In flight

(i) At intermediate throttle settings differing combinations of airspeed and ambient air temperature may cause the selected rpm to vary. It will then be necessary to adjust the throttle to maintain a constant figure.

(ii) With the throttle fully closed, the idling rpm increase with altitude and with increasing airspeed.

(iii) Negative G must not be applied for more than 15 seconds.

(iv) It is recommended that movement of the throttle should be smooth and progressive. However, rapid throttle movements can be made at any altitude if necessary, but the ACU will automatically control the rate of rpm acceleration.

(v) At low altitudes maximum power can be obtained within 5 seconds if rpm are above a minimum of 4,500 and at high altitudes within about 10 seconds from idling.

59. Management of the engine anti-icing system

(a) With the system in use, all throttle movements must be made smoothly. There is a loss of thrust and fuel economy and jpt usually rises by about 20°C. At full throttle rpm may then fall if the top temperature control comes into operation.

(b) Level flight

If icing conditions are met in flight, climb or descend out of the icing level (see (c) and (d) below).

(c) Climbing

Open up to not less than 7,200rpm and set the anti-icing switch to OPEN. Climb at the maximum practicable rate.

(d) Descending

Throttle back to not less than 5,500rpm and set the switch to OPEN. Descend at the maximum practicable rate; maximum anti-icing protection is obtained at the highest practicable rpm.

(e) When clear of icing conditions, set the switch

to SHUT and wait for about 10 seconds before making any rapid throttle movements. If icing conditions persist down to airfield level keep the switch at OPEN and maintain engine speed above 5,500rpm until finally committed to a landing.

60. Management of the flood flow system

For all normal conditions of flight the FLOOD AIR-FLOW CONTROL switch should be at AUTO. To prevent windscreen misting MANUAL should be selected wherever necessary. During take-off the increase in airflow is considerable.

61. Testing AC inverters on the ground

(a) To test No 2 inverter first check that the inverter circuit breakers are in, then set the battery master switch ON, and the starter master switch ON. The standby inverter will start up, the artificial horizon will erect and the magnetic indicator will show white.

(b) Then to test No 1 inverter and auto-changeover move the RADAR TEST switch to TEST. The inverter will then start up and take over the supply to the instruments. The indicator will change to black. When the test is complete return the radar test switch to NORMAL.

(c) To shut down after the test, switch off the battery master switch (otherwise No 1 inverter will continue to run) and, unless an immediate engine start is to be made, switch off the starter master switch. The battery master switch should only be set ON again when No 1 inverter has run down.

STARTING TAXIING AND TAKE-OFF

62. External checks

The outside of the aircraft should be checked systematically for signs of damage and for security of panels, filler caps, doors and the like. The engine intakes and boundary layer ducts must be free from obstruction and the jet pipe free from distortion. The main and nosewheel oleos should be checked for extension, the tyres for cuts and creep, and the brake leads for security and leaks. The elevator accumulator gauge in the port side of the fin should read 1,575lb/sq in and the ailerons accumulators gauge in the starboard wheel bay should read 900lb/sq in. Remove the pressure head cover.

63. Cockpit and pre-start checks

(a) On entering the cockpit

(i) Check that the drogue gun safety lock and emergency oxygen bottle pins are removed, the flight instruments and radar supply circuit breakers are in and the test switch is at NORMAL FLIGHT (see para 18(f) (iv)). Check that the hood locking pointers are in line with their spigots.

(ii) Strap in, adjust the rudder pedals and then make the main and emergency oxygen and anti-G suit connections, have the radio plug connected and both of the seat safety pins removed and stowed.

(b) Cockpit checks

Note: The starred items are directly connected with engine starting.

* Battery master switch	ON – ensure that turn and slip indicator starts up
Undercarriage emergency air pressure gauge	Pressure 2,000lb/sq in
Flaps emergency air pressure gauge	Pressure 2,000lb/sq in
Wheel brakes accumulator air pressure gauge	Min pressure 750lb/sq in
Tailplane actuator and hood motor circuit breakers	In
Radar test switch	Switch to TEST

Check aurally that the main inverter starts up, Art Hor flag black, and the oil pressure needle moves off its stop to zero. Reset switch to NORMAL, and the battery master switch OFF then ON.

Cockpit pressure warning	Switch to TEST. Cockpit test switch pressure warning light comes on. Reset switch to NORMAL
Flood air flow switch	AUTO
Cockpit pressurisation switch	ON
Cockpit temperature control switch	AUTO
Cockpit temperature selector	As required
External stores jettison switch panel	OFF
*LP cock	ON
VHF	Channel selectors off
VHF set selector switch	As required switch
Relight switch	Hold to RELIGHT for 2 seconds. If HE ignition system serviceable a clicking sound is audible. Switch battery master switch OFF then ON again.
Throttle damper	Adjust as required
Hood control switch and clutch selector	Clutch FREE, check manual operation of hood. Clutch LOCKED, switch as required
*HP cock throttle	Set 1 inch open
Top temperature controller switch	ON
Aileron and rudder trim indicators	Neutral
Undercarriage emergency selector	Ensure selector not pulled up
Radar ranging switch	OFF. Magnetic indicator black
Hood jettison handle	In
Undercarriage warning light	Out
Undercarriage position indicator	Three green lights, bulb changeover switch
Undercarriage selector AL.1 buttons	Down button in. Up button not twisted to EMERGENCY
Flaps selector lever	Up
Flaps emergency selector	Ensure selector not pulled out
Hydraulic warning light	On
Elevator and aileron power	Off
Magnetic indicators selector switches	White
Fuel level warning lights	Both out
Flight and engine instruments	Condition.
Fire warning light	Out
Cockpit altimeter	Condition
Audio warning switch	On (spring-loaded in on position)
Fuel low pressure warning light	On (out if pressure in system)
Oxygen	Main and emergency supplies connected
Emergency bottle control	In
Regulator ON/OFF valve wired	ON
Contents and delivery	100% OXYGEN selected. Check for mask leaks by pressing

	in emergency switch in central position. Return switch central, check blinker operation.
Emergency light	As required
Engine anti-icing switch	SHUT
Cockpit lighting switches	As required
Explosion suppression warning light switch	Out. If on, operate RESET
Fuel gauges	Contents (if transfer indicators are 'cross-line' a full indication will not be given)
Outboard drop tank 'empty' indicators	Black if fuel in tanks otherwise white
Booster-pump switches	OFF
Booster-pump warning lights	On (out if pressure in system)
Transfer indicators	Cross-line (if pressure in system they will be 'in-line' and contents gauges will read full)
Tank selector switches	AUTO (normally locked in AUTO)
Tank selector indicator	Point to REAR
Emergency oxygen bottle selector	Down
IFF switches	As required
Navigation lights switch	As required
Rebecca Mk 7	Off
Anti-G control	Switch ON. Pressure 1,800-2,000lb/sq in. Depress button gently to test
Butt test switch	Gated
Fire warning test switch	Operate to test light
Tank pump test switch	Off
Tank pump and starter circuit breakers	In
Lower centre console:	
Generator failure warning lights	Both on
Camera pressure head heater and starter master switches	Off
Flying controls	Full and free movement. Rotate spring feel trim fully in both directions. Check stick moves laterally in same sense. Return to neutral
Parking brake	ON. Check pressure at each wheel. If brake accumulator pressure is between 750 and 1,500lb/sq in each brake needle should read accordingly)

64. Starting the engine

(a) Starter master bar	ON. Check aurally that No 2 inverter starts up. Art Hor flag black.

Without delay press and release the starter button.

(b) When the starter fires, the engine speed should build up rapidly to 1,600rpm. As the engine lights up, the rpm increase to idling (2,500±200) and the throttle should be closed. The jpt may momentarily exceed the idling limit.

(c) Failure to start

(i) If the starter fails to fire, check the position of the switches and press the starter button again. If three attempts prove unsuccessful have the defect investigated.

(ii) If the starter fires and the engine rotates but fails to light up, set the throttle to HP cock OFF position. The starter may only be fired again when the engine has stopped rotating and at least one minute must elapse before a second attempt is made. If it is suspected that an excessive amount of fuel has collected in the engine the starter should be fired with the ignition switch off and the throttle set at the HP cock OFF position.

(iii) The starter can be fired three times at a minimum of one minute intervals but should be limited to three.

65. Checks after starting

Fire warning light	Out
RPM	2,500±200
JPT	Max 525°C
Oil pressure	10lb/sq in (min)
Generator warning lights	Out
Airbrake	Check operation with test switch. Magnetic indicator should go white, then black
VHF	Frequency selected
Rudder and aileron trimmers	Check and set neutral. Lock on.
Tailplane actuator	Test through full range on emergency, lower cover then check on normal and set at neutral. (The stick should move forward as nose-down trim is applied and vice versa.)
Tailplane switch (if fitted)	OFF. Note trim position on tp incidence indicator and trim to new position at least 1° different.
TP switch	ON. Check tailplane does not move from new position. Move stick fore-and-aft and check tailplane follows up.

If the follow-up tailplane is fitted, during the check on the emergency switch if the switch cover is lowered with the tailplane at a different angle from that when the cover was raised the tailplane will revert automatically to the original trim angle.

Hydraulic pressure	2850±150lb/sq in on both centre needle of triple pressure gauge and brake accumulator gauge
Pressure at each wheel	1,500 lb/sq in. Warning light out.
Elevator and aileron power control selectors	Switch each ON in turn and move the control column through its full traverse (using a two-handed force if necessary) to engage the controls until free movement is felt; check the indicators black.

If Mod 452 is incorporated hold the selector switch to POWER with the left hand and move the control column with the right hand, laterally for aileron engagement, fore-and-aft for elevator engagement, until the corresponding magnetic indicator goes black. Then release the switch. To engage aileron power more easily it is advisable to either:

(i) Select POWER with the stick held hard over to starboard and then move it to port or

(ii) Attempt the engagement immediately after light-up before hydraulic pressure builds up.

If Power is incorrectly engaged the controls revert to Manual on releasing the switch.

Flaps	Check operation
Instruments	Correct functioning. Erect Art Hor. Set Mk 4F compass and compare with E2
Fuel	Contents full. Set both booster-pump switches ON, check that each failure warning light goes out. Transfer indicator 'in-line'
Tank indicator	REAR
LP warning light	Out
Rebecca	Set to SB

Note: 1. To check the correct functioning of the hydraulic failure warning light and audio warning, exercise the stick laterally several times and select full flap simultaneously to exhaust hydraulic pressure. The light should come on and the audio warning sound when pressure falls below 50lb/sq in. When pressure rebuilds check that the flying control magnetic indicators are still black. In Mod 452 aircraft Manual reversion may occur during this check.

2. If a check on low-geared Manual aileron is required apply full lateral stick deflection and switch to Manual, observing a decrease in aileron angle. Re-switch to Power and check aileron reverts to its original position.

66. Taxiing

(a) Taxiing is normal for nosewheel-type aircraft. Fuel consumption at idling rpm is about 2 gals per minute.

(b) The aircraft should not be taxied at a speed which requires excessive use of the brakes as this causes overheating of the tyres and reduces their life.

67. Checks before take-off

Trim	Tailplane: Clean a/c: Neutral With 38° flap and/or inboard stores: 1° nose-up With 38° flap and outboard stores: 0.5° nose-up Spring feel trim: Neutral Rudder and aileron trim: Neutral, lock on
Fuel	Contents Booster-pump switches ON Warning lights out Transfer indicators in-line Tank selector switches AUTO Indicators at REAR LP warning light out Drop tank indicators black (if fuel in outboard tanks)
Flap	Up (38° with stores)
Instruments	Check and set
Oxygen	On 100% reaching mask blinker annunciating (emergency switch centred)
Harness	Tight and locked
Hood	SHUT (not OFF) Clutch: LOCKED Cockpit pressure: ON
Hydraulics	Pressure 3,000lb/sq in. Flying controls in power. At 4,500rpm apply full aileron and elevator and ensure that the magnetic indicators remain black. Warning light out.

Note: In bad weather, to avoid the possibility of instrument failure on take-off, the radar test switch should be set to TEST then back to NORMAL. This action overrides the undercarriage micro switch and brings in No 1 inverter After a few seconds the artificial horizon and Mk 4F compass should be checked for correct functioning.

68. Take-off

Warning: A check that the power control indicators are black must always be made immediately before take-off at not less than 4,500rpm. At any stage of a flight, if either automatic Manual reversion or any form of stick jamming occurs with POWER selected ON, immediately switch Power OFF. Do not attempt to re-engage Power. Return to base and land in Manual.

(a) Align the aircraft and roll forward a few yards to straighten the nosewheel. Apply the brakes with rudder bar central and open the throttle smoothly. If the brakes do not hold at 6,800rpm they should be considered unserviceable and the aircraft should not be flown. Release the brakes and open the throttle fully, checking the engine operation (see para 58).

(b) In cross-wind conditions, gentle braking is necessary to keep straight until the rudder becomes effective.

(c) Ease the nosewheel off at about 125 knots and hold it just off the runway, taking care not to achieve an excessively nose-up attitude. At normal loads the aircraft will unstick at 150 knots.

(d) When comfortably airborne apply the brakes and raise the undercarriage, holding the brakes on until the undercarriage is locked up. There is no noticeable change of trim as the undercarriage retracts but the nosewheel locks up with a distinct thud. It may be necessary to climb quite steeply initially as retraction must be complete before 250 knots is reached.

(e) Until experience is gained the lightness of the flying controls may lead to over-controlling in both pitch and roll.

(f) When carrying stores

Using 38° flap the nosewheel can be eased off at 125 knots and the aircraft flown off at 145-150 knots. When safely airborne immediately raise the undercarriage and then the flaps, one notch at a time, retrimming after each selection; delay in raising the flaps will result in an increasing nose-down change of trim as speed increases.

(g) Short take-off

Only a small reduction in ground run is achieved by using 38° flap; however, the distance to 50 feet is appreciably shortened. A strong nose-up change-of-trim and a small amount of sink occurs when the flaps are raised.

HANDLING IN FLIGHT

69. Climbing

Note: If pressure has been set OFF for take-off, set it ON at 5,000 feet.

(a) Best rate of climb

(i) Climb at full throttle at 0.85M. Speed should be allowed to increase to the recommended figure during the initial climb to 5,000 feet.

(ii) Below 20,000 feet the rate of climb is not greatly affected by variation in speed, provided that it is maintained between 400 knots and 0.9M.

(iii) Above 20,000 feet it is important to keep to the recommended speed; above 30,000 feet the rate of climb will fall off quickly if speed is reduced below 0.85M.

(b) Normal climb

If maximum rate of climb is not essential set 7,800rpm using the same speed as above.

70. General flying

(a) Flying controls

(i) Ailerons (in power). The ailerons are light throughout the speed and Mach number range giving a high rate of roll, but see para 74.

(ii) Elevator (in power). The force required for any manoeuvre depends on the distance the control column is displaced from the 'zero load' position and is completely independent of airspeed. It follows therefore that when large elevator deflections are required, eg at low airspeeds and very high Mach numbers, the stick forces are relatively heavy; at high airspeeds, however, since only small deflections are usually required, the stick forces are light. The control is light, effective and should be used cautiously until its characteristics are known and its effectiveness appreciated. However, elevator effectiveness is somewhat limited by jack stalling, which occurs when the air load on the elevator equals the jack output force and restricts movement of the control column rearwards. Depending on the tailplane angle and CG position, jack stalling may occur when manoeuvring above 0.93M; if it occurs the tailplane must be used as a means of control.

(iii) Rudder. The rudder is light at low IAS and becomes progressively heavier as speed increases. The application of rudder produces a strong rolling tendency.

(b) Trimmers

Note: Do not fly with the thumb on the tailplane control switch as this may cause intermittent making and breaking of the contacts, which may damage the switch. Care must also be taken not to operate the switch inadvertently during manoeuvres as this may result in excessive G being applied.

(i) The aileron trim tab trimmer must not be used whilst the controls are in Power, and it should be locked in the neutral position. The aileron spring feel adjuster should normally be used to counteract any out-of-trim forces which may occur in Power. It should not be used when flying in Manual.

(ii) The tailplane trimmer should be used in the normal manner; the angle is usually between 0.5° and 1.5° nose-down. When manoeuvring, the stick forces are light and little use of the trimmer is required. The full-power elevator tends to mask any out-of-trim forces which may be present. During sustained flying the stick forces should always be trimmed out (see para 73).

Warning: if this is not done and an inadvertent Manual reversion occurs the stick force may be too heavy for the pilot to hold. If the normal tailplane trimmer fails the standby control should be used; this operates at about one-third the speed of the normal control. If both trimmers fail, the aircraft can be flown throughout its speed range with the trim at full nose-down, but at full nose-up the elevator is not sufficiently powerful to stop the nose rising at speeds in excess of approx 420 knots.

(c) Airbrake

(i) The airbrake may be used throughout the speed range. Its use may cause the hydraulic warning light and audio warning to come on momentarily.

(ii) Selection of airbrake causes moderate buffeting and a momentary nose-down change of trim, which reverts to moderate nose-up trim change when the airbrake is fully extended. Correcting the out-of-trim forces at high IAS may lead to overcontrolling.

(d) Flaps

(i) At high Mach number

As speed is increased from 0.9M to 0.94M a marked nose-down change of trim occurs. Lowering flap also produces a marked nose-down change of trim; the degree of out-of-trim increases with the amount of flap selected and with speed. As Mach number is increased beyond 0.92 elevator and tailplane effectiveness decreases. Pilots will realise therefore that if 0.9M is exceeded inadvertently with flap lowered, or if flap is lowered inadvertently at high speeds in excess of 0.9M, longitudinal control will be very substantially reduced and in the worst condition may be lost completely. If control is lost with any degree of flap lowered, the flap should be raised immediately.

(ii) At high airspeed

If speed is increased inadvertently beyond 350 knots with flap extended, the increasing air loads decrease the flap angle, which results in a nose-up change of trim.

(e) Changes of trim

Increase and decrease
Nil power

Operation of undercarriage
Negligible

Flap down
Strong nose-down above 200 knots

Airbrake
Low IAS, negligible. High IAS and high Mach number, slight transient nose-down, moderate nose-up when extended

(f) Flying in turbulent conditions

The recommended speeds for flight in turbulent conditions are as follows:

Climbing or level flight
Above 25,000 feet, 0.83M
Below 25,000 feet, 350 knots

Descending (airbrake out, flap 20°)
Above 35,000 feet, 0.83M
Below 35,000 feet, 280 knots

(g) Flying with electric follow-up tailplane

(i) The electric follow-up tail gives an improvement in manoeuvring capabilities above 0.9M. Handling characteristics are otherwise normal, but it should be noted that the interconnection remains fully operative with the elevator in Manual.

(ii) The tailplane interconnection can be selected ON or OFF at any speed provided that the aircraft is trimmed for hands-off flight whenever either selection is made. The aircraft should be flown in the trimmed condition whenever possible.

(h) Flying with drop tanks on inboard and outboard pylons

(i) At speeds below the permitted maximum the handling characteristics are similar to those for clean aircraft, and with fuel in the drop tanks rolling manoeuvres must not exceed 360°. During combat manoeuvres with fuel in the tanks it is recommended that aileron movement is restricted to half of full movement to avoid an excessive rate of roll.

(ii) Pressure errors

The presence of drop tanks on the outboard pylons has a material effect on pressure error at high speeds, resulting in under-reading of the ASI and machmeter as compared with clean aircraft.

(iii) Practice Manual flying

Practice selection of Manual should not be made when carrying external stores other than empty inboard drop tanks since their presence increases the difficulties of reselecting Power.

(j) Flying with RPs

Aileron movement becomes restricted at 0.97M when carrying RPs. When drop tanks are carried on the inboard pylons it is difficult to exceed .98M.

(k) Flying for endurance

The recommended speed is 180 knots.

71. Flying at reduced airspeed

(a) Fly at 70-200 knots using 23° (2 notches down) flap.

(b) The hood may be opened fully below 200 knots, but at about 150 knots excessive vibration sets in. This vibration is reduced if the hood is closed to the halfway position, ie about level with the pilot's shoulders. In this position the noise level is high and rather distracting but landings can be made without undue inconvenience.

(c) Lateral rocking may occur between 200 and 250 knots with the flaps fully down. If the rocking becomes excessive the flaps should be raised. Any lateral rocking which occurs below 200 knots is easily controlled with the ailerons. Note: The hood cannot be jettisoned unless fully closed.

72. Stalling

Note: Because the rate of descent is very high and because it is possible to induce an inadvertent spin when the aircraft is fully stalled, stalling practice is not to be continued beyond the buffet stage nor below 25,000 feet.

(a) Pre-stall buffet speeds, throttle fully closed are:

(i) Full ammunition and 1,100lb fuel remaining (max landing weight approx)

Undercarriage and flap up	130 knots
Undercarriage down and flap up	130 knots
Undercarriage down and full flap	120 knots

(ii) No ammunition and 800lb fuel remaining (normal landing weight)

Undercarriage and flap up	125 knots
Undercarriage down and flap up	125 knots
Undercarriage down and full flap	115 knots

(b) Above 30,000 feet buffet occurs at approximately 135-140 knots with the throttle fully closed, undercarriage and flap up.

(c) If the extended wing leading edges are fitted, the intensity of pre-stall buffet is increased but stalling speeds are little affected.

(d) Use of the airbrake increases the buffet but does not affect the stalling speeds or other characteristics.

(e) Under typical approach conditions, the buffet speeds quoted above are not appreciably affected, but the height lost during recovery is reduced.

(f) (i) Although the aircraft must not be deliberately fully stalled in flight, the characteristics are described here to assist pilots who inadvertently enter the fully stalled condition.

(ii) If the control column is held back after the buffet stage is reached, a nose-up change of trim will occur, and though it will vary in degree from aircraft to aircraft, to counteract it may require full forward stick movement. Either wing may tend to drop but can be controlled by the ailerons. Relaxation of the forward pressure on the control column at this point will lead to a further reduction in forward speed accompanied by a very high rate of descent. In this condition the elevator is relatively ineffective and response is slow. If extended wing leading edges are fitted there is less tendency for either wing dropping or yawing to occur.

(iii) A spin or spiral may develop, and in any case considerable height will be lost. Large deflections of the ailerons near the stall will cause the aircraft to yaw in the direction of the downgoing aileron and will increase the possibility of a spin or spiral developing.

(g) Outboard stores seriously affect the stall characteristics, particularly the time taken for recovery after initiating recovery action. This feature is aggravated if inboard stores are also carried.

73. High speed stalling

(a) (i) High speed stalling is subject to the overriding restriction of para 56 whereby pilots must not exceed +7G.

(ii) At airspeeds above 0.9M between 10,000 and 30,000 feet, unless Mod 533 is fitted, an accelerometer reading +4G must not be exceeded. At airspeeds below 0.9M in that height band, G must not be applied beyond the buffet stage.

(b) Pitch-up (pre-extended wing leading edges)

(i) During turns and pull-outs adequate stall warning is given by buffeting at all heights. If the backward pressure is continued inadvertently after the stall warning, wing drop or a momentary pitch-up and sudden increase in G may occur.

(ii) Below 10,000 feet the maximum acceleration of +7G may be applied at the higher Mach numbers without stalling the aircraft; no pitch-up is therefore experienced.

(iii) Between 10,000 and 30,000 feet at speeds above 0.9M the pitch-up, if experienced, may be coincidental with the buffeting and may be severe enough to exceed +7G. It is for this reason that the restriction of +4G accelerometer reading is imposed.

(iv) Above 30,000 feet it is not possible to achieve high accelerometer readings, although pitch-up is still experienced.

(c) Pitch-up (extended wing leading edges fitted)

When Mod 533 is embodied the tendency to pitch-up is reduced and the G limitations (accelerometer readings) are +7G and -3.75G at all heights and speeds. Pitch-up may still occur to a reduced degree when G is applied, at speeds above 0.9M in the height band 25,000 to 30,000 feet. If pitch-up occurs it is still possible to maintain some degree of longitudinal control. Buffeting is considerable and wing dropping may occur.

(d) Jack stalling

Buffeting and/or limiting G can be obtained with little effort, particularly when ammunition is not carried, but if jack stalling occurs it reduces the amount of G it is possible to obtain. If the stick is held hard on the restriction stop an increase in G must be anticipated if speed is decreased, because the jack unstalls and further stick movement becomes possible as airspeed decreases. The increase in G will be sudden if the jack unstalling is coincidental with the nose-up change of trim as speed falls through 0.95M. When manoeuvring at high IAS near the ground, sufficient height must be allowed for a recovery to be effected in the jack stalled condition using the VI tailplane. Care should be taken when manoeuvring at high Mach numbers as it is easy to pull into the pitch-up inadvertently due to the light stick force.

Warning: Care must be taken to ensure that the stick force is never completely trimmed out when G is being applied at high Mach numbers because, as speed falls through 0.91M when the trim changes to nose-up and the elevator and tailplane become more effective, a sudden increase in G may result. This is particularly important below 10,000 feet when manoeuvring near limiting G and/or 'blackout threshold'.

74. Flying at high IAS

(a) When flying at high IAS all control movements must be smooth and progressive to avoid overcontrolling, particularly when flying at aft C of G and/or in turbulent air. The tailplane trimmer should be used carefully.

(b) The maximum rate of roll increases with IAS up to 420 knots; at higher speeds, however, the rate of roll progressively decreases due to jack stalling. Normally maximum rate of roll is not required.

75. Flying at high Mach number

(a) General

The maximum speed in level flight at full throttle is 0.94M. The aircraft will reach sonic speed in a 30° to 40° dive at full throttle. Transonic dives must not be started below 25,000 feet.

(b) Trim changes

As speed increases to about 0.90M there is a progressive nose-up change of trim. Between 0.90M and 0.94M a nose-down trim change followed by a nose-up trim change occurs, the aircraft being almost back in trim again by 0.96M. At higher speeds as the aircraft becomes supersonic the trim again changes to slight nose-down.

(c) Changes in stick force and tailplane and elevator effectiveness

(i) As Mach number is increased beyond 0.92 the elevator becomes less effective. This is particularly evident at transonic speeds when even large elevator deflections have a delayed and reduced response. The tailplane may be used to assist in manoeuvring, although its effectiveness is somewhat reduced.

(ii) Since the effectiveness of the elevator decreases as Mach number increases, greater deflections are required to manoeuvre. Consequently the stick forces increase.

(d) Transonic flight

Practice transonic flights should be made by putting the aircraft into a 30° to 40° dive with the tailplane trim set at 1° nose-down. At 0.97M very slight wing drop may occur, which can easily be counteracted with aileron; if the dive angle is too shallow the aircraft will reach a maximum speed of 0.97M, at which speed the elevator is not effective enough to increase the angle of dive.

(e) Recovery from transonic dives

During recovery the throttle should be closed. The airbrake may be used; its extension causes moderate buffeting and a nose-up change of trim. It is not very effective in reducing speed. Recovery can be made without using the tailplane trim, but jack stalling may occur (see para 73 (c)). However, trimming nose-up decreases the angle of deflection between the elevator and the tailplane, thereby unstalling the jack and enabling further stick movement to be applied. Normally it should not be trimmed more nose-up than 0°. This is because it is necessary to ease the control column forward and retrim nose-down to prevent a sudden increase in G caused by the nose-up trim change as speed falls through 0.95M and because of the increasing effectiveness of the tailplane and elevator.

76. Aerobatics

(a) Until experience is gained, the following speeds, in knots, are recommended:

Roll	350
Loop	425
Roll off the top	450
Vertical roll	500

(b) It is recommended that until experience is gained loops are started in the height band 10,000 to 15,000 feet.

77. Spinning

(a) Intentional spinning is prohibited. The following information is provided to acquaint pilots with the spin characteristics and recovery action in case the aircraft is spun inadvertently.

(b) When pro-spin controls are applied at the stall the aircraft rolls on to its back and enters the spin with the nose about 40° below the horizon. The attitude then becomes normal and the first turn is steady with a slow rate of rotation. During succeeding turns hesitation and marked variations in the rate of rotation occur, with associated pitching and rolling. When spinning in either direction the ailerons should be held neutral as out-spin aileron will cause the spin to become erratic.

(c) Normal recovery action is effective particularly if it coincides with the aircraft hesitations. Full forward stick movement may be necessary. When rotation ceases, the controls should be centralised immediately otherwise the aircraft may enter a spin in the opposite direction. If a spin occurs with undercarriage and flaps down they should be retracted immediately.

(d) Care should be taken during recovery action to ensure that the ailerons are central as out-spin aileron may hinder or prevent recovery. In-spin aileron is powerful in assisting recovery but great care must be taken to prevent development of a spiral in the same direction or spin in the opposite direction.

CIRCUIT PROCEDURE AND LANDING

78. Circuit procedure

Note: 460lb (60 gallons) of fuel should be allowed for a circuit, landing and possible overshoot (see para 57 (g) (iii)).

(a) Circuit speed

With the undercarriage lowered fully, 6,500rpm and 38° flap (four notches) give a comfortable speed of about 180 knots. To reduce speed for joining the circuit, flap, within the limitations, can be used successfully to augment the airbrake. Do not select more than one hydraulic service at a time and allow the cycle of each hydraulic operation to be completed before the next service is operated. The undercarriage should only be selected down with the wings laterally level.

(b) Checks before landing

Airbrake	IN, indicator black
Undercarriage	Down below 250 knots, three green lights
Brakes	Pressure, operation, off. Main supply 3,000±150lb/sq in, at each wheel 1,500lb/sq in
Flaps	As required, fully down on finals
Fuel	Contents, booster-pumps ON
Harness	Tight and locked

Note: 1. The brake lever should be held on for 3 seconds. If the system is serviceable both brake needles will remain at maximum pressure. If a hydraulic pipe is fractured, a lower than normal pressure will be indicated initially and this will decrease at a rate depending on the size of the fracture. The check should not be repeated because at each operation hydraulic oil will be lost.

2. Landing with symmetric or asymmetric stores presents no difficulty. When landing at low fuel states, with RPs retained and with ammunition expended, maximum nose-down trim is required. If inboard drop tanks are also carried there will be insufficient nose-down trim available; to avoid a moderate push force on the approach not more than 38° flap should be selected.

(c) Final approach

(i) Turn across wind at 160 knots aiming to lower full flap on the final stages. Steep approaches are not recommended.

(ii) To ensure most rapid engine response maintain at least 4,500rpm until finally committed to a landing. Under conditions of high wind or gustiness it is more comfortable if the speeds below are increased by 5 knots.

(iii) The recommended speeds, in knots, at the runway threshold are:

At normal landing weight	
No ammunition,	
800lb or less fuel remaining	130
At maximum landing weight	
No ammunition,	
1,800lb fuel remaining	135
Full ammunition,	
1,100lb fuel remaining	135

79. Landing

(a) As the touch-down point is approached, the rate of descent should be checked and the aircraft flown gently on to the ground at about 5-10 knots less than the runway threshold speeds. Holding off may result in an excessive nose-up attitude with the subsequent danger of dropping a wing; if the latter occurs, corrective aileron may be effective in raising the wing, but will cause the aircraft to yaw markedly in the direction of the down-going wing. The nosewheel can be held off at speeds down to about 70 knots, but the shortest run is achieved by putting the nosewheel firmly on to the runway and applying the brakes.

(b) Braking

Note: The effectiveness of the braking system is greatly decreased on very wet or icy runways. If wind conditions are favourable it may be advantageous to use aerodynamic braking. When the nosewheel has been lowered on to the runway the brakes can be used continuously and the maxaret units will prevent wheel locking; however, to prolong the efficiency and life of the brakes, braking should be judicious according to length of landing run available. The landing can be cut to less than half normal by using continuous full brake once the aircraft is firmly on the ground, but this procedure causes rapid brake and tyre wear and should not normally be used. The aircraft must be firmly on the ground before applying the brakes. If it is allowed to touch down with the brakes on, the maxaret units will not operate and the wheels will lock; however, if once having started turning the wheels should stop because of skid or bounce, they will not lock unless the skid or bounce continues for more than 4 seconds. After a landing involving heavy braking, ten minutes should elapse before the next landing. If the intervening period of taxiing has also required prolonged use of the brakes, twenty minutes should elapse before the next landing. Observe the same precautions in brake tests. On wet runways the landing run may be decreased by lowering the nosewheel on to the runway, applying the brakes and pulling the control column right back. The brakes must be in use while the back pressure is applied.

(c) Cross-wind landing

For cross-wind landings the 'crab' technique should be used. In light winds no difficulty should be experienced in touching down but in strong cross-winds full rudder may be required to correct the crabbing. The rudder has a delayed reaction which will require anticipation. The secondary effect of full rudder is to produce a marked roll, which must be counteracted with aileron. When the crabbing has been corrected the aircraft should be flown gently on to the ground and the nosewheel then lowered on immediately, thus reducing the tendency for the cross-wind to lift the into-wind wing. Care should be taken to centralise the rudder before applying brake. If the cross-wind is gusting strongly the approach speed should be increased by 5 knots.

80. Instrument approach

The following are the recommended airspeed, power and flap settings for an instrument approach with the undercarriage lowered:

	RPM	**Flaps**	**Airspeed (knots)**
Downwind	6,500	30°	170/180
Base leg	6,500	30°	170/180
Glide path	6,500	Full	150/160

81. Going round again

Open the throttle smoothly to the power required, raise the undercarriage, and at a safe height raise the flaps and retrim as necessary.

82. Checks after landing

Brake pressure	Sufficient
Flaps	Up
Cockpit pressure	Off
Camera master switch	Off
Both booster-pumps	OFF
Tailplane	Set to neutral
DME	OFF
VHF	One set OFF

83. Stopping the engine

Close the throttle. When the rpm have stabilised at 2,500 (min) lift the override catch and set the throttle to HP cock OFF position. Then check:

Flying control switches	Both OFF
All electrics	Off
Battery master switch	Switch off
Hood clutch selector	FREE
LP cock	OFF when engine stops rotating
Ejection seat	Replace both safety pins before leaving the cockpit

FLYING IN MANUAL

84. Selecting manual

(a) It is structurally safe to fly in Manual within the speed limitations of para 56, and in emergency Manual could be selected at any altitude and airspeed. However, until experience is gained, it is commended that selection of Manual and subsequent practice flying in Manual should be carried out above 10,000 feet. This is because of the out-of-trim forces which may be present when Manual is selected and the extreme heaviness of the controls and lack of manoeuvrability when in Manual. The out-of-trim forces are usually slight provided that the aircraft was in trim in Power, but if the tailplane and elevator are incorrectly rigged these forces may be large. Because of this possibility, when practice flying in Manual is to be carried out, the first selection to Manual should be made above 10,000 feet and at a low airspeed.

(b) Before selecting Manual, ensure:

Airspeed	250 knots or 0.80M if above 40,000 feet
Tailplane	In trim
Aileron and rudder	Neutral trim indicators
Aileron trim lock	Disengaged

85. Flying in Manual

(a) The elevator forces are high but tolerable and no difficulty should be experienced with longitudinal control. The ailerons are heavy and require some effort, but full stick deflection can be obtained with a single-handed force up to 220 knots. Reaction of the aircraft to aileron deflection is slow and delayed, therefore all necessary aileron movements must be anticipated. The rudder, the further effect of which is marked, can be used to assist control in the rolling plane but should be used with care at low airspeeds.

(b) When carrying stores

Because of the increased inertia, lateral control is less effective; this is particularly noticeable on the approach when lateral rocking due to either turbulence or over-controlling is difficult to damp out. Due to the increased turbulence below the wing and the tendency of the ailerons to up-float, likelihood of obtaining a false anchorage when re-engaging Power is increased.

(c) Trim changes

Increase and decrease of power	Nil
Operation of undercarriage	Roll in either direction (depending on which main leg lowers first). See para 84
Flaps down	Strong nose-down especially above 200 knots
Flaps up	Strong nose-up especially over 200 knots

86. Landing in Manual

(a) Until pilots have considerable experience of flying in Manual control, practice landings should be made only in ideal conditions, ie a steady wind down the runway. Because any asymmetric lowering of the undercarriage is liable to cause lateral control difficulties, the undercarriage must be lowered at a safe height. A wider than normal circuit should be made, followed by a long straight powered approach. When the airspeed is below 150 knots with undercarriage and full flap down, appreciable aileron buffet can be felt on the control column. If an overshoot has to be made, the undercarriage should not be raised until a safe height is reached. Flaps must only be selected up to a mid-position because of a nose-up change of trim, full retraction being deferred to a safe height.

(b) With external stores (see also para 85 (b))

If the drop tanks are full or partly full they must be jettisoned before landing. Landing with any asymmetric loading other than one empty inboard drop tank must not be attempted. Trials have shown that with a nominal 1,000lb weight on an inboard pylon the wings cannot be held laterally level at speeds below 180 knots. In turbulent conditions asymmetrically loaded RPs must be released if practicable; if this is not possible great care must be taken when landing and not more than 38° flap must be selected.

87. Reselecting power

Before reselecting aileron and/or elevator Power, the aircraft must be trimmed, laterally and longitudinally, in straight and level flight at the same airspeed and approximate altitude as when Manual was selected. Immediate re-engagement is more likely on the elevator than on the ailerons; it is therefore preferable to reselect elevator before ailerons. After re-selection check that the appropriate magnetic indicators are black.

88. Clearing false anchorages (ailerons)

(a) If power is re-selected below 10,000 feet and a false anchorage occurs, immediately switch Power OFF and either

(i) Return to base and land in Manual, or

(ii) Climb above 10,000 feet before making any further attempts to re-engage Power.

(b) If Power is reselected above 10,000 feet and a false anchorage occurs, it may be cleared as follows:

(i) One-way restriction

The control column should be moved rapidly over its full movement in the direction of unrestricted travel. If the airspeed is below 250 knots, this can be done without producing excessive aircraft roll.

(ii) Both-ways restriction

Lower full flap at an airspeed of 250 knots and select aileron Power ON (or hold switch to POWER if Mod 452 is embodied). The aileron buffet which occurs in this configuration will probably be sufficient to enable the pawls to engage in the slots, but if necessary move the control column laterally until the magnetic indicator turns black.

PART IV

EMERGENCY HANDLING

89. Action in the event of fire

(a) If the engine fire warning light comes on, the throttle should be closed immediately. If the light then goes out within 5 seconds, a hot gas leak as distinct from a fire is indicated. It is safe to use the engine in this condition, but a landing should be made as soon as practical. Reduced power should be used, but unless the power required is small the light will probably come on again. It is advisable, therefore, to throttle back every 5 minutes to check the light goes out in order to ensure that a fire has not started.

(b) Should the light remain on after throttling back, a fire is indicated. Should a fire occur, set:

HP cock	OFF
LP cock	OFF
Booster-pumps	OFF

and quickly reduce airspeed to a practicable minimum. Then press the extinguisher pushbutton. Should the light remain on and the fire persist, the aircraft should be abandoned.

(c) If a satisfactory extinction has occurred the warning light should go out as the circuit is broken by the cooling of the flame switches. The engine must not be restarted, due to the risk of fire with the fire-fighting resources exhausted.

(d) Carry out forced landing procedure (see para 94) or abandon.

90. Electrical system failures

(a) Single generator failure

If either generator fails, the output of the other is sufficient for non-combat flying provided that all non-essential electrical services are off. When the follow-up tailplane is in use an additional heavy electrical load is imposed; use of the interconnection should therefore be limited to flight above 0.9M.

(b) Double generator failure

If both generators fail, all electrical services will be supplied by the batteries. These, if fully charged, will maintain the following output before they fail (ie battery voltage below 16 volts).

Output (amps)	150	125	100	75	50	38	25
Time (mins)	1.5	2	3	5	10	18	40

Note: When two 24-volt batteries are fitted these times should be doubled. To assist the pilot in deciding which loads should be shed if generator failure occurs, the following table lists the major services and the current they require:

Service	Load (amps)	Service	Load (amps)
Booster pumps	35	Pressure head heater	6
Radar ranging	30	Starter control (relighting)	6
AC inverters	15	Fuel contents gauge	4.5
Tailplane actuator	12	GGS	4
Hood control	10	GGS retraction	3.5
VHF sets	7.5	All other electrics	10
IFF set	6.5		
DME	6	Maximum flight load	156

Note: The average flight load is 105 amps.

(c) Double generator and battery failure

Once the batteries are discharged, no electrical services can be operated, eg trim tab actuators, tailplane motors, electro-hydraulic selectors, etc. If the flying controls are in Power, Manual selection will not be possible unless Mod 502 is embodied, in which case Manual may be selected by pressing the emergency selector buttons. In addition the fuel gauges and electrically operated flight instruments will become unserviceable. The fuel booster-pumps will cease operation, which may entail reduction in altitude and engine rpm (see para 57(e)). No relight facilities will be available.

(d) Action in the event of double generator failure

(i) If complete generator failure occurs, switch off all non-essential electrical services to conserve the battery for as long as possible.

(ii) Switch OFF the tailplane interconnection (when fitted) and set the tailplane angle to 0.5° nose-up.

(iii) Provided that the engine is running correctly and hydraulic pressure is normal the flying controls may be left in Power. The magnetic indicators will show white when the batteries are discharged although the controls will remain in Power.

Note: 1. If any doubt as to the serviceability of the hydraulic system exists, select Power OFF before the batteries are discharged, unless Mod 502 is incorporated.

2. If it is necessary to descend more rapidly than at normal gliding speed, 15° flap may be used to increase the rate of descent, but speed must not exceed 0.9M and the tailplane incidence must be left at 0.5° nose-up. Intermediate degrees of flap can only be selected while electric power is available. The pull force needed to reduce speed to 210 knots will be considerable if in Manual.

(e) Emergency operation of the tailplane and electrical follow-up tail malfunction

(i) If the normal tailplane motor fails or 'runs away', lift the cover of the standby switch on the port shelf and use that control. Rate of operation of the latter is about one-third that of the normal control. If complete electrical failure occurs, control of the tailplane is impossible. It is important therefore to set the tailplane to 0.5° nose-up before the batteries are fully discharged.

(ii) If the follow-up tail malfunctions, either in its capacity as a flying control or in its trimming action, the standby trim switch cover must be lifted immediately. The standby actuator can then be used to obtain satisfactory trim. It is possible that the malfunction may not affect the main trim actuator with TAILPLANE selected OFF although it might have shown itself during either normal control movement or trimming with TAILPLANE ON. After raising the standby trim switch cover, select TAILPLANE OFF, then reselect the main actuator and check the normal trimming action. If the malfunction is still present, revert to the standby actuator and return to base.

(f) Aileron trim malfunction in Manual

(i) When flying in Manual should the aileron trim actuator malfunction resulting in full trim being applied with subsequent inability to rectify, carry out the following actions:

Reduce speed below 200 knots

Select elevator power ON

If above 5,000 feet select aileron power ON; if below 5,000 feet leave ailerons in Manual (but see (iii) below)

Land as soon as possible

(ii) If a one-way restriction occurs when reselecting aileron power, clear it in the normal way. If a both-ways restriction occurs, re-select Manual immediately and remain in Manual.

(iii) A landing in Manual with full aileron trim applied will be hazardous and should only be attempted if conditions are favourable.

91. Hydraulic system failures

(a) Complete hydraulic failure

(i) Periodic checks of the triple pressure gauge should be made in flight and the central needle should normally read 2,850±150lb/sq in. If the reading drops substantially below this figure when no service is being operated, then hydraulic failure should be suspected. The red warning light, and the audio warning when Mod 327 is fitted, should come on if the pressure falls below 600lb/sq in.

(ii) If the hydraulic supply pressure fails, there may be sufficient reserves in the power controls accumulators for a maximum of 3.5 full reversals of aileron and elevator, the actual reserve depending on the state of charge of the respective accumulators at the time of failure. However, even if no control movement is made, accumulator pressure will not be maintained for a long period, due to normal hydraulic component seepage. When the accumulators are exhausted the controls will revert automatically to Manual.

(iii) A failure in the power controls hydraulic circuit, as distinct from supply failure, may lead to immediate and automatic Manual reversion when the pressure at the locking pawls has fallen well below that at which the warning light becomes illuminated.

(iv) The wheel brakes accumulators provide sufficient pressure for brake operation during landing down to an accumulator pressure reading of 750lb/sq in approximately.

(b) Action in the event of complete hydraulic failure

(i) Complete failure will be indicated by a loss of pressure on the gauge followed by the red light coming on and by the audio warning.

(ii) Immediate action should be:

(1) Reduce speed to 150 knots or, if above 40,000 feet, 0.80M.

(2) Release the aileron trim safety lock and check that the trim indicator shows zero.

(3) The controls may be left in Power, providing electrical supply is normal, but when Manual reversion occurs, and in any case at a safe height before joining the circuit, switch OFF elevator and aileron Power.

(iii) Lower the undercarriage and flaps on the emergency systems.

(c) Partial hydraulic failure

Partial hydraulic failure can occur due to a type of leak which could cause complete loss of pressure during the operation of a service, but which would allow pressure to build again when the operation is complete. This type of failure can cause the power controls to disengage temporarily and re-engage with the possibility of a false anchorage.

(d) Action in the event of partial hydraulic failure

(i) Reduce speed to 250 knots or 0.80M.

(ii) Release aileron trim safety lock and check that trim indicator shows zero.

(iii) The controls may be left in Power to assist in manoeuvring, providing electrical supply is normal, but should a false anchorage occur and in any case before landing, switch OFF elevator and aileron Power.

(iv) If Mods 452 and 502 are not embodied and a partial hydraulic failure occurs subsequent to complete electric failure when the controls are still in Power, the landing is likely to be extremely hazardous, and it is recommended that the aircraft be abandoned.

92. Engine failure and flame-out

(a) Engine seizure

Engine seizure will involve both hydraulic and electric supply failure. If the engine has failed due to obvious mechanical causes do not attempt to relight. Proceed as follows:

(i) Close the throttle. Turn off the HP and LP cocks and switch off the booster-pumps.

(ii) Reduce speed and glide at 210 knots.

(iii) Switch off all unnecessary electrics and tailplane.

(iv) Set tailplane incidence to 0.5° nose-up. Release aileron trim safety lock and check that trim indicator is at zero.

(v) The controls may be left in Power, but Manual reversion must be expected.

(vi) Carry out forced landing procedure (see para 94) or abandon.

(b) Flame-out

(i) If flame-out occurs, a relight may be attempted immediately, while the rpm are decreasing, by holding the relight switch at RELIGHT for two seconds and then releasing it, leaving the throttle at its set position. A successful relight will be indicated by the rpm stabilising and then commencing to rise. The jpt gauge should be checked to ensure, by throttling back if necessary, that the maximum allowable temperature is not exceeded. The likelihood of obtaining a successful relight is increased if the height and airspeed are below the permitted maxima for relighting.

(ii) If no relight occurs within 30 seconds proceed as follows:

HP cock	OFF
LP cock	Leave on
All non-essential electrics (including tailplane)	Off
If below 35,000 feet	Relight at once
If above 35,000 feet	Switch off both booster-pumps, descend to 35,000 feet and carry out relighting drill.

Note: If above 35,000 feet the decision to descend quickly or glide at 210 knots will depend on the prevailing circumstances, eg weather conditions, distance to travel, etc. The following should be borne in mind.

1. Relights are more easily obtained at lower altitudes.

2. At best gliding speed the aircraft covers approximately 2 miles per 1,000 feet.

3. With normal services running the batteries cannot be relied on for very long. (See para 90(b).) All non-essential electrics must therefore be switched off to conserve battery power.

4. Unless Mod 502 is embodied Manual should be selected before attempting to relight in case relighting is unsuccessful and the batteries become exhausted.

5. Descending at a speed above 210 knots will increase windmilling rpm. Consequently hydraulic pressure will be higher and the generators may continue to supply power. (See para 90 (d), Note 2.)

6. If a glide is to be made at 210 knots the tailplane should be set to 0.5° nose-up.

93. Relighting

(a) Check and/or set:

Maximum altitude	35,000 feet
Maximum airspeed	0.80M
All non-essential electrics	Off (see para 90)
Throttle	HP cock OFF
Battery master switch	ON
Ignition switch	ON
Starter master switch	ON
Booster-pumps	ON

(b) Set the relight switch on for 2 seconds, then set the throttle lever to the HP cock open position. When the rpm rise to idling, increase power carefully.

(c) If no relight occurs within 30 seconds, set the throttle to HP cock OFF and switch off the booster-pumps. Allow if possible 30 seconds for the engine to dry out before the next attempt.

(d) Failure to relight

(i) If the engine fails to relight turn off the HP and LP cocks and switch off all non-essential electrical services (including the booster-pumps).

(ii) Set the tailplane incidence to 0.5° nose-up and reduce speed to the trimmed condition.

(iii) Release the aileron trim safety lock and check that the trim indicator is at zero.

(iv) Carry out forced landing procedure (see para 94) or abandon.

94. Forced landing procedure

Note: All the circumstances prevailing at the time of engine, hydraulic, or electrical failure cannot be predicted. Each system and the effects of the failure of each system must be thoroughly understood. The following information is intended to help the pilot reach the best decision.

(a) In all cases of engine failure when a relight is not possible or practicable glide at 210 knots (see (b)). If a deceleration is required to obtain the best gliding speed climb rather than use the airbrake. If used, the airbrake requires considerable hydraulic power and constitutes a hazard on landing should it remain out.

(b) The recommended gliding speed for maximum range is 210 knots. In still air conditions a distance of approximately 86nm can be achieved from 40,000 feet or about 2nm/1,000ft. The average rate of descent is approximately 2,300ft/min. These distances do not allow for the recommended overhead height of 7,000 feet. (See 94 (f).) The gliding speed of 210 knots should be maintained when gliding for range and/or in Powered control. If maximum range is not essential and a descent in Manual is being made, control is more comfortable at 175 knots.

(c) Check and/or set the following:

HP and LP cocks	OFF
Booster-pumps	OFF
Tailplane incidence	0.5° nose-up
Tailplane switch	OFF
All non-essential electrics	OFF
Aileron trim	Safety lock off, indicator zero
Elevator and aileron power	Leave on temporarily, but select OFF (1) Before electric power fails (2) If a false anchorage occurs

(d) (i) Engine failure may lead to either or both electrical and hydraulic failure. Experience shows that at 200 knots windmilling rpm will create sufficient hydraulic pressure for limited use of the power controls, but large or frequent use of the controls or selection of any of the services may cause Manual reversion. If pressure then rebuilds, false anchorages may occur.

(ii) Selecting Power off can only be accomplished whilst electrical power is available. To avoid the danger of obtaining a false anchorage with no electrical power available, and to prevent inadvertent Manual reversion at an inconvenient stage of the approach, select Power off before electrical failure occurs. When gliding at 210 knots, windmilling rpm at high altitude will be sufficient to keep the generators charging, but rpm will decrease gradually as height is lost. Frequent checks on the generator warning lights should be made, and after they come on, the batteries should not be relied on for more than 5 minutes.

(iii) If electrical failure occurs before Manual can be selected, to keep in Power, windmilling rpm should be maintained as high as possible and control movements be kept to a minimum; if a reversion to Manual then follows, reduce airspeed (not below 170 knots) to decrease windmilling rpm and use the controls excessively to avoid a rebuild of pressure. If pressure rebuilds the landing is likely to be extremely hazardous, and it is recommended that the aircraft be abandoned.

(iv) If a false anchorage occurs with no electrical power available it can only be cleared as described in para 88. If for any reason this is not possible the aircraft must be abandoned.

(e) If electrical power is still available, the undercarriage and flaps may be lowered by means of the normal systems, but the services will take longer than usual to lower. If electrical power is not available, the emergency systems can be used, but the undercarriage, once locked down, cannot be retracted either in the air or on the ground in the event of overshooting, and the flaps can only be lowered to the fully down position.

(f) (i) If a forced landing on an airfield is being made attempt to arrive overhead at approx 7,000 feet or above.

(ii) Plan a Manual approach and aim to be downwind opposite the caravan at a minimum of 4,000 feet AGL or above, depending on prevailing wind conditions, at 175 knots.

(iii) Turn across wind maintaining speed at 175 knots, and when it is certain that the touch-down point can be reached select full flap.

(iv) Maintain 175 knots on the final approach and aim to cross the threshold at 150 knots. In Manual the pull force to round out from a glide approach is large.

Note: 1. Experience suggests that it is preferable to lower the undercarriage when making a forced landing on an aerodrome or in open country. In the down position it absorbs much if not all of the initial impact, assists in retarding the aircraft, and, providing that electrical power is available and that the emergency lowering system has not been operated, it may be retracted after touch-down if necessary. With the undercarriage up, the aircraft must be lowered gently on to the ground at the normal speed; if the speed is too low a wing drop is likely to occur, and if the speed is too high the aircraft is prone to bounce, the initial impact having a damaging effect on the cockpit.

2. With the undercarriage down, the rate of descent is high, and it increases rapidly as speed is reduced below 180 knots.

3. For practice forced landings, a flap setting of 23° and 5,300 rpm (min) will give a rate of descent comparable with an engine-off glide.

95. Undercarriage and flaps emergency operation

Note: Both systems operate independently of their normal selectors. When Mod 428 is incorporated both systems are isolated from the powered control hydraulic system. Thus when either emergency system is operated the power controls will be unaffected, ie no jettisoning of hydraulic fluid will occur and the drills contained in (a) (iii) and (iv) and (b) are no longer necessary.

(a) Undercarriage

(i) If the undercarriage fails to lock down by normal selection when both electric and hydraulic systems are serviceable, repeated raising and lowering and the application of side-slip or G may have the desired effect.

(ii) If all attempts to lower the undercarriage (or any unit of the undercarriage) by the normal system fail, the emergency system should be used. However, if electrical and hydraulic power is normal and either of the undercarriage selector buttons are in, all hydraulic fluid will be jettisoned, resulting in Manual reversion. The rate of jettisoning is greater if the UP button is in. Therefore, depending on prevailing weather conditions, the following should be borne in mind.

(iii) If conditions are favourable, ie good visibility, and a steady wind down the runway, Manual should be selected before operating the emergency control and the landing made in Manual.

(iv) If conditions are not favourable, ie a strong cross-wind or if the wind is gusting, the loss of hydraulic fluid may be prevented if both undercarriage selector buttons are set to the out position. To achieve this carry out the following procedure:

1. Select up normally to ensure that the 'DOWN' selector button is out.

2. Pull out the 'UP' button, which requires little effort, and ensure that it remains out.

3. Operate the emergency air selector by holding the knob in and then pulling the handle.

Warning: It is essential that the UP button remains out. There is a tendency for it to return to the in position. If this occurs Manual reversion will follow, in which case the power control switches must be set to OFF.

(v) With both selector buttons out, both solenoids in the selector valve are de-energised enabling the slide valve to adopt a neutral position. The undercarriage hydraulic lines are thus isolated from the remainder of the hydraulic system and only the fluid in the lower end of the jacks will be jettisoned. The flying controls will remain in Power and the flaps may be operated normally. When the undercarriage up selector button is out the airbrake is isolated.

(vi) If complete electrical failure has occurred, both solenoids in the selector valve will automatically be de-energised and the slide valve will be in the neutral position. When the emergency lowering system is used only the fluid in the lower end of the jacks is jettisoned.

(vii) To raise the undercarriage on the ground in an emergency following normal selection, rotate and press the normal UP selector button. It must not be used in this way to retract the undercarriage in the air as there is a risk of the legs being retracted when not fully extended and the leg fairings may be damaged. If the UP button cannot be pressed in normally after take-off, the wheels must be left down and the aircraft landed.

(b) Flaps

(i) If electrical or hydraulic supply failure has occurred the flaps may be lowered, fully down only, by pulling out the emergency lowering control after first pushing in the central knob. If the electrical supply has failed, only the hydraulic fluid in the lower end of the jacks will be jettisoned when the emergency is used.

(ii) If the electrical and hydraulic supplies are normal, provided that wind conditions are favourable and runway length is adequate, it is recommended that a flapless landing in Power should be made (see para 98) because it is not possible to isolate the flap hydraulic lines from the remainder of the hydraulic system, and all hydraulic fluid will be jettisoned when the emergency lowering system is used. Manual reversion will then follow.

96. Landing with an undercarriage unit not locked down

Note: Experience has shown that these techniques cause minimum damage to the aircraft and none to the pilot.

(a) Both main wheels only locked down

(i) Use up as much fuel as is safe in order to move the C of G as far aft as possible. Unless circumstances dictate otherwise land on a runway.

(ii) Check harness tight and locked.

(iii) Select hood open when crossing the threshold.

(iv) Make a powered approach at the normal speed. On touch-down set HP cock OFF and maintain a moderate nose-up attitude.

(v) Trim the tailplane to give full nose-up trim and as the speed falls below 100 knots maintain a high nose-up attitude without actually touching the tail cone on the ground.

(vi) Lower the nose on to the runway at 80-90 knots and use the brakes gently to keep straight.

(b) Nosewheel and one main wheel locked down

If all attempts to lower the undercarriage satisfactorily fail and only the nosewheel and one main wheel come down, make a normal approach and landing; after touch-down hold the wings level for as long as possible by use of the ailerons. When aileron is applied and when the wing finally drops, the aircraft will swing in the direction of the unlocked wheel; this should be counteracted as much as possible by opposite brake. Experience has shown that the distance from the landing path to the point of rest averages approximately 250 yards, varying from 100 yards (min) to 400 yards (max). A runway should be chosen which has an area about 400 yards wide available in the direction of the anticipated swing.

(c) Belly landing

If it is necessary to land with the undercarriage retracted, make a normal approach and fly the aircraft gently on to the runway at the normal speed.

97. Landing with a burst tyre

No special difficulty is encountered when landing with a burst tyre(s), directional control and braking being adequate.

98. Flapless landing

A very long shallow powered approach should be made and the threshold crossed at the normal speed. Even When the throttle is closed speed decreases very slowly; the aircraft should therefore be placed firmly on the runway as soon as possible, the nosewheel lowered and the brakes applied, and if necessary the HP cock set OFF. The landing run is very much increased.

99. Wheel brakes emergency operation

(a) (i) Should the hydraulic system fail, the wheel brake accumulators provide sufficient pressure for brake operation during landing, down to a gauge reading of 750lb/sq in approximately. To conserve pressure when landing without main hydraulic pressure do not operate the lever during the down-wind checks.

(ii) Apply brake pressure in such a way that maxaretting does not occur (little pressure at high speed but increasing with decrease in speed) and avoid differential braking as much as possible. The brake lever should not be released after brakes have once been applied as this will result in a large loss of fluid.

(b) The HP cock should be put OFF on touch-down if the brakes accumulator pressure is low and the runway short.

(c) Check the accumulator pressure after landing. If it is almost exhausted, ie nearly down to 750lb/sq in, the aircraft must not be taxied. When pressure falls below this figure the brakes are completely inoperative.

100. Emergency use of oxygen

(a) Toxic fumes

If the cockpit becomes contaminated with toxic fumes, ensure the air-dilution switch on the regulator is at 100% OXYGEN and deflect the EMERGENCY switch to either side to prevent inward leaks on the mask.

(b) Blinker failure

(i) If the blinker mechanism ceases to operate check that the main tube is correctly connected, that the pressure on the regulator gauge is normal, and that the main oxygen contents gauge is indicating that oxygen is still available.

(ii) If the above indications are satisfactory, depress the regulator EMERGENCY switch when in the central position. A supply of oxygen under increased pressure indicates that the regulator is serviceable but that the blinker mechanism is defective. If the regulator is unserviceable, as indicated by no flow, use the emergency bottle and descend immediately to a safe altitude and check all oxygen tube connections.

(c) Use of emergency bottle

(i) If it is necessary to use the emergency oxygen bottle through failure of the regulator or because of exhaustion of the main supply, pull up the emergency control on the right of the seat, and disconnect the mask tube from the main supply point.

(ii) The main supply tube is automatically released and the emergency bottle brought into use on ejection.

(iii) The duration of supply from the emergency bottle is 10 minutes only.

101. Hood jettisoning

(a) The hood is jettisoned by pulling the handle on the port shelf. A strong pull force is required. To ensure a clean jettison, the hood must be closed and the speed should be in excess of 140 knots. The aircraft may be flown at speeds up to 450 knots with the hood jettisoned, but above this speed suction on the pilot's helmet may be disconcerting.

(b) The hood is also jettisoned when either the ejection seat blind or the seat pan alternative handle is pulled, but the GGS is not automatically lowered.

102. Abandoning the aircraft in flight

(a) Reduce speed as much as possible, set parachute container fully back, and lower the seat fully.

(b) Retract the GGS. If electrical power is available, the GGS should automatically lower on jettisoning the hood. The hood is jettisoned immediately the ejection seat blind or seat pan firing handle is pulled, ejection following one second later. (See para 101(b)).

(c) After ejection, the drogue gun will fire automatically.

(d) If ejection takes place above 10,000 feet automatic separation will not occur until that height is reached. If ejection takes place at or below 10,000 feet automatic separation will occur within 3 seconds.

(e) If the automatic system fails to operate after

ejection, proceed as follows:

(i) When forward speed is sufficiently low, discard the face screen and disconnect the main oxygen tube from the seat.

(ii) Pull the override D-ring to isolate the parachute auto device.

(iii) Operate the harness quick release.

(iv) Disengage the seat restraining straps.

(v) Lift the flap over the rip-cord D-ring and grasp the handle.

(vi) Push clear of the seat and pull the rip-cord handle.

(f) The minimum recommended heights for safe ejection in straight and level flight at speeds above 120 knots are:

Mk 2H seat	
60ft/sec gun	200ft
80ft/sec gun	100ft
Mk 3H seat	
80 ft/sec gun	100ft

103. Ditching

(a) Model tests of a clean aircraft indicate that a ditching in any but ideal conditions would be very hazardous.

(b) It is recommended, therefore, that except in calm sea and air conditions combined with good visibility, the pilot should abandon the aircraft rather than attempt a ditching

(c) If ditching is inevitable:

(i) Jettison the hood, select 100 per cent oxygen and deflect the emergency toggle sideways.

(ii) Lower full flap and make the approach at the lowest forward speed and rate of descent compatible with good control.

(iii) Ditch along the swell, or, if the swell is not steep, into wind.

Note: The airbrake must be retracted.

103A. Drop tank jettisoning

Drop tank jettisoning must be carried out in straight and level flight without yaw or sideslip in the following speed ranges.

(a) Drop tanks without fins	250-300 knots
(b) Drop tanks with fins	
From inboard pylons	200-450 knots
From outboard pylons	200-400 knots

103B. Cockpit pressure failure

(a) Pressurisation failure will be indicated by the warning light.

(b) If the aircraft is above 30,000 feet select EMERGENCY on the oxygen regulator and descend as rapidly as possible to below 30,000 feet, when the oxygen emergency lever may be set to OFF.

PART V

OPERATING DATA

104. Loading and CG data

(a) CG limitations

(i) The CG limits, undercarriage down, are as follows:

Forward limit	2in aft of datum
Aft limit	14.5ins aft of datum

(ii) Ballast must be carried if any of the following items are not carried:

Radar ranging, IFF, DME

(b) Effects of consumption of expendable stores

(i) Ammunition: Firing the guns causes the CG to move aft

(ii) Fuel: Consumption of fuel from the rear tanks causes the CG to move forward fairly rapidly. Consumption of the wing tank fuel causes a slight aft CG movement. As fuel is used from the front tanks the CG moves progressively aft reaching a similar position to that obtaining with all internal tanks full when about 900lb (450lb per side) remain. Consumption of fuel from inboard drop tanks has a negligible effect on CG.

(c) Typical service loads

The figures given below are approximate and are intended only as a guide. Reference should always to be made to RAF Form 4801 or to AP.4347F-Vol 1 Sect 2 Chapter 3.

	Configuration	Approx AUW (lb)
(1)	Clean, full internal fuel and ammunition	17,650
(2)	As (1) plus 24 x 25lb RPs	19,300
(3)	As (2) plus 2 x 100-gal (full) drop tanks	21,300
(4)	As (1) plus 2 x 100-gal (full) drop tanks	19,800
(5)	As (1) plus 4 x 100-gal (full) drop tanks	21,680
(6)	As (1) plus 2 x 1,000lb bombs	19,750
(7)	As (6) plus 2 x 100-gal (full) drop tanks	21,670

Note: Full ammunition = 676lb

105. Pressure error corrections

(a) The ASI sea level pressure error corrections are as follows:

IAS (knots)	200	300	400	500	600
Clean or with any stores combination excluding a port outboard drop tank	+1	+3	+4	+3	0
With any stores combination including a port outboard drop tank	0	+4	+6	+9	+12

(b) The machmeter pressure error corrections are as follows:

Configuration	Height (feet)	Level flight – IMN					Diving IMN			
		0.7	0.8	0.85	0.9	0.93	0.94	0.985	1.0	1.1
Clean, or with any stores combination excluding a port outboard drop tank	All	–	–	Negligible	–	–	–	+0.015	+0.03	+0.13
With any stores combination including a port outboard drop tank	SL	+0.02	+0.03	+0.03	+0.05	–	–	–	–	–
	40,000	Negligible	+0.02	+0.04	–	–	–	–	–	–
	–	–	–	–	–	+0.07	+0.09	–	–	–

(c) Both pec tables apply irrespective of whether extending leading edges are fitted or not. The only factor which affects pecs to any degree is the presence of a port outboard drop tank.

106. Fuel consumptions

The approximate fuel consumptions in lb/hour, for various rpm and altitudes, are given below.

Height (feet)	8,000 rpm	7,600 rpm	7,000 rpm	6,750 rpm	6,500 rpm	6,250 rpm
Sea level	13,200	10,020	5,300	3,930	3,140	2,790
10,000	9,420	7,920	4,660	3,530	2,590	2,050
20,000	6,600	5,820	4,010	3,080	2,230	1,650
30,000	4,680	3,960	3,140	2,590	1,980	1,400
40,000	3,060	2,760	2,120	1,810	1,440	–

107. Flight planning data

(a) The tables on the following pages show the flight planning data for:

(i) Climbing

The climb table gives the data for climbs in ISA conditions using the speed recommended in para 69 (repeated in the table).

(ii) Cruising

Each separate altitude block in the cruise table shows:

(1) The speed for maximum range, the approximate ANM/100lb and the approximate fuel consumption for the particular height. In addition a speed band is given, use of any speed within which should not cause more than a 5% reduction in range.

(2) The range obtainable for various amounts of available fuel when flying at the best range speed for the height. The range given is to the point of let-down, allowance being made for the descent fuel required.

(3) The range obtainable for various amounts of available fuel including the distance covered on the climb, if a climb is made to another altitude. In this case the climb must be made at the speed given in para 69 and the flight continued at the new altitude at the best range speed for that height.

Note: The range at any altitude is independent of temperature, but dependent on the weight of fuel carried.

(iii) Descent

The descent table gives data for descending from one height to another.

(b) Use of the tables

(i) Pre-flight planning

Enter the cruise data table in the sea level block at the fuel state applying immediately after take-off. Select the height at which maximum range is available at that fuel state. The distance available includes distance covered on the climb, but not on the descent. (Absolute maximum range is obtained by adding on the descent distance provided that the let-down is commenced at that distance from the destination.) For short-range flights inspect the sea level block and select the height at which the distance to be covered requires the least amount of fuel. This is the best altitude for the flight.

(ii) In-flight planning

At any stage of a flight the available range may be ascertained by applying the fuel state to the level flight range in the particular altitude block. If increase in range is required, or if a climb has to be made, the new available range may be obtained by entering the existing altitude block at the particular fuel state and moving vertically downwards within the block until the new altitude is reached. Figures in heavy type indicate the best altitude for the maximum increase in range. Above these heights no further range increase is possible. If a descent is necessitated the new range is shown by moving direct from the existing altitude level flight range for the particular fuel state to the new altitude level flight range.

CLEAN AIRCRAFT

Fuel contents	381 gallons, 2,933lb AVTAG
Taxi and take-off	270lb
Landing	60 gallons, 460lb

Climb data

From (ft)	To (ft)	Fuel (lb)	Distance (nm)	Time (mins)
Sea level*	10,000	375	6	1.75
	20,000	480	13	2.75
	30,000	585	23	4
	40,000	705	40	5.75
	50,000	930	89	12
10,000	20,000	105	7	1
	30,000	210	17	2.25
	40,000	330	34	4
	50,000	555	83	10.25
20,000	30,000	105	10	1.25
	40,000	225	27	3
	50,000	450	76	9.25
30,000	40,000	120	17	1.75
	50,000	345	66	8
40,000	50,000	225	49	6.25
% increase per 10°C temperature rise from ISA		+10%	+16%	+15%

Climb at full throttle at 0.85M

* Fuel and time figures in the sea level block are from start-up and wheels rolling respectively.

Descent data

From (ft)	To (ft)	Fuel (lb)	Distance (nm)	Time (mins)
50,000	40,000	8	7	1
	30,000	23	14	2
	20,000	58	21	3
	10,000	103	29	4.25
	Sea level	183	37	7.25
40,000	30,000	15	7	1
	20,000	50	14	2
	10,000	95	22	3.25
	Sea level	175	30	6.25
30,000	20,000	35	7	1
	10,000	80	15	2.25
	Sea level	160	23	5.25
20,000	10,000	45	8	1.25
	Sea level	125	16	4.25
10,000	Sea level	80	8	3
Airbrakes	OUT			
Flap	30°			
RPM	6,500			
IAS	280kts			

CRUISE DATA: CLEAN (17,500lb)

Fuel state (lb AVTAG)		**2,663**	**2,500**	**2,000**	**1,500**	**1,000**
Sea level	Range	275	260	200	135	**70**
ANM/100lb 12.8	10,000ft	340	315	235	150	65
lb/hr 2,790	20,000ft	413	385	275	165	60
Speed 360kts	30,000ft	500	470	330	190	40
95% range 260-400kts	40,000ft	**585**	**535**	**365**	**190**	–
	50,000ft	540	485	315	145	–
10,000ft	Range	–	325	245	160	80
ANM/100lb 16.7	20,000ft	–	405	300	190	**80**
lb/hr 2,230	30,000ft	–	495	355	215	**70**
Speed 320kts	40,000ft	–	**565**	**395**	**225**	45
95% range 30kts-0.70M	50,000ft	–	515	345	175	–
20,000ft	Range	–	420	310	205	**95**
ANM/100lb 22.1	30,000ft	–	515	375	235	90
lb/hr 1,775	40,000ft	–	**590**	**420**	**250**	75
Speed 290kts	50,000ft	–	545	375	205	–
95% range 230kts-0.77M						
30,000ft	Range	–	–	395	255	**110**
ANM/100lb 28.5	40,000ft	–	–	**450**	**280**	**105**
lb/hr 1,475	50,000ft	–	–	400	230	60
Speed 0.73M						
95% range 200kts-0.83M						
40,000ft	Range	–	–	**475**	**305**	**130**
ANM/100lb 34.4	50,000ft	–	–	425	255	85
lb/hr 1,330						
Speed 0.81M						
95% range 0.7-0.85M						
50,000ft	Range	–	–	**465**	**295**	**125**
ANM/100lb 34.06						
lb/hr 1400						
Speed 0.84M						
95% range 0.79-0.88M						

Note: Values of ANM/100lb are based on a mean cruising weight of 16,000lb. These tables are based on estimates.

With 2 x 100-gallon drop tanks

Fuel contents	581 gallons, 4,473lb AVTAG
Taxi and take-off	35 gallons, 270lb
Landing	60 gallons, 460lb

From (ft)	To (ft)	Fuel (lb)	Distance (nm)	Time (mins)
Sea level*	10,000ft	400	7	1.75
	20,000ft	530	16	2.75
	30,000ft	660	28	4.25
	40,000ft	815	48	7
	47,000ft	1,025	90	12.25
10,000ft	20,000ft	130	9	1
	30,000ft	260	21	2.5
	40,000ft	415	41	5.25
	47,000ft	625	83	10.5
20,000ft	30,000ft	130	12	1.5
	40,000ft	285	32	4.25
	47,000ft	495	74	9.5
30,000ft	40,000ft	155	21	2.75
	47,000ft	365	63	8
40,000ft	47,000ft	210	42	5.25
% increase per 10°C temperature rise from ISA		+10%	+16%	+15%
Climb at full throttle at 0.85M				

* Fuel and time figures in the sea level block are from start-up and wheels rolling respectively.

Descent data

From (ft)	To (ft)	Fuel (lb)	Distance (nm)	Time (mins)
47,000ft	40,000ft	6	5	0.75
	30,000ft	21	12	1.75
	20,000ft	56	19	2.75
	10,000ft	101	27	4
	Sea level	181	35	7
40,000ft	30,000ft	15	7	1
	20,000ft	50	14	2
	10,000ft	95	22	3.25
	Sea level	175	30	6.25
30,000ft	20,000ft	35	7	1
	10,000ft	80	15	2.25
	Sea level	160	23	5.25
20,000ft	10,000ft	45	8	1.25
	Sea level	125	16	4.25
10,000ft	Sea Level	80	8	3
Airbrakes	OUT			
Flap	30°			
RPM	6,500			
IAS	280kts			

CRUISE DATA: WITH 2 x 100 GALLON DROP TANKS (19,400lb)

Fuel state (lb AVTAG)			**4203**	**4000**	**3500**	**3000**	**2500**	**2000**	**1500**	**1000**
Sea level		Range	440	425	365	305	245	185	125	**65**
ANM/100lb	11.9	10,000ft	550	525	450	375	295	220	140	60
lb/hr	3,080	20,000ft	690	660	560	460	360	255	155	50
Speed	365kts	30,000ft	850	805	680	555	425	295	**160**	–
95% range	260-400kts	40,000ft	**985**	**935**	**785**	**630**	**475**	**315**	155	–
		47,000ft	905	840	685	530	375	220	–	–
10,000ft		Range	–	535	460	385	310	230	155	**75**
ANM/100lb	15.5	20,000ft	–	675	580	475	380	275	175	70
lb/hr	2,470	30,000ft	–	835	710	585	450	320	190	55
Speed	320kts	40,000ft	–	**990**	**840**	**690**	**535**	**375**	**215**	–
95% range	250-380kts	47,000ft	–	920	765	610	455	300	145	–
20,000ft		Range	–	–	595	495	395	295	190	**85**
ANM/100lb	20.3	30,000ft	–	–	730	605	475	345	215	**80**
lb/hr	1,995	40,000ft	–	–	**850**	**700**	**545**	**385**	**225**	60
Speed	300kts	47,000ft	–	–	800	645	490	335	180	–
95% range	230kts-0.77M									
30,000ft		Range	–	–	745	615	490	360	230	**95**
ANM/100lb	25.8	40,000ft	–	–	**875**	**725**	**575**	**415**	**255**	90
lb/hr	1,690	47,000ft	–	–	830	675	520	365	210	–
Speed 95% range	0.75M 220kts-0.83M									
40,000ft		Range	–	–	**900**	**750**	**595**	**440**	**285**	**120**
ANM/100lb	31.1	47,000ft	–	–	855	700	545	390	235	80
lb/hr	1,380									
Speed	0.81M									
95% range	0.7-0.85M									

Fuel state (lb AVTAG)		**4203**	**4000**	**3500**	**3000**	**2500**	**2000**	**1500**	**1000**
47,000ft									
ANM/100lb	30.75								
lb/hr	1,500								
Speed	0.83M								
95% range	0.78-0.87M								
Range		–	–	**875**	**720**	**565**	**410**	**255**	**100**

Note: Values of ANM/100lb are based on a mean cruising weight of 17,250lb. These tables are based on estimates.

108. Take-off distances

The approximate take-off distances, in yards, at sea level for a clean aircraft and with drop tanks, are given below.

(a) Clean aircraft (17,500lb)

Temperature °C	0	+15	+30	+45	
Zero wind	Ground run	590	720	840	970
	To clear 50 feet	980	1,150	1,310	1,470
30 knot wind	Ground run	340	420	490	570
	To clear 50 feet	660	770	880	990

(b) With 2 x 100-gallon drop tanks (19,400lb)

Temperature °C	0	+15	+30	+45	
Zero wind	Ground run	770	940	1,100	1,270
	To clear 50 feet	1,250	1,470	1,680	1,880
30 knot wind	Ground run	450	555	640	740
	To clear 50 feet	830	980	1,120	1,250

For every 1,000 feet the aerodrome altitude is above sea level, increase the above distances by:

12 per cent for ground run

10 per cent to clear 50 feet

Appendix 9

References

Naturally, a popular and historic aircraft like the Hunter has received a huge amount of coverage on the internet. There are some excellent websites that include interesting information and/or photographs, but unfortunately (like so many other subjects covered in cyberspace) the Hunter is also featured in a large number of sites that have been produced by amateurs who often have no knowledge of the subject. Consequently, great care should be taken when using websites and it is important to remember at all times that just because something appears on the web, it is not necessarily true or technically correct! From the seemingly endless website coverage, the following sites are certainly recommended:

http://www.canit.se/~griffon/aviation/text/34hunter.htm The Hunter in Swedish service

http://www.hunterflyingclub.co.uk/ Hunter Flying Club at Exeter

http://www.vicflintham.co.uk/ Excellent aviation reference site

http://www.bharat-rakshak.com/IAF/ Indian Air Force reference site

http://www.hunterteam.com/ HHA Aviation at Scampton

http://www.nlcas.ca/ Northern Lights Canada

http://www.deltajets.com/ Delta Jets at Kemble

http://belmilac.wetpaint.com/ Belgian Air Force reference site

http://www.radfanhunters.co.uk/ Ray Deacon's outstanding site devoted to Hunter operations in the Middle East

Books on the Hunter are also widely available and (like websites) vary in quality. Recommended titles are:

Hawker Hunter – 50 Golden Years (ISBN 0-9540666-0-X)
Ray Deacon is a respected aviation historian, with a particular interest in the Hunter. His excellent book is currently out of print although copies can still be purchased through specialist dealers.

Modern Combat Aircraft 15: Hawker Hunter (Ian Allan, 1982; ISBN 0 7110 1216 4)
Although this is an old book and now out of print, copies can still be found. Written information is good (but brief – the book is not particularly large) and there are some good images of the Hunter.

Hawker Hunter – The Operational Record (Airlife, 1989; ISBN 1 85310 048 X)
Author Robert Jackson also produced Ian Allan's MCA book on the Hunter, but this later volume (although containing material found in the Ian Allan book) is a more comprehensive and interesting read, with plenty of information not found in generic Hunter history publications.

Hunter Squadrons of the Royal Air Force and Fleet Air Arm (Linewrights, 1985; ISBN 0 946958 13 0)
Richard Ward is a greatly respected historian and, as might be expected, this book is an excellent reference source packed with photographs (some in colour) and informative accompanying text.

Hawker Hunter (Crowood Press, 1998; ISBN 1 86126 0830)
Written by Barry Jones, this is a satisfactory look at the Hunter's history, but contains little that has not been published before. The photographs, however, are well reproduced and some are fairly uncommon.

Hawker Hunter in Action (Squadron/Signal Publications, Aircraft Number 121, 1992; ISBN 0-89747-273-X)
As with most Squadron/Signal reference publications, the format is aimed at modelling enthusiasts and written information is kept to a minimum. Author Glenn Ashley has collected an interesting and varied selection of Hunter images that are combined with a set of colour profiles.

Hawker Hunter – Biography of a thoroughbred (PSL, 1981; ISBN 0 85059 476 6)
One of the best Hunter books to have been produced, Francis Mason provides an interesting and readable account of the Hunter's history, combined with a good selection of images, many from the Hawker archives. Sadly, the book is currently out of print.

Warpaint Series No 8: Hawker Hunter (Hall Park Books, 1997; ISSN 1363-0369)
Another reference publication aimed at modellers, this volume provides a mix of photography combined with a general history of the Hunter. Unfortunately, some of the information is less than accurate, but the photographic selection is good, even though print quality is sometimes rather poor.

In The Red (Pentland Press, 2000; ISBN 1 85821 754 7)
Author J. T. Kingsley's excellent account of his experiences as an RAF pilot, including coverage of his association with the Hunter.

Best of Breed (Pen & Sword, 2006; ISBN 1 84415 412 2)
Nigel Walpole presents a fascinating description of Hunter reconnaissance operations.

Cold War Hot Wings (Pen & Sword, 2006; ISBN 9 78184 415 5750)
A first-hand account of RAF combat flying by author Chris Bain, including excellent coverage of Hunter flying.

Vliegend in Nederland – Hawker Hunter (Flash Aviation, 1990; ISBN 90 71553 09 4)
Although the text for this detailed reference booklet is all in Dutch, the photography – which covers all aspects of the Hunter's operations in the Netherlands – is excellent, and is accompanied by a complete disposal history for each of the RNethAF Hunters.

RAF Hunters in Germany (Kipp & Lindsay, 2003; ISBN 0 9544069 0 7)
A fascinating account of Hunter operations in Germany, complete with many rare photographs and first-hand accounts from former RAF Hunter pilots.

Hawker Hunter 1951 to 2007 (Lulu.com, 2007; ISBN 978 1 4303 0593 4)
Author David Griffin joined the Royal Air Force as an armourer and worked with the last two squadrons of Hunters in operational roles while serving at RAF Wittering. After leaving the RAF he went to work for Airwork Services Ltd, on contract with the Royal Air Force of Oman where he remained for the next 21 years, serving at Thumrait and working until 1993 with the Hawker Hunter as a line armourer and later as armament instructor. His book is undoubtedly the most comprehensive publication yet published on the individual histories of the Hunter, providing current details of each individual Hunter aircraft operated by military and civilian operators over the fifty-six years since its introduction. It runs from the early prototypes right through to the two latest Hunters with the UK Military, ZZ190 and ZZ191. Many unpublished and some very rare photos are included, with copies of the original specification and contract documents. Part Two of the book – *Hawker Hunter 1951 to 2008* – was published early in 2008, and is a totally updated volume with all new photographic content and a number of new chapters covering aspects of the Hunter's construction and use.

Index